THE FUTURE OF AERONAUTICS

THE FUTURE OF AERONAUTICS

EDITED BY JOHN E ALLEN & JOAN BRUCE

Foreword by HIS ROYAL HIGHNESS THE PRINCE PHILIP DUKE OF EDINBURGH KG KT

HUTCHINSON OF LONDON

HUTCHINSON & CO (*Publishers*) LTD
178–202 Great Portland Street, London W1

London Melbourne Sydney
Auckland Bombay Toronto
Johannesburg New York

First published 1970

This book has been set in Times, printed in Great Britain
on Antique Wove paper by Anchor Press, and
bound by Wm. Brendon, both of Tiptree, Essex
ISBN 0 09 100920 0

Contents

(Editor's note: Please note that all references to Second Century refer to the Second Century of the Royal Aeronautical Society which was founded on 12th January 1866)

Foreword

Predicting the future has exerted a peculiar fascination over man ever since he discovered how to think about it. Predicting the future of aeronautics is particularly tempting because there seems to be so much to predict. There are also pitfalls. The rate of progress and development is so fast that events have a nasty habit of overtaking predictions. Furthermore the developments themselves change the environment and what may seem desirable or attainable in terms of today can easily be irrelevant in terms of tomorrow.

It may well be possible to predict technological developments with considerable accuracy but anyone trying to predict the economic climate has a more difficult task. A relatively small change in the price of some material or technique may have a far greater influence upon what is economically possible than some predictable technological development.

The value of this book lies not so much in any of its predictions, but rather in its attempts to foresee the size and complexity of the problems which are more or less inevitable at the present rate of progress. These estimates are vital if the process of discovering solutions is to be set in motion with a sufficient margin of time.

Much of this book is speculation but it is speculation by people with a very thorough understanding of their subjects and it is well worth paying attention to what they have to say.

Philip

Preface

Air Commodore F. R. Banks

In 1966 the Royal Aeronautical Society celebrated its first centenary and published a 300-page anniversary volume which surveyed the achievements of this period; recalling the groping and early speculations, and the experiments leading from gliders and kites, and the study of birds, to balloons and, somewhat later, at the beginning of this century, to the manned aeroplane and airship.

The advent of the gas turbine engine, which allowed greatly increased speeds for the subsonic aircraft and made supersonic speeds possible, has led to a revolution in aviation in all its forms, and particularly in civil aviation; and it is therefore a fascinating exercise to attempt to forecast the next hundred years in the second century of the Royal Aeronautical Society.

The Royal Aeronautical Society, at the suggestion of J. E. Allen, invited a number of distinguished engineers, researchers and scientists, educators and lawyers, to give their views on the progress that might be foreseen for aerospace in the next hundred years, and principally to forecast what might be towards the end of that period; and this book is the result of their endeavours, being a collection of the Second Century Papers written by these distinguished people for the Aeronautical Society-after the Centenary Year. The twenty-two chapters cover the most impor tant scientific and technological disciplines relevant to aeronautics, but they also contain new lines of thought on the place of aerospace in our world of the future. The book should not only be valuable to scientists and technologists, but also to economists, sociologists and, later, for historians, since it covers a most important and even vital period in civilisation's evolution with the impact of technology on mankind becoming a critical issue.

Biographies of the Contributors

Chapter 1 *John E. Allen,* BSc(Eng), CEng, FRAeS, FIMechE, AFAIAA, is Chief Future Projects Engineer, Hawker Siddeley Aviation. He was formerly with A. V. Roe, Weapons Research Division as Head of the Aerodynamics, Projects and Assessment Department. Previously he was at MAEE and RAE, Farnborough.

Chapter 2 *Dr. Dietrich Küchemann,* CBE, FRS, FRAeS, FAIAA, FIMA, has been Head of the Aerodynamics Department, Royal Aircraft Establishment, Farnborough, since 1966. From 1936 to 1946 he did research work at Aerodynamische Versuchsanstalt, Göttingen, and from 1946 at the RAE, Farnborough.

Chapter 3 *Air Cdre. Allen H. Wheeler,* CBE, MA(Cantab), FRAeS, joined the Royal Air Force in the General Duties (Flying) Branch direct from Cambridge University in 1925. When he retired in 1954 he was Head of the A & AEE, at Boscombe Down. He is now a Consultant in Aviation and is Trustee in charge of the Shuttleworth Collection of Historic Aircraft.

Chapter 4 *Bryan Thwaites,* MA, PhD, FIMA, has been Principal of Westfield College, University of London, since 1966. Previously he was Professor of Theoretical Mechanics at Southampton University. He initiated and still directs the School Mathematics Project, is Chairman of the Council of the Church Colleges of Education, and was a magistrate for some years.

Chapter 5 *A. V. Cleaver,* OBE, CEng, HonFAIAA, HonFBIS, FRAeS, is currently Director and General Manager of the Rocket Department, Rolls-Royce, Ansty Division. Before joining Rolls-Royce in 1957 to take charge of the engineering on the Blue Streak propulsion system, Mr. Cleaver spent twenty years at de Havillands working on variable pitch aircraft propellers and building up their Rocket Division.

Chapter 6 *F. O'Hara,* MA, CEng, FRAeS, is Chief Superintendent of the Royal Aircraft Establishment, Bedford. He was previously at the Marine Aircraft Experimental Establishment, the Airborne Forces Experimental Establishment and the A & AEE, Boscombe Down, before transferring to the Aerodynamics Department, RAE Bedford in 1954.

J. Burnham, BSc, is Head of the Gust Research Section, Flight Division, Aerodynamics Department, Royal Aircraft Establishment, Bedford. He joined the RAE in 1954 where he began his work on atmospheric turbulence and its affect on aircraft.

Chapter 7 *W. G. Molyneux,* DSc, CEng, FRAeS, is currently Assistant Director of Scientific Research (Air) in the Ministry of Technology by whom he has been employed since 1942. He has published some 30 technical papers and holds a like number of patents for inventions.

Chapter 8 *Beverley S. Shenstone,* MASc, CEng, HonFRAeS, HonFCASI, FAIAA, retired from the position of Technical Director, BOAC, in 1966. He is now an Aeronautical Consultant. He was previously Chief Engineer, BEA. Mr. Shenstone was President of the Royal Aeronautical Society in 1962–3, and was Editor of OSTIV and *The World's Sailplanes* from 1958 to 1965.

Chapter 9 *Professor M. W. Thring,* ScD, FIMechE, FIEE, MIChemE, FInstP, FInstF, FRAeS, has been Head of the Mechanical Engineering Department, Queen Mary College, University of London, since 1964. He was formerly Professor of Fuel Technology and Chemical Engineering, Sheffield University. Professor Thring has been President of the Institute of Fuel, and was a member of the Advisory Council on R & D, Ministry of Power, 1960–7.

Chapter 10 *L. G. Dawson,* BSc(Eng), MIMechE, MAIAA, AFRAeS, is Director of Advanced Engineering, Aero Engine Division, Rolls-Royce. He joined Rolls-Royce as a technical assistant in 1939. Mr. Dawson was Chief Engineer–Military Engines, from 1958 to 1963 and Chief Engineer–Projects, from 1963 until 1965 when he was appointed to his present position.

Chapter 11 *Dr. G. Melvill Jones,* MA, MB, BCh, AFRAeS, FCASI, is Director of the Defence Research Board Aviation Medical Research Unit and Professor in the Department of Physiology, McGill University, Montreal, Quebec, Canada. After obtaining his medical degree in 1948, and spending two years as a House Surgeon at the Middlesex Hospital and Addenbrooke's Hospital, Cambridge, he was drafted into the Royal Air Force and joined the staff of the Institute of Aviation Medicine, Farnborough. He also undertook full flying training to RAF Squadron standard. In 1956 he joined the External Staff of the Medical Research Council to study problems of orientation in the context of aviation.

Chapter 12 *Richard Stanton-Jones,* MA, DCAe, CEng, AFRAeS, is Managing Director of British Hovercraft Corporation Ltd. He was formerly with Saunders-Roe Ltd., joining as a Senior Aerodynamicist in 1950 and was appointed Chief Designer in 1959.

Chapter 13 *P. A. Hearne,* DCAe, DLC, FRAeS, is Director of Elliott Flight Automation. He joined Elliott Brothers in 1959 as Divisional Manager, Weapons Division, and became Assistant General Manager, Guided Flight Group, in 1960 and Joint General Manager in 1965.

Chapter 14 *Raoul Hafner,* CEng, FRAeS, retired from the position of Technical Director (Research), Westland Aircraft Ltd., in 1969, and is now a consultant to the Company. Mr. Hafner, who was born and educated in Austria, produced his first helicopter design in 1927. He worked on rotating wing research for MAP during the 1939–45 war. Later as Chief Designer, helicopters, he designed the range of Bristol Helicopters.

Chapter 15 *A. J. Kennedy*, DSc, PhD, has been Director of the British Non-Ferrous Metals Research Association since 1966. He was previously Professor of Materials and Head of the Materials Department at the College of Aeronautics, Cranfield, and before that held appointments at the British Iron and Steel Research Association, the Royal Institution and University College, London.

Chapter 16 *Professor J. Black*, PhD, FRAeS, is Head of the School for Engineering of Bath University of Technology. He is a member of the University Grants Committee and is Chairman of the Sub-Committee on Educational Technology, in which capacity he is responsible for the allocation of television, language laboratories and learning resources in the UK Universities.

Chapter 17 *John Stamper,* MA, CEng, FRAeS, is Technical Director, Hawker Siddeley Aviation. He joined the Blackburn Aircraft Company in 1947 as a postgraduate apprentice, eventually becoming Chief Designer in 1962. Following the merger of Blackburns with Hawker Siddeley, Mr. Stamper continued at Brough as Executive Director and Chief Designer until 1966, when he was transferred to the Head Office at Kingston, and in 1968 he moved to Hatfield as Executive Director and Deputy Chief Engineer–Civil.

Chapter 18 *Harold Caplan*, DCAe, CEng, FRAeS, is Joint General Manager of International Insurance Services, and is principally concerned with the legal problems of international insurance. He was previously an Assistant Manager and head of the legal department of the British Aviation Insurance Co. Ltd. Before that he was for ten years engaged on accident and claim investigation, risk analysis, repair and salvage control for aviation insurers. Mr Caplan studied aeronautical engineering as an apprentice with the Bristol Aeroplane Co. Ltd., and was subsequently one of the first students at the College of Aeronautics, Cranfield. He was called to the Bar by the Middle Temple in 1955.

Chapter 19 *Alan H. Stratford,* BSc(Eng), CEng, FRAeS, FRSA, formed the firm of Alan Stratford and Associates in 1965 and is engaged in air transport and aviation research for airlines, manufacturers and airport authorities in the UK and overseas. Previously he was Aviation Research Manager, Hawker Siddeley International, and before that he was Manager of the Transport and Economics Department of the Whitworth Gloster Aircraft Company and Superintendent of Performance Analysis in BEA.

Chapter 20 *Andrew Stratton*, MSc, CEng, FInstP, FIEE FInstNav, FIMA, is Director of the Defence Operational Analysis Establishment, Ministry of Defence. From 1966 to 1968 he was Professor and Head of the Department of Mathematics at the College of Aeronautics, Cranfield, and before that was at the Royal Aircraft Establishment.

Chapter 21 *Ronald Smelt,* MA, PhD, FAIAA, FRAeS, is Vice-President and Chief Scientist, Lockheed Aircraft Corporation, California. Previously he was Chief of the Gas Dynamics Facility of the USAF Arnold Engineering Development Center, Deputy Chief of the Aeroballistic Research Department of the Naval Ordnance Laboratory, and Chief of the Guided Weapons Department of the RAE Farnborough. He is President of the American Institute of Aeronautics and Astronautics, and is Chairman of NASA's Research and Technology Advisory Committee on Space Vehicles.

Chapter 22 *Lord Kings Norton*, PhD, DTech, DIC, DSc, LlD, FIMechE, FRAeS—is Chairman of Berger, Jenson and Nicholson and a director of other companies. He was Chairman of the Metal Box Company from 1961 to 1967. He is President of The Royal Institution, the Institute of Marketing and the National Council for Quality and Reliability. He is Chairman of the Council for National Academic Awards, Chairman of the Governors of the College of Aeronautics and of the Council of the Air Registration Board. From 1924 to 1929, Harold Roxbee Cox (as he then was) worked on the construction of the Airship R101. Except for a short spell as Chief Technical Officer at Cardington in 1931, he worked from 1929 to 1937 at the RAE. From 1937 to 1939 he was Chief Technical Officer at the ARB. In 1940 he was appointed Deputy Director of Scientific Research at the Ministry of Aircraft Production, and later Director of Special Projects. In 1946 he became Director of the National Gas Turbine Establishment and two years later was appointed Chief Scientist of the Ministry of Fuel and Power. He was knighted in 1953 and was made a Life Peer in 1965.

1 Looking ahead in aeronautics

John E. Allen

I INTRODUCTION

> 'We should accept the challenge of predicting how management and design will develop towards the end of this century. It could be a stimulating problem which might inspire the younger members to greater thoughts and open up wider horizons than treading the somewhat narrow path defined by current development of techniques.'
>
> A. A. LOMBARD, *June 1967*

The Centenary of the Royal Aeronautical Society[1] in Great Britain in 1966 drew attention to the aeronautical happenings of the previous hundred years in a vivid way. It was, therefore, not difficult to arouse interest and sympathy to the idea of balancing the historical view with an attempt to look ahead over a roughly similar time scale. Forecasting, predicting or prophesying has been an attractive intellectual activity over the centuries and there has been a renewed interest in such matters generally during the 1960s, particularly in technological forecasting encouraged by the growing availability of large sized computers which could handle hitherto unattainable masses of data. Such preoccupation with the prospects for the future was also stimulated by the abnormal situation of living in a period of rapid change. The great novelty of the mid-twentieth century was the replacement of the individual prophetic writers like Sir Thomas More, Bishop Wilkins, and Jules Verne by the research teams in Government and Industry specifically charged with technological and sociological forecasting. In 1964 the Rand Corporation developed the Delphi technique of predicting technological breakthroughs and in 1965 Dr. Robert Jungk founded the Institute for Questions of the Future[2]. By 1967 a world-wide survey showed that several hundred private industrial and governmental institutions were working on some aspect of 'future research'.

The studies of Utopias—societies of the future—and the future possi-

Originally published in the *Aeronautical Journal of the Royal Aeronautical Society*, January 1968.

bilities for inventions are indeed fascinating subjects in their own right and many surveys are now available[3-8]. But the history of prediction forecasting gives many examples of highly bizarre situations and worse still, fads in thinking such as that called 'Utopiology', practised by German sociologists in the nineteenth century and more recently the dangers inherent in 'Futurology'. When the chapters of this book were being written there was some apprehension lest the authors could be accused of slipping into such dangerous paths. Perhaps the best definition of the way in which the authors approached their tasks was the notion of futuribles of Bertrand de Jouvenel[3]. A futurible is defined as a futurum that appears in the mind as a possible descendant from the present state of affairs. The task set was directed particularly to aeronautical affairs, taking appropriate account of possible changes in world activities likely to have a bearing on the future of aeronautics but without paying too much attention to deciding which of the possible Utopias should be the framework within which all authors should confine their thoughts. In fact there was only scant co-ordination between authors but, as there were sufficient of them to ensure many shades of opinion the book reflects typical attitudes and beliefs of the time. However, each author started his speculation from a professional baseline.

So, the book does not attempt to visualise what the world could be like in AD 2067. In some matters big steps will have occurred before the end of the twentieth century. In time many hitherto separate technical disciplines may combine, some will probably disappear altogether and meanwhile novel areas will arise which have yet to be recognised. Each chapter in the book treats a selected topic, each author aiming to describe important developments in his field which are likely to emerge during the next century.

Aeronautics is a particularly fascinating subject for long-term studies of this type. There is still a magic and a fascination in flight through the air. Many of the attainments are quite unbelievable and the book will show there will be no lack of further opportunities for invention and imaginative grasp of technology in the future. But aeronautics will serve a vastly changing world and as always will attract men of skill and vision to support it and so before embarking on the future this chapter surveys the future of aeronautics as seen through the eyes of great aeronautical thinkers of the past and in so doing sets the scene for the remaining chapters.

2 SOME AERONAUTICAL PREDICTIONS

> 'In 1901, I confess that I said to my brother Orville that man would not fly for fifty years. Two years later we made flights. This demonstration of my impotence as a prophet gave me such a shock that ever more I have distrusted myself and avoided all predictions.'
>
> WILBUR WRIGHT, *1908*

The Aeronautical Society started with a small group who believed that manned flight would result from the correct application of scientific method although the world was generally sceptical of such realisation. 1866 is a quite arbitrary date in the history of the aeroplane, which has a more important beginning in the theoretical studies of Sir George Cayley dating from 1796 and his successful glider experiments. Cayley made three attempts to form an Aeronautical Society and died eight years before the Society was founded by others, most of whom knew of Cayley's work and some had known him personally. I have, therefore, started my review of previous aeronautical predictions from Cayley and proceed from the beginning of the nineteenth century by six predictions to the immediate present. The six aeronautical predictions I have selected are:

On Aerial Navigation. Sir George Cayley, Bart. *Journal of Natural Philosophy, Chemistry and the Arts*, by W. Nicholson. Published in London. November 1809, February 1810 and March 1810[9].

Proceedings of the International Conference on Aerial Navigation, Chicago. August 1–4 1893[10].

The Future of Aerial Transport. Dr. A. P. Thurston. British Ministry of Reconstruction[11] 1919.

F. W. Lanchester—Various publications from 1907 to 1940.

Looking Forward. Prolegomena for a Detailed Study of the Future of British Aviation. Dr. H. Roxbee Cox. *Journal of the Royal Aeronautical Society*[12]. September 1940.

SST Prospectus—Economics of the United States Supersonic Transport. Lee R. Howard and Dr. J. I. Williams. *Lockheed Horizons*. Autumn 1965, page 2[13].

These are my personal choice; perhaps there are better examples but each of them seemed exceptional in some way and each seemed to be a collector's item.

Cayley

It is remarkable that this gifted man, the Father of Aerial Navigation and of Aerodynamic Science, should have published so many brilliant anticipations. It is also of significance that Charles Gibbs-Smith has delivered to us such a thorough study of his work extending over the period 1796–1855[14]. The papers referred to represent Cayley's first basic work from

1799 to 1809 and he was urged into print by a false report of manned mechanical flight by a Viennese, Jacob Degen, in 1809. Cayley was forthright about the status of aviation—'I cannot conceive, that in stating the fundamental principles of this art (aerial navigation), together with a considerable number of facts and practical observations, that have arisen in the course of much attention to this subject, I may be expediting the attainment of an object, that will in time be found of great importance to mankind; so much so, that a new aera (*sic*) in society will commence from the moment that aerial navigation is familiarly realised.'

Cayley believed that the secret of flying was to separate the provision of lift from that of thrust, i.e. unlike the bird, although he studied bird flight intensely in both flapping and gliding motion. In explaining the latter he separated air resistance into vertical and horizontal components analogous to lift and drag (although he did not use these terms). The beneficial effects of camber, ratio of lift to drag and reduced pressure over the wing were all known to Cayley. He proposed dihedral for stability and the need for a tailplane and controllable surfaces at the tail.

Interpreting some of his ideas in modern parlance we may cite:

Fluid motion a 'slender filament of current is constantly received at the anterior (leading) edge of the (wing) surface'. Surely the quoted idea is the forerunner of the streamline.

Structural design 'principle (of) this wing is that of the poles being couched within the cloth, so as to avoid resistance. The poles (spars) are not allowed to form the (leading) edge of the wing . . . and the thick part is purposely conveyed considerably within the edge.'

Aerodynamic design 'the shape of the hinder part of the spindle is of as much importance as that of the front, in diminishing resistance. This arises from the partial vacuity created behind the obstructing body . . . may probably be obviated . . . as this portion of the spindle . . . (may be) . . . a simple cone.'

Aerodynamic trade-off 'Let it be remembered as a maxim in the art of aerial navigation, that every pound of direct resistance that is done away, will support 30 pounds of additional weight without additional power.'

Airworthiness 'A hundred necks have to be broken before all the sources of accident can be ascertained and guarded against.'

It was many decades before better sense was added to the sum of the world's aeronautical knowledge. So much for his essential grasp of fundamentals founded not on guesswork but on calculation and experiment. For prediction he felt 'perfectly confident, however, that this noble art will soon be brought home to man's general convenience, and that we shall be able to transport ourselves and families, and their goods and chattels, more securely by air than by water, and with a velocity of from 20 to 100 miles per hour.' The speed of flight vehicles always seems to have been underestimated until we reach the predictions of earth satellites!

It is really quite astonishing how much profound prediction Cayley was able to write in a single sentence, e.g. 'Force can be obtained by a much lighter apparatus than the muscles of animals or birds and therefore in such proportions may aerial vehicles be loaded with inactive matter' (payload?) . . . 'but by increasing the magnitude of the engine 10, 50 or 500 men may equally be conveyed; and convenience alone, regulated by the strength and size of materials, will point out the limit of size of vessels in aerial navigation.' Perhaps the latter betrays a hint of optimism but the largest payload he quotes is a fair prediction of the Russian An 22 and the Boeing 747, 160 years later. Another hint of optimism (in 1810) —'It appears to me that *nothing more is necessary*, in order to bring the following principles into common practical use, than the endeavours of skilful artificers, who may vary the means of execution, till those most convenient are attained.'

In retrospect three essentials were missing (*a*) the propeller*, (*b*) means of lateral control, and (*c*) the means of disseminating knowledge of Cayley's ideas quickly and convincingly to those who would be receptive to them. I am sure Cayley would have been surprised that the first successful manned flight did not occur for another 94 years. Cayley was wise enough not to put dates to his predictions, but we are left convinced of his intuitive skill and his passionate belief in the future of aviation. But is there a hint of something still to be found—'the act of flying when properly adjusted by the Supreme Author of every power requires less exertions than, from the appearances, is supposed.'

Of course Cayley left behind him much more than speculations and predictions in aeronautics for he conducted experiments in which he measured airflow and the aerodynamic forces on moving bodies. He was also highly inventive and, for example, is credited with the discovery of the tensioned spoked wheel, Fig. 1. Starting with small gliders in which conventional wing and tail units were represented together with wing camber he continued experiments into full scale by two man-carrying gliders, one of which is acknowledged to have been flown (glided) across a Yorkshire valley, Fig. 2. The occupant was his coachman who was a rather reluctant passenger in the frightening novelty of the first unpiloted aeroplane.

International Conference at Chicago, 1893

There is no reason to think that this particular aeronautical meeting was special and it has been chosen to indicate the wide interest and varied

*Cayley did introduce the propeller in later writings.

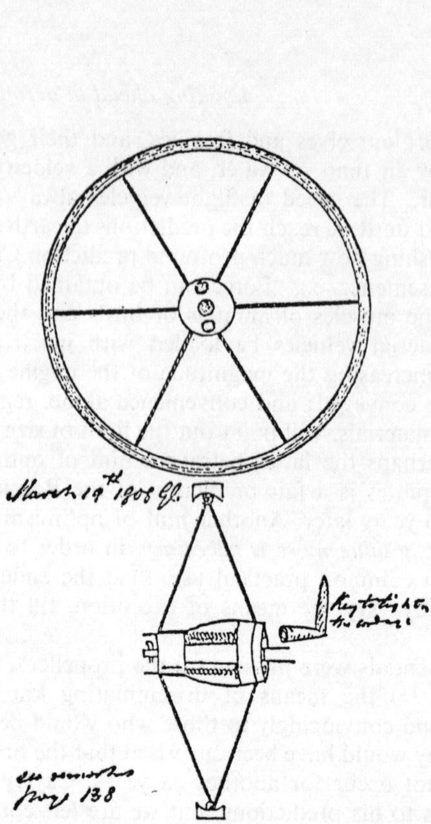

First design for the tension (or suspension) wheel for aircraft under-carriages: (a) side elevation; (b) cross-section, with sketch of the key for tightening the cordage: 1808.

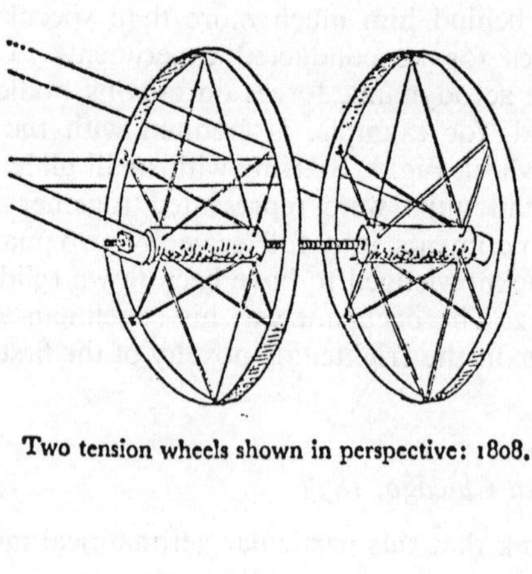

Two tension wheels shown in perspective: 1808.

Figure 1

Mechanics' Magazine,

MUSEUM, REGISTER, JOURNAL, AND GAZETTE.

No. 1520.] SATURDAY, SEPTEMBER 25, 1852. [Price 3*d.*, Stamped 4*d.*

Edited by J. C. Robertson, 166, Fleet-street.

SIR GEORGE CAYLEY'S GOVERNABLE PARACHUTES.

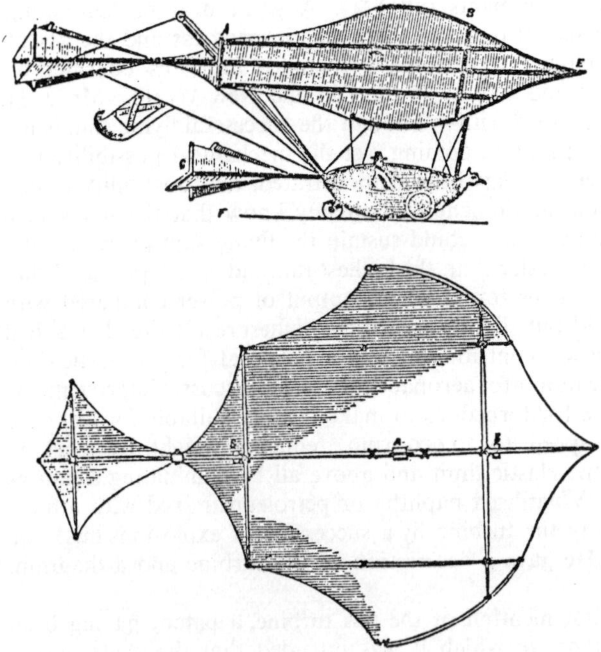

Developed design for a man-carrying glider, called a "governable parachute"; facsimile of the illustrations as they appeared in the *Mechanics' Magazine*: 1852.

Figure 2 From 'Sir George Cayley's Aeronautics 1796–1855'. C. H. Gibbs Smith. HMSO. London 1962

topics which were the subject of speculation a decade before the Wrights flew. It was held to coincide with the Chicago World Fair of that year and the participants were predominantly English and American with some French, Austrians and Danes; they included famous names such as Chanute, Langley and Zahm from the USA, Wenham and Baden-Powell from Britain and Lawrence Hargrave from Australia.

Thirty-six papers were presented showing greatest interest in balloons (20%), of which over half were presented by one author, design of heavier-than-air flying machines (14%), gliding (11%), propulsion, wind and upper atmosphere experiments (all 8%). A great deal of ballooning, gliding, kiting and model powered flight was in progress and the assembling of such a large international conference indicates considerable professional interest and activity in the Old and New Worlds. Mr. J. H. Dow started his lecture,[10] with—'Though the successful flying ship is not yet built and may be long in coming, yet the mechanical possibility that the air may be navigated has been demonstrated, and possibility is sure prophecy of eventual attainment. We already know that the screw propeller aided by the aeroplane could sustain the flying ship in the air and scale [propel] it ahead faster than the highest railroad speed, provided that a reliable motor could be found whose output of power compared with its own weight would but slightly exceed the highest result already reached experimentally.' Dow's contribution was a proposal for an elastic-fluid turbine as a possible motor for aeronautical use. He discusses improvements to his steam turbine for torpedoes to make it more suitable for airborne use by increasing speed to 20,000 rpm, reducing weight by working materials close to the elastic limit and above all by eliminating steam as the working fluid. 'Volatilised naphtha or petroleum mixed with atmospheric air and driving the turbine by a succession of explosions might be highly successful.' He gave a description of the turbine and a diagram, Fig. 3.

This is not the first mention of the gas turbine, a patent having been taken out before 1800, in which it was intended that the shaft output should drive a propeller. Nevertheless it is an interesting example of far-sighted engineering. In conclusion he made two other prophetic observations—'civilisation will undergo a crucial test of its moral stability in the tremendous economic revolution that must attend the successful introduction of aerial navigation, when everybody the world over shall become neighbour to everybody else, and neither fortifications, nor battleships, nor Chinese walls, nor cordons of customs officers can offer any barrier to enemies or intruders. But the world is safe for the present—great inventions do not appear until the moment when civilisation becomes ripe for them. The flying ship is a certainty of the future, and with equal certainty must come the moral climax of universal brotherhood.'

At this 1893 conference a surprising amount of data on materials was given. Steel was considered to be the best aeronautical material although

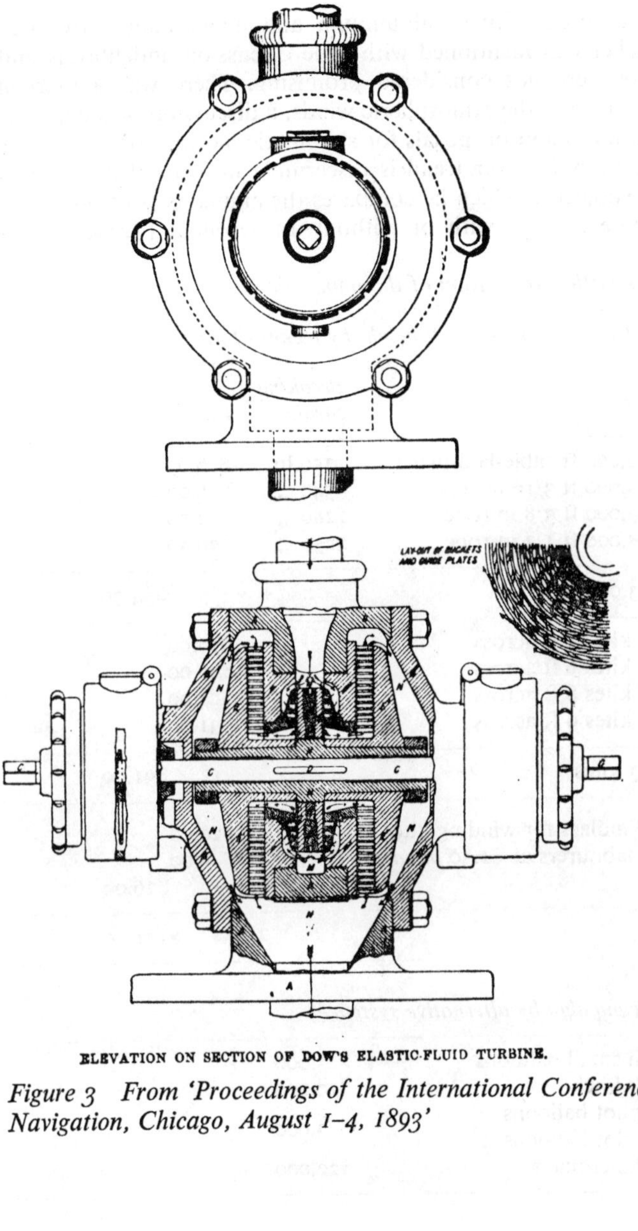

ELEVATION ON SECTION OF DOW'S ELASTIC-FLUID TURBINE.

Figure 3 From 'Proceedings of the International Conference on Aerial Navigation, Chicago, August 1–4, 1893'

the light-weight potentiality of aluminium and magnesium alloys was recognised. Nickel was mentioned with little discussion and fibrous and textile materials were not considered promising. There was a marked interest in the nature of the atmospheric winds, a discussion of the atmosphere up to ten miles and proposals for systematic meteorological studies up to this height. A familiar trend is discernible in Mark Harrington's paper[10] on this subject in which he compares the annual cost of operating a balloon or kite system, with or without an aeronaut (Table 1). He

TABLE 1 *Systematic exploration of the upper air*

Cost to carry instruments to 15,000 ft height by means of kites[10]

		Breaking Strain	*Cost*	
8 highest kites	12,000 ft cable-laid twine	250 lb	$ 6.00	
8 next ,,	3,000 ft 3/16 in rope	540 ,,	$ 6.00	
8 next ,,	3,000 ft 3/8 in rope	1280 ,,	$12.20	
9 next ,,	5,000 ft 1/2 in rope	2250 ,,	$30.50	
	23,000 ft			$54.70
Kites	8 kites 5 ft across		$16.00	
	8 kites 6 ft across		$20.00	
	8 kites 8 ft across		$24.00	
	9 kites 9 ft across		$31.00	
	33 kites			$91.00
	Windlass for winding kites		$20.00	
	4 labourers at $1.50 *per diem*		$ 6.00	
				$26.00
Total				$171.70

Cost of a years' campaign by alternative systems[10]

To 3,000 ft with small balloons	$ 3,000
14,000 ft with kites	$10,000
20,000 ft 52 pilot balloons }	$ 3,000
50,000 ft 12 pilot balloons }	
20,000 ft with aeronaut	$20,000

favoured a 'man in upper atmosphere' programme costing $20,000 per station per year, and the meeting unanimously agreed that Congress should be urged to approve expenditure to carry out this proposal.

None of these predictions are of tremendous significance, nevertheless we note comparative cost estimates, scientifically planned weather studies, appeals for government backing for such programmes and the idea of the intrusion of moral judgment in aeronautical affairs. It was, however, the first occasion at which the aeronautical world saw Hargrave's box kites which were to have a profound effect on the early designs of aeroplanes.

There is an obvious parallel between the proceedings of this 1893 Aeronautical Conference and the International Astronautical Congresses from 1950 onwards. Whereas in aeronautics the theoretical deliberators were replaced ten years later by the practical sporting aeronauts, the space flight pioneers were in their time taken over by the great national astronautic organisations.

Thurston, 1919

Gibbs-Smith has remarked on the luck needed to trace gems in aeronautical history: I consider myself fortunate to have discovered Thurston's 1919 paper. It is not well known and, as an official report, does not quote him as its author. As A. D. Young has said[15] 'It is remarkable both for its sound common sense and its vision.' It was the Plowden Report of its day, done by one man in 8000 words and for a fee of twenty guineas. In spite of its small size one cannot recall another paper of such scope, so simply written and which has so well stood the test of history. Although he did not consider air transport within Great Britain to hold much promise his caution relates to the assumed speeds of 100 mph. For overseas flight of passengers, freight and mail he had great hopes but he realised the financial implications and discussed State assistance to private enterprise and State ownership of airlines. Reflecting current opinion of the day he believed there was a future for both the aeroplane and the airship. His paper dealt adequately with legislation for safety, meteorological services, research and education in which he saw an increasing demand for well qualified aeronautical engineers and pilots. A good example of his imagination was the idea that the fuel supply in Britain might be revolutionised by the momentous effects which might result from the discovery of oil in Derbyshire and other parts of England.

Thurston's prediction of inventions is surely the most outstanding:

'The following lines of invention may be indicated as being of great importance to commercial aeronautics: the gyroscope and its incidental uses, the turbine system applied to combustion engines, and directional wireless. Developments may be anticipated in the servo-motor which will make large machines easy to control, and in automatic landing gears, which will make safe the difficult operation of landing at night. The metal aeroplane is still in its infancy and promises large developments in systems of construction.

'The possibility of revolutionary inventions should always be kept in mind. Thus a new form of motor utilising the radio-active energy of the earth, or the internal energy of the atom, might be evolved at any moment as our knowledge of matter and ether progresses. At the end of the 19th century, before the discovery of Hertzian waves, the possibility of wireless telephony or telegraphy was on the extremest limits of imagination, and the evolution of this motor may not be more distant than wireless telephony and telegraphy were less than a generation ago. We know that the store of energy is there: we have but to discover how to tap it at its source.

'Nor is this all, for beyond arises the possibility of levitation or some form of auto-gravity as the reward of deeper knowledge of the construction of matter and universe. Already model electric cars have been driven through the air without support and without rails by the repulsion of electric currents. But this system is costly in energy. Some day a cheaper system of levitation may be discovered.'

I suspect that Thurston would not have claimed originality for any of these for all were mentioned elsewhere, but in the context of this official document they take on greater significance. One suspects that his list did not exhaust all speculative ideas of his day—his skill lay in choosing the more advantageous devices even though their execution was beyond the engineers of the time. However, one of his ideas which does not seem to have succeeded was 'an all round improvement in efficiency and weight saving is to be expected and some possible combination of airship and airplane may exercise far reaching effects.'

Thurston's conclusions I think will stand as a classic example of simplicity in forward thinking and I quote them here in full:

'The general position in regard to the future of civil aviation may be summed up as follows:

1 The maintenance of an Air Force backed by a strong commercial air fleet is as vital to the safety and prosperity of the British Empire as the maintenance of a strong Navy and a strong Mercantile Marine.

2 The aircraft industry must be kept vigorous in order to respond to any possible war emergency of the future.

3 The development of civil aerial transport services will not at present be sufficient to keep the industry alive unless it receives substantial support from the state.

4 Meanwhile the commercial possibilities of aircraft will be developed, but their progress depends on:

(*a*) A growing sense of confidence on the part of the public in the safety of aircraft.

(*b*) Such developments in construction, organisation etc., as will enable aircraft to offer speedier transport than land and sea services.

(*c*) Large-scale reductions in cost.

5 At present the only classes of traffic for which any public support may be reasonably expected are:

(*a*) Limited amount of pleasure traffic.
(*b*) Special rapid business journeys, such as from London to Paris and back in a day.
(*c*) Express mail and parcels services between towns not less than, say 400 miles apart.
6 Before civil aviation can become a practicable commercial proposition it is essential to lay out considerable sums of public money on landing grounds, the organisation of aerial routes, research and education.'

One final small point: Thurston's paper was one of a series published by the UK Ministry of Reconstruction, which also included a paper on New Fields for British Engineering and another on Scientific Business Management.

The previous three examples of prediction were made in a qualitative manner at a time when there was a paucity of information and only rudimentary mathematical techniques for study and prediction were available. A certain type of mind is needed to make successful predictions in such an environment. The fifth and sixth papers are examples of predictions made on the basis of a significant amount of technical, economic and operational data and both use mathematical studies to improve the quantitative value of the analysis. A distinction might therefore be drawn between looking ahead in a speculative way and looking ahead on the basis of technical forecasts.

F. W. Lanchester

Before proceeding further, however, I would like to discuss some unique contributions to quantitative and intuitive prediction made by F. W. Lanchester whose aeronautical work started before the 1893 Conference and continued beyond Roxbee Cox's 1940 paper. Lanchester, unlike Thurston, has not left us a single, forward looking paper—for he despised the idea of 'propheteer'. He was often called a prophet: indeed, H. E. Wimperis, the Society's President of 1937, described him as 'one of the major prophets of aeronautics' a title he gracefully declined[16]. Nevertheless, although he himself considered his work, seen by others as prophetic, to be but research and the result of logical deduction from observation he certainly brought a surprising intuition to bear in posing and solving problems, many of which did not become generally recognised until ten, twenty or even thirty years later.

Two aspects of his work seem relevant to this chapter. First, a brief description of some of his 'predictive' researches and second, his introduction of mathematical techniques, particularly into operational research. In both these aspects there seems to be much sound judgment which is equally true in the very different circumstances of today.

In his *Aerodynamics* book of 1911 Lanchester[17] discussed an electro-magnetic analogy for Eulerian inviscid fluid flow. He concluded that 'In spite of difficulties, it appears probable to the author that in the near future some use may be made of existing electrical theory as an auxiliary means of investigating the aerodynamics of flight. Thus, in the general dynamics of the periptery (Lanchester's term for lifting circulatory flow), and in connection with the relations of the strength of the cyclic motion and the magnitude of the load reaction, it may be that mathematical solutions exist, ready to hand, in the analagous electrical theory, such as appropriately interpreted may some day be found to be of service.'

Malavard in 1947 presented a paper[18] on 'The use of Rheo-electrical analogies in certain aeronautical problems' describing a technique which is still employed very effectively in France.

In some respects Lanchester's predictions were modest, e.g. the maximum range of an aeroplane he calculated to be 360 miles in 1907 and 2000 miles in 1915[19]. Occasionally he was proved to be far wrong, e.g. in stating (1915) that not less than some 100 or 150 acres should be sacrificed for every important flying ground. With major present-day airports of the order of 50 times greater might we consider Lanchester in error, or that perhaps aeronautics has, of necessity, developed along an erroneous path? With research into VTOL we now seem at last to be approaching an airfield size which he advocated as sufficient, bearing in mind the economic burden of a very large airfield. Much later (1937) he recognised that flight efficiency called for wing loadings of 50 lb/sq ft (2 kN/m^2) and that the corresponding take-off problem would need either long runways, of the order of three miles (5 km) in length, or a two-stage aeroplane. On this he concluded that the runway would be the solution ultimately adopted[20].

Von Kármán[21] considered Lanchester's papers on 'Flight at High Altitude'[22] (1937) and 'Rocket Efficiency' and 'Exhaust Efflux Propulsion'[23] (1939) to 'represent a very competent discussion of the correlation between high speed and high altitude with very much foresight into the development which took place in the following twenty years.' He also noted that Lanchester, analysing stability of aircraft in 1908, foresaw aeroelastic effects decreasing the stability of wings due to flexibility[21].

In 1914 Lanchester published his *Aircraft in Warfare,* acknowledged to be one of the fundamental sources of inspiration for the science of Operational Analysis. In this Lanchester discussed the strategic and tactical use of aircraft, aeroplane bombing, armament and armour in the aircraft, airborne torpedoes, rockets, aircraft carriers etc. The major mathematical contribution was his '*N*-square Law' which considers the numerical strength in weapons or weapon systems necessary for annihilation of the enemy. Like all Lanchester's mathematics the law is simple, showing that the fighting strengths of the two forces are equal when the squares of the numerical strengths multiplied by the fighting values of the individual units are equal, i.e. $\eta_1 N_1^2 = \eta_2 N_2^2$. He gave several worked examples of this law and analysed some classic battles in terms of it: he attributed 'the Nelson touch' to an intuitive application of this law.

I have introduced this mid-way discussion of Lanchester because he spans across the period of generally non-mathematical intuitive speculation to the latter period of strong mathematical support for technical forecasting. In all his researches Lanchester gives quantitative evidence and derives quantitative laws, and yet he keeps his mathematics outstandingly simple[24]. At the conclusion of his 1937 paper on skin friction he cautioned:

'I am inclined to give a word of warning on the subject of the application of mathematical elaboration when discussing real problems in aerodynamics. Although the arithmetic concerning the economics of flight is more precise as depending upon the first principles and less upon empirical data than is the case in other modes of locomotion, it is possible to overweight the subject with elaborate mathematics. It is rare that anything more than the simplest algebraic expressions, scarcely worth dignifying with the name of mathematics, serve any useful purpose. In this connection the most elaborate investigation I have found necessary is that relating to the differentiation of an expression resulting in two simultaneous equations, in order to obtain an expression giving the condition of least resistance, taking account of the weight of the aerofoil.'

Roxbee Cox, 1940

This famous paper has several similarities to Thurston's. It was an overall programme for civil aviation written at a time when aviation was overshadowed by war and its purpose was sharpened by it: and it was comprehensive. Comparison of the fine structure of the text clearly shows the advances made in British aeronautics in two decades. By 1940 the question of the airship for the long-haul routes had been settled, the future of the flying boat was full of promise but trans-Atlantic range was difficult to attain so that efforts were being made with flight refuelling and two-stage aeroplanes. The paper made use of what we might now call market survey techniques and the need for greater size and speed was urged. The author settled for a 300 mph (130 m/s) 5000 mile (8000 km) range air liner: whether for 50 or 90 passengers was still to be decided. Construction would be preceded by an overnight mail plane as a less expensive prototype. The idea of the special mail plane which was the order of the day in the twenties still lingered into the 1940s but was not to survive.

Following his experiences on the structural design of the large airships in the twenties Roxbee Cox studied the application of the square-cube law to aircraft which was currently thought to limit their size. Cayley had introduced this law to aviation over a century before, showing how increasing size from birds to man-carrying aircraft favoured the carriage of payload. Roxbee Cox quoted design studies for large propeller-driven aircraft up to one million pounds in weight. (Fig. 4). Such thinking led to the all-wing arrangement, as Junkers had concluded in 1922, and because of airfield size limitations this favoured the flying boat (Fig. 5).

Although Roxbee Cox's outline designs were overtaken by the development of concrete wartime runways which sent the flying boat to obsoles-

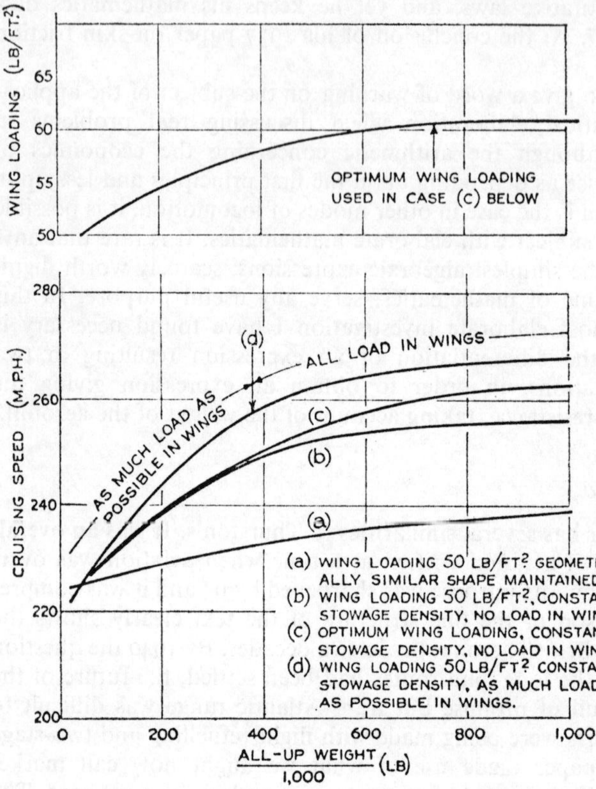

Figure 4 Cruising speeds of aeroplanes of different proportions. From 'The Journal of the Royal Aeronautical Society', September 1940

cence and the fact that the coming of the jet engine altered the thrust-weight ratio, speed and range radically, his ideas were nevertheless characterised by boldness:

'Without . . . a progressive civil aviation policy . . . our civil design will become pedestrian, afraid to dare, making the big jumps—like the increase of wing loading due to flaps, the change in technique due to nose wheels—only when these have been successfully cleared by the leaders.' Although he could not say so at the time Roxbee Cox knew very well the secret progress made with Whittle's jet engine and must have appreciated its potential for civil aircraft.

The history of aviation has shown that there are unexpected discontinuities in technique (e.g. the demise of the airship; jet engine; sweepback; space flight; etc.) and that these can occur within a decade. Recognising

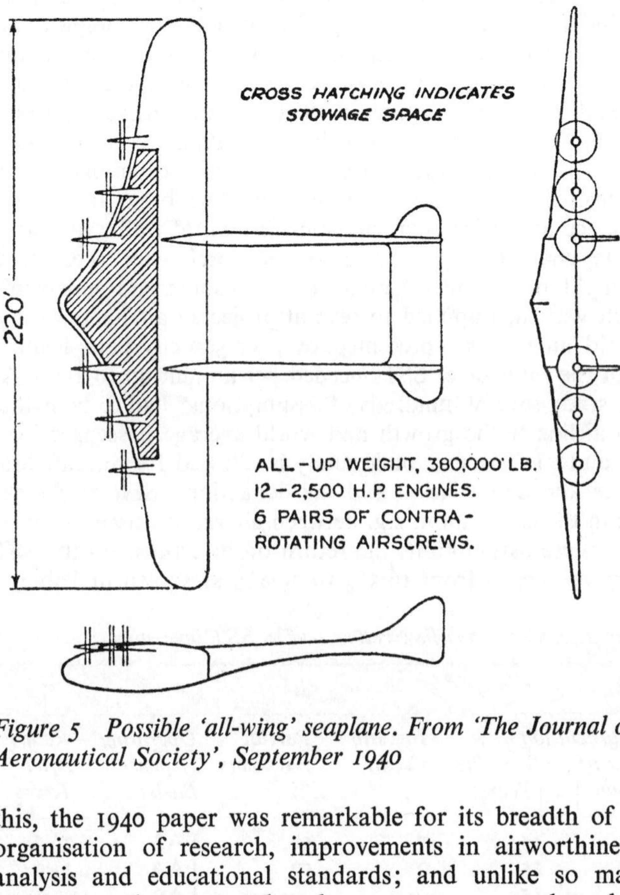

Figure 5 Possible 'all-wing' seaplane. From 'The Journal of the Royal Aeronautical Society', September 1940

this, the 1940 paper was remarkable for its breadth of vision into the organisation of research, improvements in airworthiness, into design analysis and educational standards; and unlike so many predictions which have been overtaken by events, no aeroplane has yet reached 1,000,000 lb (450 Mg) weight, although Jumbo jets now building at near 700,000 lb (318 Mg) approach this closely.

Howard and Williams, August 1965

This study of the likely market and finances of the SST is a fair example of prediction in the present era of aeronautics. It may be called a prediction for it looks ahead to the year 1990, and yet it appears wholly unlike the previous examples and is more like a complex piece of accounting. A lot of changes had taken place since 1940. Differential analysers were very recently available at the time of Roxbee Cox's paper but not for aeronautics. Large computers were introduced for nuclear physics in the 1940s, and during the 1950s the use of analogue simulators to predict performance of guided weapons was commonplace. Then came space flight, the massive digital computer and its deployment both for complex performance analysis and market research studies. The paper by Howard and Williams considers both aspects[13]. Its object was to assess the economic effect of a US-built supersonic transport (SST) on the airlines and also on the economy of the USA. The vast scale of this undertaking is underlined by the seven year evolution of a computer-based economic model of airline systems. This programme allows for five basic inputs—(*i*) aircraft performance and price characteristics, (*ii*) world traffic forecasts, (*iii*) airline route characteristics, (*iv*) passenger preference (i.e. fares, frequency and flight time) and (*v*) mixture of subsonic and supersonic traffic. The result was then applied to several projected air liners over a selection of world air routes represented by over 500 city pairs leading to an estimate of the number of SSTs needed per annum up to 1990. At this stage of the study several hundreds of assumptions* had to be made, including world airline traffic growth and world average passenger load factors (Figs. 6 and 7). The economic analysis allowed for aircraft first cost, maintenance and depreciation and tax deductions. Several choices had to be made in the calculation, e.g. between alternative fiscal policies. Depending upon these assumptions the return on investment in the SST after tax deductions ranged from 16·5% to 29·8% as shown in Table 2.

TABLE 2 *Airline discounted cash flow return on US SST investment*[13]

Assumptions						
Discounted Cash Flow Schedule	*Depreciation Tax life (Years)*	*Book life (Years)*	*Investm't Credit %*	*Partial Payments %*	*Operating Expense Basis*	*Return After Taxes*
Basic	6	15	7	50	FAA	26·3 %
Alternate 1	15	15	0	50	FAA	21·9
Alternate 2	15	15	0	0	FAA	29·8
Alternate 3	15	15	0	50	FAA +20%	16·5

The final stage in the economic analysis was the difference in the USA balance of payments resulting from the decision to build an SST.

*'assumption' may imply insufficient accuracy; each would be the subject of an 'auxiliary prediction study' and in some cases agreed predictor formulae were used e.g. supersonic transport model ground rules.

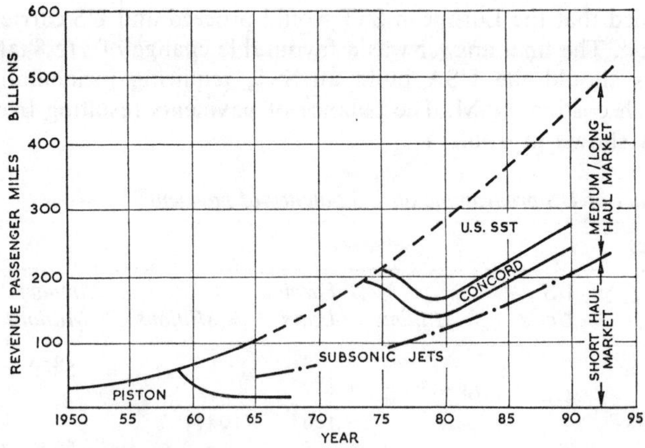

Figure 6 (above) Division of Free-World Passenger Traffic

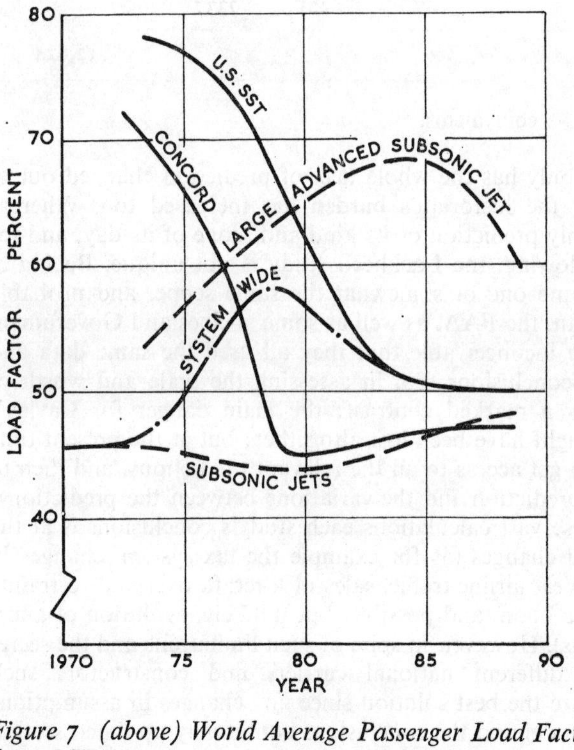

Figure 7 (above) World Average Passenger Load Factor. Both figures from SST Prospectus—Economics of the United States Supersonic Transport. 'Lockheed Horizons', Autumn 1965

It was assumed that the European SST would proceed and US carriers would buy it too. The final answer was a favourable change of $12,842M in US reserves should the USA build an SST, requiring incidentally, 500 aircraft each costing $33M. The balance of payments resulting from this forecast is shown in Table 3.

TABLE 3 *Impact of US programme on US balance of payments*[13]

Airline Purchases

	US Units	*$ Millions*	*Foreign Units*	*$ Millions*	*Impact $ Millions*
Without US SST					−5829
Concordes	410	6876			
US Subsonic Jets			126*	1047	
With US SST					+6995
Concordes	74	1239			
US Subsonic Jets			108*	897	
US SST			221	7337	
TOTAL IMPACT					12,824

*Numbers of 707/DC-8 equivalents.

We see that not only has the whole task of prediction changed out of all recognition but the historian's burden has increased too. Whereas Cayley's was the only prediction of its kind and scope of its day, and for several decades following, the Lockheed study is not unique. Boeing at least, must have done one of somewhat the same scope, and probably also North American, the FAA, as well as some airlines and Government agencies. It is quite inconceivable that they all used the same data and came to the same conclusions. So, in assessing the scale and worth of predictions there is a marked contrast: the main danger for Cayley's work was that it might have been lost altogether; but at the present time the major task is to get access to all the relevant predictions, and then to discover the mean prediction and the variations between the predictions! And in spite of these vast calculations each study's conclusions is at the mercy of peripheral changes (as for example the tax system, changes in the share of East-West airline traffic, sales of aircraft, overland restraints because of the sonic boom and possible, but unlikely, evolution of alternative travel systems). However, in spite of such limitations and the secret interplay between different national carriers and constructors such prediction studies are the best solution since any changes in assumptions can be put in the model and the conclusions continually up-dated so that trends can be anticipated and, given reasonable warning, they will usually be allowed for.

I have emphasised in these examples the considered study as this probably gives the best guidance for the future. To conclude I have collected a few brief predictions apparently made in all seriousness by well-informed people which show how dangerous forecasting can be and will certainly emphasise the virtues of the other paper I have referred to.

'The demonstration that no possible combination of known substances, known form of machinery and known forms of force can be united in a practicable machine by which we will fly long distances through the air, seems to the writer as complete as it is possible for the demonstration of any physical fact to be. May not our mechanicians . . . be ultimately forced to admit that aerial flight is one of the class of problems with which man can never cope, and give up all attempts to grapple with it?' Simon Newcomb, American scientist, *circa* 1900.

Also, W. H. Pickering, Professor of Astronomy, Harvard University, in 1908—'It is doubtful if aeroplanes will ever cross the ocean, and despite the Wright success they offer little menace to warfare. The public has greatly overestimated the possibilities of the aeroplane, imagining that in another generation they will be able to fly over to London in a day. This is manifestly impossible, because every time the speed of an object through the air is doubled the resistance is quadrupled. The popular conception of a machine being perfected in a few years which shall be capable of carrying twenty or thirty people is nonsense. The aeroplane will be valuable for reconnoitering the enemy, also for dropping explosives. That is all.'

3 CONCLUSION

> 'The human mind is often so awkward and ill-regulated in the career of invention that it is at first diffident, and then despises itself. For it appears at first incredible that any such discovery should be made, and when it has been made, it appears incredible that it should so long have escaped men's research.'
>
> FRANCIS BACON

Reviewing past predictions in this way reveals some persistent ideas that have ultimately been realised after a great elapse of time, some of them even after over a century. The jet engine, the 500-seater aircraft and automatic landing systems are such examples. Other concepts such as auto-gravity or combined aircraft-airships seem to have got nowhere and many other inventions and techniques have emerged with little prior anticipation. In one sense we tend to think that at the present time there is much more scope for possibilities than ever before, but even Cayley, writing to Henson in 1846 felt that 'as to new principles there are none; of practical expedients there will soon be an endless variety and to select the best will be the point at issue.' A very special problem of our day however is the narrowing of the time interval between discovery and application in the physical sciences, for example the interval was 65 years for the electric motor, ten years for the nuclear reactor, five years for radar, three years for the transistor and two years for the solar battery. Such an acceleration in the pace of development seems to bring speculative ideas rather closer than might be expected.

So much for the past. At the present time several developments are proceeding that seem certain to come to fruition during the next century, such as city centre to city centre VTOL, hypersonic civil and military flight, very large subsonic transports, and so on. Can we be as confident that nuclear propulsion, an operational laminar flow transport, or man-powered flight will be achieved? Avoidance of congestion at airports —both on the ground and in the air—seems likely to become a matter of ever-increasing concern—who can foresee how this may be resolved in even the next thirty years? Might solutions to such transport system problems place unfamiliar restraints on the otherwise natural developments of subsonic and supersonic transport aircraft? Although banished from the main stream of aeronautics, airships and flying boats are still receiving some attention. Is it possible these may make unexpected come-backs? The possibilities for new materials both for structure and internal systems offer tremendous scope for future research, the outcome of which can only be dimly visualised. The whole field of aeronautics is still richly strewn with an unending list of exciting questions.

So much for the questions of the present. Even though they may not be answered for some time, nevertheless the future possibilities and promises for aeronautics look very rich indeed.

References

1 *The Centenary Journal of the Royal Aeronautical Society*, January 1966.

2 *Science Journal*, October 1967.

3 JOUVENEL, BERTRAND DE. *The Art of Conjecture*. Weidenfeld and Nicolson, London, 1967. Translated from the French by Nikita Lary.

4 ARMYTAGE, W. H. G. *Yesterday's Tomorrows*. A Historical Survey of Future Societies. Routledge and Kegan Paul, London, 1968.

5 BROWN, H., BONNER, J., and WEIR, J. *The Next Hundred Years*. Weidenfeld and Nicolson, London, 1957.

6 JANTSCH, E. Forecasting the Future. *Science Journal*, p 40, October 1967.

7 JEWKES, J., SAWERS, D., and STILLERMAN, R. *The Sources of Invention*. Macmillan, London, 1962.

8 DARWIN, CHARLES GALTON. *The Next Million Years*. Rupert Hart-Davis, London, 1952.

9 CAYLEY, SIR GEORGE. On Aerial Navigation. *Journal of Natural Philosophy, Chemistry and the Arts*. W. Nicholson. London. November 1809, February 1810 and March 1810.

10 *Proceedings of the International Conference on Aerial Navigation, Chicago*. 1–4 August 1893.

11 THURSTON, DR. A. P. The Future of Aerial Transport. British Ministry of Reconstruction 1919.

12 ROXBEE COX, DR. H. The 28th Wilbur Wright Memorial Lecture. Looking Forward. Prolegomena for a Detailed Study of the Future of British Aviation. *Journal of the Royal Aeronautical Society*, Vol **44**, No 357, September 1940.
13 HOWARD, LEE R. and WILLIAMS, DR. J. I. SST Prospectus—Economics of the United States Supersonic Transport. *Lockheed Horizons*, p 2, Autumn 1965.
14 GIBBS-SMITH, C. H. *Sir George Cayley's Aeronautics 1796–1855*. HMSO, London, 1962.
15 YOUNG, A. D. Dr. A. P. Thurston, A Review of His Contributions to Aeronautics. *The Newcomen Society*. 6 April 1966.
16 LANCHESTER, F. W. The Part Played by Skin-Friction in Aeronautics. *Journal of the Royal Aeronautical Society*, p 67, Vol **41**, February 1937.
17 LANCHESTER, F. W. *Aerodynamics*, p 425, 1911 Edition, Constable, London.
18 MALAVARD, L. The Use of Rheo-electrical Analogies in Certain Aeronautical Problems. *Journal of the Royal Aeronautical Society*, p 739, Vol **51**, September 1947.
19 LANCHESTER, F. W. The Flying Machine from an Engineering Standpoint. James Forrest Lecture, 1915 and book Constable and Co. Ltd. London, 1916.
20 LANCHESTER, F. W. The Part Played by Skin-Friction in Aeronautics. *Journal of the Royal Aeronautical Society*, p 125, Vol **41**, February 1937.
21 VON KÁRMÁN, TH. The First Lanchester Memorial Lecture. *Journal of the Royal Aeronautical Society*, p 89, Vol **62**, February 1958.
22 LANCHESTER, F. W. Flight at High Altitude. *Journal of the Royal Aeronautical Society*. Parts I and II, Vol **41**, May and June 1937.
23 LANCHESTER, F. W. Exhaust Efflux Propulsion. *Flight*. 16 and 23 November 1939; 7 and 14 December 1939.
24 LANCHESTER, F. W. Sustentation in Flight. The 14th Wilbur Wright Memorial Lecture, 27 May 1926, *Journal of the Royal Aeronautical Society*, p 587, Vol **30**, October 1926.

2 An aerodynamicist's prospect of the Second Century

D. Küchemann

I INTRODUCTION

This chapter is restricted in many ways. It deals only with *aircraft*, that is to say, the means of transporting people and goods through that 'uninterrupted navigable ocean that comes to the threshold of every man's door', as Cayley put it. This means that orbital vehicles and spacecraft are not discussed and, also, that the emphasis is on civil aircraft. The discussion is further restricted in the main to *aerodynamic* aspects. It is also limited to aerodynamic shapes which are firmly based on present concepts in fluid mechanics, on the assumption that the physical principles implied will not be refuted during the time under consideration. Hence, the discussion is centred on the possible further exploration of these principles and not on invention or imagination. Moreover, the opinions expressed must of necessity be strictly personal and, since such a long view is to be taken, no proof can be offered for any of the conjectures and statements made.

The task of reviewing the prospects over a whole century may be regarded as being at once impossible, presumptuous, and not without a touch of the absurd. Also, as far as the author is concerned, he cannot be taken up on what he says, if the world turns out differently. However, he may assume that human society still exists in a hundred years' time and this in turn allows him to assume that the necessary steps will have been taken by then to make this possible. Further, a responsible working man will want to take his mind off his day-to-day jobs, occasionally, and look at the prospects before him. It is in this state of mind that some thought has been given to the matter, with results which are now offered in this chapter.

Some privileges should be implied in being asked to write such a chapter. The long view should entitle the author to make wide overall assumptions

Originally published in the *Aeronautical Journal of the Royal Aeronautical Society*, February 1968.

and to present results to a 'zero-order' accuracy only. These privileges are hereby invoked.

This chapter should be read within the context of the other chapters, especially with regard to the selection of the material. The chapter follows closely a lecture given during the Centenary year[1] and the material it is based on can all be found in ref. 2. There is, however, no discussion here of the 'total transport system' into which the aircraft considered should, no doubt, fit; this important matter is regarded as being beyond the scope of this chapter.

There are three main sections. One is concerned with past developments and present conditions so as to set the scene for the future. Another is concerned with the technical prospects, where the aerodynamic aspects receive special attention. Since these prospects are closely tied to what we might want these aircraft to do, some remarks are made in the final section about possible future changes in the external motivation for developments in aviation and on how its aims may be put on a more rational basis.

2 INITIAL CONDITIONS

Before we start taking a long view forward, we may be well advised to take at least a glimpse backward over past developments. There may be some lessons to learn, and we may be able to define some initial conditions for the journey into the future.

The art of flying through the air has, of course, been developed first in the animal world, with man as the envious on-looker for many thousands of years. Flying has been invented independently four times by different kinds of animals at different stages of evolution, by the insects, the saurians, the birds, and the bats (*see* e.g. ref. 3, chapter 11). The same primary aim has been achieved at different times by a great variety of different means. But all the different species had to adapt themselves to the laws of aerodynamics and, in the present context, it seems most remarkable that during a gradual development which has taken millions of years practically all of them have adopted the same aerodynamic solution, namely, what we now call the Kutta-Joukowski flow over *classical aerofoils*. They usually employ the forces nearly normal to the relative wind direction, which act on lifting surfaces at what we call small angles of incidence. It is not surprising, therefore, that we find the same solution employed also in human engineering and that we so often encounter troubles when we dare to go beyond this type of flow.

The reason why it took so long for human flight to come about must be seen in the fact that animal flight cannot (yet), and need not be, imitated for man-carriage: the imitators all failed. What was needed was first some *understanding* of the basic physical principles involved in animal flight and then a *translation* of these into what was feasible in terms of human engineering, involving in some cases also the lifting of some restrictions to which the animals are subjected (*see* e.g. ref. 4). It may be fairly stated that the first successful 'interpreter' in this sense was Sir George Cayley (*see* e.g. ref. 5), who, in 1809, developed the concept of what we now call the *classical aircraft type,* a wing-fuselage combination with separate and largely independent organs for providing volume, lift, propulsion, and controls.

Cayley's concept was enormously successful* and dominated the technical evolution up to about 1950, when Gabrielli and von Kármán[6] could rightly include among the various means of transport only one 'commercial airplane'. On the technical side, it is possible to trace developments in such a way that progress appears smooth and continuous, resulting from a steady stream of 'improvements', brought about by the diligent and steady efforts of scientists and engineers and of pilots who realised from the beginning that they must become an integral and living part of the aircraft (even though it might appear that they are now being slowly removed from the 'loop', in some cases). If we look at specific aircraft characteristics, such as weights and speeds as they have changed, we find that single curves (or two curves, if we distinguish between civil and military aircraft) can be drawn through sets of slightly scattered points (*see* e.g. ref. 7). We shall try to show later that there is no reason to believe that the development of the classical type of aircraft has come anywhere near its end: the continued application of diligent and steady efforts is likely to lead to further significant improvements so that the development of this type is likely to extend well into the Second Century.

From time to time, 'barriers' seemed to have appeared in the course of this development which led some to believe the end had been reached. Typical of this were apparent speed limitations. On the propulsion side, it was difficult to imagine piston engines operating on an Otto cycle able to produce the power required for operation at high speeds and altitudes, and this would have led to absurd weights and sizes. On the aerodynamic side, such compressibility effects as the drag rise and unsteadiness of the flow beyond the critical Mach number appeared to limit the flight speed to some value well below the speed of sound. Strictly, this limitation still applies to propeller-driven, straight-winged aircraft, and these branch off the main line of the classical aircraft and level out at about $M=0{\cdot}7$ (*see*, e.g. ref. 7). But these barriers were soon overcome by the development of the turbojet engine and of the swept wing, and these two new concepts alone led to the evolution of a new type of aircraft, the *swept wing*, which, of course, is still closely related to the classical type of aircraft but extends its speed range beyond the speed of sound. The evolution of this type of aircraft has gone on for some 20 years only and

*So successful, in fact, that most people still think every aircraft must have its fuselage, wings, and engines and that even most science-fiction writers tacitly assume that the men from Mars must follow Cayley.

may thus be regarded as being still in its early stages, with substantial developments to come.

The most important development during the past decade or so of the First Century may have been the realisation of the fact that aviation is not restricted to one major type of aircraft only, in its classical and swept-winged forms, but that there are at least two more: the *slender wing* and the *waverider*, which are primarily suited to flight at supersonic and at hypersonic speeds respectively. In the long view, this may turn out to have been a major step in the evolution of aircraft, widening enormously the field of what aviation can do for mankind, and providing work and fruitful applications for a long time to come, possibly for the whole of the Second Century. For, if we have not yet exhausted the possibilities of the classical aircraft which has occupied aviation for the First Century, who would dare forecast the day when aviation will have exhausted the possibilities of a whole *spectrum of major types of aircraft*?

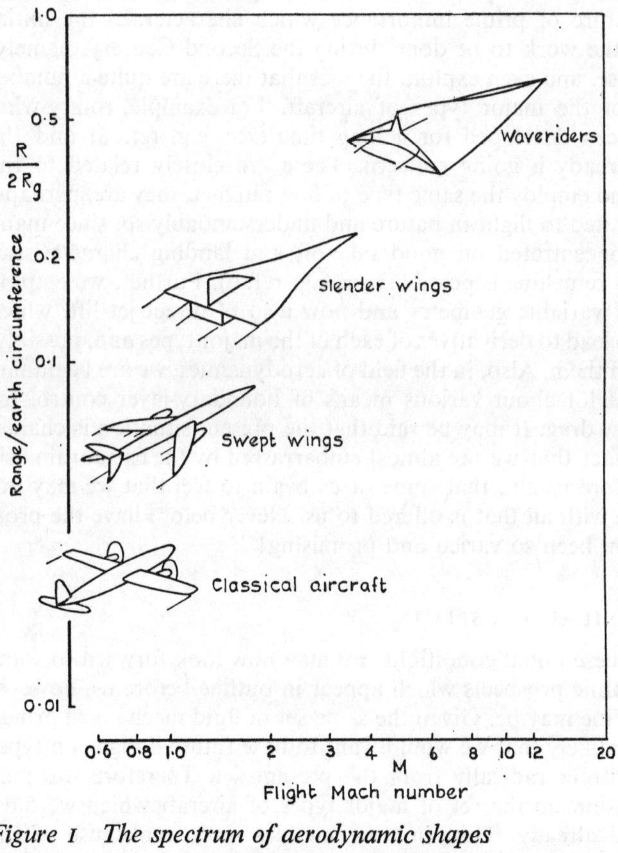

Figure 1 The spectrum of aerodynamic shapes

Unlike the classical, propeller-driven, type of aircraft, the new types are 'man-made' and have no counterpart in nature. But like the classical type, their shapes have been evolved by the researches of aerodynamicists, and their first distinguishing feature lies in the different *types of flow* they employ: leading-edge vortex flow for the slender type; and attached strong-shock-wave flow for the waverider type. We can just begin to perceive the major trends underlying the main properties of these types of aircraft and it looks as though they differ in their non-aerodynamic aspects also. An overall pattern is just beginning to emerge; to explore it, to consolidate it, and to exploit it in practical engineering applications will be obvious tasks for the Second Century.

This overall spectrum as it appears to us now in the form of aerodynamic shapes, is sketched roughly in Fig. 1. Some attempt to elaborate on this theme has been made in refs. 1 and 2, and it may be that primary ranges of application can now be assigned to each of the types in terms of suitable speed ranges coupled with suitable flight ranges. But there is yet another feature of prime importance which characterises the initial conditions for the work to be done during the Second Century, namely, that we recognise, and even exploit, the fact that there are quite a number of *derivatives* of the major types of aircraft. For example, rotary-wing aircraft have been developed for a long time (*see*, e.g. ref. 8) and the helicopter is already a going concern. These are closely related to the classical type and employ the same type of flow; in fact, they are, perhaps, most closely related to flight in nature and understandably so, since many species have concentrated on good take-off and landing characteristics so that frequent refuelling is possible (*see*, e.g. ref. 9). Further, we employ the concepts of variable geometry and now also of direct jet lift, which may turn out to lead to derivatives of each of the major types and, possibly, to links between them. Also, in the field of aerodynamics, we are beginning to learn quite a lot about various means of boundary-layer control for high-lift and low drag. It may be said that the present situation is characterised by the fact that we are almost embarrassed by the extraordinarily wide choice before us and that some of us begin to feel that we may not be able to cope with all that is offered to us. Never before have the prospects in aviation been so varied and promising!

3 SOME TECHNICAL PROSPECTS

Starting from these initial conditions, we may now look forward to some of the aerodynamic prospects which appear in outline before us, however blurred the outline may be. Given the same set of fluid mechanical principles, it seems unlikely that we would want to base future designs on types of flow which differ radically from the present set. Therefore, we may base the discussion on the set of major types of aircraft which we have roughly defined already. This, in turn, enables us to make use of the results of the analysis in ref. 2, but the time scale involved entitles us to take an even broader view.

We must assume that, during the Second Century, the economics and costs involved in transporting people and goods by air still matter and, further, that roughly the operating costs are related to the hours flown, that the revenue is related to the miles covered, and that both are related to the seat-miles covered in a given time (*see*, e.g. ref. 7). The parameters that interest us here are, therefore, the payload fraction W_P/W, the flight range R/R_g, the product $(W_P/W)\,(R/R_g)$ as a measure of the seat-miles produced, and the flying time t, where we use as standards a given overall weight W of the aircraft and half the earth's circumference, R_g, as the ultimate global range. In the present context, we may use Breguet values for R and t.

It is convenient to separate out several different aspects: *aerodynamics* (of the airframe and of propulsion); *weights;* and *fuels.* This can be done on the basis of the relation

$$W_P/W = k_1 - k_2 W_F/W$$

(*see* eqns. (3.1), (4.6), (5.9) of ref. 2), where k_1 and k_2 are related to the weights of various parts of the aircraft and W_F is the fuel used during the flight, which is determined by

$$W_F/W = 1 - e^{-(R/R_g)\frac{R_g}{H\eta_p L/D}}$$

(*see* eqn. (2.4) of ref. 2), where H is the calorific value of the fuel, η_p the propulsive efficiency, and L/D the lift-to-drag ratio during cruise flight. The parameter characterising the aerodynamic aspects is, therefore, the combined propulsive-lifting efficiency $\eta_p L/D$.

We now need some numerical values to indicate the spread between what we think would be attainable now and what may be attainable in the future. Not too much significance should be assigned to the actual values taken: the results would look much the same for a variety of sets of values. On the basis of the analysis in ref. 2, and in the present context, we need not make any distinction between the different types of aircraft as far as the aerodynamic parameter $\eta_p L/D$ is concerned and we may take 3 as a 'present' value and 5 as a possible 'future' value. The weight parameter k_1 (which depends on the wing weight, the cabin weight, the weight of the installed engine, the reserve fuel, the weight of the undercarriage, of the services and equipment, and of the furnishings and other operators' items) is the payload fraction at zero range, and we take 0·25 as a 'present' value and 0·35 as a possible 'future' value for all types of aircraft. The weight parameter k_2 (which depends on the cabin weight and the weight of the furnishings) may be taken as 0·45 for the 'present' and as 0·55 for the 'future', for all types of aircraft. We must bear in mind here that not all the types of aircraft exist 'at present' and that for these even the 'present' values are hypothetical. The long view asked for allows us this liberty. For the calorific value H of the fuel, we may take as a 'present' value the value for kerosene, $H=2{\cdot}7\times10^3$ statutory miles for all types of aircraft, and as a possible 'future' value for a high-energy fuel the value for hydrogen, $H=7{\cdot}3\times10^3$ statutory miles, probably for waverider aircraft only.

With these two sets of values, curves for W_P/W and for $(W_P/W)\,(R/R_g)$ as they depend on R/R_g can be drawn and are shown in Fig. 2. An additional scale is shown in the middle, which gives the Mach number at which the aircraft must fly to make the *flying time* two hours. Together with that scale, possible *ranges of application* for the various types of aircraft can also be indicated. This should speak for itself. We would only draw attention to what may be regarded as 'inhabited areas', at this stage, the longish one being that inhabited by classical and swept-winged aircraft (in that case, the Mach number scale does not, of course, apply), and the small circle being that to be inhabited by slender wings such as the Concorde. Comparing these inhabited areas with the extent of the as yet unexplored regions, we are tempted to conclude that aviation is still in its infancy or, at best, just growing up.

We must make it clear here that *all* the elements of the spectrum of aircraft, and also their derivatives, equally offer a wide range of as yet unexplored and promising possibilities. 'All is flux' applies to all of them and there are no signs whatsoever that any of them converges towards, and settles down in, a particular groove. For example, we can be reasonably confident that classical and swept-winged aircraft are here to stay for flights over short and medium ranges and that they will continue to develop into a whole family of aircraft of the same kind, from small private aircraft to large airbuses. But we are far from being able to say even what the pattern of their relative sizes will turn out to be, quite apart from the details of their aerodynamic shape, propulsion, construction, and operation. We note, in particular, that the question of size is not only a matter of reaching an equilibrium between engineering possibilities and traffic handling, but also of the aerodynamics of the design of the airframe and its propulsion as well as of the aerodynamic handling and safety characteristics. In this context, novel aerodynamic concepts such as the all-wing airbus (*see* ref. 2, section 4) must also be mentioned, as well as the possibilities of engine/airframe integration which have not even been charted.

We are interested here particularly in the aerodynamic aspects and we can see from Fig. 2 that the *direct* influence of the aerodynamic aspects of airframe and propulsion on future prospects is very substantial. But there are also very important influences of the aerodynamics on the weight and fuel aspects, which are *not* shown in Fig. 2. A few examples may illustrate this point.

At short and medium ranges and for classical and swept-winged aircraft, in particular, aerodynamics can strongly influence the wing size and hence the wing weight by providing, with safety, higher lift at the same weight or the same lift at less weight. Here lie some of the immediate promising prospects for a combined attack in the fields of aerodynamics and *pilot engineering* and the outcome is likely to have a significant influence on the cost of flying. The same applies to any *simplifications* in

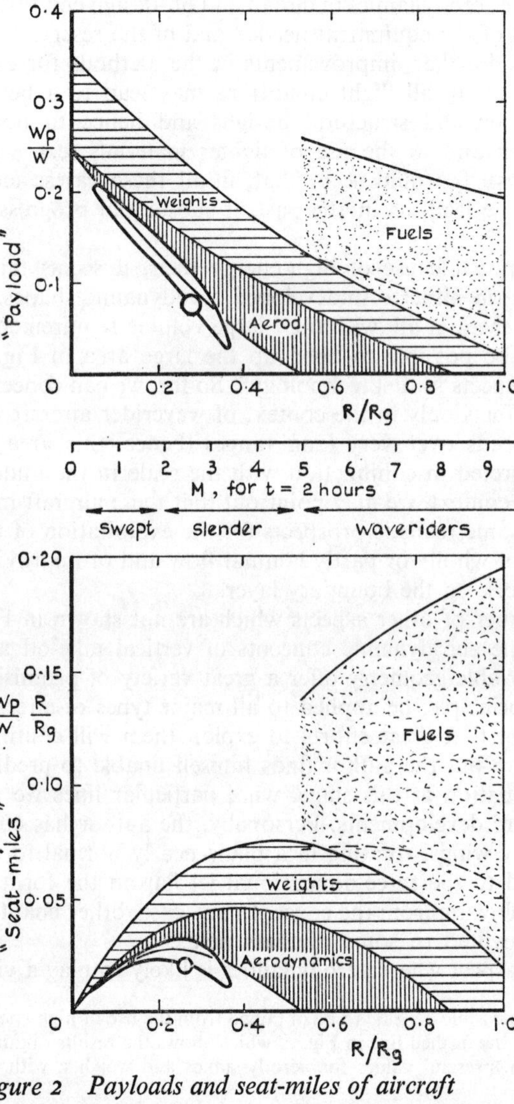

Figure 2 Payloads and seat-miles of aircraft

design which can be brought about by aerodynamic *refinements*. Aerodynamics can also strongly influence the performance of the engine and hence the installed engine weight, and the latter may further be reduced by providing aircraft shapes with less drag. Very typically, aerodynamic developments may lead to new and much more compact layouts, such as the slender all-wing aerobus of Gates for flight over very short ranges, and hence to substantial weight reductions. More subtly but nevertheless equally important, better aerodynamics in design and off-design conditions may reduce the weights of the equipment needed and of the reserve fuel, for all types of aircraft. Further, improvements in the methods for estimating aerodynamic loads in all flight conditions may lead to a better matching of aerodynamic and structural designs and hence to novel methods of construction and to the use of lighter materials (e.g. with unidirectional properties). It would seem that, in all these cases, aerodynamics should give the lead, as in the past, if substantial progress is to be made.

With regard to the possible use of high-energy fuels, it seems fairly clear that we must rely on hitherto undeveloped aerodynamic shapes to make it possible to use them at all, with their large volume requirements. Otherwise, it may not be possible to open up the large area in Fig. 2 which indicates the prospects in fuel technology*. So far, we can conceive the use of high-energy fuels only in the context of waverider aircraft for flight at hypersonic speeds over very long ranges (hence, this area in Fig. 2 should be interpreted in conjunction with the scale in the middle of Fig. 2). In this same context, we may point out that these aircraft may offer for the first time some realistic prospects for the exploitation of the aerodynamic concepts of wholly or partly laminar flow and of improving propulsion by doing work on the boundary layer.

There are still a number of other aspects which are not shown in Fig. 2. As already stated, the aerodynamic concepts of vertical take-off and landing and also of variable-geometry offer a great variety of promising prospects and can, in principle, be applied to all major types of aircraft. There seems little doubt that great efforts to exploit them will continue to be made in the future but the author finds himself unable to predict, in view of the fluid situation at this stage, what particular lines we are likely to select for future developments. Personally, the author has committed himself to rotary wings, rotating in a plane nearly normal to the direction of flight[9], and to the three-dimensional jet flap in the form of slender wings with jet-blowing from the edges. In any case, other chapters in this book will be devoted to some of these fields.

One very important aspect where aerodynamics is likely to play a vital

*Another indication of the possible benefits to be obtained from the use of high-energy fuels is given by comparing the dashed line in Fig. 2, which shows the results obtained for hydrogen fuel but with 'present' values for aerodynamics and weights, with the lowest line.

part is *flight safety* with all its many implications. The 'precipice' surrounding the flight envelope of the classical and swept-winged types of aircraft is an obvious case in point. But aerodynamic handling and associated safety aspects are vital for all types of aircraft and here we hope, with reasonable confidence, that substantial progress can be made. We are seriously involved here also with the fluid mechanics of the atmospheric environment and of its effects on aircraft handling and safety; their relative importance as compared with performance and economic matters must not be under-estimated, and commercial pressures must not be allowed to dominate. These aspects will be discussed in a later chapter.

Yet another aspect is aerodynamic *noise*, be it engine noise, jet noise or sonic bangs. There is no doubt that one of the major problems that faces the aerodynamicist in the Second Century is to achieve quieter air transport—we are much less successful than nature in this respect at present. As traffic density builds up, the noise aspects will vitally affect the choice of type of vehicle to be used.

Finally, our long view would not be complete without returning once more to flight in nature. There, we have a demonstration of the fullest exploitation of the aerodynamic principles of the classical type of aerofoil flow (including V/STOL), of elastic structures, of energy storage, and of a fully automatic system of operating the flight mechanism (i.e. one that seems to require no 'learning'). There is little doubt that the classical Cayley concept is only one of the possible engineering adaptations of flight in nature: there may well be others.

In concluding our discussion of some of the technical prospects before us, we must try to state some of the conditions which must be fulfilled if we are to have at least some confidence that some of the goals before us may be reached and realised in practice. In the author's opinion, it will be necessary for us to be less over-confident in our ability to cope with any engineering problem before us, provided only large funds will be made available; that we must be more selective and discriminating (for a long time to come) and concentrate on what we have reason to believe will turn out to be simple and 'natural' solutions (these, in turn, are much more likely to be also commercially viable and profitable); that we must be much less in the dark and must have done much more preparatory work before we proceed with actual projects; in short, that we must do much more *fundamental* and *aimed research* than we have been accustomed to doing. This in itself poses the problem of how to organise large-scale aimed research effectively, and here we are still far from knowing a solution. It will have to be solved in steps: in this country, in Europe, in the world. To be able to do all this, we need much clearer aims to our work, and this leads us to discuss, at least briefly some of the prospects for possible changes in the motivation of future developments in aviation.

4 THE MOTIVATION AND AIMS OF AVIATION

Before we begin to discuss these matters, it must be understood that all scientific and technological developments are strongly based on natural instincts of man: to be curious and to make and use tools. The technical prospects discussed in the previous section lead to work to be done, which can be expected to be interesting enough in itself to give full satisfaction to those who do it and to still the craving for knowledge which, according to Aristotle, all men possess by nature. Therefore, we do not need to invent pretexts for wanting to improve natural knowledge. What we want to discuss here is something quite different, namely, the *external motivation* of aviation and the jobs to which this tool can usefully be applied. If we can clarify these matters (which is likely to take us quite some time, well into the Second Century of the Royal Aeronautical Society), we can take better aims and this alone should greatly benefit our work. Apart from this, however, we are obliged to take up our share of *social responsibility* and to help to clarify these matters: because of the size of the job to be done; because of the (largely public) funds required to carry out the job (*see*, e.g. ref. 10); and, above all, because of the social implications of this work and of its outcome, the existence of which now begin to become clearly apparent.

If we look at the present situation first, it is fairly clear what we are rather fond of just now: it is market research and forecasting; emphasis on economic aspects and cost-effectiveness (even if we find it difficult at times to define the terms of the effectiveness); it is the 'great modern game of allotting priorities' (Lord Florey[11]). At present, commercial considerations, cost and timing, tend to dominate, and revenue and profit aspects are beginning to come in more strongly. This may well be a reasonable approach at the present juncture. So far, the 'market' has always been there to absorb the technical developments offered, and the consideration of 'market trends' and 'forecasts' made sense and still is, to some extent significant. On the present approach, we can be reasonably confident that there is a potential market for all the major types of aircraft discussed above: for classical aircraft at subsonic speeds for short ranges; for swept aircraft at high subsonic or low supersonic speeds for short and medium ranges; for slender aircraft at supersonic speeds for trans-Atlantic and similar ranges; for hydrogen-fuelled waverider aircraft at hypersonic speeds for longer ranges up to global distances. There should also be a market for some of their derivatives and here, the short-hop bus able to operate in highly populated areas may develop into a class of aircraft by itself rather than a possible derivative. More specifically, it appears that the *potential* domestic US market (in terms of seat-miles per year) for short and medium ranges is very large; that the potential domestic market in Europe is several times larger still, especially for aircraft of the airbus types to fly over very short and short ranges (*see*, e.g. ref. 7); and that the potential world market for aircraft to fly over very long ranges from about 5000 statutory miles up to global distances may be even larger (*see*, e.g. refs. 12 and 13), if we make the (necessary) assumption that the gap between the developed and under-developed countries will be closed during the Second Century; indeed, aviation can help to bring this about. We can also be reasonably confident that there will continue to exist a very large potential market for aircraft to fly over trans-Atlantic and similar ranges. In addition, we may expect that there will remain very strong pressures, for economic reasons and also simply because of human nature, to cut the *flying time* down to a few hours. On this assumption, the spectrum of major types of aircraft has been arranged in the way indicated in Fig. 2. In all this, we must of course, make the assumption that air transport will be matched by equally good ground transportation.

In describing the present approach, we must be aware of the fact that this differs considerably from that adopted in the past. Aviation began with dedicated pursuits of individuals (and this still is—and, one would hope, always will be—an essential ingredient of aviation!). Later, military incentives provided some driving force and civil aviation tended to become to some extent the exploitation of 'fall-out' from military aviation, dictated largely by what was technically feasible. Thus the *incentives* have definitely changed over the years, and there is no reason to believe that they will stop where they happen to be right now and that there will be no further change. We must, therefore, resist the adoption of 'instant traditions' and the compulsion for 'absolutising the present', which is being recognised as no more than a mode of conduct typical of large organisations that can, of course, be corrected. In considering the Second Century, we must assume that the incentives and motivations will continue to change. The question is, what the new trends are likely to be.

It is significant in this context that some would say that market forces are essentially *blind* and that we live in what has been called the 'Accidental Century' when our environment is transformed more by hazard than intent. It would seem fairly clear that we cannot afford to make our Second Century another accidental one. A deeper understanding of these matters could give some direction to new trends, possibly away from blind market forces and towards a greater social awareness (*see* e.g. ref. 14).

There is also an urgent need for such a change in direction. It would seem to be a case of Darwinian 'adapt or perish' to find a workable solution to the *population problem*, as formulated by Lord Florey[11]. To quote, 'there is now overwhelming evidence that rapid population growth is bringing with it dire consequences, not only in the great Asiatic countries, but probably even here. Evidence is slowly accumulating that the question is not simply whether food can be supplied for an ever-increasing population, but whether overcrowding *per se* does not lead to obscure and so far ill-defined difficulties of mental and social adjustment to a crowded and rapidly changing environment. Observations on animals are certainly pointing to some such possibility. Perhaps we should be paying more attention to the generally unpleasing form that life is assuming in great cities. It may be that *to relate population to environment optimally* is the greatest technological task of the end of this century. Its solution will not depend on biological science or medicine alone. It will involve every discipline from economics to psychology'.

In view of the enormous impact on our environment and on the way we live made by the available means of *communication* and *transport*, we may add here that a vital part in solving this problem must be played by transport technology in general and, in particular, by aviation which offers the prospect of providing a global network of transport operations, from very short to global ranges, where any two places may be no more than about two flying hours' apart. This is where aviation should come in during the Second Century, fulfilling an essential function within the whole system and providing a service.

From the point of view of a scientist, it would seem desirable to try to put these developments on a more rational rather than a political basis and to tackle the problems by the exercise of reason aided by imagination, rather than by relying on any dogma or assertion, since 'the whole universe of science is built on the belief that this is a rational universe' (*see* ref. 15). Our whole argument would be more complete if a more rational approach would turn out to be feasible and practicable. Fortunately, there is now at least some reason to believe that this may indeed be the case. This is based on the expectation that we shall be helped by advances in our knowledge of the *nature of man*: in ecology, by studying peoples and institutions in relation to environment; and in ethology, by studying the behaviour of man and his natural make-up (*see* e.g. refs. 16 and 17). It is conceivable that this may lead to the evolution of a system of organisation which is by no means 'ideal' but at least sufficient in the sense that it is so designed that our inborn instincts—once we know what they are—and our natural make-up are at least not violated. For example, we could then begin to think of new and appropriate ways of how we want to live and work together and completely revise our present concept of cities which really came about because people once wanted to congregate there for easy communication (and defence). We develop already, and could then sensibly proceed further towards linear cities and from there to the concept of 'strung settlements' of H. Rau, spreading over continents, which in turn become feasible and effective only when aviation is brought in.

In this long view over another century of developments in aviation, we may then finally envisage the ultimate external motivation for aviation to emerge. Knowing what it wants and realising how far aviation can serve these purposes, society can set the aims.

5 CONCLUSIONS

We have found that the Second Century of the Royal Aeronautical Society begins at a stage in the evolution of aircraft where an overall pattern is beginning to emerge, with four major types of aircraft and several possible derivatives. The first, the classical aircraft, is based on flight in nature, suitably adapted to human engineering, and has dominated developments in aviation during the First Century; the other types are 'man-made' and, in their early stages or still, hypothetical.

We have seen that there are real prospects for major improvements in *all* types, to be brought about by advances in our knowledge concerning aerodynamics, weights, fuels, and flight operation and safety. Aerodynamics, in particular has led the evolution of aircraft shapes and propulsion and offers substantial direct improvements. Indirectly, aerodynamics will in many ways create the conditions necessary for improvements in weights, fuels, noise, and flight safety to be effected. The overall area of as yet unexplored prospective improvements is vast compared with the present 'inhabited area'. Since the development of the classical type has taken the whole of the First Century and has not reached its end, we must expect that the development of a whole spectrum of aircraft will occupy aviation for the best part of the Second Century at least.

We have come across some important lessons from the past and also some conditions which should be fulfilled so as to give us a good chance of reaching our prospective goals. Among these are: we must apply ourselves diligently and steadily to research in all fields, to understand the basic physical principles involved; we must try to find feasible and natural engineering solutions and become less over-confident and reliant on funds being available, i.e. we must not hope to be able to beat the problem in an *ad hoc* manner by brute force; we must not repeat past mistakes and be deterred by apparent 'barriers' or smugly imagine we have already reached 'the ultimate'; we must be less in the dark when we start on an actual project in earnest, which means: we must do much more large-scale aimed research beforehand, in good time, and we have yet to learn how to do this effectively; to do all this, we need clearer aims.

We have recalled that the aims and the external motivations of aviation have changed during the course of time. We have suggested, therefore, that they are likely to continue to change and, in particular, that the present emphasis on foreseeable commercial profitability may give way during the Second Century to a greater appreciation of the social implications of aviation as a means for providing a global network of transport operations, where any two places are no more than about two flying hours' apart. Specifically, aviation as a vital part of a general system of transport and communications can help to solve the overridingly important population problem, as formulated by Lord Florey, which may be regarded as the greatest technological task of the Second Century. There is even some hope that all these developments may be put on a more rational rather than a commercial or political basis.

We may, therefore, conclude that aviation can look forward not only to a vast expansion on the technical side into most promising and as yet unexplored fields, but that it may also be built into a more rational social system with the aim of serving mankind as a whole.

Acknowledgment

The author acknowledges the permission of the Chief Scientist to publish this chapter.

References

1 KÜCHEMANN, D. Possible Types of Flying Vehicle in the Future. *The Skyward Urge—Aviation 1866–2016*, RAeS Centenary Symposium. London 1966, RAeS 1967.

2 KÜCHEMANN, D. and WEBER, J. An Analysis of Some Performance Aspects of Various Types of Aircraft Designed to Fly Over Different Ranges at Different Speeds. *Progress in Aeronautical Science*, Vol **9**, p 329, 1968.

3 KÜCHEMANN, D. and WEBER, J. *Aerodynamics of Propulsion*. McGraw-Hill, 1953.

4 VON HOLST, E. and KÜCHEMANN, D. Biologische und Aerodynamische Probleme des Tierfluges. *Naturwissenschaften*, Vol **29**, p 348, 1941. Abridged trans. *Journal of the Royal Aeronautical Society*, Vol **46**, p 39, 1942.

5 GIBBS-SMITH, C. H. *Sir George Cayley's Aeronautics 1796–1855*. London 1952.

6 GABRIELLI, G. and VON KÁRMÁN, TH. What Price Speed? *Mechanical Engineering*, p 775, October 1950.

7 COCKBURN, SIR ROBERT. A New Phase in Aviation? *The Aeronautical Journal of the Royal Aeronautical Society*, March 1968.

8 HAFNER, R. Quo Vadimus? *Proc AGARD Symposium*, Göttingen 1967.

9 KÜCHEMANN, D. Some Ways of Obtaining an Integrated Aircraft. *Journal of Helicopter Society of GB*, Vol **12**, p 121, 1958.

10 SHAW, R. R. Technological Progress and the Airlines. *Journal of Royal Aeronautical Society*, Vol **71**, p 597, 1967.

11 FLOREY, LORD. Address of the President at the Anniversary Meeting of the Royal Society, 30 November 1965. *Proceedings of Royal Society (A)*, Vol **289**, p 141, 1966.

12 STACK, J. The 55th Wilbur and Orville Wright Memorial Lecture, The Prospects for Hypersonic Flight. To be published in *The Aeronautical Journal of the Royal Aeronautical Society*.

13 WORTH, W. Making the Tenfold Transition—Jets to the Hypersonic Transport. *Astronautics & Aeronautics*, Vol **5**, p 62, 1967.

14 BRAGG, S. L. The Mechanical Engineer of 1987. *New Scientist*, p 302, 10 August 1967.

15 PYKE, M. Scientific Understanding and the Chief End of Man. Cambridge Meeting of the British Association for the Advancement of Science 1965, *Nature*, Vol **208**, p 921, 1965.

16 VON HOLST, E. Probleme der Modernen Instinkforschung. *Merkur*, Vol 15, p 913, 1961. Reprinted in *Zur Verhaltensphysiologie bei Tieren und Menschen*, Piper, 1969.

17 LORENZ, K. *Natural History of Aggression*. Methuen, 1966.

3 The future for agricultural aviation

A. H. Wheeler RAF (retd.)

INTRODUCTION

Predicting the future for Agricultural Aviation is rendered even more uncertain than predictions for most forms of aviation because in this case the future depends on two entirely separate sets of unknowns. These are the normal unknowns affecting aeronautical development but, more important, there are the unknowns affecting the development of agricultural chemicals. Also, in predicting the future evolution of any activity one must normally look back over the rate and the trend of developments in the activity, so far as they are relevant, from the past and then project this rate and trend into the future, bending and extending the line of projection in accordance with known or foreseeable influences. Here again we get a third set of unknowns related to future farming techniques, although these unknowns are perhaps less significant because the main expansion in agricultural aviation will probably lie in the vast undeveloped regions of the world where the farming techniques will merely be brought up to present standards, or just introduced where there was literally no agriculture before.

Starting from a selected base line in 1968 and looking back one becomes aware that the line of known rate and trend is very short: agricultural aviation as a practical and economic operation is little more than 20 years old, although thoughts and attempts at it go back a little further. One can almost say that the use of sophisticated chemicals in agriculture hardly pre-dates aviation so far as practical applications go. The properties of lime and, to a lesser extent, nitrate have been appreciated for much longer but only in primitive forms. If one looks back in aviation for half a century one finds that the aircraft were available in those far off days at a price and with flight characteristics which were entirely satisfactory for the purpose of spraying against pests or spreading fertilisers. What was not available was the type of chemical which was needed either for pest control or for fertilisation in a form which could be delivered from the air.

Originally published in the *Aeronautical Journal of the Royal Aeronautical Society*, March 1968.

AGRICULTURAL CHEMICAL DEVELOPMENT AS FARMING TECHNIQUES CHANGE

To illustrate this one can quote two cases, one from pest control chemicals and the other from fertiliser chemicals.

Twenty-five years ago there were virtually no selective weed killers at all and the most common weed killer used—and almost the only weed killer available—was a highly poisonous mixture which was used on garden paths in very large volume for the area treated. After the application of these weed killers the ground became virtually sterile for an unacceptably long time.

Pests such as mosquitoes in the swampy areas of the world were dealt with fairly effectively by spreading considerable quantities of petroleum products on the stagnant water where they bred.

Chemicals such as the now common DDT were undeveloped or certainly only in the laboratory stage. Today very efficient selective weed killers are available at an application rate of only 16 fluid ounces per acre (112 mm^3/m^2), representing a reduction in application rate from 100 gallons per acre (112,000 mm^3/m^2) in the past 20 years. If the present selective weed killer had been available half a century ago the aircraft were already available to deliver them economically at such a small dosage rate.

Chemical fertilisers make a less impressive story so far as recent development goes. One of the commonest active agents required in fertilisation is nitrogen: in 1920 ammonium sulphate was available as a chemical fertiliser with a nitrogen content of 20·6%, but it was available in a form which would have made dispersal from aircraft almost impossible. By 1940 urea was available with a nitrogen content of 46% and by 1958 this fertiliser was available in pellet form which allowed it to 'run' almost like a liquid and it could, therefore be spread easily from the air by either a centrifugal or venturi-operated mechanism. With this form of fertiliser the aircraft available in 1920 could have operated economically but it is doubtful if farmers would have availed themselves of it for other reasons concerning crops, cropping techniques and wage rates. A superficial picture of this trend in development is given in Table 1, showing the cost and speed of applying fertilisers between 1900 and 1960. One of the reasons for farmers not using aerial fertilising as against surface vehicles, even if the modern chemicals had been available 40 years ago, is also shown in Table 2, which illustrates how crop yields per acre have increased since then, thus making it more important to avoid wheel damage to the crop. This applies particularly to crops like potatoes when the operations must be done when the crops are nearly full grown. Even the latest highly developed forms of grass are too precious to be subjected to serious wheel damage which can delay development of up to 10% of the crop.

Another factor which is only recently beginning to bear on the popularity, if not directly on the economy, of aerial operations also concerns wheel damage. Today we have moved very much more into the habit of using tinned, preserved frozen or otherwise 'packaged' foods which includes vegetables. When we open a package we expect to find the contents, peas, beans, etc., all at the same state of ripeness. Wheel damage tends to delay the growth, and therefore the ripening, of all vegetation: this would necessitate selective harvesting which in modern mechanised farming would be quite unacceptable economically. It is therefore essential that the crops should not be damaged in any way which might retard development of a part of them. To achieve these requirements spraying can only be done from the air.

TABLE 1 *Cost and speed of applying fertiliser—UK*

Year	*Method of delivery*	*Speed in acres per hour (m^2/h)*	*Cost in shillings per acre (pence/m^2)*
1900	Dumped by cart, spread by hand, harrowed in	¾ (3000)	3s. 0d. (0·009)
1919	Spreading only by hand	1 (4000)	2s. 0d. (0·006)
1960	Tractor spreader and	1½ to 3 (6000–12000)	10s. 0d. to 15s. 0d. (0·03–0·045)
	Trailer spinner	6 to 10 (24,000–40,000)	10s. 0d. to 12s. 0d. (0·03–0·035)
1960	By aircraft	26 to 40 (105,000–162,000)	21s. 6d. (0·064)

TABLE 2 *Average crop yield—good and bad farms in UK*

Years	*Wheat*	*Potatoes*
1900–1909	17·4	5·9
1939–1948	18·8	7·1
1950–1954	22.5	7·9
1955–1959	26·1	7–9

One strangely stagnant line of possible development in the fertiliser range of chemicals is *lime*. Very high on the list of importance, particularly on high marginal lands where rocks and broken ground make surface transport difficult or impossible, lime is required in large quantities of the order of four tons to the acre as against some four cwt per acre for nitrogenous fertilisers. Until now this sort of application rate has ruled out aerial delivery, although very recently in forestry areas a scheme has started where an application rate of only three cwt per acre has promise of being effective: at present this quantity is being delivered by helicopter at an economic rate but only if the growth improvement comes up to expectations. The obvious question with lime is,' Why cannot the active agent in lime be concentrated per unit weight and volume in the same way as has been done with other chemicals in agricultural use?' The answer is that lime (calcium hydroxide) has two functions—namely flocculating (i.e. clumping together) very small clay particles, making the soil less like cement and more easily crumbled, and raising the pH, i.e. reducing acidity*. Potassium hydroxide would raise the pH very efficiently and raise the potassium level at the same time, but it would have a disastrous effect on the soil crumb. Aluminium hydroxide, however, might be excellent for the soil crumb but inefficient for raising the pH. A mixture of the two would seem a compromise, but would probably be no better than lime and it would be a lot more expensive. I make no apology for this rather detailed chemical analysis of the lime problem because it helps to illustrate the importance of chemical development which was emphasised at the beginning of this paper: if anyone interested in agricultural aviation could increase the lime-effectiveness/weight factor they would achieve more for the art than any miracles which might now be done with the handling characteristics of an aircraft. At one stroke it might double the acreage in UK over which agricultural aircraft could operate economically: at the same time it would create a demand for a very much larger aircraft.

Apart from the improvement in chemical formulae by concentrating the active agents there are other improvements on the way such as producing fertilisers which decay over a very long period: with these available the spreading of fertilisers could be done at any convenient season, thus simplifying the utilisation problems of the aerial operator.

All the above is intended to give a superficial background to indicate some factors which have a strong bearing on the development of agricultural aviation, but which are entirely unrelated to developments in aeronautics. Most other forms of aerial transport are not subject to this changing pattern of load. Passengers may tend to get heavier as our standard of living increases or motor cars may get smaller as our roads get more crowded, but the change will be small and these types of loads do not have to be reloaded every few minutes and then pumped out through a nozzle†.

*I am indebted for this information to Mr. R. C. Amsden of Fisons, Chesterford Park Research Station.

†See also *Journal of the Royal Aeronautical Society* November 1961, Vol **65** p 758. 'Future Developments in Agricultural Aviation' Air Cdre A. H. Wheeler.

LOOKING TO THE FUTURE

We may assume therefore that the factors given above have an almost over-riding effect on the development of agricultural aviation. Without further developments in chemistry, biology, botany, mechanical distributors and farming techniques, the effect of aeronautical development will be small. But this does not mean that the increase in the whole activity over the past few years has not been startling or that the increase in the future may not also be startling. The rate of increase of the activity is shown in Fig. 1 and from the chart one could extrapolate to give an almost infinite figure for 1980, limited only (in acreage) by the extent of the land surfaces of the world and even then we must bear in mind that sea farming has already started in some parts of the world, so one might also add in those areas of the oceans of the world down to (say) the 100 fathom (182·88 m) line. This gives quite an area for expansion! Fig. 1, of course, only really proves the unreliability of extrapolation from a

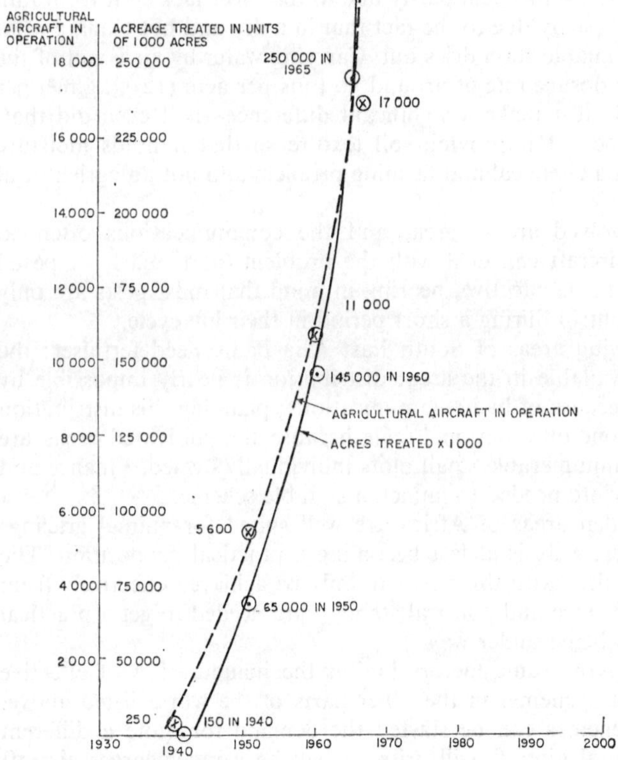

Figure 1 Rate of increase of agro-aerial activity

relatively short known period. But the implications from these figures are significant and when one bears in mind the fact that almost the only parts of the world in which aerial farming is practised extensively today are the more advanced parts, from an agricultural point of view, then the significance of the figures becomes almost unbelievable. But the fact remains that, for financial, educational, political, and other reasons, by far the largest part of the world which is farmable is still either underdeveloped or not developed at all. These areas exist in South East Asia, India, Africa, South America and any number of smaller areas and islands. The reasons for this lack of development are various, but a few typical instances may be given to indicate how future development may proceed and how, in this geographical expansion, aeronautical development will become more significant.

The sub-continent of India is perhaps one of the most advanced of the areas listed. Over the whole area pest control and fertilisation is lacking in varying degrees—chiefly for financial reasons. The standard of crops and therefore the standard of living could be improved considerably. Water is, of course, a problem partly due to the sheer lack of it from rain or irrigation and partly due to the fact that in a dry and hot climate what water may be available soon dries out. Carrying water by air is out of the question, since a dosage rate of around 50 tons per acre (12·5 kg/m^2) per week would only just make a significant difference—if it even did that. There are methods of improving soil texture so that it holds moisture better but this is a chemical and farming problem and not an agricultural aviation one.

The areas involved are so great and the communications often so bad that only aircraft can deal with the problem (particularly of pests) quickly enough to be effective, bearing in mind that most pests are only susceptible to control during a short period in their life cycle.

The rice growing areas of South East Asia badly need fertiliser; the chemicals are available in the area; distribution is nearly impossible by surface means because of bad communications; planning this distribution could only be done on a national scale because the cultivated areas are broken up into innumerable small plots individually owned. Finance and political stability are needed to launch a suitable scheme.

The tsetse-ridden areas of Africa are well suited for animal grazing; control of the tsetse fly is at last becoming a practical proposition. The speed needed to deal with this pest can only be achieved effectively from the air. Again, finance and political stability are needed to get a practical and rewarding scheme under way.

Approximately the same factors inhibit the inauguration of effective—and rewarding—schemes in the other parts of the world listed above.

When such schemes can be started the demand for quite a different type of agricultural aircraft will arise. It will be a much larger aircraft than those now most commonly in use, perhaps as large as present air liners and it may be predicted that the first (experimental) schemes will be tried out by air liners or military transports converted to agricultural use. As operations with these converted types become more extensive and as the source of supply of these types dries up—by about 1980—the demand for functionally designed types will arise and the demand may become considerable as under-developed countries appreciate the advantages accruing. These types will not replace the existing fleets of smaller agricultural aircraft; they will actually open up opportunities for greater use of them.

SPECIALISED AIRCRAFT FOR THE FUTURE

This subject can only be dealt with by taking cases individually since the requirements are so diverse.

Aircraft for operating in enclosed and highly developed areas

The first case selected will be the type of agricultural aircraft which will be needed for the existing highly developed farming areas of the world and which will largely replace the existing range of agricultural aircraft. It will also probably be the first development away from existing designs. The requirement for it should be undeterred (relatively) by financial consideration since the rewards are foreseeable and unlikely to be affected by political instability.

The problem before the present day operators is the wage bill—nothing new or unique in this. A typical team for operating one agricultural aircraft carrying a payload of approximately one ton consists of one pilot, one loader/tractor driver/refueller, and two markers (able-bodied men who mark each succeeding swathe to be dealt with). That makes four men in all. Some teams increase this number by adding a team manager, which makes the operation only marginally economical: some teams try to decrease the number by making the pilot do work on the ground as well as in the air; this makes the operation more than marginally dangerous. If all teams followed exactly the letter of the regulations recently written by an authority, concerning mapping all obstructions adjacent to the operating area before starting work, the whole operation would become uneconomical and probably no safer.

Most of the members of the team described are available at wage rates which are not unduly high, but the supply of pilots is rapidly drying up. There is, in fact, a world shortage of pilots with commercial licences which now exceeds 3000 and the shortage is increasing steadily. As a result of this expert pilots in agricultural aviation are being paid as much as £7000 pa for twenty weeks work in the year: inexpert pilots are carrying an insurance premium as high as 35%—if they can get cover at all. For some time now authorities have been urged to provide for a separate type of commercial pilots' licence tailored to the peculiar needs in agricultural flying since the requirements are very specialised and not common to many of the present standard commercial pilots' qualifications: if this requirement were accepted, as it is in some of the major aviation countries of the world such as USA, France, Holland, there would be an immediate alleviation of the pilot shortage. It has not so far received favourable consideration in UK, so that the shortage of pilots is a serious factor in agricultural aviation operations. Perhaps this is the major factor which is focusing attention on the desirability and (now) practicability, of developing a pilot-less 'Robot' flying machine which could operate under ground control.

Specification for small pilot-less agricultural aircraft of the future

An impending major technical break-through in agricultural aviation will be a form of 'Robot' flying machine which will supersede the existing generation of agricultural aircraft, of which there are today some 18,000 in use, but the difficulties and problems involved must not be underrated. To start with, the craft must be able to operate not more than five feet above the crops and be able to move from one swathe to the next with an overlap error not exceeding (approximately) five feet (1.5 m) in a swathe width of approximately 35 feet (10 m). This must be done in an area which may be an irregular shape, may not exceed five acres in size and may be surrounded with high obstructions: larger areas may have obstructions in the middle of them. Guidance for this craft should be provided by one 'station' adjacent to the reloading/refuelling point with perhaps one other slave 'station'. Both 'stations' must be mobile on all types of ground and of a width not exceeding the width of normal farm gates; the 'slave station' should be operable unmanned. The craft should be capable of operating at night and in daylight conditions of visibility as low as 30 yards (27 m). It should also be able to operate in wind conditions up to at least 30 knots (55 km/h) and carry a payload of at least 200 lb (90 kg) of liquid spray. If used for spreading fertiliser, thus increasing possible utilisation, a payload of at least 500 lb (230 kg) would be required. It must be capable of vertical take-off and landing, with an operating speed of at least 20 knots (37 km/h)—preferably 50 knots (93 km/h). The craft must have a form of fail-safe quality built in whereby it would immediately land if control or guidance were lost. Within this specification it might be possible to reduce the operating team to one controller/technician and one loader/driver, although a third member of the team would probably be needed.

Airworthiness certification for such an un-manned craft should be a simpler problem than for aircraft carrying a pilot and insurance premia should be correspondingly reduced. The craft should be able to operate all hours of the day and night—subject to no Ministerial veto in the interests of public amenities—and thus untilisation could be very much increased.

Cost of future agricultural (Robot) unit

From the foregoing it will be appreciated that an operating unit of this type comprising three highly sophisticated items, will be a very expensive affair and—hazarding a guess—the total cost of one (triple) unit could be in the region of £150,000 or possibly £250,000, but it would achieve a very high utilisation since the three items comprising it would be easily transportable and could be operated in various parts of the world according to the required seasonal applications. The personnel operating the unit, although highly trained, would not need any special inherent aptitude such as is usually required in pilots. To achieve maximum utilisation each unit would probably need manning with three or four complete team shifts of two men plus one on-the-spot manager qualified in all operating functions. It is assumed that contractors operating these units would have a central planning headquarters from which the daily operating area allocations would be issued and all administration matters would be dealt with.

If one assumes that such a unit becomes a practical and economical proposition and that its payload is such that spreading fertiliser can also be done by it, then an enormous market would be open to it all over the world.

Although a unit of this kind would cover a greater area every 24 hours than the present conventional aircraft, it would open up possibilities for utilisation greater than that now available and it could be predicted that by 1980 approximately 500 of these machines would be required: thereafter the demand for them would increase rapidly until by the turn of the century approximately 120,000 of them would be in regular use. This rather startling prediction becomes understandable when it is appreciated that the great majority of the tedious farming operations could, by then, be done by this type of machine as indicated further on in this chapter. One therefore assesses and guesses at the probable number, not by counting the number of agricultural aircraft now in use but by counting the number of farm tractors!

Specification for large agricultural aircraft of the future

As indicated before there will be areas of the world, forests, swamps, jungles, and even sea areas, which will be so vast but botanically homogeneous that they could not be dealt with economically by the smaller Robot type of aircraft described. A much larger aircraft would be needed and, it is suggested that, at first, converted obsolescent air liners or military transports could probably be used. Although the operations undertaken by these, mostly flying high enough to be clear of obstructions and at any normal speed suitable to the craft, would be fairly straightforward it is not considered that the later breed of swept wing jets could be used except in very exceptional circumstances.

The essential requirements for such craft would be:

For spraying A payload capacity so large that it would permit continuous work for at least two hours, but preferably for a period lasting for the planned endurance of the aircraft. This would, of course, involve a balance between engine fuel and spray liquid, bearing in mind that the refuelling/reloading base may have to be as much as 100 miles (161 km) distant from the operating area.

For fertiliser spreading A payload capacity as high as possible, but it is unlikely that this could provide for a sortie period exceeding one and a half hours and the base might need to be further away than the specified 100 miles for spraying operations.

For landing and take-off These aircraft should be able to operate from runways 7500 feet (2300 m) long at the most, but the operations would be much more economical if they could use some of the old wartime landing sites of only 6000 feet (1830 m). In operations of this kind a greater risk of engine failure on take-off would probably be permissible so long as quick dumping of the load were possible.

There will always be the difficult decision to make in these operations as to whether to use small aircraft from small landing grounds with small loads, but close to the operational area, or whether to use large aircraft from more remote and bigger airfields. The decision on this will mostly be decided by the surface communications which are available.

Loading, discharge and dumping Quick turn round time is essential to the economy of the whole operation. The spray or fertiliser loading operation should be capable of completion in the same time taken for refuelling and between flight inspection.

It will be necessary to adapt the wing structure so that it can support the necessary spray gear or spreading gear at full economical cruising speed.

It should be possible to dump 30% of the total payload in one minute without serious change of trim.

Crew The aircraft should be able to operate with a crew of two—one pilot and one navigator/wireless operator/engineer, but a third crew member should be allowed for special operations. An automatic pilot will be essential.

Navigational facilities In all this work, however large the area treated, it will be essential to provide the aircraft with the most accurate navigational equipment available: if it were possible, navigational accuracy within a few feet should be provided along a swathe run of several miles. Without this order of accuracy chemical spray or fertiliser will be wasted or large strips may be left untreated. In pest control this inaccuracy might almost nullify the effort expended. Probably in the future many crops will be planted in a line pattern which will render swathe selection easier, as it is now in operations on such crops as cotton and vines where the rows are recognisable and countable during the turn. Although this will help the larger aircraft with crews, the pilotless 'robots' will depend entirely on electronic guidance.

OTHER MISCELLANEOUS CRAFT OF THE FUTURE

Between, before, and among, the two main types specified above there will undoubtedly be a number of specialised types of agricultural aircraft.

Air cushion vehicles

Air cushion vehicles (sometimes called hovercraft) have already been tried for agricultural work and at the present time developments are in progress. An obvious application for air cushion vehicles will be on such work as weed and pest control in lakes, rivers, and swamps. They could also (in the future) be used for fertilising feeding grounds and edible sea weeds in tidal areas. Limited use of them would be for harvesting, fertilising and (perhaps) planting in rice growing areas; they could also be used for pre-emergent weed control and fertilising on normal crop areas, although the high speed airflow around the craft's skirt would tend to disturb the top soil during dry seasons. For the same reason this high speed airflow might tend to damage crops during any but the earliest stages of growth.

It seems therefore, that air cushion vehicles will be useful complementary craft operating in specialised roles among the other aerial craft, and it would be a pre-requisite for their use that they must operate more economically.

Smaller piloted aircraft

Certainly, before the successful development of fully automatic pilotless aircraft, there will be an intermediate stage wherein aircraft of fairly conventional design may be used, but with highly sophisticated guidance systems able to direct the pilot on accurate swathe runs without the (rather expensive) use of two able-bodied men as swathe 'markers'. When this is done, guiding the pilot down successive swathe runs 30–50 feet (10–15 m) wide within an error margin of less than five feet, then a major breakthrough in economy will have been achieved. BUT here one must sound a warning: it will be no economy to replace the two able-bodied, but relatively unskilled, 'markers' with two highly skilled technicians operating two highly sophisticated and expensive stations directing the pilot through an equally expensive airborne 'black box'.

In spite of these considerations the manned, but sophisticatedly guided, aircraft may prove to be a necessary step towards the unmanned 'robot', even though it may be only a (rather expensive) passing phase.

Photographic aircraft

In all farming operations photographic surveys are becoming essential because of their inherent economy. Apart from the well known photographic survey operations in dry weather to locate uncharted land drains and all other survey requirements, a new and promising application is the early detection of blight in (mostly) potato crops by the use of infra-red photography which senses areas of thin foliage in the crop days before this would become noticeable to normal human observation.

Another photographic operation already in limited use is that of crop assessment and even animal counting.

No special aircraft would be needed for this activity beyond those already available for normal photographic survey work.

Weather report and control

It is just worth mentioning here the well known possibility of inducing rain by cloud seeding and also the considerable use now made in compiling accurate weather reports available from data obtained by both aircraft and satellites. It is, however, not anticipated that any special aircraft would be developed especially for the agricultural requirements within this category; although the large agricultural aircraft mentioned before would be ideally suitable for cloud seeding. This would tend to increase their world-wide utilisation.

AGRICULTURAL OPERATIONS WHICH MAY IN THE FUTURE BE TAKEN OVER BY AIRCRAFT

Farming operations which are already, to a significant extent, done from the air are:

Spraying against weeds and pests of all kinds;
Spreading fertilisers;
Detecting and controlling forest fires;
Delivering farming materials in inaccessible places;
Seeding crops over difficult terrain;

Assessing and inspecting crops and animals by air photography.

In the near future we may see much higher percentage of all these operations done from the air as the various factors already mentioned develop and also as farm wages increase in order to try to counteract the considerable drift of workers towards factory and office employment. There are also a number of possible developments which, at any time, might quite abruptly make aerial work the most economical of all. Typical of such 'local' break-throughs would be changes in seeding techniques and developments in new forms of virus control.

In seeding

The main reason why cereal seeding is not now generally done from the air—particularly when ground conditions render surface seeding impossible—is because it is impossible at present to render the seed unpalatable to birds for long enough to allow it to germinate and start growing. When that becomes possible, seeding from the air will be so quick that it could easily compete with surface sowing which, at present, has the advantage that it not only spreads the seed but also buries it in drills. In fact broadcasting seed is in many ways preferable since each seed then tends to have an area of ground to itself.

Doubts have been expressed for years in the farming world about the wisdom of compacting the top soil with wheeled traffic and turning in the top soil with a plough. In some parts of the world already pastures are being sown from the air after the previous herbage has been killed from the air, and the new crop has then been cleaned from the air by selective weed killers. There is no reason to suppose that this whole operation could not be practical on normal crops sown in normal farming areas. It will then only remain to devise methods of harvesting from the air—which is not unforseeable—and the land will never have to suffer haphazard compacting, as at present.

In virus control

At present control of such viruses as potato blight is achieved by spraying. Although the spray volume required with ultra low volume sprays is very efficient and economical, it is within the realms of possibility that equally effective measures against a virus could be taken by flying over the crop with a very powerful ultra-violet ray. If this could be done the aircraft would only have to land at the infrequent intervals necessary for refuelling, since a small generator would provide the current necessary for the source of ultra-violet light. Development work will be needed on such rather revolutionary methods of virus control as this, or any others which may be considered. At present almost all research on virus control is done by the big chemical companies, but they would naturally not be interested to the same extent in developing a technique which eliminates the need for chemicals.

NEW TECHNIQUES

There must be many other entirely new techniques which will be developed to increase the economy and efficiency of growing and harvesting all types of crops, from normal everyday cereal and root crops of established farming to the more exotic crops like seaweed in Japan, timber in the forest areas of the world, fish in the oceans of the world, and many others. In all such new techniques developed, speed will be an important factor because of the short period available during many crops' growing stages when the particular operation must be done: speed also cuts down labourers' time. Many of these new techniques will be aimed at developing the more remote and inaccessible regions of the world where communications are bad or impossible, thus where only aircraft can operate. This will especially be the case in areas of jungle, mountains and swamps.

NUMBERS AND TYPES OF AIRCRAFT IN THE FUTURE

Earlier in this chapter the case has been made for three essentially different categories of aircraft. These are:

1 *The large manned aircraft* modified from existing air liner types which will be needed to fertilise, control pests and forest fires and generally survey the vast botanically homogeneous areas of the world land surfaces with sideline employment on rain-making, ocean fertilising, transportation and such temporary jobs as may crop up.

Owing to the inherent inter-continental mobility of such aircraft, their utilisation should be very high so long as such factors as national prestige and international finance do not inhibit their world-wide operation. This high utilisation will tend to reduce the numbers needed, but it will also reduce the operating costs which in turn, will extend the range of activities which can be done economically. Bearing all these factors in mind and particularly how vast are the areas of the world over which they can operate, one could predict that by 1980 between 1000 and 4000 of these aircraft could be in regular use: the higher figure of 4000 depends, not on technical factors at all, but on the political stability of under-developed countries and therefore, the readiness of developed ones to invest in them. As mentioned before, the rewards from fertilising under-fertilised crops can be enormous. Taking the same factors into consideration, the numbers of such aircraft operating in AD 2000—no longer modified air liners but functionally designed craft—will be between 10,000 and 40,000—again depending almost solely on political factors.

2 *The small unmanned 'Robot' craft* which will operate over the highly

developed farming areas of the world. Unlike the larger aircraft in category 1, the numbers of these operating in the future will, in this case, depend almost entirely on technical factors and not on political ones. If a practical and reliable agro-aerial system (as described earlier) can be developed which can spray and spread fertiliser to the fine accuracy required, then the figures given for future numbers in operation will be realistic. There could well be 500 in operation by 1980 and 120,000 by AD 2000, but during this period one must bear in mind that in 1980 there will be approximately 60,000 small conventional manned aircraft operating and that this figure will tend to fall to around 10,000 by AD 2000, except for such aircraft remaining which are referred to in category 3 and which are only of marginal interest in this chapter.

3 *A rather mixed assortment of craft* ranging from air-cushion vehicles to small manned craft of fairly conventional design by present day standards. The latter will be in use for such 'fill in' complementary jobs as crop inspection, aerial counting, personnel transportation, very occasional crop spraying and spreading of fertilisers, and a number of other diverse operations none of which will be on anything like a massive scale.

Depending largely on where one draws the line around the somewhat vague term 'agricultural employment', the numbers in this sub-division should be between 20,000 and 60,000 in 1980 and between 60,000 and 100,000 in AD 2000. But almost all this sub-division will be to designs not specifically for agricultural use.

This comment will also apply to such air-cushion vehicles as are used for purely agricultural purposes. A few may be specifically designed for such work, but since the layout of such craft lends itself to fairly diverse types of operations, it seems unlikely that a specialised craft will be built. Predicting numbers for the future in this case is difficult but use can be foreseen for approximately 1000 air cushion vehicles—chiefly over swamp and water—by 1980 with that figure rising at least fourfold by AD 2000.

These predictions are summed up in Table 3; from this it will be seen that we arrive at a (predicted) total for functionally designed, or functionally fully modified, agricultural aircraft operating in 1980 of approximately 45,500. This, as indicated, is assumed to be a *maximum* figure likely: but at the most pessimistic rate of expansion the figure is unlikely to be less than 41,500. Out of this total we have the great majority (in 1980) as fairly standard types, essentially to present-day designs, representing a total investment of something of the order of £120 million. The large (converted but essentially standard) type represents an investment of the order of £800 million (or £200 million minimum). The very small number of absolutely revolutionary design, of automatic 'Robot' (triple) units just coming into operation as almost a pioneer venture, will represent an investment of the order of another £100 million.

If predictions are true, by AD 2000 the pattern will have changed. the (by then) standard 'Robots' in operation will represent a total investment of the order of £24,000 million: the large (by then) functionally designed type will represent an investment of the order of £20,000 million; and the older manned agricultural aircraft of today will represent an investment of the order of only £30 million and this figure might be much

TABLE 3 *Maximum predicted numbers of different types of agricultural aircraft AD 1970–2000 (world-wide)*

	1970	*1980*	*AD 2000*
Similar to Type of Agricultural Aircraft in 1968 at £3,000 each equipped	22,000 (£66 million)	40,000 (£120 million)	10,000 (£30 million)
Large Converted Air liners Type at £200,000 each equipped	100 (£20 million)	4,000 (£800 million)	40,000 functionally designed at £500,000 (£20,000 million)
Robot (unmanned) Type at £200,000 per triple unit	0	500 (£100 million)	120,000 (£24,000 million)
Miscellaneous Aircraft not used for normal direct agricultural purposes—but closely associated purposes	2,000	10,000	40,000
Hovercraft	0	1,000	4,000
Total world-wide (all types)	24,100	55,500	214,000

less since, during the 'run down' of obsolescent types a large number in use are always relatively old, most of them in use in backward countries, where either lack of education or political instability (or both), has inhibited development in modern farming techniques.

Of the figures given above approximately 30% will represent the expenditure on many forms of electronic guidance.

But, summing up all these predictions, the probability is that there will be operating in the world functionally designed (or truly functionally modified) agricultural aircraft which represent an investment of some £1000 million by 1980 and some £44,000 million by AD 2000.

4 The future role of computers with special reference to aerodynamic design

Bryan Thwaites

INTRODUCTION

Of all the paradoxes of our present age, one of the most curious is the apparently world-wide phenomenon whereby intelligent young people are turning their interest away from the sciences towards the humanities. Science, in the sense of knowing how and why, has been with us since the dawn of human history; technology, too. The Iron-age Man, rubbing his stick on a stone to light his fire, was a technologist if not a scientist. The Ancient Civilisations, in that they began to put knowledge into a structure, were demonstrably scientific, and arguably not so technological. Through the Middle Ages, the Renaissance and the Industrial Revolution, science and technology seemed to go hand in hand, as the steadily increasing organisation of knowledge of the physical world was manifested in inventions which had profound effects on the development of the human society. Now, in the second half of the twentieth century, it is difficult to compare the relative significance of our exploding knowledge of fundamental physical mechanisms and of our mighty capacity for creative applications of that knowledge.

Young people, especially those who are destined (at least according to our present concepts of education and training) to fill positions of power in our society, now seem to be reacting against this historic evolution of science and its applications. This reaction may, of course, be a wholly insignificant perturbation in the smooth exponential growth of scientific understanding and awareness; it may be that we are far too close to it in time to draw any valid conclusion from it. What is more, insofar as we see it as the first discernible shift in men's thought away from science, it may be suspect on the argument that the probabilities are heavily stacked against the first shift occurring just at the present moment.

But the shift of interest *is* discernible and if we are to peer a hundred years into the future of any realm of science it is relevant to assess briefly

Originally published in the *Aeronautical Journal of the Royal Aeronautical Society*, April 1968.

some general human factors as manifested by the intellectual attitudes, the approach to community life, the comparative importance of human problems, the possibilities for a world-order (or otherwise)—factors such as these which will affect the intellectual, political and economic factors controlling scientific developments in the future. To make such an assessment, however crude and however brief, one can look profitably for instruction to our more intelligent young people, that is, to the generation which is now in the tertiary sphere of education, or in the first few years of their professional careers.

Their disillusionment with science stems, I suggest, from two main observations. First, as they look at the society in which they are beginning to take a creative place, they identify as the major problems those involving, not science in any obvious way, but humanitarian, sociological and moral considerations. They see the wars or potential wars that are raging as clashes of ideas and ideals, not as manifestations of scientific uncertainties. They see the manifest inequalities among human life as failures of humane man, not of scientific man. They see the illiberality of our national society—either an overdone permissiveness or a too-rigid conformism (to have it both ways)—as a failure of communication rather than as a phenomenon susceptible to scientific analysis.

Second, they are sceptical (to put it mildly) of the advantages of some of the most notable scientific and technological achievements. There is no colour in the spectrum to their liking—a spectrum which ranges from the nuclear bomb at one end to the desecration by the automobile of civilised city life at the other; or from the space-race to the artificial creation of monstrous new living organisms. Gone is the excitement and curiosity which we like to think science has generated in people of past ages; in their place, there is cynicism and disillusion.

THE COMPUTER

What, you may now be asking, has this to do with the impact of computers on aerodynamic design over the next hundred years?

The relevance lies in the attitude which the highly-educated section of society adopts to the computer. On the one hand, the computer could be regarded as yet another component of science or of technology. In this case, there is a real danger that its development and applications will be limited so severely that the benefits to mankind, in general as well as in the scientific spheres, of which (in my opinion) it is capable, will not be realised. On the other hand, the computer (by which must be meant both hardware and software) can take its place as a provider of manipulative brainpower on the grand, indeed truly superhuman, scale appropriate to every intellectual discipline under the sun.

Let us try to be a little more precise as to this conjunction of attitude. If we accept that the invention of the computer was a scientific feat of the first rank, let us compare it, or at least its effects, with such feats as the realisation of atomic energy, the successful design of an aircraft jet engine, or the practice of antisepsis or asepsis. It has to be admitted that the three last examples not only are magnificent intellectual achievements but also carry profound implications for the character of society. Atomic energy promises virtually limitless power at minute cost (if not now, at least within the next hundred years) which, even if manifested in practice only as electrical energy, will be one of the foundations of future civilisation. The jet engine and its successors have already ensured a speed of physical transportation which will reduce the whole of our worldly life to that of an old city whose walls are bounded not by a watery moat but by the void of space itself. Advances in medicine, again, have upset and will continue even more violently to upset what we still think of as the natural balances of life and death.

Yet despite the indisputable depth of the impact of these three sciences or technologies on the evolution of civilisation, each has only a limited area of impact. They do not, for example, overlap each other to any significant extent. The realisation of atomic energy does not depend in any appreciable way on the prolongation of the human span of life; or future advances in medicine do not depend crucially on the speed of intercontinental travel. None has a universality of application to the whole of human activity.

The computer, on the other hand, seems to possess this property of universality which is the prerogative of only a minute number of scientific inventions. It is as difficult as it is invidious to attempt to enumerate these inventions, but if one mentions the thousands-of-years-old idea of the wheel as being the basis of all organised movement on earth, and the centuries-old idea of printing as being the basis of all organised communication, there seem few other candidates for the short list of truly universal scientific or technical inventions. To this pair, I would certainly add the computer, for it is now difficult to imagine any human activity which cannot be aided (if enriched is too strong a word) by the computer. Certainly every activity involving quantitative thought is susceptible to computerisation. Thus whether the quantitativeness arises from the intrinsic use of numbers as in mathematics, or from the enumeration of samples as in data processing, or from the governing of systems describable in numerical terms as in traffic control, or from the description of elements of structures as in language translation, the computer will wholly transcend the capabilities of the individual unaided human worker.

But the quantitative mode of thought is not the only one for the computer. True, the inner mechanisms of a computer are essentially quantitative on the microscopic scale; but methods of working, or styles of programming, are now such that the computer at least appears to work in a qualitative fashion. That is to say, it appears to exhibit the capability of independent thought and judgment.

In saying that, let us have none of the nonsense about the computer evolving into a truly superhuman position in which men are powerlessly subservient. Let us, too, not be bothered overmuch by the niceties of semantics, by the problem as to whether a computer is actually behaving qualitatively if it appears to behave qualitatively. Let us here be content with the observation that the elementary computer facility of comparison of two numbers can clearly be developed into such a complex structure of decision making that a computer can adopt a heuristic role—indeed, a learning role as well, in the sense that its decision procedures can be modified in the light of experience. A chess-playing computer has become almost an uninteresting triviality since it is now clear that, given further programing effort, it will be able to beat any human competitor; nearer our aeronautical sphere of interest is Professor Donald Mitchie's computer at Edinburgh University which has learnt to balance upright a rod which is free to fall when placed on a randomly-moving trolley.

Further away from these regions of physically-deterministic systems which are susceptible to heuristic methods are the great regions of the humanities into which the quantitative appears not to trespass at all. Even of these regions, it could be said that the activity of the human brain as it contemplates them might one day be modelled within a computer and it would be a rash man who would assert with assurance that computers will never be programmed to make aesthetic judgments or value decisions, provided that these are based on describable criteria.

After that small detour away from the main line of our argument, we can reassert the conviction that the invention of the computer holds potentialities for civilisation which transcend certainly in scope, and probably in ultimate importance, any of the others which are recognised as major inventions. Thus it necessarily follows that it has a central role to play in any one particular science. We now therefore pause to outline briefly and in general terms, this role in the science of aeronautics before going on to consider in more detail the role in the smaller field of aeronautical design.

THE COMPUTER IN AERONAUTICS

Computers mostly, of a relatively simple kind, are already built in to the operational structure of aeronautics.

In commercial aircraft, the 'black-box'—whether it be a simple recorder of flight-data or an automatic control such as an autopilot—will soon develop, even if it has not already done so, into a computer which will act either in real-time or as a control or as a data-processer.

Acting in real-time, it will, for example, keep a constant watch on the optimisation of the whole flight process which includes the aircraft altitude, speed, height, taking into account the variables and constraints such as atmospheric winds. outside air temperature, passenger and fuel loads, meteorological conditions ahead and final destination. As a control, it will feed information to the pilot, to the aircraft's systems and to ground authorities, and receive feedback information from all three. As a data-processer, it will, for example, receive data from the aircraft radar as to the presence of other aircraft, or from ground position indicators.

The future development of the aircraft 'black-box' computer seems clear, at least up to the point at which an aircraft doors close on the incoming passengers to the moment that doors open again at the destination. Beyond that, it will be necessary for the total world activity of commercial aircraft to be treated as a single entity within which the individual aircraft will be just one element whose flight will be optimised with respect not only to its own local and personal conditions but also to all other aircraft in its neighbourhood—the word neighbourhood here being defined in a space in which the three-dimensional physical space is a sub-space.

At this stage of development (round about the turn of the century), the central processing computer will probably be one or more linked machines in permanent orbit, in equally permanent communication with each aircraft's own satellite computer. By then, too, there will be a wide variety of aircraft ranging probably from the hypersonic vehicle for the longest journeys (that is, semi-circumnavigations), through the supersonic aircraft for trans-ocean flights to subsonic machines for national routes and helicopters (or their successors) for airport-city connections. As to the optimisation process, it will extend over a scale from the universal set of aircraft through the individual aircraft to the passenger himself. Passengers will arrive as they will at city passenger terminals, record the necessary details of their intended journey themselves on a keyboard with CRT display attached so that the computer can ask whatever questions it wants, and the passenger will then be assigned to a path involving a number of aircraft which has been optimised with respect to all the known conditions.

If that is a picture which can be seen clearly for the period beginning at about 2025, then less clearly can be seen the further sophistication of the personal vehicle—the successor to the family car—which, with aerodynamic and propulsive properties still undreamt of, becomes yet another element in the total computer-controlled structure of aeronautical transport.

Within the second quarter of the twenty-first century, the world system just described will itself be just a part of a bigger system of astronautical travel. By that time—certainly well within the next hundred years—the moon will have been successfully colonised and there will be commercial services between the Earth and the Moon. Space travel is, of course, absolutely dependent even now on massive computing power devoted not only to the technicalities of the operating machinery in the space-craft but also to the calculation and timing of inter-planetary orbits. All this, therefore, will be integrated with the earth-bound system.

Much as one might hope that the military demands on the sciences of aeronautics and astronautics will relax as the years go by, one should perhaps adopt a pessimistic stance, at least when trying to predict what may happen and admit that military applications will remain a chief motivation in future technological developments. Thus we might note that already many military devices have a degree of scientific sophistication which far surpasses what is in common use in the civil field. It is this fact as much as any other, which encourages one's confidence that what has been said so far, as regards the capabilities of computers, is actively within the bounds of possibility. Indeed, most of the technical know-how for the computing side of the developments is already to hand: what will delay the production of the requisite hardware and software will be the physical development of the aeronautical and astronautical vehicles, and shortage of money.

In short, the computer will be behind the aeronautical engineer right where he wants it.

THE PROCESS OF AERODYNAMIC DESIGN

We must now begin to narrow the scope of the discussion, and look in more detail at the special area of aerodynamic design. What is meant by 'aerodynamic design'? This is not a simple question.

One might describe it in such general terms as these: 'the design of a total aircraft which minimises the passenger-per-mile cost'. By 'cost' is meant an all-inclusive cost of design, research, development, depreciation and operation. By an 'aircraft' is meant a machine which can take off and land on whatever ground facilities are available, carry a quota of luggage per passenger, give the passenger certain creature comforts, and fly prescribed stage-lengths. In proceeding on this traditional basis, we may be guilty of lack of foresight; it is as least possible that the intercontinental traveller in 2070 will be happy to be anaesthetised and packed sardine-like into a canister which a space-rocket ejects early in its flight to free-fall or glide under automatic control through the atmosphere to its destination.

The general problem of aerodynamic design under the traditional constraints just mentioned is further idealised again in an entirely traditional way, to the problem of designing an aircraft for a given payload and range, which will take off and land safely at low speed and which will have the minimum drag at its cruising speed.

This simplified problem is still further reduced, in current practice, by isolating certain geometrical or operating areas which are susceptible to attack by existing analytic methods—for example, infinite swept wings in transonic flow, or complicated flap-systems in two-dimensional incompressible flow, or swept-wing fuselage intersections in subsonic flow. The 'art' of design, if it is an art, has lain more in the ultimate fitting together of these isolated pieces of the design jigsaw than in the excellence of the basic physical theories.

Thus, when one comes to predicting the future course of design processes, it seems safe to say that it will be analagous to the building-up of a large computer program: small groups of simple operations, some treated no doubt as loops or iterations, will be combined to form larger sub-routines; and in their turn these sub-routines, some in loops, will form larger blocks of program which, as time goes on, will be themselves relegated to the status of sub-routines.

In discussing the role of the computer not by analogy, but as an integral element of modern and future design, we may first usefully describe the situation as it exists at present. Then by referring to some of the more sophisticated work at present being developed at the interface of computers and mathematics, we may guess at likely developments in techniques in design over the next quarter of a century. Finally, we shall grope a little further into the darkness of the unseen.

THE MATHEMATICAL MODELLING OF FLUID DYNAMICS OF THE 1960S

The most important feature of fluid dynamics, at least as it impinges on aeronautical design, is its determinism; in other words, it is assumed that the real physical situation can be modelled mathematically in such a way that a solution exists to every 'well-posed' problem. Newton founded this tradition by modelling physical situations in gravitation, mechanics and fluids by differential equations, and he was followed in the fullness of time by those who established canonical equations for well-defined phenomena, such as Maxwell for electro-magnetic continua and Stokes for fluids. These great mathematical physicists undoubtedly took a deterministic view of the phenomena they were modelling, a view which is still held in our macroscience of aeronautics.

Perhaps the phenomenon of turbulence calls for special mention at this point. A certain disillusionment seems now to have settled upon us as to the relationship between the Navier-Stokes equations of fluid-dynamical flows and the observed character of turbulence; the guarded optimism which was encouraged by the earlier work of scientists such as G. I. Taylor and von Kármán, and which was maintained after the last war by Batchelor and his generation, has now given way to a tacit acceptance that no direct solution of the Navier-Stokes equations is in sight for any realistic turbulent flow. Nor does there appear any likelihood of accurate predictions for the onset of turbulence in general circumstances. Yet this is not to say that a turbulent flow does not satisfy the governing Navier-Stokes equations; there is no acceptable evidence which would support the contention that a turbulent flow does not satisfy the equations —or even that turbulence is in any deep sense indeterminate. Certainly the statistical theories are not acceptable as a valid alternative to the Navier-Stokes equations.

So, if we include the border-line case of turbulence, we are in the position of having a set of differential equations which, we are confident, describe all the flows of interest to us.

These canonical equations, however, are too complicated for general analytical solutions to be found; indeed, such exact solutions as do exist can be described as trivial. Approximations, therefore, have to be made, and the computer immediately jumps into line with all the other more traditional techniques of approximation.

Two approaches to approximation can be distinguished, motivated by mathematical and by physical considerations respectively. In the first, alterations to the canonical equations are determined according to mathematical criteria and Nature is then asked kindly to conform; in the second, Nature's own tendencies are observed and translated into mathematical approximations.

Fluid dynamics happens to supply an excellent large-scale example of this dichotomy. The nineteenth century witnessed an extraordinary flowering of mathematics based on the wholesale neglect of the second-order derivatives in the Navier-Stokes equations. Yet this approach failed completely to account for the possibility of, for example, the sustained level flight of an aircraft, an adequate explanation for which had to await the work of Blasius[1] who, pursuing the other line, discovered the boundary-layer effect in differential equations.

It is fashionable to attach the greater prestige to the second approach. I doubt whether this is justified. The mathematical physicists of the nineteenth century developed, *en passant*, a vast amount of mathematics —indeed of what, round about the turn of the century, was regarded as pure mathematics. The title of Whittaker and Watson's famous book[2] of 1902, *Modern Analysis*, may now ring absurdly on our ears but modern equivalents such as Kantorovich and Krylov's fine *Approximate Methods of Higher Analysis*[3] have the same underlying physical motivations. Then again, Nature can sometimes be most accommodating to a mathematical simplification, as for example in Nonweilers' concept[4] of a hypersonic 'wave-riding' lifting surface. Either approach to approximations to the canonical equations, therefore, carries great potentiality of one kind or another.

But, whichever approach is favoured, there remains the fundamental question which any theoretician must, at some stage, answer—to what extent, if any, is he interested in the final production of *numbers*? It is often argued that research into fundamental physical mechanisms, the discovery of effects which are strongly coupled, and the interpretation of natural observations all require more than merely the production of numbers. It can be further argued that mathematical analysis can achieve certain vital objectives that computers cannot—in particular, the nature of singularities, or the exact variation of a solution with a certain parameter over the whole of its domain.

As the first point, one may remark that these activities of the research scientist have strong affinities to those of the pure mathematician and I hope to indicate later how computers, through essentially numerical means, are already active as laboratory assistants to the pure mathematician. As to the second, it may be observed that singularities are abhorred by Nature and that they arise only as a result of analytical approximations or of the inadequacy of the basic mathematical model. As to the third, an exact functional relationship usually arises only by virtue of earlier analytical approximations; thus it seems inherently sounder to derive such a relationship as an approximate interpretation of exact results rather than as an exact interpretation of approximate results. Remembering therefore that the speed, storage, methods of output and methods of analysis of output of the computers of the present generation are all still at a primitive level, we may surely expect great advances in the future in the 'approximate interpretation of exact results'. And finally, it seems conclusive that the theories of the applied mathematician are not sufficient unto themselves: whatever their sophistication, they must still be tested against observation. And since the essence of scientific observation is quantitative measurement, numbers are, in the end, inescapable.

As an aside, it is intriguing to consider the course mathematics would have taken if Newton had had a computer. It is ironic that differential equations which are derived in the first instance by taking the limit of finite-difference formulations are now recast in finite-difference form for computer solution. In these circumstances, what is the role of the differential equation? Is differentiability a concept which will be as significant in the future as over the past two hundred years? Would the great structure of special functions and the associated functional analysis ever have been built if the computer had existed in the sixteenth century? These are fascinating questions and a serious examination of them might help us to see more clearly the future of mathematical methods.

For myself I am prepared to be so bold as to assert that within fifty years we shall not be interested in explicit analytical formulae for solutions of differential equations and that computers will operate directly from the differential equations themselves. And I shall slant the following remarks in that direction.

PRESENT APPLICATIONS OF COMPUTERS TO FLUID DYNAMICS

Let us exemplify the potentialities by considering a boundary-value problem posed on one of the very simplest differential equations, the two-dimensional Laplace equation. Despite the intensive theoretical treatment to which this has been subjected over the past century, the class of boundary shapes which yield explicit solutions in closed form is highly restricted.

In Fig. 1* is shown a shape falling outside this class, that of the front-half of a body of revolution. The broken line shows the fluid velocity on its surface as calculated by a standard method, and this is to be compared with experimental points. We notice that the theoretical method is in fact so crude that it entirely smooths out the effects of a surface bump even as gentle as this one. In contrast to this, a straight-forward computer calculation, based on 51 discrete source-rings in the

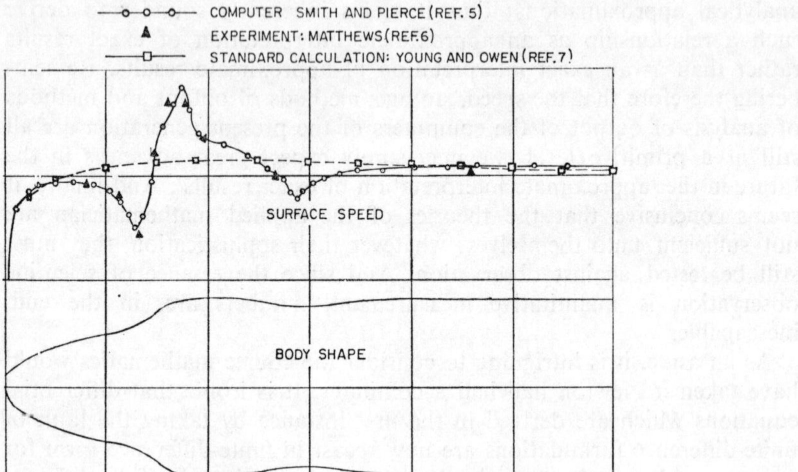

Figure 1 The computer obtains solutions for a wider class of physical problems than can be solved by classical methods, as is illustrated by this comparison of three attempts to find the fluid speed on the surface of a body of revolution submerged in a stream parallel to its axis

surface, produces a velocity distribution which clearly reflects the details of geometrical shape and, incidentally, is in close agreement with experiment. Such a computerised method of solution is, in effect, a definitive method applicable to a generality of surface incomparably wider than can be successfully tackled by other methods; it renders obsolete analytical procedures for solving Laplace's equation for such a class of problem. Of course, this particular method makes use of a fundamental result concerning the representation of harmonic functions by distributions of singularities on the boundary; and this illustrates the truth that the efficient setting-up of a procedure of computation may require a due amount

*Figures 1, 2, 3 are taken from Thwaites[16] and are reproduced by the courtesy of the Institute of Mathematics and its Applications.

of initial mathematical analysis. But in that the presupposed theorem is classical, this example is probably not typical; in the future the analysis preparatory to computation usually will itself involve novel theoretical methods.

The next example, relating to the Bénard problem[8], indicates one of the ways in which the sheer speed of the computer makes possible new ways of examining aerodynamic situations. Let us consider the two-dimensional case of fluid contained between two horizontal strips, the lower of which is maintained at a temperature higher than that of the upper. The fluid will not stay at rest. The heated lower layers become bouyant and feel the urge to rise; thus rising spouts of hotter fluid form and elsewhere, down-draughts of cooler fluid. The character of the flow depends on the Rayleigh number R which may be interpreted as the ratio of the destabilising buoyancy to the stabilising viscous forces. For low values of R, a regular pattern of cellular motion establishes itself, such as that shown in Fig. 2(*a*); under these conditions, the problem can be treated in a linear

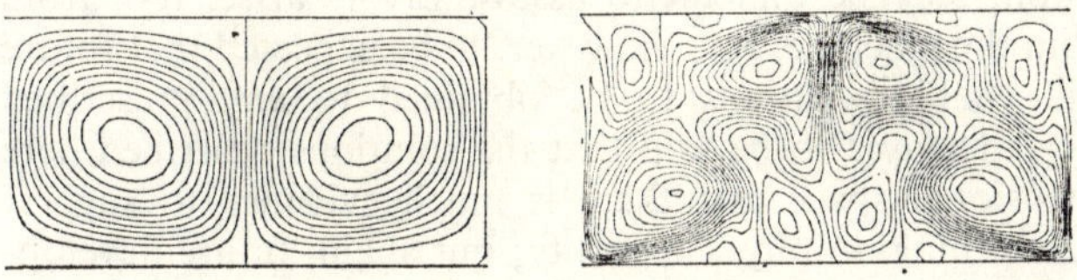

(*a*) (*b*)

Figure 2 (a) Conventional analysis could yield the steady-state streamline pattern in a Bénard problem for $R=3\times 10^4$; (b) Only the computer could reveal the details of the unsteady flow at $R=3\times 10^6$, of which this is a typical streamline pattern. Both figures were plotted by computer[9]

fashion. For high values of R, the flow is unstable; the analysis then becomes non-linear and untreatable. Fromm[9], however, has computed solutions of this unstable time-dependent situation. Fig. 2(*b*) shows a typical plot of the instantaneous streamlines for the high value $R=3\times 10^6$.

In this computer solution, there were no initial theoretical manipulations; the governing equations were immediately cast in finite-difference form and the time-dependent solutions calculated by a straightforward step-by-step method. At each time, the two-dimensional space field was solved by relaxation-type techniques involving up to 10^2 iterations over a mesh of some 10^3 pivotal points. Altogether some 10^3 time intervals were calculated; the whole work thus involved some 10^8 pivotal calculations and so is as far beyond the reach of human calculations as is the non-linear solution beyond the reach of mathematical analysis.

Fromm's program went beyond a typed output on a high-speed line printer. It displayed the instantaneous streamline patterns on a cathode-ray tube (the two Figs. 2 are photographs of CRT displays) and these were photographed and run as frames on a film. Those who have seen this film would, I think, agree that its visual impact is as physically suggestive as the direct observation of a real flow; indeed the possibility that such solutions can display fine details, such as small separated regions, in a way that conventional experimental techniques cannot, suggests that computer solutions of this kind will be most valuable adjuncts to, if not substitutes for, laboratory experiments.

This idea of programming output for physical phenomena in such a way that a direct visual comparison can be made with either full-scale or laboratory observation is one which is bound to be highly developed in the future. It may not be superabundant to give here yet another example, again produced by Fromm[10]; Fig. 3 speaks for itself.

It needs emphasising that, in solutions such as these, the complete differential equations are being solved, with no simplifying approximations made. Perhaps a little qualification of this is needed. For example, errors are introduced by the finite-difference formulation whose magnitude, in the final solution, may be difficult to assess. Errors arise, too, from rounding-off within the machine. It is, however, to be expected that future developments will enable solutions to be obtained to any prescribed degree of accuracy. It may also be said that the solutions described here refer to two-dimensional space and there is much evidence to suggest that there is no such thing, in physical reality; but again, future developments will certainly embrace the full four-dimensional situation of time dependent solutions in three-dimensional space. But if we are confident—as we are in these cases—that the basic equations do indeed give a complete description of the physical situation, then the computer is doing what mathematical analysis cannot do, namely produce definitive solutions for very general classes of problems.

SHORT-TERM DEVELOPMENTS IN FLUID-DYNAMICAL COMPUTING

It has to be admitted that, at present, the range of computed solutions is very limited; limited in fact to problems of two independent variables. Solutions such as those of Fromm described before, barely break beyond these limits: the introduction of the time as a third independent variable carries computing difficulties much less severe than would occur with a third space variable; also these solutions, for all their value and interest, are not definitive—the magnitude of the errors is not known. There are thus substantial advances in present numerical techniques still to be made before reliable solutions are obtainable for general three-dimensional unsteady flows. Perhaps the greatest strides forward are being taken already by Russian workers, if one may judge from such remarkable papers as those by Babenko *et al.*[12] or by Belotserkovskii *et al.*[13].

The hyperbolic character of supersonic and hypersonic flow poses less

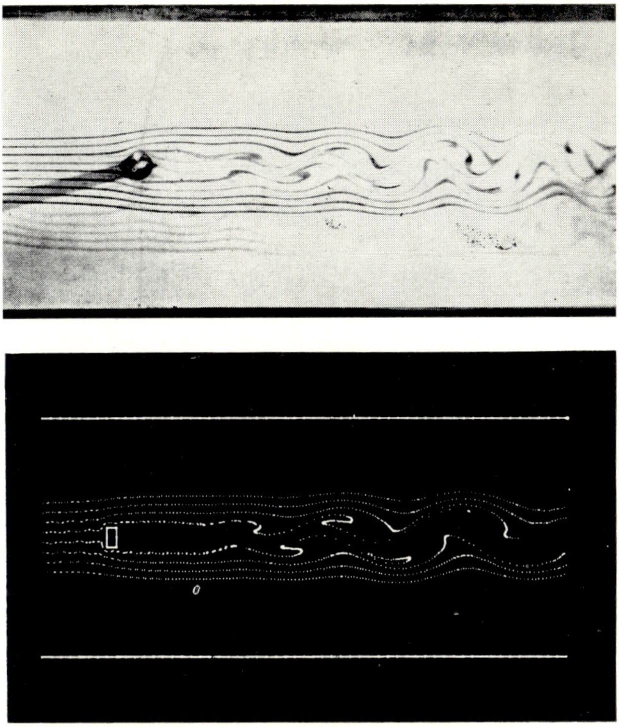

Figure 3 The computer in its role as a laboratory. The top diagram is a photograph[11] *of a real flow past a bluff body. The bottom is the CRT output of a computer calculation*[10]

formidable difficulties to numerical analysis than the elliptic character of subsonic flow. Thus in the following brief discussion of the problems which call for solution over the next quarter-century, we restrict ourselves to the subsonic case, so as to give any predictions a pessimistic rather than an optimistic air.

In the research programmes in this field over the next two decades, there may be two major impediments. First there may be an increasing divergence between the industrial designer and the academic researcher. The constraints on a creative industry are becoming biased ever more strongly towards the financial requirements, while the academic's search for knowledge-for-its-own-sake holds monetary considerations in contempt. It is probable that the country will be steadily less and less able to afford the luxury of the 'pure' researcher in a field so highly applied as aeronautics, and that he will find himself increasingly under pressure to choose problems relevant to the current technology. Secondly, there is the problem of turbulence. This is of acute importance to the designer who urgently needs to be able to predict, to a good degree of accuracy, the turbulent boundary layer on a swept-wing/fuselage combination; yet (as we have already mentioned) the fundamental study of turbulence seems to have run up against a brick wall. Turbulence, therefore, is a topic in which a theoretical break-through is required; it will, undoubtedly, come one day; meanwhile, perhaps we should look to the computer to provide the key.

There are, however, some important types of flow in which rapid oscillations, not of a strictly turbulent character, are of great significance and which, though very difficult of treatment, may yield to the power of future computers. Such flows in which the typical scales of variations in time and length are of a higher order of magnitude than those of fully-developed turbulence, occur in some buffeting situations, or in the oscillation of a shock-wave or in the general stability of a very high-lift flow in the landing condition. They are invariably beyond the limits of analytical solution even under very ideally-modelled conditions, so that the computer can be expected to aid major advances in the study of such flows.

It seems probable that the most rapid advance on such problems will be in the low Reynolds number case—say, for $R<10^4$. For $R>10^4$, and especially for the much higher Reynolds numbers typical of large aircraft, the large values of derivatives associated with a thin boundary layer pose problems for computer calculations whose solutions even in principle are not yet clear. It may turn out that a combination of boundary-layer theory on the one hand and more direct numerical analysis for the external flow on the other will be best; alternatively, the size and speed of computers may increase so dramatically that the number of mesh points in the grid can be increased by one or two orders of magnitudes so as to cover adequately the boundary layer.

In either case, it is now thought that, even though the ultimate goal must be the solution of unsteady and turbulent flow, a necessary preliminary is the development of satisfactory numerical methods for steady laminar flows. These flows then fall naturally into two categories, each carrying its own uncertainties as to the validity of the conventional mathematical model.

The first is categorised by the thick wake, or by the fact that the classical first approximation of potential flow past the body is not adequate to describe, even roughly, the real flow outside the boundary layer and wake. This category includes the practically important case of an aerofoil at high incidence with a long bubble. The short-comings of all the methods which have, from time to time, been proposed for this category of flow are well-known—all such methods involve paying the penalty of indeterminateness for the apparent advantage of workable analysis. Almost certainly the escape from this impasse lies in the ability to calculate definitive solutions over the whole field, including the interior of the wake and bubble.

The second category is defined by the classical problem of high Reynolds number flows past thin streamlined bodies at low incidence with thin wakes. Although much is known about the individual components of this flow as a result of many decades of intensive research—about the inviscid external flow, about a laminar boundary layer, and about a thin laminar wake—it remains astonishing that the fitting together of these components in anything but an extremely crude fashion, is still quite beyond our capabilities. In the very short term, say during the next ten years, therefore, advances even with the aid of computers are likely to be confined to the establishment of routine methods of high accuracy for those individual components of the flow. One might hope that, after 1980, some real progress will be made with genuine whole-field solutions and these seem bound to be computer solutions.

The rate of advance in the study of any of these types of flow will depend on whether they are posed in two or three space dimensions. The preceding few paragraphs were written with the two-dimensional case in mind. The programming difficulties seem much more severe, and the time scale considerably longer, for the general three-dimensional case of a lifting body with separations from all its edges, possibly with long and short bubbles as well, and at high Reynolds numbers. Once again, however, it seems that in the computer lies the only hope of solving this general problem. And the time scale? It would be rash to expect a standard program for the problem to be available before 1985.

MEDIUM-TERM DEVELOPMENTS IN THE SOLUTION OF CANONICAL EQUATIONS

It is possible now to draw some tentative conclusions for the fifty years in the middle of the next hundred years—that is from about 1995 to

2045. The previous section looked realistically at the daunting difficulties of definitive solutions of the Navier-Stokes equations; in particular, it tried to see the situation through the eyes of today's active researcher whose attention is focused primarily on the next small step forward and whose work is circumscribed by existing computing machines.

But in looking ahead to these same problems in 2000 and beyond, one should try to anticipate the development of at least the simplest traditional features of the computer—namely the size of the store, and the internal machine speed of store transfers, of arithmetic and logical operations, and of input. Perhaps it would not be optimistic to make the very rough prediction that by the year 2000 stores will be of a virtually unlimited size and that if access times are to be roughly independent of the size of store, all basic operating speeds within the machine will be about two orders of magnitude faster than at present. Operating speeds of about three orders of magnitude greater than at present are probably technically feasible, but at such speeds some of the physical distances are becoming molecular and the speed of light is inhibiting the overall size.

Thus we may expect that soon after the turn of the century, we shall be able to 'solve'—in whatever way that word will be interpreted at that time—problems in unsteady three-dimensional laminar flow in a time of the order of one second. We shall then have passed a critical point in the history of theoretical analysis, beyond which the time required for the input of the data exceeds the time required for the actual processing of the solution. The emphasis will have shifted away from the technicalities of the solution of the canonical equations towards the means and character of input and output.

The wide discrepancy which will then exist between times of input and times of solution—and also between the times involved in the human contemplation of some problem—may be sharply illustrated by the following rough orders of magnitude:

Computer solution:	10^0 units of time (of the order of one second)
Manual input of numerical data:	10^3-10^4 units
Mental formulation of problem:	10^6-10^7 units

From this emerges that the computer will no longer be regarded, as it is now, as a miracle computational device performing numerical tasks beyond the capacity of mid-twentieth century mathematical analysis, but as an adjunct to the human mind in the conceptual formulation of problems.

While, therefore, times of input and output will be reduced by two or three orders of magnitude either by methods of man/machine transcription not yet invented or by far greater use of linked computer stores, the major development will lie in the reduction of the time of formulation to (shall we guess?) the order of 10^4 units, or one hour.

Perhaps we should try to be a little more precise in quoting conceptual formulation times of the order of one hour. What is meant by this? First, we must enlarge now the scope of our discussion beyond the confines of the external aerodynamics. The ideas discussed so far apply equally to the canonical equations governing the other aeronautical disciplines which at present remain in their own more or less self-contained compartments. The internal aerodynamics of jet engines, or the propulsion characteristics of rockets are, or will be, soluble in the same way as those of the external aerodynamic problems. The same is true of the structure of an aircraft, although a complication involving a factor of some orders of magnitude is introduced by the fact that a complete flying machine will be made up of many individual structural components to each of which may be applied, as required, the canonical equations of continuous structural analysis. The factor of complication is not, however, of the same order of magnitude as the number of individual components since, in the consideration of their interactions within the total structure, only certain macroscopic features of each are required. And so the list may be extended to all other aspects of a machine's design—aeroelastic properties, control systems, navigational techniques, and so on; all these pose, in some sense, deterministic problems of an inherently numerical character which will be susceptible to the general developments already envisaged for the Navier-Stokes equations.

In talking, therefore, of problem formulation times of the order of one hour, one is certainly thinking of total aeronautical problems, that is of problems involving the integration of many different current disciplines into a single overall scheme or objective. One is not thinking of problems as simple as that of, for example, the design of a two-dimensional aerofoil to have certain lift–drag properties over the whole of its operating ranges of speed and incidence, for which an aerodynamicist would need, at present, a time of the order of one hour for the formulation of the concepts and desiderata. Half a century ahead, that hour will be sufficient for the designer to develop his detailed concepts for a complete machine, since his computer will feed him with immediate information in a form appropriate to his task.

Such future methods of concept-formation depend on various computer techniques which are additional to those involved in mathematical solution of canonical equations and the remaining sections of this chapter are devoted to a summary of some of the ideas involved.

DIALOGUE WITH COMPUTER—INPUT AND OUTPUT

No doubt the computer of the future will always possess the basic characteristics of the first few generations of machine, in which the sequence of operation is: program, data, run, output. No doubt, too, there will always be input facilities involving cards, punched tape or magnetic tape, and output facilities of the same kind as well as high-speed line printers and graph plotters. But these basic facilities will be supplemented most powerfully by the teleprinter and keyboard, and by the cathode-ray tube and light pen. A word about each is required.

The teleprinter keyboard will be more than the means of conventional input: it will be the medium of conversation between the scientist and the machine. Familiar already is the idea of building up a program on a keyboard attached to a multi-access machine, and of checking each block of program as it is written into the machine. Less familiar are the programs which are already developed, and which will evolve rapidly into much greater sophistication, for accepting and manipulating mathematical expressions typed on the keyboard. This, of course, requires a language of conversation and it is clear that traditional mathematical symbolism with its composite symbols, suffices and indices is manifestly unsuitable for mechanical input and output through a teletype. Instead a symbolism is required which is serial in the sense that any conceivable mathematical expression can be written as a series of symbols on a single line. Such languages already exist, for example that developed by Iverson[14], and in due course the mathematical and scientific community will have to come to an agreement on such a universal language.

One use of the cathode ray screen will be to display, in traditional notation, mathematical expressions typed in a new universal language. This can be exemplified by a program already in existence[15] by which, for example, the equation

$$\ddot{x}+\omega^2 x=\epsilon x^3,$$

when typed in as

```
(DRV(:T,2,X(:T)) + OMEGA↑2*X(:T) = EP*(X(:T))↑3)⫽
```

appears on the screen as

$$\frac{d^2}{dt^2}X(t)+\omega^2\cdot X(t)=\epsilon\cdot X(t)^3$$

The light pen is essentially an input device, identifying some part of the total CRT display which is then regarded as a data input for whatever program is running at the time. Thus in the manipulation of mathematics, the light pen may identify a particular term of an equation for further operations on it, as would be required, for example, for the deliberate replacement of $\sin 2x$ by $2 \sin x \cos x$; or it may point, for example, to two power series for multiplication and display in power-series form. All this can even now be done, and there seems no limit to the computer's technical competence in the manipulation of mathematics.

The cathode-ray tube will also be used in many other ways, but probably the most common use will be the geometrical. It will be used for the input or output of shapes of all kinds, both two- and three-dimensional; with the light pen, parts of such shapes will be identified or modified. Results of calculations will be presented in graphical form and the speed of future computers will be such that displays of time-dependent situations will be seen as films in real-time, each 'frame' being the result of a calculation performed in the instant between that 'frame' and the preceding one.

In these and many other ways, therefore, the computer will enlarge our capacity for technical manipulations of all kinds, and they will be essential constituents of the scientific methods of the future.

LIKELY DEVELOPMENTS IN THEORETICAL AND EXPERIMENTAL TECHNIQUES IN AERONAUTICAL DESIGN

It will be helpful in this section to have in mind a specific problem. Let us take the problem of designing a two-dimensional aerofoil shape for a prescribed pressure distribution in low speed flow. In the inviscid model, this is a relatively simple problem which can be treated more or less successfully by a variety of methods. In the realistic viscous situation, the problem is at present intractable despite its apparent simplicity, except possibly in the special case of unseparated flow at high Reynolds number. In looking to the future when we expect to be able to solve far more complicated design problems than this, we must assume that this basic two-dimensional problem will be solved virtually at the touch of a button. Nevertheless, the methods which may be applied to much more complicated situations in the more distant future can be illustrated briefly by reference to this simple aerofoil problem.

First, there will be statistical approaches in which are analysed the trends given by large numbers of earlier results. This approach assumes an experimental attitude to a class of problems associated with the given problem. In the future, the adjective 'experimental' will refer to theoretical or numerical experiments as well as to physical laboratory experiments; the essential character of this experimental approach is that it yields a large amount of data. The traditional characteristic of laboratory experiments, namely the methodical variation of a small number of carefully chosen parameters—which was necessary primarily for the efficient extraction of the maximum amount of information from what was usually a lengthy, laborious and expensive process—will no longer be of such significance. Thus, in our example, one might envisage aerofoil shapes being first defined in discrete form in various different ways (of which the usual cartesian co-ordinate representation would be only one) and then statistical correlations established between the shapes and the pressure distributions.

A more direct approach would be trial-and-error, in which, for example, an aerofoil shape is drawn by a light pen and the associated pressure distribution is immediately shown on the screen as the output of a program solving the governing equations. Modifications by hand and eye are then continuously made until the desired state is reached. The apparent crudity of this approach, against which there may be some psychological resistance on the part of the 'purer' scientist, conceals the sophistication inherent in the derivation of the pressure distribution.

Next, it may be assumed that laboratory experiments (in our case, wind-tunnel runs) will be highly automated; thus, at the least, the control of the tunnel speed, of the aerofoil incidence and other test quantities will be in the hands of the computer which will itself decide the parameter values for the next run so as to fit in to the total scheme of the experiment. In such a linked system, all readings would also be automatically recorded and fed direct to the computer. The computer could then keep a constant account of the comparisons between the calculated pressure distributions and the scale experiments; in subjecting these comparisons to the kind of statistical analyses already suggested, it would make valid extrapolations or accurate allowances for physical effects (for example, for turbulent boundary layers) for which full allowance had not been made in the theoretical solutions.

But the idea of a linked computer/wind-tunnel system should be extended to include an automated workshop. Thus if, in the design problem we are illustrating, it is admitted that the theoretical basis is insufficient and that purely theoretical predictions, however obtained, must be backed up by direct physical measurements, then the computer itself should be programmed to choose critical cases for wind-tunnel tests and to fabricate the aerofoils; all the workshop technician would have to do is to fit the model in the tunnel.

The methods outlined above, and their combinations, have the important merit of being independent of the restrictions imposed by mathematical idealisations. For example, aerofoil theory has been, and wing theory still is, dominated by linearised approximations of various kinds and it may well be that genuine effects are masked by such approximations. Thus we might have specified for our problem a pressure distribution which falls outside the class of valid linearised solutions and, in these circumstances some other approach is necessary.

Or again, most practical problems are highly constrained and the objective is some kind of optimum within these constraints. For example, if to our aerofoil design we add only one simple condition, namely, a lower bound for the area enclosed by the contour, coupled with an upper bound for some error-norm on the pressure distribution, conventional methods are at once inadequate, whereas the computer-based approaches are complicated to the relatively trivial extent of an additional iteration loop.

All these ideas, then, have obvious relevance to the design of a complete flying machine and in this wider context a final remark is particularly relevant. When optimum states are being sought, it may be relatively easy to find local optima by systematic variation of parameters about some designed mean. In simple systems, it often happens that such a local optimum is also an overall optimum, which comes about as a result partly of the small number of parameters and partly of the curious but convenient fact that most physical systems exhibit smooth behaviour under variation of parameters. But things do not happen so conveniently in more complicated systems when genuine optima can occur which lie well outside the region of conventional analysis within which some local optima may be found. It is likely that such optima can only be established by the wide-ranging searches which computer methods make possible.

THE COMPUTER AS A HEURISTIC INTELLIGENCE

We now turn attention to the researches over the past few years which have firmly established the capabilities of computers to learn from their own past performance and to adopt heuristic attitudes to problems which may be set them. 'Machine intelligence' may be a glib phrase but it now fairly describes these future capabilities; and this section outlines very briefly such capabilities as have already been realised in the realm of mathematics. A fuller account has been given by Thwaites[16].

First, then, methods are now well-established whereby abstract quantities and operations on these quantities can be coded for computers. It is possible to establish an axiomatic system, and to derive limitless results in it; or to postulate a logic and establish any number of true statements within it. The theoretical concepts involved in the computerisation of such systems—networks, graphs, trees and so on—are now well known. Theorems can be proved by automatic traversing along the lines of a graph between two nodes. Results can be generated which, if not theorems (for theorems are characterised by being statements about an infinite class), at least have great power in that they may refer to large, although not infinite, classes.

These developments prompt the remark that, although mathematical machines of various kinds have existed in the past, the special preserve of the mathematician has always been the determination of the strategy of his thinking. But now, things are different, as Collins and Mitchie have vividly explained[17]: 'the distinctive aim of machine intelligence within the general field of computer science is artfully to encroach on this preserve, annexing to the machine's domain more and more of the human's elusive aptitudes—his ability to handle non-numerical data structures of all kinds, to generate and apply data-descriptions schemata, to learn from past experience to abstract features of common to different problems, to devise and apply criteria of "relevance" to use approximate and intuitive arguments, to construct general rules from particular instances and to define concepts by way of examples.'

As an example of how these ideas have already been applied to the more mundane processes of applied mathematics, the two keystone techniques of differentiation and integration of real functions of a real variable may be cited.

They apparently involve two quite different algorithmic techniques. Differentiation is a completely determinate process. Thus there is no difficulty of principle in computing a derivative of any expression involving functions whose derivatives are known. The form in which a derivative is required depends, however, upon the context so that even this apparently determinate process will in fact involve heuristic considerations.

By contrast, integration appears undeniably heuristic. Yet paradoxically, it has been found that a large class of integrands is susceptible to algorithmic treatment: in fact a program[18] developed at the Massachusetts Institute of Technology integrates at a speed which is often two or three orders of magnitude faster than a heuristic program[19]. Thus indefinite integrals involving the elementary functions can now be performed in a few seconds—which means instantaneously on the machines of the twenty-first century.

Two main conclusions may be drawn from these brief remarks on the present state of 'machine intelligence' as applied to mathematics. First, the art of programming computers for heuristic intelligence is still in its infancy, and there seems no reason to suppose that computers, within the next hundred years, will not surpass human capacities at least in some areas of the intellect's activity. Second, the methods now being developed for specifically mathematical purposes have immediate and obvious applications to all other sciences which are fundamentally mathematical in character. Thus the methodology of engineering, and in particular of aeronautical design, is bound to be radically changed by computers programmed for heuristic intelligence.

ECONOMIC CONSTRAINTS

If a possible deduction from what has been said so far is that more or less any process within a logical structure will be susceptible to successful treatment by the computer of the future, the time-scales involved, and even the nature of problems which are deemed significant or worthwhile will be strongly influenced by economic considerations. Here one is thinking not so much of the current or developing states of the economics of individual countries, or trade blocs, or continents, but of the sheer total cost of the development and running of computer systems of the kind envisaged.

On capital costs of machine hardware, it might be said that bigger and faster machines will certainly be built in any case and that their unit costs (of each basic machine operation or of each bit in store) will decrease sharply over the years as computer technology progresses.

There are, however, capital costs of another kind in connection with the computer applications to any particular area of science, for example to aerodynamic design; and these relate to the development of the appropriate software. It seems likely that each science will, in the future, require highly specialised computer software just as it now requires its own distinctive laboratory or experimental equipment. Thus the programs and displays and input methods for future aerodynamic design will be as different from those for the design of future power stations as is a wind-tunnel from an atomic pile. The major component of the total cost of experimental computing facilities for aerodynamic design will undoubtedly arise from the software—and it is, incidentally, in the development of appropriate software that this country is already seriously lagging behind others.

Financial considerations cannot, however, be ignored; indeed they will be an essential ingredient in the planning of design methods. For example, is it reasonable to aim at completing the design of a whole aircraft within an hour or two (as was suggested earlier) if it would be just as effective for one's macroscopic goals to complete the design in a day or a week, and at a much lower cost? Such a question poses formidable problems concerning the deployment of human and material resources.

CONCLUSION

An extraordinary situation therefore seems to have arisen with regard to the development and applications of computers. Within the incredibly short space of twenty-five years from their invention, they have developed to such an extent that the firm forecasts which can be made of their potentialities lead to the question, 'How can we effectively use the power they put into our hands?'

In asking such a question one has instinctively returned to the paradoxes of human attitudes described at the beginning of this chapter. Computers will be capable of almost any process which possesses a structure describable by the human intellect. How does this capability fit in to one's vision of human society; how is it to be reconciled with the *essentially* human components of the world community of the year 2070? The scientist or engineer may say that it is not for him to seek to answer such questions. But if he does not, the humanist will, for they are compelling questions demanding answers.

My final remark, therefore, is not one of scientific prophecy, but of exhortation to the scientist, whether he be an aeronautical designer, geneticist, economist or what you will. Let him acknowledge the power over the evolution of the human society which his science gives him and his successors, and let him with conscious deliberation render compatible this power with the fulfilment of the highest aims of humanity.

References

1 BLASIUS, H. Grenzschichten in Flüssigkeiten mit kleiner Reibung. *Zeitschrift fur Mathematics und Physics*, Vol **56** pp 1–37, 1908.

2 WHITTAKER, E. T. and WATSON, G. N. *A course in modern analysis.* Cambridge; Cambridge University Press, 1902.

3 KANTOROVITCH, L. V. and KRYLOV, V. I. *Approximate methods of higher analysis.* Groningen: Noordhoff, 1958.

4 NONWEILER, T. R. F. Aerodynamic problems of manned space vehicles. *The Journal of the Royal Aeronautical Society*, Vol **63**, No. 585, pp 521–8, 1959.

5 SMITH, A. M. O. and PIERCE, J. Exact solution of the Neumann problem. Calculation of non-circulatory plane and axially symmetric flows about or within arbitrary boundaries. Douglas Aircraft Co. Inc., Report ES26988, 1958.

6 MATTHEWS, C. W. A comparison of the experimental subsonic pressure distribution about several bodies of revolution with pressure distributions computed by means of linearised theory. Report, National Advisory Committee on Aeronautics, Washington, RM–L9F28, 1949.

7 YOUNG, A. D. and OWEN, P. R. A simplified theory for streamline bodies of revolution and its application to the development of high-speed low-drag shapes. Report and Memoranda, Aeronautical Research Council, London, 2071, 1943.

8 CHANDRASEKHAR, S. *Hydrodynamic and hydromagnetic stability.* Clarendon Press, Oxford, 1961.

9 FROMM, J. E. Finite difference methods of solution of nonlinear flow processes with application to the Bénard problem. Report, Los Alamos Scientific Laboratories, LA–3522, 1967.

10 FROMM, J. E. A method for computing nonsteady, incompressible, viscous fluid flows. Report, Los Alamos Scientific Laboratories, LA–2910, 1963.

11 THOM, A. The flow past circular cylinders at low speeds. *Proceedings of the Royal Society*, **A141**, pp 651–69, 1933.

12 BABENKO, K. I., VOSKRESENSKIY, G. P., LYUBIMOV, A. N. and RUSANOV, V. V. *Three-dimensional flow of ideal gas past smooth bodies.* Moscow: 'Science' Publishing House, 1964. (National Aeronautics and Space Administration, Technical Translation F–380).

13 BELOTSERKOVSKII, O. M. and CHUSHKIN, P. I. The numerical solution of problems in gas dynamics. 1–126 in *Basic developments in fluid dynamics* (Editor, HOLT, M.). New York: Academic Press, 1965.

14 IVERSON, K. E. *A programming language.* New York: Wiley & Sons, 1962.

15 MARTIN, W. A. Syntax and display of mathematical expressions. Massachusetts Institute of Technology Memoranda, MAC–M–257, 1965.

16 THWAITES, B. 1984: Mathematics <=> Computers? *Bulletin of the Institute of Mathematics and its Applications*, Vol **3**, No. 6, pp 133–59, 1967.

17 COLLINS, N. L. and MITCHIE, D. (Editors) *Machine Intelligence I.* Edinburgh: Oliver and Boyd, 1967.

18 MOSES, J. Symbolic Integration. Massachusetts Institute of Technology Memoranda, MAC–M–310 and MAC–M–327, 1966.

19 SLAGLE, J. R. A heuristic programme that solves symbolic integration problems on freshmen calculus, Symbolic Automatic Integrator (SAINT). PhD Thesis, Massachusetts Institute of Technology, 1961.

5 Astronautics—its development during the Second Century of the Royal Aeronautical Society (1966-2066)

A. V. Cleaver

1 INTRODUCTION

The Royal Aeronautical Society is now 102 years old, and astronautics has already existed as a technical discipline for some 70 of these years, even if one disregards its long pre-history of mythology, science-fiction, and vague speculation. Tsiolkovsky began to make his amazing and entirely serious scientific contributions just before the dawn of the twentieth century.

This suggestion that aeronautics and astronautics are much of an age would surprise many laymen, who are apt to accept literally the dating of the popular Press; this usually argues that the 'Space Age' began with the launching of Sputnik I in 1957, disregarding even the first successful large rocket, the V2, which flew 15 years before Sputnik I. However, whether one argues that astronautics is already a septuagenarian, is in its vigorous and youthful twenties, or is not yet even a teenager, one must admit that its amazing development has occurred mostly in the last decade. One can also be quite sure that this development will continue far beyond even the second century of the Royal Aeronautical Society, and that what we have seen will seem insignificant by comparison with what our children, and their children's children, will see. Like Man's achievement of flight within the earth's atmosphere but with an even greater eventual potential for influencing his whole life and outlook, spaceflight is more than just an applied science. It is, or will be, a vast field of human activity, with commercial, military, social, and philosophical overtones as well as scientific ones, and not least (as has often been said) it is 'The greatest adventure awaiting mankind'.

Now, in 1968, we can already be fairly sure that men will first set foot on another world within the next five years; the official schedule of the NASA Apollo programme calls for them to be Americans, landing on the Moon in 1969 or 1970*. Even if this date is not kept, and even if the

*Editor's note: The first manned landing on the Moon was on 21st July 1969.

Originally published in the *Aeronautical Journal of the Royal Aeronautical Society*, May 1968.

Russians succeed first in this historic achievement, it is still almost certain to fall within the first decade of the Society's second century.

During this period, we shall also see a very considerable extension of those 'practical' or 'useful' space applications which cannot be ignored even by those who dismiss lunar expeditions and interplanetary travel as mere scientific stunts, or as extravagant and anti-social displays of national showing-off, comparable to Roman circuses or the Pharoahs' pyramids.

Already, telecommunication satellites are offering a commercial service across the Atlantic and Pacific, sometimes relaying TV programmes in real time; others are providing military telecommunications. Yet other satellites are helping ships and especially the nuclear submarines of the US Navy to navigate, and meteorologists to make better long-range weather forecasts, with warnings of impending hurricanes and other natural disasters. Satellites are already a primary and probably decisive factor in military reconnaissance, and therefore are making a vital contribution to keeping the peace of the world. The deterrent effect of the great inter-continental ballistic missiles, in this writer's opinion, had already made such a contribution even before this; these were the forerunners of the space-launching vehicles (in fact, both in the USA and the USSR they were themselves, after adaptation, the first such vehicles), and they are really space weapons. If anyone regards this as stretching a point, he must surely concede such a description to the recent Russian revelation of what the Americans (in typical jargon) have called a 'Fractional Orbital Bombardment System'. We shall see other military developments in space. In the field of pure scientific research, satellites and interplanetary probes have already extended human knowledge (the van Allen radiation belts, conditions on the Moon, Mars and Venus), and once-sceptical physicists and astronomers would not now care to be deprived of this invaluable new tool. All nations with any pretension to culture now subscribe to the view that pure scientific knowledge has an inherent value (apart from its often more spectacular value in good or bad applications), so this space activity is also likely to develop along the lines already initiated.

Quite apart from any controversial questions of manned interplanetary travel, or the exploration and eventual exploitation of other worlds, there is in consequence every reason to predict a steady and vigorous development of space technology in the future. If we were concerned here only with the next 20 years, instead of the next 100, these factors of what I have called the 'practical' or 'useful' applications in space would ensure this. Once-sceptical quarters have become convinced, even in the short period since Sputnik I, of their value; the Pentagon and the Kremlin would hate to be deprived of their reconnaissance satellites; even the notoriously-conservative British GPO (which once resisted the introduction of telephones into the City of London because their own excellent service of messenger boys made Dr. Graham Bell's new-fangled invention superfluous) now wants to use American communications satellites, although it remains to be convinced that we should develop our own European ones.

Elsewhere, hard-headed business interests have realised that satellite telecommunications are a new and developing multi-million dollar operation. Very soon (perhaps already, within Russia or China) politicians will grasp the fact that, especially with TV, they can be used as an extremely powerful propaganda weapon. Educators have already appreciated their similar, but more admirable, uses (especially for emergent countries) in education. Perhaps the direct-broadcast TV satellite will have the biggest effect in controlling the excessive birth-rates of Asia and Africa, just as the works of the civil engineer (at least as much as those of medical science, until antibiotics) have operated to control the death-rate.

We shall return to the longer-term problem, with all its vital and fascinating questions of motivation, but let us first take a look, in rather greater detail, at the short-term (about 20 years). This must inevitably be wise, in attempting a long-term forecast, since the course of the next 100 years will, very largely, be indicated by the trends during the next 20, just as the latter are rooted in those of the present.

2 THE FIRST TWENTY YEARS (1966–1986)

During this period, the most practically important space application is likely to remain the communication satellite, of one sort or another, but others will emerge. All will share a great deal of common technology; indeed, since all satellites are concerned with the transmission of some kind of information, one might pedantically argue that they are all 'communication satellites'.

Using the term 'comsat' in its usual narrower sense, these devices will progress from their initial provision of telephone channels and a TV relay service between large ground stations, into more sophisticated forms. These will demand increasingly advanced on-board auxiliary power sources, provided by solar cells, nuclear isotopes, or even small fission reactors (with direct conversion to electrical power). TV relays from a large base station to many community centres, with small antennae, will soon be provided and in turn will develop into direct-broadcast systems, capable of reception by domestic roof-top aerials. It is with these later systems (and especially the last) that the full potential of the TV comsat for cultural and propaganda purposes, as distinct from entertainment, will become apparent. They will be an extremely powerful social influence within the next generation. Apart from TV, the simpler technique of sound radio broadcasting from satellites should not be ignored, especially in its uses for the dissemination of political propaganda and educational material.

In parallel with these services, comsats will be used for transmitting all sorts of other information: facsimiles of documents, links between computers, and so on. One consequence of all this must surely be some reduction in social needs for the actual physical transportation of people and, to some extent, of mail. Apart from holiday travel, for its own sake, there may well be a significant lessening of the necessity for men to travel long distances to attend short business conferences or interviews, when all those concerned can be linked by telephone (and television, if desired), and provided with fascimiles of all relevant papers. Such facilities will probably become cheap (calls from London to Sydney will then come to be regarded as no more extravagant than those to Bristol are today), so that this is a factor which should be taken into account by all those concerned with long-term forecasts of the growth of airline traffic. (Maybe this is an added reason for developing the air freight market.) Forecasters in another field are already aware that, at any rate over long distances, satellite links are soon likely to be preferred to any major expansion of cables or direct surface-to-surface radio.

In the same period, specialised satellites will also be used increasingly as aids for navigation and traffic control for ships and aircraft, and for the gathering, dissemination and prediction of weather data. The latter function (already a very important one) foreshadows yet another, which may become one of the most valuable uses of space technology: that of the 'earth resources satellite'. Both are close descendants of the military reconnaissance satellite, using infra-red, TV or photographic sensors (in the latter case, with recovery of films if required, as already practised by the Americans and the Russians for the military application). Preliminary experiments have already shown that a satellite can reveal an amazing amount of data, not only about the earth's atmosphere (as viewed from on high), but also about its surface: not only about military and industrial installations, but also about geological formations, the depths of seas and waterways, ocean currents, the migrations of fish and animals, the health of crops, forests and vegetation, and so on.

Most satellites for all these purposes, if unmanned and fully automatic, will probably continue to be of less than 1 tonne (1000 kg) in mass. Depending on the radio-techniques employed, perhaps they may all lie within this range, but probably some may eventually require to be as large as 2 or 3 tonnes (2000 or 3000 kg) in mass. Most, but not all, will find it advantageous to occupy the geo-stationary orbit (at 26,000 miles (42,000 km) radius) which may become rather crowded, by the normal standards of 'empty space'.

This particular synchronous orbit represents a mission launching requirement more onerous, for example, than the attainment of escape velocity, or impact on the Moon. The range of payloads specified for it (as above) should therefore encompass most likely requirements for advanced scientific research satellites or space probes. There is no doubt that these will continue to be sent to the surfaces, or at least to the vicinity, of all the planets (including Jupiter), and deep into space, or that they will be of increasing ingenuity and complexity. However, the same launching vehicles will suffice for both the applications and the needs of the pure scientists. Suitable ones exist already, and are smaller than the largest of those available even today.

That this is so, of course, is a tribute to the amazing development of modern electronics, with its miniaturised and then micro-miniaturised circuits, transistors and still later solid-state devices, all of high and increasing reliability. This development has both been stimulated by the development of missiles and space vehicles (for their guidance and payload systems), with widespread 'fall-out' into ground-based computers and other fields, and then has in turn been a potent factor in making it possible for these spacecraft to achieve increasingly more impressive tasks.

It is nevertheless certain that by 1986 there will have been developed a whole new range of launch vehicles; because we already have some with the required performance, it is unlikely we shall remain satisfied with them, when improved designs have become feasible. However, arguments of cost-effectiveness would impede such progress, if the only demands in sight were those of the automatic research and application satellites. The necessary R and D is already well under way to build the improved vehicles (less clumsy for a given duty, more efficient) but the overwhelming case for having them becomes apparent only when we want to send men into space.

A man is a bulky, heavy, and inconvenient payload to send into space; for all but the earliest, most experimental, and least ambitious of missions, he also likes company—there will be two or three or more of him. They will all demand air to breathe, and food and drink, complicated temperature and humidity controls and arrangements for the disposal of physical waste products. He and his fellow-men back on earth will always expect all astronauts to be safely recovered, together with the relatively large capsule of captured Earth-environment which has accompanied them into space; all this results in further demands for heat-shields, retro-rockets, etc. All these requirements drastically increase the size and weight of the spacecraft (Sputnik I weighed 184 lb (83·5 kg) but Vostok was already 10,400 lb (4717 kg)).

Why, then, send him at all, when clever automatic devices can already do so much, and promise soon to do even more? There is no doubt that a large element of romanticism, of desire for personal or vicarious adventure, has played a part in the answer to this question, which nevertheless continues to be asked by many people. There are, however, other answers of a more practical nature, more likely to go at least some way towards satisfying the questioners. No automatic device yet invented, or perhaps ever likely to be invented, can demonstrate the skills of judgment, initiative and improvisation which are characteristic of any highly-trained human beings, such as test pilots, experimenters or explorers. For some tasks, therefore, we are likely to continue to send men into space for the foreseeable future, certainly throughout a period extending well beyond 1986.

If manned spacecraft are therefore developed for certain purposes, it follows that we must also develop the larger vehicles to launch them on their way. Given the availability of both, it becomes a fascinating speculation to consider whether the future of even 'practical' (or workhouse) space applications belongs quite so much to the robots, as is often currently believed.

Granted the high and increasing reliability of modern electronics, it still goes wrong occasionally, and is always likely to do so. Might not a service engineer sometimes be a valuable protection for a huge capital investment? And certain duties—military, meteorological, or 'earth resources' reconnaissance—suggest that they might find the special human qualities already mentioned to be of great value. We may yet see something like the large manned space stations beloved of the early astronautical pioneers and science-fiction writers. Current interest by the USAF in 'manned orbiting laboratories' (MOL), with astronomical telescopes peering down at earth and up to the heavens, suggest this.

My guess is that we shall see the beginnings of such a trend even within the relatively near future (up to 1986) which is all that we are so far considering. If correct, this will soon exert a significant effect on the trend of spacecraft and launch vehicle development. Space operations will attain a scale on which it begins to seem worthwhile to recover and re-use the vehicles employed. As envisaged a long time ago by the pioneers, these will be required to operate a ferry service, first to erect and later to service the large manned orbiting stations. There is always an aesthetic appeal to engineers in the recoverable, re-usable vehicle; an expendable one somehow seems to be a sin against the crafts of complex design and precision manufacture. However, the hard economic arguments do not support these feelings until the required frequency of operations becomes very large, because only then can the huge development costs of even more complex (and individually still more expensive) recoverable vehicles be amortised (that is, themselves recovered) over a sufficient number of missions. Until then, it is cheaper to build a few simpler expendables, and throw them away each time.

Such eventual Earth-based recoverable vehicles may be of the ballistic variety, using retro-rockets aided only by parachutes or other forms of air-brake, in landing, or they may rely more on aerodynamics, with use of wings or lifting-body shapes. My own guess is the latter. They may also make some use, in their first stages, of some form of air-breathing engines, or of air-collecting types (first using air in the gaseous phase, in a ramjet, then at higher speeds liquefying it for use in a rocket engine); my guess is not. I think they will still be of the pure ballistic rocket type, with VTO, but will land aerodynamically, probably with some use of auxiliary air-breathing turbojets to cruise back to base. However, I could be wrong!

Apart from such developments as these, the development of other advanced systems—and in particular of propulsion systems—is already well under way. Here it is worth emphasising the fundamental character of propulsion for astronautics. This is not to say that other technical disciplines, such as guidance or structures, are less 'important' (they are of course vital), but in a very special sense, propulsion is fundamental. Man could not travel in outer space at all without a propulsion system which worked independently of an atmospheric environment; so far, one form or another of rocket engine is the only solution of any practical significance. (You could, in theory, launch a projectile into space from a giant gun, and then change its orbit by the radiation pressure on solar sails—but we need not concern ourselves with that.)

It is also interesting to note that all ideas for improved forms of rocket propulsion have found their genesis in the interest of men in space-flight—and manned space-flight, at that. The original inventors (or discoverers-by-accident) of the solid (gunpowder) rocket are unlikely to have been motivated by an interest in anything more esoteric than fireworks, or in crude bombardment weapons—but certainly most (and probably all) early thinking about liquid-propellant types resulted from a desire for interplanetary travel. It is of course true that such lofty ambitions have frequently been channelled into more prosaic, and sometimes less worthy, ends (e.g., the history of V2). A recent example is electric propulsion, which was first conceived as a possible efficient means for sending large spaceships to the planets, but is now often advocated for transferring a very high proportion of the possible payload launched by a booster into a low orbit up to a higher one, such as that for a synchronous communications satellite.

Von Braun's V2, the first (1942) successful large rocket, achieved an exhaust velocity of only about 2km/sec (i.e., a Specific Impulse (SI) of approximately 200 lbf of thrust per lb mass ejected per sec: exhaust velocity or SI is the most important single parameter of all in determining rocket vehicle performance). Incidentally, its structural factor, i.e.

$$\frac{\text{empty mass less payload}}{\text{mass full of propellants}}$$

was about 0·3; this is the next most important influence, and the V2 figure compares with values of 0·05 to 0·10 achieved today. This is a measure of the enormous contribution already made by the structural designer and the suppliers of high-strength, lightweight, materials. Their colleagues in the field of guidance have made even more striking advances—from V2's doubtful ability to land in the Greater London area, when launched from the Netherlands, only 200 miles (300 km) away, to our present errors of only a mile (1610 m) or so when aiming for a target at ICBM range, or even at one on the Moon.

Figure 1 The BAC 'Mustard' project study for a recoverable, reusable space launching vehicle. Three parallel, substantially similar, stages. The two outers are boosters, which peel off and return to base, turbojet-assisted, while the central core vehicle continues out to orbit.

Figure 2 A very large wheel-shaped space station of the future, rotating slowly to produce artificial gravity for its occupants, who are accommodated in the rim. One 'wheel' is still under construction, from material ferried out from Earth

(Picture from current MGM-Stanley Kubrick film '2001: a Space Odyssey'; screenplay by Kubrick and Arthur C. Clarke)

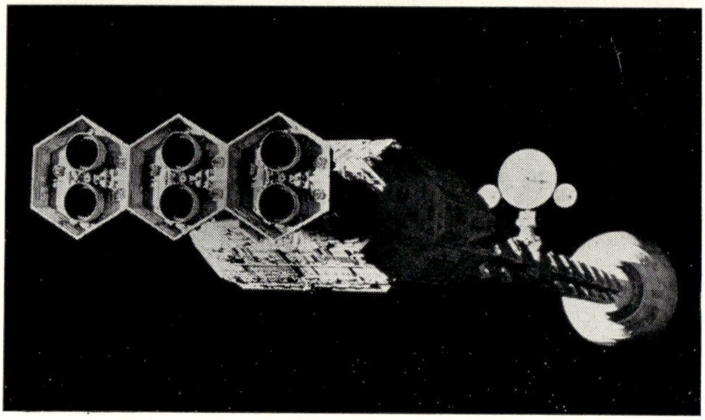

Figures 3 and 4   A deep-space exploration ship of the future, with nuclear propulsion unit at the rear, spherical crew-cabin at the front, and radio-radar antennae amidships. In Figure 3 (top) a small one-man 'space pod', used for minor excursions from the main ship, is emerging from the sphere

(Pictures from current MGM-Stanley Kubrick film: '2001: a Space Odyssey'; screenplay by Kubrick and Arthur C. Clarke)

Today, with the use of better propellants (liquid oxygen/kerosene, or nitrogen tetroxide/hydrazine, instead of liquid oxygen/alcohol-water), in more efficient engines working at higher pressures (ca. 35 to 70 atmospheres instead of less than 20), an exhaust velocity of 2·5 to 3·0 km/sec is normal. (SI of about 250 (280 *in vacuo*)). However, the even more powerful liquid oxygen/liquid hydrogen propellant combination has already demonstrated effective exhaust velocities *in vacuo* well in excess of 4 km/sec; it is used in the 'Centaur' upper stage, and in the upper stages of the huge Apollo/Saturn V. The latter also uses five first-stage (liquid oxygen/kerosene) engines of 6700 kilonewtons (1,500,000 lbf) thrust each—some 30 times larger than those of V2. Rocket engines, during the next couple of decades, will continue to become more efficient more reliable, and still larger (if necessary). They will operate at higher pressures (perhaps up to 200 atmospheres), will use more efficient, topping turbopump cycles, or centre-body nozzles, where appropriate, and also new cooling techniques (such as hydrogen film cooling). However, a restriction to chemical propellants will limit vacuum exhaust velocities to less than 5 km/sec; only fluorine/hydrogen, or perhaps something like beryllium/oxygen/hydrogen, or lithium/fluorine/hydrogen, will secure any further improvement over the oxygen/hydrogen combination now coming into use—and that only a relatively small one.

We must, and can, do better than this. A nuclear fission reactor, heating liquid hydrogen as a working fluid, has already enabled an experimental American rocket engine to demonstrate an exhaust velocity of over 8 km/sec (SI=800), while electrical rocket propulsion systems can show values of between 10 km/sec (for arc-jets or resisto-jets), to as much as 50 to 100 km/sec (for electrostatically-accelerated ion motors), with intermediate performances for many colloid or MHD types.

All this work will come to fruition, and find increasing application, by 1986. It has been stimulated mostly by the demanding goal of more efficient manned space flight—though much of it will find an application also, once it has attained a completely successful status, in other applications.

So far, we have said little about the achievements of man in space by 1986. By then, his first landing on the Moon will lie further back in his past than Sputnik I does in ours today. It seems highly probable that he will also have landed on Mars (or at least have circum-navigated it, and Venus too). Nuclear rocket engines are likely to have been employed for this mission.

The detailed discussion of the probability of these forecasts, however, begins to bring us up against those difficulties which are inherent in the completion of the rest of our task—for we contracted to discuss the second century, and so far have mentioned only the first twenty years! This is rather as though one of the original RAeS members, trying to project his mind forward to the era when Concorde was taking shape, had got as far as 1886.

We will not abandon the task which the editor set us, but let us digress for a long moment to consider the basic problems which it involves.

3 THE BASIS OF TECHNICAL PROPHECY

This considerable digression is, I fear, quite necessary to make any sense of what will follow, when we try to throw our mind forward in time five times further than we have so far done. It is also not without interest for its own sake, and is very relevant to other papers in this series, though particularly relevant (indeed, inseparable) from the theme of astronautics.

In the best book I have ever read about the future[1] Arthur C. Clarke (who also first proposed communication satellites, in 1945) wrote that most failures in prediction could be described as either failures of nerve, or failures of imagination. In the former, the prophet does not dare to think about (or if he does, dares not mention for fear of ridicule) some things which are clearly possible. We shall try to avoid this failing. The second pitfall is more difficult to avoid, for it results from one's inevitable conditioning by the contemporary environment; certain things may become possible, because of the indirect consequences of certain factors which never occurred to one at all. We shall do our best.

There are two other categories which are extremely useful when considering any technological future. They are: what *could* be done, and: what may *actually* be done.

3.1 *What could be done—technical feasibility*

This aspect is, of course, quite fundamental (if we could be sure that something was technically impossible we would simply rule it out). Until about 50 years ago, this criterion that a project must be technically possible was really the only one that mattered, because if even a major new device was theoretically feasible, it was almost certain that some individual would then, sooner or later, become sufficiently interested to prove this in practice. Whether there was a widely appreciated social demand for it, or even whether it was generally agreed to be beneficial, were at least secondary considerations. The primary factors were the conviction and determination of the pioneer; the resources required to satisfy his ambitions were not beyond those at his own disposal, or at least were no greater than those which he could muster, after infecting a few other private backers with his own enthusiasm. (Sir Geoffrey de Havilland built his first aeroplane after persuading his grandfather to anticipate the date by which he would in any case have inherited a legacy of £1000: Count Zeppelin, his first airship with his own fortune, later augmented by public subscriptions.)

If technical feasibility still remained the only criterion, the task of the technical prophet would be much easier, but still not an easy one. It would still be subject to failures of both nerve and imagination.

An excellent example of the former is provided by the history of the rival forms of flight within the atmosphere—lighter-than-air and heavier-than-air. From 1910 until about 1935, it was fashionable to believe that the airship was the answer for long-distance air travel, and the aeroplane for short or medium distances. This view changed only gradually over this long period. To begin with, it was almost universal; most of the ardent advocates of the aeroplane, even, subscribed to it at first, and only towards the end did they dare to claim that the aeroplane would soon triumph in all spheres. (And then most of them agreed that the flying boat, rather than the landplane, would be the best long-distance vehicle!) This was because they did not dare to envisage the cumulative effect of successive advances in many specialised and detailed fields: aerodynamics, materials and structures, concrete runways, flaps, retractable undercarriages, supercharging and VP propellers, pressure cabins, swept wings, gas turbines and jet propulsion.

Failures of pure imagination can be even more of a pitfall, however, because they can neglect (or even deny the possibility of) a complete breakthrough. This has become an overworked word, and is sometimes used to describe a mere improvement in existing techniques. Here, I am using it in the more justified sense of an application of completely fresh basic knowledge, which is of course much rarer.

This particular occupational hazard for technical prophets can be illustrated vividly by three examples. James Watt would not have found much difficulty in comprehending Sir Charles Parsons' steam turbines, but he would have been completely mystified by the electrical generators which they often drove. Parsons, in turn, would have found a nuclear reactor quite incomprehensible, only one stage (if that) removed from black magic. Any of the great aeronautical pioneers at the time of the Society's foundation (1866)—say, F. H. Wenham or John Stringfellow or even Sir George Cayley before them—would not have found too much difficulty in understanding a Boeing 707 and even its jet engines. However, the electronics of such a modern aircraft (its radio, radar and any computer in its inertial navigation system) would have been meaningless to them.

Over any period as long as 100 years, one must allow for the probability of new discoveries in basic physics.

3.2 *What may actually be done*

We have said that once this was practically synonymous with what was technically possible, or at least (allowing for the true breakthroughs) with what was technically conceivable. However, there has been a gradual departure from this simple situation.

First, there was a period when the backers of the pioneers had come to be large companies, rather than just relatively wealthy individuals; or alternatively, during which the speedy initial development of the invention or activity became dependent even on the support of national governments (e.g., the rapid development of aviation in the First World War, or of radar and the gas turbine in the Second).

The process was not arrested at this point, since the increasing complexity and the difficulty, and hence the rapidly escalating costs, of major new developments began to demand support on a scale which was obtainable *only* from national governments. Examples were the development (during and since the Second World War) of nuclear weapons and reactors, and of large rockets (first for war and now for space).

The laboratories, large test installations, and skilled labour forces required for these are so expensive that it is not remotely likely that they would, or could, ever have been provided, even gradually over a long period, by the private enterprise of even the largest corporation, let alone of individuals. (In passing, let it be noted that the interest of the individual pioneer still remained as an essential main-spring, to trigger off the gigantic State effort: men like Fermi and von Braun were still the first begetters.)

From now on, therefore, we must consider not only whether a major project is feasible, but whether someone (which in practice means some government or group of governments) is likely to be prepared to pay for it. There will continue to be exceptions to this rule for the smaller projects, e.g. hovercraft, at least in their earlier phases—but this chapter is essentially concerned with major ones. The question of the backers' willingness and ability to foot the bill is inseparable from the question of why they should be prepared to do so, and hence any useful discussion of the matter must include some consideration of non-technical aspects, such as politics, economics and even basic human psychology.

In an effort to put the whole question in true perspective, it is again useful to look back for a moment to the beginning, in 1866, of the Society's first century. At that time, railways, steamships, and the electric telegraph were recent and contemporary technical marvels. Britannia ruled the waves, much of the world was coloured red (not Communist Red, but that of the British Empire, the greatest the world had ever seen and on which the sun never set, etc.). Russia and China were huge backward nations, Japan a mediaeval one, Africa a dark continent for which hardly anyone ever envisaged racial problems of one-man-one-vote, and the United States (which had just emerged from a devastating Civil War) was a country of brash, if enterprising, ex-colonials who seemed to have difficulty in running their own affairs.

A lot has happened since then, much of it very relevant to our present concern. The continuing industrial and scientific revolutions have vastly increased human productivity, the world population has grown enormously, and the Super-States (of which the USA and USSR are the only present examples, but which are the precursors of several others) are now capable of mounting national efforts on a relative scale which would have been impossible even for rich Victorian Britain. There have even been tentative and so far largely unsuccessful forays in the general direction of a World State (the League of Nations and the United Nations) which have nevertheless, in their less-political activities, managed to demonstrate something of the potential for still larger global organisations. The founder-members of the RAeS would find the modern world surprising in more than merely its technical aspects, and some of the political and social changes have largely determined the scale of the technical achievement.

After this, it may seem a *non sequitur* to point out that the Ancient Egyptians thought it worthwhile to devote a large proportion of their national resources to the building of tombs and monuments, and the Europeans of the Middle Ages to the erection of cathedrals. Since then, modern man (with less originality) has continued to believe that it is a sensible and practical activity to pile up vast stockpiles of destructive armaments, which, at best, benefit him most if they never have to be used.

These observations are not really irrelevant to our theme, or even unconnected. The message is that the capability of nations changes drastically with time and with their organisational structure, and that the ways in which they exercise this capability, at any given time, are determined in the last resort by the caprice of the collective human spirit.

One last illustration of this argument will return us closer to our own theme. In 1953, I wrote a paper for the British Interplanetary Society entitled 'A Programme for Achieving Interplanetary Flight'[2]. Believing then what I still believe today, I wrote, 'If we believe (as I certainly do) that spaceships *could* be built within the next few generations, albeit at the great expense of overcoming immense technical difficulties, then the question of whether they *will* be built at that time depends entirely on such non-technical considerations as motives and availability of resources.' Following this, there were predictions of artificial satellites by 1965, manned orbital flight by 1975, and that a lunar expedition (using nuclear rockets) might be mounted in about 2000 AD, with a landing on Mars some few years later. With hindsight, these prophecies seem fainthearted, but not many people thought so then. I had not neglected the possible influence of the cold-war competition either, because I also wrote, 'To put it crudely, it might be possible to scare the Pentagon into establishing a lunar base by convincing it that the Kremlin had similar plans—or vice versa . . . national prestige might remain as a factor, even if military usefulness did not.'

However, I had drastically under-estimated it. In the middle 1950s, the first steps in space flight were technologically feasible for both the USA and USSR. The Russians decided that such an activity offered a heaven sent opportunity to score points in the game of international prestige. They may also have decided, on more philosophical grounds, that it was basically worthwhile—at any rate, I am sure their prime mover pioneers were of this opinion, like their fellows elsewhere. In 1957, a humiliated USA had to follow suit in spite of indignant references by a few dissenting Senators to 'Moon-doggles'. If this was not when the Space Age began, it was certainly when it emerged from infancy, and it certainly marked the beginning of the Space Race.

This has continued ever since, basically because the two Super-States have decided to compete in this particular arena, for reasons far removed from those of most of the dedicated engineers and scientists involved, and who made the whole thing possible. However, the currently somewhat wavering Congressional support for the NASA programmes (which is often exaggerated—budget $\$5{\cdot}9 \times 10^9$ in fiscal 1966, compared with $\$4{\cdot}6 \times 10^9$ in fiscal 1969) is again indicative of the fact that the *rate* of technical progress in practice is conditioned by the level of support which society is prepared to give.

With all these background considerations in mind, let us revert to the problem of what may happen in astronautics during the rest of the second century.

4 THE FOLLOWING 80 YEARS (1986–2066)

4.1 *Technical probabilities*

These will include an extrapolation of all the developments discussed in section 2. That is, even by the early 21st century, it seems more than likely that large manned space stations with various operational duties will be circling the Earth. One of these duties may have progressed from mere weather surveillance, and predictions, to the beginnings of local weather control. Bases of some sort will probably also have been established on the surfaces of the Moon and Mars and perhaps on those of some other planets or their moons.

Certainly these achievements will have become technically feasible by then. In their accomplishment, use will have been made of improved nuclear rocket engines with solid fission reactors (exhaust velocity $\simeq$ 10 km/sec), and of various advanced electrical propulsion systems, of low thrust but even higher exhaust velocity (about 50 to 100 km/sec). Space operations in general will have attained at least a sufficient scale for the employment of recoverable and re-usable vehicles to have become economically desirable, and therefore probably quite the normal practice.

However, the actual scale of space operations, and the purposes for which they are carried out, will be influenced by the success or otherwise of certain technical possibilities beyond any discussed so far.

As Kipling clearly pointed out, engineering is a profession which encourages, more than perhaps any other, a respect for truth and a corresponding disposition to state the facts of a case in blunt and straightforward terms. Nevertheless, even engineers and scientists are not above practising a certain deception, by the use of somewhat devious terms and methods, on occasions. I have always thought this to be amusingly illustrated by their choice of favourite performance parameters. The designers of many electrical machines, propellers, compressors or pumps usually quote efficiencies without hesitation, since their brain-children achieve values respectably close to the possible limit of 100%. Those responsible for heat engine prime movers with rather low thermal efficiencies, however, are more reluctant to do so; they prefer to quote a specific consumption (in lb per hp or lb thrust per hour), with no such obvious limit to which actual performances can readily be compared. Faced with the consumption of oxidant as well as fuel, even this practice alarmed the rocket engineer, whose device might sometimes be compared with earlier air-breathing propulsion systems. He therefore invented the term 'Specific Impulse' (or specific thrust), which looked quite different from a specific consumption although it was simply its reciprocal.

Rocket vehicle constructors are quite shameless in this respect; they use the term 'payload' in very cavalier fashion. For the air liners now departing from London Airport every few minutes, this means the passengers, their baggage, and any mail or freight which is also carried; for a military aircraft, the crew and weapons which are aboard. In either case, the payload, as a percentage of the vehicle mass at take-off, is of the order of 10 to 20%. Anything less than 5% would be considered unacceptable, even for a supersonic transport.

For a manned space launching vehicle, however, the 'payload' quoted usually includes the mass of the space capsule (or crew cabin), and various equipment essential to fulfilment of the mission, even that of upper stages which will be used in outer space or for return to earth. As a case in point, the Saturn V (take-off mass=approx. 6,000,000 lb (2,700,000 kg)) is often said to have a payload for the lunar mission of around 95,000 lb (43,000 kg)—but this includes the mass of the control, service and landing modules. The true 'payload' (in aircraft terms) consists of three brave (if, no doubt, somewhat nervous) astronauts, plus a modicum of observational scientific equipment; say at a generous estimate, some 1000 lb (500 kg), or 0·02% of the take-off mass. For launch vehicles placing large robot payloads in close-Earth orbits, this factor has already risen to a more respectable 1% or 2%, and of course the developments already discussed will improve both figures, though not by as much as an order of magnitude.

In short, we shall need to do better still, if space operations are not always to be restricted to an essential minimum of highly desirable tasks—broadly those of the kind that are already authorised: initial research and exploration, or vital services demanding in any case only relatively small 'payloads', such as reconnaissance or communications.

It is extremely difficult to predict with any precision how these further demands will be met, but there are a great many possibilities, and this prophet at least would be very surprised indeed if many of them have not been proved feasible before AD 2066.

4.2 *Technical possibilities*

One of the most interesting possibilities (or so it seems at the time of writing) is one on which I personally would not bet—but of course I could be wrong, and many 'experts' with at least equal claims to that dubious title would disagree with me. It is the use of air-breathing engines in the first stage (for lift-off from Earth), and possibly the use also of wings for this phase of flight, with horizontal take-off from a track or ramp; perhaps with some auxiliary launching carriage (e.g. rocket-propelled, or electromagnetically accelerated along a 'hover-train' track). The reasoning behind such proposals is elegant in its simplicity, and at least superficially convincing. We have to depart through Earth's atmosphere: why not made active use of it, instead of merely passively combating its adverse effects of aerodynamic drag, stresses and heating?

A recent and more sophisticated American suggestion for an air-breather was LACE (Liquid-Air-Cycle-Engine), though project studies in some depth have not yet confirmed even its technical feasibility. This device would function as a hypersonic ramjet while it accelerated round the earth within the atmosphere; however, just as the bee makes a little honey on the side, it would also collect some excess air, liquefying this in a refrigeration cycle with the help of its liquid hydrogen fuel; probably it would separate the liquid oxygen and store it in tanks, rejecting the liquid nitrogen as a film coolant into its ramjet. At some point, the whole apparatus could convert to a rocket mode of operation, using its stored liquid oxygen, and leap into an orbit in outer space.

If air-breathing engines ever make any significant contribution to space-launching vehicles, I suspect that they will need to assume some such radical form as this. The related device of a nuclear ramjet is also of doubtful practicality; the upper-atmospheric rarefied air, even after compression at hypersonic speeds, is not an attractive heat-transfer and working fluid for a solid nuclear reactor.

However, my own basic (and qualified) scepticism is concerned with a more general objection. I suspect that even if these winged, air-breathing, first stages succeeded in their objective of raising the payload percentage of take-off mass, they would be found to have defeated it after all by increasing their costs. It is, after all, the latter economic objective which is the target, not just the achievement of a higher payload fraction as a demonstration of technological virtuosity. The development costs of such engines would be enormous, because they would demand particularly expensive ground testing facilities. Even beyond this, in production, rocket propellants are relatively cheap, while air-breathing engine hardware and airframe structure designed to carry heavy loads, are not.

On the other hand, I suspect that air-breathers (turbo-jets) and wings (lightly-loaded, because the vehicle will then have used up most of its propellants) *will* play a part in the subsonic return to base of recoverable, re-usable, space vehicles. This is the classic solution (a vertical, ballistic, rocket launch, with recovery as a possibly power-assisted glider) envisaged by many of the pioneers, and I still think they were right.

The alternative, currently advocated by Phil Bono of McDonnell-Douglas, is a ballistic stern-first recovery, with rocket braking assisted by a minimum of aerodynamic devices (parachutes, drogues, or rigid air brakes). Even with such a technique, auxiliary light-weight lift turbo-jets might find a role. It is too soon to reject such a solution, but my own bet would still be placed on the winged, more highly-manoeuvrable, vehicle, landing horizontally on a prepared runway.

It seems very unlikely that there will be any major, further advances in the power of chemical rocket propellants; advances, that is, comparable to the one from hydrogen as a fuel. It may be found feasible to use tri-propellant combinations, such as those in which Li or Be metallic fuels are burned with O_2 or F_2, and used to heat H_2 as a diluent or working fluid; such systems, if they can be made to work, would raise the possible exhaust velocity of a chemical rocket engine to around 5 km/sec.

Beyond this, however, the only possibility (which seems very remote) is the use of 'unnatural' chemical propellants, such as free radicals, which have a fleeting existence in some ordinary chemical reactions. For example, the reactions: $H+H \rightarrow H_2$, or $H_eH \rightarrow H_e+H$ would realise enormous energies, and produce exhaust velocities in the 10 to 20 km/sec range. But how to make, and store under stable conditions, such materials? Perhaps at temperatures approaching absolute zero? There has been some speculation about this, but none of the work has yet given much encouragement that it will ever find any practical application. In principle, of course, the idea is attractive, since such fuels might be made in large nuclear ground installations, with huge power inputs, for storage and subsequent use in a flight vehicle.

Another highly conjectural proposal for the use of large powerplants not on the space vehicle itself involves the transmission of energy to it by microwave radiation or laser beams. The original power generation might then be in nuclear installations on the surface of the earth or the Moon, or in a very large space station. Received by the vehicle, it might then be used in electrical propulsion units, or even, after conversion into heat in a super arc- or resisto-jet, to accelerate a propulsive exhaust of H_2. The scheme seems improbable, but has already been given some quite serious attention (e.g. in the microwave version, by Raytheon, Inc.).

In another development, however, one can feel a very high degree of confidence: long before AD 2066, and probably well before AD 2000, very considerable advances will have become possible in high-thrust nuclear rocket engine technology. The contemporary prototypes, which have already run for periods approaching an hour and demonstrated exhaust velocities of over 8 km/sec, will have been made to look as primitive as, say, Whittle's W1 turbojet does today (though they are just as significant).

Solid reactor nuclear rocket engines may eventually be developed with much higher exhaust velocities of at least 10 km/sec, but already there is also a fair amount of theoretical study of variants in which the heating of the H_2 working fluid is by a nuclear fission reaction in the gaseous phase, containment of the plasma being achieved by some vortex pattern, rotation of the system, or by electro-magnetic fields. A basic problem, of course, is to avoid ejecting the fissile material with the reaction mass of the H_2 working fluid; various solutions have been proposed, but none appears practical as yet. This should not lead us to assume that none will be found by AD 2066; very much to the contrary, there is a good possibility that by then such systems will provide us with exhaust velocities of around 20 km/s, if something even better has not become available in the meantime.

Another possible propulsion system, which might provide SI's of the same order (and at extremely high thrusts, also permitting lift-off of very heavy vehicles from the Earth), is that of the fantastic Orion vehicle studied for several years for the USAF by General Dynamics. In this, small nuclear bombs are ejected from the vehicle itself, to be exploded beneath a 'pusher-plate' at its base, the resulting pressure providing a large forward (or upward) thrust. Many calculations and design studies, and model tests using chemical explosives, were carried out over several recent years, and convinced many people that this propulsion system would be feasible. Its use even in very close orbit, however, seems likely to contravene any test ban treaty concerning nuclear fall-out, while (unlike other forms of nuclear rocket) its permissible operation from the earth's surface seems even less likely.

On the score of minimising radioactivity of their exhausts, a most attractive possibility seems to be the use in rocket engines of the much more energetic nuclear fusion reactions, instead of fission reactions. Since fusion involves light elements (i.e. hydrogen and its isotopes), and these also provide the ideal rocket working fluids, there also seems to be a natural marriage just asking to be arranged here. The snag is, of course, that at the time of writing (1968) we do not even know how to generate and contain a sustained thermo-nuclear reaction, producing more energy than we put in; we can only make hydrogen bombs, employing the reaction for destructive, explosive, purposes. However, we seem close to overcoming this problem; a possibly more serious consideration is that its solution seems likely to involve the construction of very large, bulky and heavy machines, incorporating huge electro-magnetic coils.

Such techniques as super-conductivity may help here, and anyway this is probably the sort of problem which appears insuperable to one generation, but fairly easy to the next, or the next-but-one. (I remember hearing an eminent mechanical engineer, just before the Seond World War, declaiming that one application which the gas turbine was unlikely ever to satisfy was aircraft propulsion, because of the size and weight of the necessary compressor, in relation to the turbine itself. Whittle rose and disagreed with him.)

I would feel fairly confident, therefore, that nuclear fusion rockets will have appeared by AD 2066, and probably displaced those of the fission variety, with considerable consequent benefits to the SI produced and to the radiation hazards involved. Immensely higher working temperatures will be employed, contained by electro-magnetic fields and what in effect will amount to film-cooling techniques, adjacent to solid chamber and nozzle walls.

It is perhaps just remotely conceivable that we might also have learnt to construct synthetic stable 'atoms' which, giving an even lower molecular weight than H_2, would form a still better working fluid for a rocket exhaust.

There may, of course, be even more sensational possibilities. There probably will, since 100 years is a very long time indeed at the current pace of advancing science and technology. Most people today have heard of Einstein's classic $E=mc^2$ equation, explaining the equivalence of energy and mass, with a conversion factor equal to the square of the velocity of light. It is a sobering thought that in the fission reactions of the heavy elements (uranium or plutonium), only some 0·09% of the reacting mass involved is converted into energy, while even for the fusion reactions of the light elements (hydrogen and its isotopes) the figure is still only 0·50%. May we not, some day, learn to achieve the complete annihilation of matter (perhaps of *any* element) with production of energy—as when a particle of 'natural' matter collides with its anti-matter equivalent, such as an electron with a positron? Even beyond this, there are already hints from the more esoteric realms of theoretical and experimental high-energy physics, and from astrophysics and cosmogony, that there may be even newer, stranger and much larger sources of energy to be tapped in Nature's hidden stores. . . . Within the next century, we may have learnt to rifle them, and to control what we have stolen, or borrowed.

Such speculations are entirely reasonable, and not 'unscientific' at all, but they bring us close to another observation made by Arthur Clarke. 'Any sufficiently advanced technology,' he said, 'would probably seem to us to be indistinguishable from magic.'

The authors of science fiction are less inhibited than engineers and physicists when writing or speaking in their strictly professional roles. (The qualification is necessary, because the best science fiction authors often come from these very same professions!) They have imagined even more exotic possibilities, some of which may well be realised by AD 2066.

Even writing as a scientist, for example, one of the recognised astronautical pioneers (Eugen Sanger) discussed the possibility of 'photon rockets', in which the complete conversion of mass into energy would produce a 'jet' of pure radiation, with an exhaust velocity equal to that of light! With such propulsion by radiation pressure, and using relativistic mechanics, he worked out the problems of space travel far beyond the boundaries of our own Solar System.

Another such dream, perhaps a little less fantastic, is 'anti-gravity'. This has figured in a great many science fiction stories, the best-known of which is probably *The First Men in the Moon* by H. G. Wells (who, incidentally, shared the centenary of his birth with the Royal Aeronautical Society.) Wells imagined an impossible substance called 'Cavorite' which would be repelled, instead of attracted, by other matter. It would be impossible (unless our most fundamental physical beliefs contain some very subtle error) because it would enable us to defeat the laws of conservation of energy (or, in the most modern terms, of mass-energy). However, there are absolutely no known laws which say that cancellation or reversal of gravitational forces are impossible—provided the effect were achieved by an energy input at least equal to the external work being done in the operation. (The ubiquitous Arthur C. Clarke is fond of pointing out that essentially it should require only about 1000 kWh of energy, worth less than a fiver at 1968 prices, to lift an average man to the Moon.)

We simply do not know how to do it, or even whether it really *is* even theoretically possible (there might be a physical objection not yet appreciated). Gravity is the most mysterious physical phenomenon in nature, although everyday familiarity with its effects tends to disguise the fact. Text books for school and university gloss over the fact that we are still completely ignorant of its physical cause or mechanism, and unable to show any connection whatever between it and any other physical phenomenon, except in a purely abstract, mathematical, sense (e.g., we can relate the detailed mechanism of electricity and magnetism and optical effects, but none of these to gravitation). The most modern accepted theories in mathematical physics, such as the work of Einstein, really say nothing about the physical nature of gravitation; they merely describe it in terms of a highly sophisticated mathematical convention, expressing a possible geometrical structure of space. They suggest that it is propagated at the speed of light—a concept which would have been regarded as revolutionary little more than a generation ago, when Newton's ideas of 'action at a distance' were reluctantly accepted in the absence of any knowledge to the contrary. A search in the more esoteric literature of contemporary mathematical physics reveals a considerable amount of speculation on the subject; suggestions about gravitational radiation, explanations linking mass to weight by mechanisms of neutrino bombardment.

In the 1950s, this speculation led the USAF to finance quite a number of pure scientists working in this field, to their great delight, and, in some cases, to smug, even contemptuous, expressions of opinion that at least some of the excessive budget of defence dollars was thereby being expended on a worthy cultural objective. The Pentagon's hope was, of course, that some fundamental discovery might be made which was capable of an early application to astronautics. (And, of course, even to aeronautics—and to a great many other fields; cranes? lifts? escalators?)

Nothing came of it; no practical experiment* has yet been devised which would prove, or disprove, any of these highly conjectural theories, or carry the speculation any further. If it had been otherwise, it would have been the first time in scientific history that a fundamental advance in basic science had been made to order, as a result of financial subsidy, and with fairly specific practical applications in view. This is not to say, however, that some unknown researcher, anywhere in the world, may not provide the solution tomorrow, next year, or sometime before 2066. . . . Perhaps he (or she) will conceive a revelatory new theoretical approach, leading to the design of crucial experiments, or perhaps some experimental fact will be stumbled upon, even in the course of research on a different subject. Perhaps further work will dispose of the possibility altogether, by placing it in the category of perpetual motion, but it certainly cannot yet be relegated to such a place.

If some method of manipulating gravitational fields has been developed by AD 2066, with the weight of a given mass being not merely nullified, but reversed in direction by several orders of magnitude, then various sweeping possibilities might emerge. Such control, by the input of energy in some suitable form, could presumably be exercised over every individual atom in the mass concerned; there would be no theoretical limiting accelerations, save only those imposed by practical considerations of the available power input, and further advances in the release of nuclear energy (as already noted) should surely remove most of the latter. This would be true both as regards the physiological effects on the human payload, and the structural effects on the vehicle itself. Any sort of rocket (chemical, electrical or nuclear) would probably become, very rapidly, as obsolete as the dirigible already is in today's world of jet propelled aeroplanes.

An even more 'far-out' speculation of the science-fiction author has sometimes been called 'teleportation' or 'matter-casting'. Mass is convertible to energy, energy can be transmitted over a distance by radiation (e.g. radio), so why not de-materialise passengers and freight in some sort of transmitter, beam them to the moons of Saturn or Jupiter at the speed of light, and re-assemble them there in a corresponding receiver? Presumably this will first have had to be transported to the desired destination

*Subsequently (March 1969), Dr Joseph Weber of the University of Maryland reported the experimental detection of gravitational radiation.

by other, less economical and convenient, means—rather like building an airfield with materials transported by sea, after which ships become superfluous for repetitions of such a journey.

With anti-gravity or teleportation, even inter-stellar travel might eventually become practicable; one should not rule out the possibility that, with further advances in fundamental knowledge, even the speed of light may not be seen as an absolute limitation. Perhaps it is a barrier only within our known frame of physical dimensions. . . .

However, I think one can be fairly sure that such things as inter-stellar, or inter-galactic, travel belong to a future even more distant than AD 2066, but anti-gravity or even teleportation, or their beginnings, might not. Think again about the idea of a resurrected Henson or Stringfellow, being mystified by a radio or radar system, or nuclear reactor, after the intervening period of the Society's first century. . . .

If such things actually do come, how shall we use them? Where shall we go, and for what purpose? Will the wider exploration of space ever lead to anything different in kind from today's occasional expeditions to the Polar regions or the summit of Everest or to the depths of the sea? On a very long-term view, extending forward a century or more, I think that it will—just as these other terrestrial exploits of the present and recent past will do: eventually we shall mine in the Antarctic and on the ocean's bed. It may be that we shall discover some cargo worth transporting back across the interplanetary void, or perhaps future children of man will be born and will live out their lives in colonies beneath pressurised air-conditioned domes on the Moon and on alien planets. Their contemporary world society may think it worthwhile to do so because, for example, of the unique opportunities so afforded for the pursuit of research in some branch of knowledge; it is not inconceivable that they may even like the environment. (The denizens of Surbiton often find it strange that some of their fellows prefer to live on the African veldt.) The advanced technology which was instrumental in transporting them into it may also be capable of feats of planetary engineering which could transform the atmosphere and climates of their new world. There may be interplanetary transport for reasons of health or recreation. The Victorians would have thought it highly improbable that many of the British working classes in the mid-20th century should enjoy Mediterranean holidays, even that the rich should make similar excursions to the Bahamas.

4.3 *Probability of realisation*

Arguments have now been presented in support of many wonderful things which advancing science and technology may well have made possible by AD 2066—but will these possibilities actually be realised? This must mainly depend on whether mankind decides that it wants them to be so realised. The point has already been made that it will no longer be sufficient for a few men to desire this; it is unlikely that all men will ever do so, but a majority (or at least a very considerable, influential, and powerful minority) must decide in their favour, if they are to come to pass. This, I think, is the significant new factor affecting the prediction of the technological future, and it introduces much greater uncertainties than the technical factors themselves.

Barring the obvious intervention of a major nuclear war, the continuous and steady development of space activities in close-Earth orbit is now assured for the indefinite future. The economic and practical benefits (civil and military) have already been demonstrated so convincingly that the world will unquestionably be prepared to pay for this; the only doubt that remains is whether, and for how long, the USA and USSR will continue to enjoy something approaching a monopoly of these activities. While this latter question is one of great concern to many of us on this side of the Atlantic, it is very secondary to the main theme of this chapter, and is mainly of importance only insofar as it may indicate a significant trend. If, for example, Europe already hesitates during the 1960s to make more than a token entry in the Space Race, may not even the Super-States pause for thought before authorising its indefinite continuation into the 21st century and beyond? Many people would see the recent downward trend in NASA's budget in the USA as confirmation of this suggestion. Those who have always thought that astronautics was a fantastic extravagance would even welcome it, as evidence of returning sanity.

Before accepting such ideas too easily, since they are somewhat depressing in relation to the larger possible future of astronautics, let us get a few facts straight. First, even European space expenditures are actually rising, in spite of our relatively weak motivation, even in the UK—but all this is concerned with close-orbital operations. Secondly, while the NASA annual budget had admittedly been decreased by rather more than 20% over the past three years, the sum total of those for NASA and DoD has remained more nearly constant. It is the American expenditure on such civil manned space-flight programmes as Mercury, Gemini and Apollo, and on advanced interplanetary research with unmanned scientific spacecraft, which has suffered some reduction. This has so far been relatively slight (in spite of anguished protests to the contrary from the space lobby in Washington and the Press), but it may or may not be significant in the long-term. Thirdly, it has in any case certainly not been the result of America's having discovered that she cannot afford her space programme; at the most, it indicates some possible doubt, perhaps only temporary, about whether she wants it quite as much as she once thought she did. Huge as it is, it still costs her less than a dollar (say, 40NP (8s.)) per week per head of adult working population; a total annual national expenditure about the same as that on cosmetics and beauty culture, less than that on tobacco or drink, and hardly prohibitive, if she so chooses, for a great nation aspiring to world leadership. Fourthly, there is no reason to suppose that Russian space expenditure is declining. Indeed, there are some hints to the contrary.

It may be that the Soviets have a firmer constancy of purpose behind their space programme than the Americans; one more fundamentally based on an ideological judgment of the inherent value of man's knowledge of, and powers to control, his whole material environment. (This is not by any means to say that they have not also been quick to appreciate and to exploit the political, military and prestige aspects of space.) They may, if this is true, press on, even if the USA falters, but at any rate it is of considerable interest, in the context of the long-term future of astronautics, to examine the current American reluctance to authorise major programmes ensuring a continuity of space exploration beyond Apollo.

Since mankind has always assigned a moderately high priority to projects of pure exploration, one might reasonably expect expeditions of this character to be mounted at least occasionally, now that they have become possible. However, if all the planets and their moons of our own Solar System, and much later, perhaps some of those belonging to other stars, are ever to be fully explored and even colonised and exploited, then the scale of such an operation will be truly immense. Very soon, its proportions will be such that only the Super-States, or groupings of states, *will* be able to afford it without real sacrifice, for the necessary costs will continue to increase rapidly as the projects concerned become even more difficult of attainment. It will help, of course, that mankind as a whole is becoming continuously richer, as advancing technology increases productivity, so that we shall be able to 'afford' to embark on progressively larger undertakings as time goes on. This is again on the considerable assumption that our progress is not summarily interrupted, or even concluded, by Armaggedon.

There are at present only two such Super-States, but during the period being discussed here (up to AD 2066) others will almost certainly emerge. A United Europe is still a possibility, and might within this period attain comparable status to that of the USA and USSR; it would in this context have the advantage of the longest technological tradition in the world, and some existing capability in space on which to build any future programmes. I would still hope, therefore that Europe may yet become a real contender in the Space Race, though even in my most optimistic moments I would not give the possibility much more than a 50/50 chance.

Of China's position however, there can be little doubt. With the largest population in the world, new-found political reasons for national cohesion, and a strong military motivation for having already embarked on vigorous development programmes in rocketry and nuclear technology, she can hardly fail to have become both a Super-State and a space power by the end of the century. By AD 2066, she may well be the most important of all. If the USA and USSR have ceased to compete between themselves so vigorously, then perhaps either or both of them may then regard China as their main adversary.

Japan shows signs of becoming almost a Super-State in her own right; within our time-scale she should certainly have acquired a larger potential capability by herself than, say, the UK, France, or Germany individually. She has already embarked on a modest, but expanding, space programme; in due course, she may become associated with other Asian nations, perhaps even with China itself, which would provide a truly massive potential for any conceivable activity, not least in space. Within the same time-scale, other Super-States may have emerged: India (perhaps re-united with Pakistan, if religious differences have begun to seem less important in a more scientific age?), Pan-Africa, or some South American Union. Most of these possibilities, however, may take even longer than a century to be realised.

By the dawn of the 21st century, nevertheless, there should be at least three Super-States capable of mounting major astronautical programmes extending throughout the Solar System and not confined to near-terrestrial space; there may be as many as half-a-dozen. Why should they do so, and will they do so? I have not forgotten of course, the other more idealistic approach, which also cannot be ruled out—namely that such an effort might be sponsored by the world as a whole, perhaps under the auspices of the United Nations. This would be admirable, but I suspect it belongs to an even remoter future. In stimulating progress, there are also advantages in competition. There is a good story, which I believe to be true, of a leading Russian space scientist to whom an American was lamenting the fact that their two countries competed, with wasteful duplication of effort, in space. Said the Russian—'If we could co-operate we should *neither* of us have a large space programme.'

On this fundamental question of motivation, it seems to me that there are, broadly speaking, three possibilities, though in practice the distinction between them could of course become blurred to a greater or lesser extent:

1 As a result of gradually accumulating experience in close-Earth orbits, and from preliminary expeditions, some very compelling military or economic benefits from inter-planetary travel may be revealed; compelling, that is, in terms of presently-accepted social standards. This is possible, but seems to me to be unlikely.

As far as commerce is concerned, it seems improbable (though nothing should be described as impossible, until we have been there) that there will be anything available in large quantities on other worlds that we could not obtain more easily (even if still with great difficulty, from the Antarctic or the ocean bed) on our own. Interplanetary freight charges, of course, are always likely to be higher than those over terrestrial distances.

On the military side, the same basic laws of transport economics must apply, but there might conceivably be eventual deterrent advantages from securing military bases elsewhere within the Solar System. The threat of retaliatory missiles launched from such a location, with allowance for the transit times in their delivery, would certainly upset the balance of strategic thinking on first and second strikes.

2 The competing Super-States may continue to regard the demonstration of their capabilities in space as being a worthwhile activity, even if a very expensive one, on a prestige basis. They may continue to use it to impress each other, and the smaller nations of the world, especially those emerging to the status of enhanced importance as political partners or potential customers. The technology displayed will be similar to that of advanced weapon systems, so that a broad hint is conveyed of potential military power; also to that of many exportable industrial products, so that in this sense space exploits are a form of national advertising. The technological stimulus from space programmes undoubtedly provides, in current jargon, a 'fall-out' or 'spin-off' which helps to offset their cost.

One is irresistably reminded of a common practice of boys, performing feats of strength or agility in preference to fighting one another. But what parent would not prefer such showing-off to having them return with tears, torn clothing, cuts and bloodied noses? The analogy is rather a good one; participation in what has been called 'the Space Olympics' is certainly preferable to fighting actual wars, even of a fringe nature, or even to piling up immense stocks of armaments which (hopefully) will never be used in anger. Also, expensive as it is, it is still far cheaper than these old-fashioned forms of rivalry; their space programmes cost the Americans about a fifth as much as the war in Vietnam, and very much less than one-tenth of their total defence budget, even without the cost of that war.

Right through the period we are discussing, I would expect this kind of motivation to be the one most likely to sustain the momentum of astronautical progress. As noted earlier in this chapter, it was the one which really brought about (with Sputnik I, in 1957) the first major advances in its long story, which will now probably continue for as long as mankind exists. It may, I suspect, become even more effective, as the advantages of space exploration over military preparations and adventures become more widely appreciated. As a result, even more men-in-the-street may gradually come to welcome the necessarily increasing expenditures on space, regarding them as complementary to, rather than merely competitive with, other necessary budgets for social welfare, foreign aid, or education, etc.—competitive, rather, with the even larger budgets accepted throughout past history for armaments and wars.

3 Eventually, the world at large and (which is perhaps more important) its top political administrators may come to accept rather more of the basic motivation of the astronautical pioneers themselves.

It cannot be over-emphasised that, beyond all the 'practical' arguments, this is really what started the whole vast enterprise off. It is also (and always will be) of critical importance in sustaining it, and if comparable views become more widespread, its continued progress will be assured. Prince Philip's observation at the Society's Centenary Congress in 1966 that aeronautics 'is as much an enthusiasm as a scientific, technological or commercial undertaking' is, if anything, even more true of astronautics.

Men like Tsiolkovsky, Goddard and Oberth, and those like von Braun and Korolev who have followed in their footsteps, did what they did because they were entirely under the spell of the *idea* of space-flight and of Man's physical exploration of other worlds; captivated by this idea, in its own right and because of its intrinsic value, in their own arbitrary judgment. (It will be apparent to any reader, I expect, that I agree with them.) They understood the arguments for short-term 'practical' applications of space flight—in fact, they invariably took the initiative in suggesting them—and fully recognised the undoubted validity of such proposals. This, however, was only a small part of their motivation; scientific research, in its purest sense, was another part, and also a very important one, but they understood (as some eminent pure scientists have not) that astronautics cannot be discussed or justified merely in what (within this context) must be described as narrow scientific terms, any more than it should be in terms only of its early commercial or military applications. Ultimately, astronautics must be thought of as affecting man's whole philosophy of life and view of his place in the Universe. Like the navigation of the oceans and the air, with all the opportunities they brought about for developing our experience, knowledge, and control of our environment—but eventually, to a much greater extent—space travel will constitute a whole chapter in the adventure story of mankind. It promises our best chance of discovering what life really is, how it came into existence, whether it has done so elsewhere (and if so in the same or other ways); perhaps, on one distant day, of comparing ourselves with beings evolved on other worlds.

It is not too far-fetched, I think, to imagine that one day such views may be widely accepted. Quite a short period of objective thought will confirm the arbitrary and questionable nature of many of the value-judgments made by society in the past. If, one day, the world of men supports a great sustained programme for the exploration and exploitation of interplanetary space, regarding this as a main aim of society, self-evident in its justification, would this at any rate be less worthwhile, any more bizarre, than some of our past activities? There are already many men who would not think so.

5 CONCLUSION

In this short concluding section, I should like to summarise briefly what has been said, and to express some purely personal views on the probabilities hidden among all the possibilities which have been discussed. If by chance this chapter should be read by anyone around AD 2066, then no doubt it will at least provide some amusement, for the future will inevitably provide some surprises, so that any guesses we can make will appear quaint in hindsight. This, indeed is one of the main points I have tried to make.

With this reservation, therefore, I would feel fairly confident that the Second Century will see a steady continued progress in astronautics, though no doubt with the minor perturbations, due to both technical and political setbacks, which have also occurred in aeronautics during the First Century and which seem very important at the time. Economic and social considerations will exert far more influence on the technical progress of astronautics, however, than they ever have done on aeronautics.

The practical benefits of exploiting near-terrestrial space will nevertheless assure support for rapid progress in this field for at least the first 20 years of the Second Century. Many more applications satellites will be established for telecommunications, TV, educational and propaganda purposes; for data links and various other services, for air traffic control and navigational aid, weather forecasting, military reconnaissance and the survey of 'earth resources' (geological, agricultural, etc.).

All this will ensure the continued development of improved rocket vehicles; in particular the increased use of high-energy chemical fuels such as liquid hydrogen, and the development of advanced non-chemical propulsion systems and power sources, such as the ion rocket and nuclear power. Nuclear-thermodynamic rocket engines, however, are more likely to be mainly associated with the larger and more ambitious vehicles required for manned space flight.

Since the contemporary world is likely to demand continued advances in all these space applications, even beyond the status achieved by the 1980s, it seems to me likely that (in spite of the increasing potential of automatic devices) fairly large manned space stations will be established after all at about this time. Their use will stimulate the development of recoverable, re-usable, launch vehicles, in the main for use as ferries.

Space research in pure science will undoubtedly have continued in the meantime, with probes despatched to the vicinity of the outer planets, and landings of unmanned spacecraft on the surfaces of Mars, Venus and Mercury. It is likely that astronomy will have made many advances as a result of observations made in space, not only from satellites but very probably also from manned space stations and the lunar surface.

Within a decade, or little more, of the first manned landing on the Moon, it is likely that men will reach Mars by the mid-1980s. The continued progress of the manned exploration of space, however, will be crucially dependent on social and economic factors. It is by no means as assured of a smooth and unbroken passage as are the various close-orbital space applications, or the pure scientific research aspects.

Nevertheless, I feel fairly optimistic that there will be no major hiatus between the late 20th and early 21st centuries; if this is so, progress thereafter should be continuous, unless mankind then encounters some extremely difficult technical barrier, becomes disillusioned with the discoveries so far made, or becomes more interested instead in applications of the biological and psychological sciences. The political difficulties and objections are most likely to be serious between now and the mid-1970s; I believe they will be overcome, in the first place, because of a continuing prestige competition between the Super-States, with China emerging as a major contender. Thereafter interplanetary flight will increasingly become accepted as a natural, inevitable and necessary activity, even if a very expensive one. (A gradual merging will occur, in effect, of the motivations categorised as 2 and 3 in the preceding section.)

For this progress to continue, the first requirement will be the employment of nuclear energy in some form, for the propulsion of the interplanetary vehicles. The most probable systems will still be describable as rockets, either of the thermodynamic or electrical varieties (in the latter case, with primary power sources in which nuclear-thermal energy is directly converted into electricity). By about the end of the 20th century, I would expect that some of these nuclear rockets would utilise the fusion rather than the fission reactions. It would not surprise me if, soon after that, even more fundamental nuclear processes were involved—or even if, early in the 21st century, the rocket principle itself had been rendered obsolescent by some novel form of drive. In fact, I would personally even describe this as probable—but that would be a pure guess, with no assured basis in science, let alone technology.

By AD 2066, I would expect to find many hundreds, and probably thousands, of men and women living semi-permanently on the Moon, Mars, lonely outposts on the moons of Jupiter and Saturn, and in large orbiting space-stations.

Inter-stellar and inter-galactic travel will take a little longer.

References

1 CLARKE, ARTHUR C. *Profiles of the Future*. Gollancz, UK and Harper & Row, USA, 1962.

2 *Realities of Space Travel*, selected papers of the British Interplanetary Society. Putnam, UK, 1957.

6 The atmospheric environment and aircraft—now and the future

F. O'Hara and J. Burnham

I INTRODUCTION

This chapter is about the relationship of aircraft to the atmospheric environment in which they operate, and about the way in which this relationship may change or develop in the future. The chapter starts with a fairly detailed survey of such effects on aircraft design and operation at the present time, and perhaps some justification for this should be given. No one will dispute that it is difficult, or in fact impossible, to make sensible predictions about the future without an understanding of the present and, in the particular area with which we are concerned, little attention seems to have been given in recent times to taking an overall view of the situation. Indeed, only one reference[1] has been found which gives any kind of survey of the present position, and that a brief and, for our purposes, an inadequate one. Starting, therefore, from the current factual position, and confident that the effects of the atmosphere on aircraft will continue to be predictable by Newton's laws of motion, however much erudite mathematics may be applied fruitfully to some aspects of it, we have a basis for assessing what might, or perhaps with more certainty, what might not happen in the period in question. New types of aircraft will doubtless appear which apply as yet unknown principles in the generation of lift, thrust and drag; and within the next hundred years we expect that man will control significant aspects of the weather. The effect of such on the interaction between aircraft and the atmospheric environment in which it flies is likely to result in differences in degree and emphasis from the present situation. However, we can see no rational reason to expect anything essentially new in kind.

Doubtless the future holds problems in store which are not yet thought of, either because imperfect understanding leads to particular aircraft designs being unduly sensitive to some aspects of the environment as, for example, in the past when intake icing troubles unexpectedly occurred

Originally published in the *Aeronautical Journal of the Royal Aeronautical Society*, June 1968.

on one aircraft; or because aircraft affect the atmosphere in some unexpected way. We have not attempted to 'dream up' particular things of this sort, although we are not unaware that it has been successfully done in the past, so that there seems, for example, to be a prophetic link between *No Highway* and the Comet structural fatigue accidents. The justification for a novel, however, is different from our purpose, as it is probably irrelevant whether the novel is scientifically right or wrong. Our position is a bit more akin to that of the weather forecaster, who would prefer to be partly right rather than wholly wrong.

Those of us in the aviation business, as in other lines of activity, are in danger at times of regarding aircraft development as an end in itself, and of seeing the future as one of infinite extrapolation. In a limited world, this clearly cannot be so, and leaving aside space travel as outside the atmospheric environment, we must reach a levelling out stage, after which objectives may change from striving after broad advances, to progress by more subtle refinements. The exercise of judgment in this respect would seem to be one of the principal demands on authors in this Second Century series. We have certainly endeavoured to keep it in mind, but the outcome must inevitably be a personal view.

2 THE ATMOSPHERIC ENVIRONMENT AND AIRCRAFT—THE PRESENT SCENE

Until quite recently, aircraft design practices to deal with the effects of the atmospheric environment, whether these be from gusts, hail, certain problems concerning handling qualities, and many others, have advanced by a process of evolution; in fact, there has been a strong vein of pragmatism in the design philosophy for these aspects of aircraft construction. Thus, new designs tended to incorporate those features which experience showed did not lead to trouble and to reject those which did. The design criteria used did not directly specify the atmospheric environment itself but represented a generalisation of those aspects of design which had proved satisfactory in the past.

This kind of extrapolation from past experience has become increasingly difficult in recent years, with new designs for which the dynamic behaviour and speeds are very different from those of past types, and attempts are continuously being made to refine design criteria in an attempt to reduce structural weight and to improve performance. The past ten or fifteen years have, therefore, seen a great increase in research aimed at understanding the atmospheric environment as it affects aircraft. A few years ago the first 'atmospheric environment standard' for aircraft design was produced[2], as part of the Anglo-French airworthiness standards for supersonic transports.

Aircraft have never been designed to withstand the very worst conditions which might exist in the atmosphere. Operational weather limitations have always been imposed, in one way or another. There is, therefore, a very complex 'trade-off' situation between the severity of design criteria and the desire and ability of the aircraft operator to avoid really bad conditions. Until recently the balance between the two has been struck in an essentially intuitive way, on the basis of experienced judgment, but in some areas, at least, recent research allows this to be done in a more quantitative way, on a probability basis. Demands to increase safety and improve operating economics are likely to increase still further in the future. However, if operating procedures are fixed, a large improvement in the accident rate due to atmospheric effects, resulting from increased severity in the design criteria, might well be associated with an increase of operating costs of 10% to 20%. In consequence the ability to refine operating procedures and so to reduce costs, while still retaining a high degree of regularity and punctuality of service, is likely to offer the greatest rewards. We expect therefore, to see research on atmospheric effects on aircraft directed particularly to the optimisation of operating practices in the next decade or so.

Before looking further into the future, however, let us look at the operating environment as we know it at the present time, together with the associated aircraft design aspects, observing where we can, trends of development.

2.1 *The atmospheric environment*

The atmospheric environment includes motions on all scales, from the global to the molecular. Apart from the smallest, in practice scales less than a foot or so, all motions directly affect aviation in one way or another. On the largest scale, we have the familiar trade wind belts which circle the earth about the tropics, and winds which blow around and between high and low pressure areas at high latitudes. Wind speed tends to increase with height up to the tropopause level, which varies from around 60,000 ft (18 km) at the equator, to about 25,000 ft (8 km) at the poles; the tropopause, that is the zone dividing the lower part of the atmosphere in which temperature decreases with height from that, at greater altitudes, in which temperature is roughly constant, (*see* Fig. 1), is associated with a number of major meteorological phenomena. The strongest winds in the lower atmosphere are the jet streams, which occur at about tropopause height, and in the boundary region between conditions of warm and cold air (Figs. 2 and 3). Wind speeds up to 150 knots (278 km/h) are fairly common in the jet stream cores and speeds up to 250 knots (463 km/h) have been recorded occasionally[3].

Shear in the large-scale motion generates turbulent motion on much smaller scales. At low altitudes, in what amounts essentially to the earth's boundary layer, turbulence is generated by shear and by the surface roughness of trees, buildings, etc. At higher altitudes, shear in jet streams

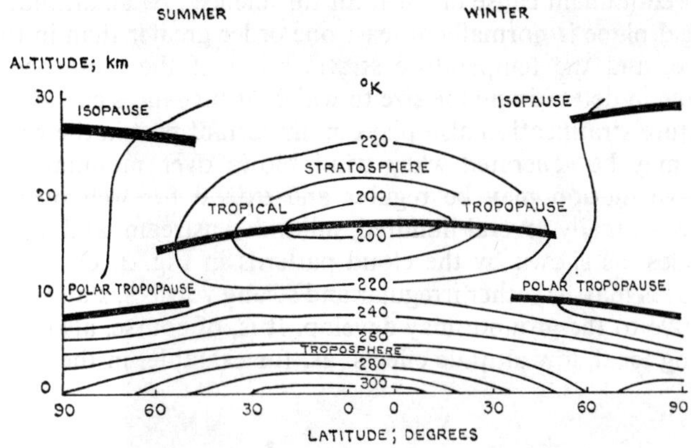

Figure 1 Mean meridional cross section of atmospheric temperature (after V. R. Dubentsov)

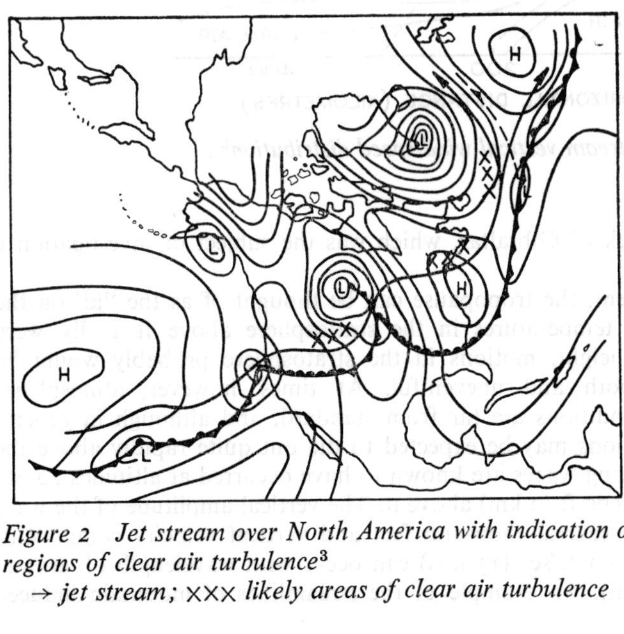

Figure 2 Jet stream over North America with indication of likely regions of clear air turbulence[3]
→ jet stream; XXX likely areas of clear air turbulence

wave motion in the stratosphere, reconstructed from recent flight test data, is shown in Fig. 5[5].

Although these wave motions are often smooth, they can sometimes be extremely turbulent, or perhaps 'rough' would be a better description since the 'turbulence' may not be of the random kind which occurs, for example, in the earth's boundary layer. Very large horizontal and vertical gusts, and large and rapid fluctuations in air temperature can be experienced by aircraft in these conditions. The origin of these disturbances can, perhaps, be interpreted as arising from a motion development, equivalent to a 'breaking' of the waves, but the actual physical processes are ill-understood, and this is an area in which further research is required.

Thunderstorms are the atmospheric phenomenon which probably causes the greatest single weather hazard to aviation, due to gusts, hail and very heavy rain. All parts of the world are affected to some extent, but they are a very frequent and particularly troublesome phenomenon

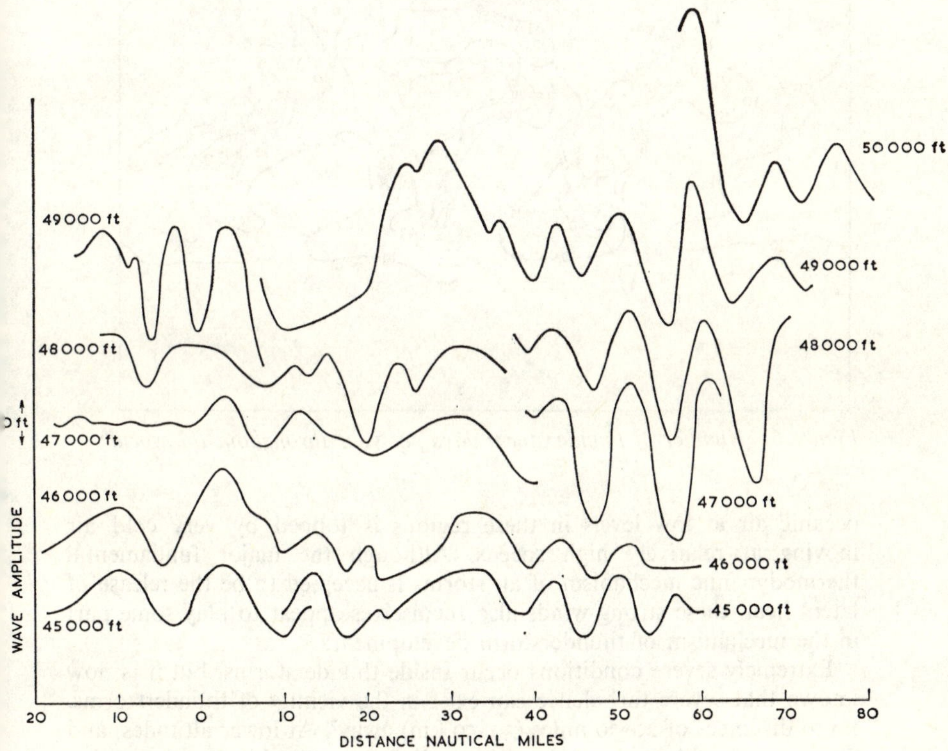

Figure 5 Streamlines of flow in mountain waves over the Western USA, based on constant potential temperature surfaces

in some areas as shown in Fig. 6[6]. A typical thunderstorm has a diameter of 5–10 miles (8–16 km) and so affects a relatively small area, but in some regions storms may form in almost continuous lines several hundred miles long.

In general, the tops of thunderstorms do not penetrate far above the tropopause, but in some regions sufficient rising energy is available for their tops to reach 10,000–20,000 ft (3–6 km) above the tropopause, to altitudes certainly in excess of 60,000 ft (18 km). Two such regions are the midwest of the USA and North East India. Very warm and very moist

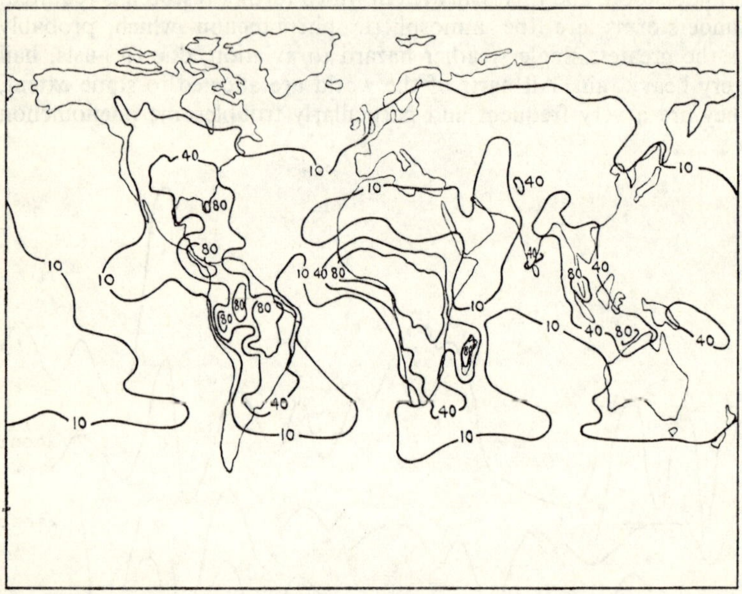

Figure 6 Number of thunderstorm days per year throughout the world[6]

oceanic air at low levels in these regions is topped by very cold air moving at relatively high speeds. Although the major fundamental thermodynamic mechanism of all storms is accepted to be the release of latent heat, these strong winds also themselves appear to play some part in the mechanism of thunderstorm development.

Extremely severe conditions occur inside thunderstorms, but it is now known that severe turbulence can exist in the vicinity of thunderstorms, up to distances of 20–30 miles (30–50 km) away[7]. At lower altitudes, and down to ground level, advancing squall fronts associated with moving storms can produce sudden extreme changes of weather at airfields. At the upper end, there is evidence that severe disturbances can exist up to several thousand feet above the tops of storms: as well as large-scale turbulence, there can be other undesirable conditions for aircraft, such as up and down draughts, and very steep temperature gradients. More is said of these conditions and their avoidance later.

Other important atmospheric phenomena, such as fog, rain and other forms of precipitation, cosmic radiation, ozone, etc., are also considered, with reference to their effect on aircraft operation. The overall range of relevant atmospheric phenomena is illustrated in Fig. 1.

2.2 *Aircraft operations in relation to the weather*

In the days when 'dead reckoning' navigation was the main means whereby an aviator found his way around the world, knowledge of wind speed and direction was all-important in ensuring that the aircraft arrived at its intended destination. Improvements in navigation aids and increases in aircraft cruising speeds have decreased its importance but the effects of increasing speed are not quite so great as, on the face of it, one might suspect. While speeds have increased, so have operating altitudes and the probability of encountering strong winds increases with altitude up to the tropopause. As an extreme example, a case is known where jet transports (which cruise at around tropopause height) were unable to fly, for several days, from Hong Kong to Bangkok due to headwinds of about 200 knots (370 km/h) throughout the height band in which the aircraft could cruise and have sufficient fuel to fly the sector. During the same period, piston and turboprop engined aircraft, designed to fly at lower altitudes, operated normally.

The effects of wind and temperature on take-off performance are too well known to need much discussion here. Except for supersonic aircraft, where the installed thrust is usually defined by cruise rather than take-off considerations, cases where take-off weight has to be restricted by considerations of aerodrome altitude, temperature and wind will continue to occur. It is more economic to accept this weight limit on a small minority of occasions than to install sufficient take-off thrust to meet every possible eventuality.

The effects on take-off performance of runway surface conditions, and in particular, slush, can fairly be attributed to atmospheric conditions. In recent years a number of engineering improvements, the development of chined tyres and the grooving of runways to improve drainage have alleviated the problems of slush, but not removed them completely. The use of coarser aggregates in runway construction may improve surface friction conditions further, but in the future one would expect to see slush difficulties overcome by under-surface heating and adequate drainage.

Bad weather landing has become possible through the development in the past decade, of automatic landing systems, which make use of modern advances in autopilot and radio guidance systems. Already, semi-automatic landings are being made in routine service in visibilities in which operation would have been out of the question only ten or so years ago. The next few years should also see the proving in flight of complementary director or automatic take-off systems, and tests on semi-automatic taxiing in very restricted visibility have already been made. In future, airline operations are likely to be little affected by poor visibility. As far as the passenger is concerned, the greatest problem seems likely to be how to get to and from the airport.

2.3 *Structural viability*

2.3.1 *Gusts* There are two different mechanisms by which atmospheric gusts may break an aircraft. Repeated encounters with relatively small gusts may reduce the strength of the structure by fatigue so that it breaks in a situation in which a load markedly less than the design proof load is exceeded. Or the proof load itself may be exceeded during an encounter with a single very large gust.

Gusts are by no means the only source of fatigue loads and, except for low altitude, high speed military aircraft, and a few rather unusual civil operations such as pipe-line surveying, they are not the most significant one. The introduction of power spectral methods, during the past 15 years, has allowed dynamic aeroelastic effects on the structural gust loads to be taken into account in a rational manner. Such methods, which assume the gusts to be members of a stationary random process, lend an aura of science to a subject which, in the past, has been almost entirely empirical. While the results of recent research make the assumptions of the spectral approach appear of doubtful universal validity, there is no doubt that, judging purely by its results, this method is a considerable advance on what has gone before. A tremendous volume of gust data is now available on a world-wide basis from recordings taken on routine aircraft operations and this appears to give an adequate picture from which gust effects on the fatigue lives of almost any future design can be assessed. Improvements in this respect are far more likely to come from greater understanding of the physics of fatigue than from refinements in knowledge of the atmosphere.

The place of gust effects in the definition of the ultimate strength of aircraft structures, presents a more complex problem. Loading actions arising from gusts are only of prime significance for aircraft which have a restricted manoeuvre envelope such as that associated with transport aircraft. There have been 22 known cases in the past 20 years in which such aircraft have broken up as a result of encounters with gusts, 17 of these occurring during flight in or near thunderstorms. It is quite clear that the latter may sometimes contain gusts well in excess of current design values and the results shown in Fig. 7 of recent research[8] allow the risk of encountering them to be assessed on a quantitative basis. As noted in section 2.1, such large gusts can also occur in other meteorological situations, particularly in association with mountain and lee waves.

It is known that in some of the cases in which structural failures occurred during flight through gusts, the latter were not the direct cause of the failure. This occurred because the pilot lost control, to some extent, of the aircraft during the gust encounter, and applied, or induced, a load exceeding the manoeuvre strength of the aircraft during attempted recovery. It is impossible to be precise about the proportion of the accidents which were in this category, since, in many cases, accident recorders were not carried, but it is certainly at least a quarter, and may be as much as a half. Also, it is now becoming clear that, even if control of the aircraft is not lost completely during an encounter with severe turbulence, loads which

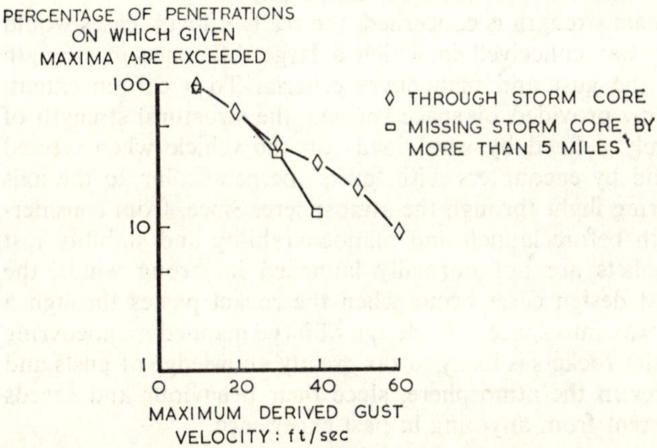

Figure 7 Probability of encountering given maximum derived gust velocities during thunderstorm penetrations at altitudes between 23,000 ft (7 km) and 28,000 ft (8·5 km)

occur as a result of the pilot's control actions are likely to be significant. There are a number of reasons why it is not possible to assess what a pilot may do in severe turbulence, by scaling up from his behaviour in light or moderate turbulence. First, results from both operational and research flying show that, so far as aircraft handling in rough air is concerned, the pilot is a highly non-linear system. Thus his actions in light and severe turbulence are likely to be very different. Second, severe turbulence does not consist only of moderate gusts of larger amplitude. As already suggested, the physical processes in the atmosphere which give rise to severe gusts are different in kind as well as in degree from those responsible for the much more frequent, less severe, gusts. Improvements in knowledge of the atmosphere are not, therefore, likely to lead to a rational improvement in gust ultimate strength criteria until such time as pilot control actions in severe turbulence and, the effects of aircraft handling qualities on them, are better understood. While in many cases of transport aircraft, the critical design case for part of the structure is the gust, rather than the manoeuvre case, the difference between the two is relatively small. In a typical example, the vertical gust case is a 66 ft/sec (20 m/s) gust and that for a symmetrical manoeuvre an acceleration of 2·5 *g*. The former is the more critical because the 66 ft/sec (20 m/s) gust is equivalent to a manoeuvre loading of 2·7–3·0 *g*. With differences as small as this, current knowledge suggests that experience does not really 'prove' that the gust criterion is 'correct'. Indeed, it is possible that the number of structural failures due to gusts would not have been very different if there had been no ultimate strength gust design criterion at all, provided that the manoeuvre criteria remained the same.

So far as ultimate strength is concerned, the really difficult cases would arise if a design was conceived in which a large difference in strength was implied by the gust and manoeuvre criteria. To a certain extent, such cases are now provided by space rockets, the structural strength of these being largely defined by wind loads on the vehicle when erected before launch and by encounters with 'gusts' perpendicular to the axis of the rocket during flight through the atmosphere. Since, from considerations of strength before launch and manoeuvrability and stability just after it, such rockets are not normally launched in strong winds, the critical flight gust design cases occur when the rocket passes through a jet stream on its way into space. The design of large manned manoeuvring recoverable booster rockets is likely to tax greatly knowledge of gusts and other disturbances in the atmosphere, since their behaviour and speeds will be very different from anything in past experience.

2.3.2. *Water* The most obvious example of the importance of atmospheric water from the structural strength viewpoint, is damage due to hail. The damage producing mechanism consists quite simply of the structure being bombarded with more or less spherical chunks of solid ice. Hailstones up to 4 inches (0·10 m) in diameter are known to exist in the atmosphere in occasional very intense thunderstorms, but stones of about one inch (00·25 m) diameter are relatively common in storms in many parts of the world and may be found at any altitudes up to at least tropopause levels. Weather radar can detect the storms which contain damaging hail and so can be used to avoid dangerous concentrations which only exist in thunderstorms. However, although so far as is known, there has been no fatal accident to a transport aircraft due to hail in the past 20 years, there have been a number of incidents (on average perhaps one or two a year on a world-wide basis) in which transport aircraft have been severely damaged by hail.

The mechanisms by which encounters with high concentrations of liquid water produce structural damage are rather more complex, and again, from a practical point of view, damaging concentrations are only likely to exist in thunderstorms. Water, encountered at high speeds, can simply erode the surface of materials, particularly radomes and transparencies, but it can also enter minute cracks in joints and under rivet heads under very high pressure. This can be sufficient to peel back joints and break off rivets. Good detail structural design can go a long way in alleviating this problem.

2.4 *Handling qualities*

The effects of pilot control actions, and by implication, of the handling qualities of the aircraft, on loads encountered during flight through gusts, have already been mentioned. Structural failures have occurred due to attempts to recover after control of the aircraft was lost during flight through gusts. In many of the latter cases, recovery was made difficult by deficiencies in the handling qualities of the aircraft concerned.

There are, of course, many more direct ways in which atmospheric conditions, and, in particular, gusts and turbulence, define handling qualities requirements. Indeed it seems fair to say that these requirements would be much less stringent than they are, if the atmosphere was everywhere uniformly smooth and non-turbulent, even though statements concerning atmospheric conditions only rarely appear directly in the specified requirements. For example, lateral stability and control on the landing approach is essentially defined by the ability of the aircraft to land safely in a turbulent cross-wind.

The interaction between atmospheric conditions and handling qualities can sometimes be much more complicated than might appear likely from a superficial appraisal. This may be illustrated by the following example of a problem which resulted in a number of accidents and incidents. These occurred on a particular type of light aircraft, which is fitted with powerful flaps, giving a rather pronounced nose-down attitude on the approach. If the aircraft is flown 5–10 knots (9–18 km/hr) too fast it is probable that the nose-wheel will touch down first. Usually this would not matter very much. However, on this particular type the nose-wheel steering was connected directly to the rudder pedals, with the result that if the aircraft landed too fast in a cross-wind, with the rudder, and therefore also the nose-wheel, significantly deflected, the result was likely to be an uncontrollable veer off the runway.

2.5 *Systems and engines*

Most aircraft contain a number of systems designed to combat various atmospheric effects. Cabin pressurisation, anti-icing and windscreen rain removal systems are perhaps the best examples of those designed primarily for this purpose.

Lightning strikes cause relatively minor structural damage but have been known to play havoc with the electrical systems of some small military aircraft. So far as is known, these effects do not occur on larger aircraft, which is fortunate since civil transport aircraft are struck by lightning relatively frequently. However, one is inclined to wonder whether the more complex avionic systems and particularly electrical control signalling of the future, may not be more sensitive to such effects than has been the case in the past.

Some jet engines dislike ingesting large quantities of water; concentrations around 10 gm/m³ which occur quite often in thunderstorms, can produce compressor stall and flame out. A further possible intake-engine

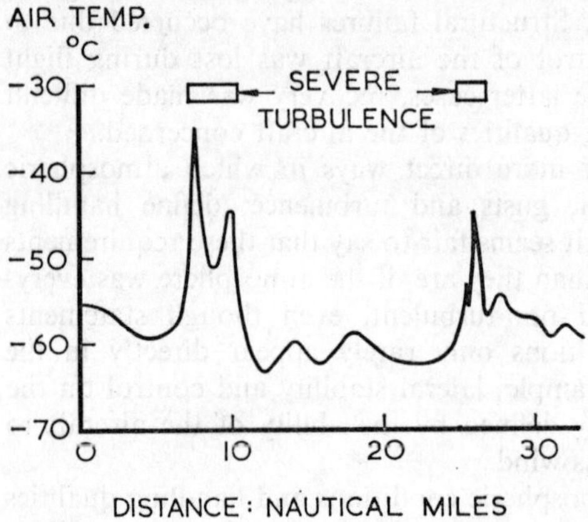

Figure 8 Time-history of air temperature measured in flight in mountain wave conditions at a height of 46,000 ft (14 km)

problem which has recently come to light[5] concerns the effects in supersonic flight of large and rapid temperature fluctuations, which sometimes occur in the stratosphere in association with thunderstorms and mountain waves, an example of which is shown in Fig. 8. Such fluctuations produce rapid changes in the Mach number, and so in the shock-wave and airflow pattern in a supersonic intake, and account must be taken of them in the design of the intake and its control system.

2.6 *Ozone and cosmic radiation*

Finally, to round off this discussion of present day effects of the atmospheric environment, mention must be made of possible problems due to other causes, such as ozone and cosmic radiation effects, on the passengers in an SST. Ozone, which exists at a mild level of concentration, in a layer of the stratosphere in which supersonic transport will operate, was thought at one stage to be a potential hazard to passengers. It is now recognised that it will be 'cracked' in the cabin air conditioning system and is no longer considered a worry. With regard to cosmic radiation, on the other hand, there is no universal agreement among the experts as to how much of a problem will really exist for the first generation SST. It does, however, seem to be generally agreed that in certain conditions associated with solar flares it will be necessary, to ensure safety of the passengers and crew particularly from possible genetic effects of the radiation, for an SST to descend into the more dense regions of the atmosphere, where the radiation is fairly heavily attenuated. For the future, it seems likely that cosmic radiation will be one of the factors which limits the length of time man can spend in space.

3. WEATHER FORECASTING FOR AVIATION AT THE PRESENT TIME

Weather forecasting can be thought of as consisting essentially of a detailed knowledge of present and immediately past weather and some means of predicting how this will change. The basis of the latter is usually a knowledge of the surface pressure field and of the heights of a few constant pressure surfaces. Present values are used, often with the help of a large digital computer, to forecast values 12, 24, 48 hours and so on, ahead. A rough guide to the present claimed predictability of weather[9], is that on a 5000 km scale, it can be forecast on a five day basis, whereas on a 5 km scale, forecasts can only be on a two hour basis. In fact, looked at on the scale of Europe, or perhaps even on that of the British Isles, the agreement between the forecast and subsequent actual pressure patterns is usually surprisingly close. But the position of major features, for example the positions of the centres of high and low pressure areas and of fronts, may be wrong by a hundred miles (160 km) or so on occasions. The effects of such errors are usually in terms of errors in the time at which particular weather gets to a particular place. However, there are some situations in which quite minor errors in forecasting the pressure field can result in radical errors in forecasting the weather; for example, an error of 20 or 30 degrees in wind direction can make the difference between a cold windy but possibly clear day in East Anglia in winter, and a blizzard there. Another kind of weather in which timing errors in forecasts can sometimes be large is fog, forecasts of the lifting of which are not usually reliable to better than an hour or so and can sometimes be wrong by several days.

3.1 *Compilation of weather data*

No forecasting can be more accurate than the data on which it is based and it is worthwhile looking briefly at what is available at the present time. This consists of a network of ground stations which measure pressure, temperature, rainfall, wind and sometimes humidity, together with a more widely spaced network of stations which send up balloons measuring temperature and pressure and wind. These stations report to some central office, usually by teleprinter, reports from surface stations often including a note of any significant weather since the last report, e.g. fog, rain, thunder, etc., and of the cloud cover and cloud heights.

The distribution on a world-wide basis of these reporting stations at the present time, has some very large gaps in the oceans and in the under-developed parts of the world. In such areas the meteorologist has to rely on reports from aircraft and ships but in recent years the coming of the weather satellite has been a great help. Although, at the present time, considering data available for forecasting rather than for research, these satellites indicate only cloud cover, and that at a resolution which is just about adequate to show a large thunderstorm at least, large-scale weather systems such as hurricanes and cyclones can no longer exist undetected for days. Satellite pictures sometimes permit a quite accurate determination of the position of jet streams and reveal the existence of mountain and lee waves as shown in Fig. 4. A further tool in determining present weather over quite large areas (e.g. a radius of 200 miles (320 km)) is weather radar which allows the position and intensity of precipitation and, of particular importance to aviation, of thunderstorms, to be determined.

3.2 *Problems of weather forecasting for aircraft operations*

An aspect of weather forecasts in which an improvement in accuracy would be appreciated by the aircraft operator concerns the prediction, around three hours ahead, of the wind at take-off. Moderate and even strong winds do not usually provide much of a problem from the flying point of view, but at some airfields where the runways are short, take-off weight has to be restricted in light wind conditions, particularly on hot days. If really accurate forecasts were available, heavier loads could safely be carried on most such occasions. Unfortunately, it is on just such hot, still days that forecasting of wind is least reliable, and a worthwhile improvement in accuracy seems a long way off. It is not suggested that improvement is impossible but, in terms of useful return on research effort, there are many areas where such effort would be better spent.

As well as forecasts of weather, as this would be apparent to an observer on the ground, the aviator expects forecasts of winds and temperatures aloft and of any hazardous conditions likely to be encountered *en route*. The accuracy of wind and temperature forecasts is generally satisfactory for current aircraft, the standard deviation of forecast errors in route mean wind on the North Atlantic for example, being 12 knots (22 km/h) at present altitudes. An improvement could result in an improvement in operating economics, but there is no problem from a safety point of view. Some fairly severe limitations on present abilities to forecast *en route* weather hazards must be admitted. The presence of thunderstorms and of turbulence at high altitudes can be predicted with fair reliability several hours ahead insofar as statements can be made to the effect that such phenomena are likely to be encountered somewhere within an area a hundred or more miles square and an indication can be given as to roughly what proportion of the area is likely to be affected. Ground weather radar can be used to pinpoint the positions of thunderstorms once these have developed but it is not at present a reliable (aviation) forecasting tool. A reasonable inference can sometimes be made as to what is likely to happen up to an hour ahead but one could not often go further than that. Forecasts of icing conditions are usually adequate but problems can occur from time to time in particular parts of the world. Even in their present limited form however, forecasts of *en route* weather hazards are invaluable in giving warnings to the pilots and in flight planning; for example, airlines flying over the Western USA often route flights so as to avoid forecast mountain wave conditions, and the associated turbulence.

3.3 *Communication and appreciation of weather information*

To make reliable and accurate weather forecasts is one thing; to transmit the forecast to the pilot in such a way that he can understand and use it, is another. In fact, if one looks at flying accidents over the past 25 years in which the pilot's appreciation of weather played a significant part, the impression is gained that it is in the latter area rather than that of basically inaccurate forecasts that the problems most often lie. For example, the pilot may have, and understand, a perfectly adequate forecast for his planned route, and then may decide to divert from it. He may 'extrapolate' the forecast, on the basis of experience and what he can see out of the cockpit, to his new route in a way which would probably be right in general, but happens to be quite wrong on that particular occasion, and possibly for a reason which was known to the man who prepared his original forecast. Or, and this is particularly the case in rapidly deteriorating visibility, the time taken between the change in conditions being noticed by the meteorologist and this information being transmitted to the aircraft may be too long.

In flight the pilot normally checks, from his general experience, how well the forecast he has been given tallies with what he can see outside the cockpit and with his own measurements of wind and often of temperature. Airborne weather radar fills what is probably the largest gap in the forecaster's capabilities, by permitting the pilot to keep a safe distance from thunderstorms. For turbulence not associated with thunderstorms, which is usually referred to as CAT, the situation is not so happy. As mentioned earlier, the present situation is that the position of such turbulence can only be forecast as somewhere in a large area, forecasts of its intensity are not always accurate and there have been a fair number of cases of severe CAT being encountered in a region where none had been forecast. Debatably, it often could have been forecast on the basis of the best current knowledge. But even over the next many years it would not be surprising occasionally to find cases of 'unforecastable' severe turbulence. At the present time the most reliable guide which the pilot has to the position and intensity of CAT is reports from other aircraft and in most parts of the world such reports are transmitted by the ATC system with commendable speed. Reports received in flight are of even greater importance for actual forecast conditions at the destination and alternate airfields. Since such forecasts usually need cover only the next hour or so, they are accurate enough on the whole, but may not always be of the right form, nor properly describe the important aspects of the weather. Two problems which remain, in this respect, concern the development of a really meaningful measure of visibility when this is marginal, and of something better than the currently transmitted readings of the airfield anemometer for the occasions when turbulence and wind shear on the approach might lead to piloting difficulties. Increases in air traffic in recent years have led at times to near saturation of radio communications. Ground-to-air teletype is easing this situation and should help to provide more detailed and timely forecasts in the future.

3.4 *Atmospheric conditions and their effect on aircraft safety*

There can be no doubt that bad weather is conducive to accidents. Such accidents include not only those in which something happening in the atmosphere (e.g. extreme turbulence) was the direct cause of the accident, but also those in which adverse weather was a factor. For example, the probability of an accident occurring on a particular landing increases noticeably as the visibility in which the landing is made diminishes; this is shown by the statistics in Fig. 9[10]. Again, flight within 50 miles (80 km) of a thunderstorm, even if the aircraft never enters cloud, or flight through an area of strong mountain wave activity, is likely to be measurably more dangerous than the average. One can think of many more examples. Quite generally, however, a pilot can almost always 'get away with' a single failure (of a non-catastrophic type), or an isolated mistake or error of judgment. Much more rarely can he 'get away with' two or more. Flight into adverse weather is often the second step in a chain of events leading to trouble.

It has happened also, that fatal accidents have resulted from flight into known severe weather, when a better appreciation of the dangers would have led the pilot to avoid the area altogether and accept the resulting delay or diversion. To attribute such accidents to 'severe weather' or 'pilot error' is an over-simplification of the situation. In practice, the pilot may have been required to balance in some way and probably subjectively, punctuality against safety; the result could have been due to his over-valuing the first or under-appreciating the diminishing margin in the second. In such a situation, commercial pressures may be involved, but so also may be personal aspects. If the pilot's assessment of the risk involved was over-optimistic, this may be because his personal experience of flight in severe weather is limited. A given pilot is likely to meet severe turbulence on the average once in every five or ten years and to land in near minimum visibility once or twice a year at most. Although in some cases the airline may make the decision as to what are the worst conditions which can be accepted (e.g. visibility and cross-winds), the advice the pilot is given about other conditions may be rather vague. Such vagueness is often the best that can be done in the current stage of knowledge.

There are now signs that possibly within the next few years, and certainly on a time-scale not going far into the next century, improvements in knowledge in some areas (e.g. turbulence) will allow the risks of flight in particular weather conditions to be assessed with realistic accuracy, on a probability basis. This will place those responsible for airworthiness and operational safety requirements in an entirely different position from that in which they now are. It is only in recent years that overall safety requirements have been expressed in numerical terms. The idea of assigning a number to the maximum acceptable risk on any particular flight is even more recent.

If air traffic is to grow at anything like the rate which extrapolation of present trends suggests, it must become a means of travel for a far wider section of the population than is the case at present, and mainly for personal rather than for business reasons. (The authors may perhaps be forgiven for hoping that other inventions, such as business conferences by private television satellite link, will lead to reduction rather than to growth in the amount of flying necessary for business pruposes.) The 'man in the street' probably knows little about the numerical figures of aviation safety, but almost certainly has the impression that at the present time flying is more dangerous than travel by car, train or ship. Even if future developments make the risks more equal, or even if flying can demonstrably be made the safest form of transport, a single accident to a Jumbo jet or SST killing 300 to 500 people, being fully publicised, as it will be, as a spectacular event, would result in an equally spectacular sinking of public confidence. Very few people outside the insurance companies assess risk on an actuarial basis, and there is no gainsaying that many who travel by air today are not really happy about it, but are themselves balancing the risks against the convenience and time saving of travelling by air.

It is clear, however, that if air transport is to grow in the future in a really satisfactory way, and as those who are involved in aviation predict and hope for, current levels of safety must be improved in the future. In view of the aviation growth implied, this means that the chance of an accident on any particular flight will have to be reduced, possibly by orders of magnitude. If this does not happen, the reluctance of the travelling public may be a greater brake on development than the forward pressure of technological advance. More is involved in this than just the effects of the weather, but the situation is such that no area of possible improvement can be overlooked. In the next hundred years, advances in knowledge of the atmosphere will lead to improved weather forecasting and, in some cases, to weather modification; to improved aircraft design and operating techniques to combat adverse weather. It will be partly a matter of choice whether these advances are used to improve safety or to make air travel more punctual or cheaper. From experience with other forms of transport, there is no historical necessity for transport systems to become more punctual and reliable with time. It can be argued that lack of success in these directions indicates some inadequacy in operational control, which clearly bears on safety. This cannot be lightly accepted in the same way for aviation, with its exposure to higher levels of risk.

4 POSSIBLE DEVELOPMENTS IN THE PATTERN OF THE RELATIONSHIP OF AIRCRAFT TO THE ATMOSPHERIC ENVIRONMENT

The pattern of aircraft operations can be affected by developments in aircraft design, which extend their operational capabilities to, for example, VTOL or orbital flight, and also by associated systems developments, in say, autocontrols or gust alleviators; the pattern can also be influenced by advances in weather prediction and weather mapping, and by in-flight bad weather detection aids; and lastly, it can clearly be influenced by developments in weather modification techniques, which make possible dispersal of unfavourable weather conditions, locally or on a larger scale. We shall consider each of these aspects in turn, looking in particular at the possibilities where the greatest technological advances might be expected, and then try to sum up what the integrated effect might mean for air operations.

4.1 *Aircraft design developments*

The likely improvements and developments to be expected in aircraft design have been outlined in other surveys. So far as we are concerned, major developments like sub-orbital operation, providing the ability to operate efficiently above the weather, can significantly affect the importance of atmospheric turbulence in relation to aircraft design. It is possible also that there might be associated reduction in the sensitivity of such aircraft to atmospheric effects if the ascent to altitude is in ballistic-type flight, with retracted wings, following power-supported take-off and subsequent landing. However, we still have to consider all advanced types of aircraft, subsonic, supersonic and hypersonic, as well as orbital, and the differences they may produce in relation to the atmospheric situation.

Before considering developments in relation to the operating role of aircraft, it may be useful to note first that although continuing advances in structural design are to be expected, fundamental changes in the mode of construction of aircraft, which might basically alter their sensitivity to the atmospheric environment, are thought unlikely. There will, however, be refinements of design practice to meet increasing aerodynamic requirements, such as thinner wings, and higher kinetic heating temperatures; there will also be full exploitation of new materials, both metallic and non-metallic. Some of these developments should make aircraft more resistant to environmental damage; they may also produce structures with novel dynamic properties but probably not with significant increase in the inherent structural damping, which could reduce the effects of turbulence. The latter is more likely to be achieved by the systems developments discussed in the next section.

On the operational side, we know, from current experience, that ever greater altitudes will be involved. Even in the near future it will be necessary for weather information to be provided for altitudes in excess of 60,000 ft (18 km) on a reliable world-wide basis, rather than for the little over 40,000 ft (12 km) now reached by radiosonde ascents in many parts of the world. Altitudes of over 60,000 ft (18 km) are currently reached in the USA and Europe and the problem of wider coverage is not only technical but one of organisation and finance. At altitudes between those of the SST and satellites, rockets will probably be used for weather sounding rather than balloons, and current research technology will have to be applied on an operational basis.

V/STOL operation has been slower in development than has variously been forecast, but it would be surprising if V/STOL aircraft are not engaged in inter-city operation on the time-scale we are considering. There can be special low altitude turbulence problems in this type of operation, arising from proximity to large buildings. To offset this, many types of VTOL aircraft are much less sensitive to turbulence, either in terms of flight path control or passenger comfort, than conventional aircraft; other types, particularly in the STOL class, if lift is generated aerodynamically by air control schemes on the wings, like sucking and blowing, or induced circulation, are likely to be sensitive enough to raise problems, but none which adequate control power and auto-stabilisation cannot deal with. It has to be noted that the STOL type is the more attractive at present from the point of view of noise level, but advances in noise suppression may make this a less significant factor.

To reduce the problems of air disturbances in city precincts, VTOL operation could take place more generally from the tops of tall buildings, as is done at present in some cases of helicopter operation. This tends to avoid the direct effect of eddies from surrounding buildings but, unless operations are from a building considerably higher than those surrounding it, the roughening effects of the buildings and the general increase in wind speed with height, may result in the turbulence at the building top being greater than in open country. There may, of course, be the possibility of artificial screening by physical or aerodynamic means, to reduce the severity of turbulence in the vicinity of such operating sites. There may even be the further possibility, considered later, of local weather modifications to prevent extreme conditions arising. To consider another possible line of development, there have been suggestions for covered cities, and operating sites on the roof level in this case would be less subject to additional turbulence effects.

4.2 *Associated aircraft systems developments*

Improvements in auto-controls on aircraft, coupled with advances in air traffic control systems, can result in much easier but also more precise flight control. This can provide the opportunity for more exact, and at the same time more flexible, delineation of air lanes, for optimum routing in relation to a comprehensively mapped weather picture.

Optimisation of route flying has to take account of performance efficiency, avoidance of structurally dangerous conditions in the atmosphere, and so forth, but has also to pay attention to smoothness of flight and the comfort of passengers. The latter can be achieved by avoiding rough air, but there are also possibilities of improving comfort by the use of gust alleviators. Schemes designed to modify the response of an areoplane to turbulence to alleviate loads and so improve passenger comfort have been considered for many years. A gust alleviator proposed for the Brabazon aircraft built in the UK just after the 1939–45 war, was tested in flight on a Lancaster aircraft at RAE[11]. At that time, knowledge of gusts and turbulence was very limited, the performance of the control systems then available was relatively poor, and the possibility of malfunction was such that the automatic control system could be allowed only very limited authority. The great improvements which have occurred in the past 10–15 years in knowledge about gusts and the response of aircraft to them, and greatly improved avionics and power controls, have led to the possibility of gust alleviation systems being reconsidered in the past few years. The older types of alleviator were mainly intended to work by sensing the gust ahead of the aircraft and then moving the controls in a suitable manner to alleviate its effects. Most systems now being considered are basically response modifiers which work by coupling structural motions through sensors to the auto-control system, thereby increasing the damping of the aeroelastic modes of the aeroplane, as well as reducing the magnitude of the response in the rigid-body modes. Such

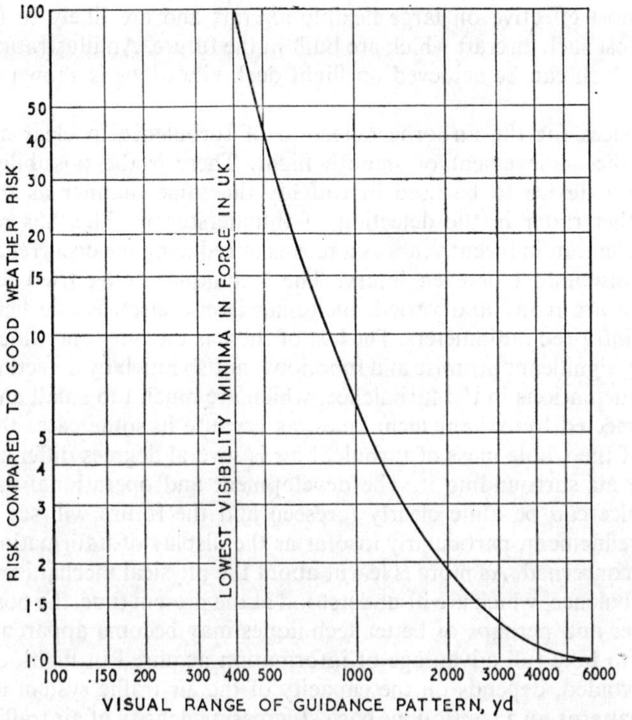

Figure 9 Relationship between risk of landing and RVR[10]

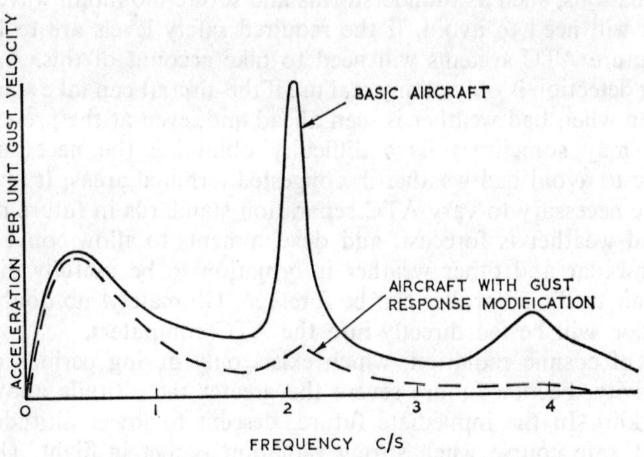

Figure 10 Possible effect of gust response modification system on normal accelerations in the cockpit of a large aircraft

systems are most effective on large flexible aircraft and are likely to be a feature of most such aircraft which are built in the future. An illustration of the effect which can be achieved on flight deck vibrations is shown in Fig. 10.

Aircraft devices for the airborne detection of turbulence in clear air can assist in the achievement of smooth flight. There is the possibility of developing a device to be used in roughly the same manner as the airborne weather radar in the detection of thunderstorms. This has received much attention in recent years as a means of reducing the occurrencc of passenger discomfort or even injury. The techniques being tried for CAT detection are many and varied, including lasers, electrostatic field detectors and infra-red radiometers. The last of these is the only one which has shown any significant promise and functions, not so much by detecting temperature fluctuations in the turbulence, which are much too small and rapid to be detected by present techniques, as because in some cases the temperature of the whole mass of turbulent air is several degrees different to that of the air surrounding it. The development and operational use of such a device can be quite clearly foreseen and the future will see a considerable refinement, particularly insofar as the display of information to the pilot is concerned. As more is learnt about the physical mechanisms of clear air turbulence, which are ill-understood at the present time, the possibility of other and perhaps of better techniques may become apparent.

The ability to take full advantage of information acquired in flight, of areas better avoided, depends on the capacity of the air traffic system to adjust traffic patterns on a continuous basis. Increasing density of air traffic will result in smaller separations for *en route* traffic and a greater degree of automation in the ATC system. In the foreseeable future there will still be weather situations, such as thunderstorms and severe mountain waves, which aircraft will need to avoid. If the required safety levels are to be maintained, future ATC systems will need to take account of this. Airborne weather detection is only of any real use if the aircraft can take some avoiding action when bad weather is seen ahead and, even at the present time, aircraft may sometimes have difficulty obtaining the necessary ATC clearance to avoid bad weather in congested terminal areas. It may be found to be necessary to vary ATC separation standards in future on days when bad weather is forecast, and developments to allow conventional weather radar and other weather information to be usefully displayed to the air traffic controller can be foreseen. Ultimately, no doubt, such information will be fed directly into the ATC computers.

The hazard of cosmic radiation, which exists only during periods of solar flare activity, becomes more severe the greater the altitude above 50,000 ft (15 km). In the immediate future, descent to lower altitudes seems the only safe course when strong radiation is met in flight. The times at which such activity will take place cannot be accurately forecast at present and the early years of supersonic passenger flying are likely to see the setting up of a network of satellites to detect high levels of cosmic radiation.

4.3 *Weather forecasting and operation planning*

Meteorology has always been strongly connected with travel, in earlier times by sea, but in recent times there have been large associated developments in both meteorology and aviation. There is sometimes a tendency to think that the meteorological organisations of the world exist primarily to issue forecasts for aviation, and that the 'general weather forecasts' are just a useful by-product, helpful in deciding, for example, whether to go to the coast or to stay at home. The services of the meteorological organisations, however, are just as important to, for example, the construction industry, and agriculture and horticulture. The cost to the airlines of a wrong forecast, such as an airport closing unexpectedly due to fog, is probably no greater than the cost to the shop and cafe proprietors of a coastal resort, when a forecast of low cloud and rain keeps the trippers away on a day which turns out fine and sunny. The economic impact of improving the accuracy of *en route* wind and temperature forecasts can be vastly over-rated. The idea that advantage can be taken of favourable winds in planning the track or altitude of a flight is somewhat unrealistic in the air traffic environment in many parts of the world today, and may be even less realistic on the more crowded airways of tomorrow. Forecasting improvements would not allow fuel reserves to be cut by much since these are carried to cope with other contingencies than wrong forecasts. In fact the cost reductions that would be achieved if present forecasting errors were halved are not very large and only comparable to the reductions that would result from obvious changes like lighter cabin furnishings and equipment. Improvements in forecasting are of greatest use, therefore, in assisting in the avoidance of dangerous or unpleasant weather conditions.

4.4 *Developments in weather forecasting*

The areas of research in which most of the attention is being given in meteorology at the present time, problems of the general circulation of the atmosphere and of the air-sea interface for example, are unlikely to be of direct interest to aviation, but can provide a more complete basic understanding of the physics of the atmosphere, which is necessary for more reliable forecasting. However, since almost all flights last for less than 12 hours, aviation is in general interested more in short term forecasts, for six or twelve hours ahead, or for the duration of the flight. Improvements in the accuracy of forecasts over this kind of time-scale are more likely to come from more accurate, and more frequent, observations than from any great breakthrough in the understanding of the physics of the atmosphere. In the forthcoming period, the number of flights made in hitherto less developed parts of the world is likely to increase substantially, and the pressure from aviation to bring the standard of short term forecasts in these areas, at least up to the best standard currently attained in Europe and the USA, will be great. So will the pressures from other users of meteorological forecasts, and it is to be expected that the spacing of the network of surface and radiosonde stations over most of the earth will be comparable to that of the current European network before too long.

Although doubt has been expressed as to the direct usefulness to aviation of forecasts for periods more than 12 hours ahead, the improvements in longer term forecasts which will certainly occur are bound to have some impact, both directly, and also through improvements resulting from better short term forecasting methods. It is therefore worthwhile saying a little about the difficulties involved, and by implication, about the likely time-scale. It is, to many, a basic postulate of science that a complete knowledge of the initial state and of the equations of motion would allow perfect prediction and that the existence of such a state of affairs is effectively attainable in a meaningful sense. This is often taken to imply that weather forecasts of indefinitely increasing accuracy can be expected to come from more tightly spaced observations (it would be illogical to say more frequent observations since increasing accuracy of longer term forecasts ought to allow observations to be made less frequently) and, from the use of more advanced computers, to take ever-increasing amounts of data as initial conditions, for the solution of equations of motion of increasing complexity. This is true, up to a point, but the gap which exists between what is currently attainable, and the making of reasonably accurate forecasts of weather (i.e. rainfall, sunshine, temperature, humidity, fog, frost, etc., as opposed to the surface pressure pattern) three to ten days ahead, is, however, truly enormous. Currently possible theoretical solutions to the aero-thermo-electro-dynamic equations which govern the behaviour of the mixture of air and water which is the atmosphere, are trivial compared to the processes which actually go on there. To give an example which does not involve thermodynamics (in the sense that no release of specific heat is involved), the equations for the motion of a stratified fluid over a three-dimensional surface, of direct application to the understanding of mountain wave phenomena as they really exist, have yet to be solved. When the behaviour of water in various forms has to be taken into account, many processes are involved which hardly are understood at all at the present time. The behaviour of condensation nuclei in the generation of rain, and the part played by man-made smoke particles in the process, may be quoted as examples. No one would claim at the present time that the problems of conventional subsonic aerodynamics are fully solved and these are of very limited complexity compared with the processes which go on in the atmosphere. The effort, over the past 50 years, which has been put into the former must be considerably greater than that spent on the whole of meteorological research. It seems likely that most of the easy problems in meteorology have now been solved and many of those which are left need a deep and wide physical insight, before practical solutions will be found.

The form in which it is possible to provide forecasts is a matter also in need of consideration. Thus it may be claimed that accurate forecasting of some meteorological event, like thunderstorms, can be made; however this is true only in the sense that thunderstorms will occur within a fairly extensive area, rather than that there will be a thunderstorm at a particular place, at a specified time. In the last analysis the precise nature of any thunderstorm occurring is likely to depend on local conditions, like the state of the crops in the fields.

Particularly from the immediacy point of view, aircraft themselves can be a fruitful source of information about the state of the atmosphere through which they are flying. This is especially true when a new type is introduced which flies through a region of the atmosphere about which forecasts were not previously made; thus for the SST during the next decade, data gathered *en route* will be of great importance to the forecaster. As implied above, it may well be the only source of data over the less developed parts of the world. Ways will have to be found to record and transmit these data in such a way that they are not out of date by the time they get to the forecaster. This is not likely to require new technological development, but organisational and financial problems may well be involved. Major airlines in the USA have found it worthwhile to establish their own meteorological services, while relying on the national network to provide most of the basic data used. The scale of most other airlines is small by comparison, and while some of them employ one or two meteorologists, they rely on the national meteorological services of the countries in which they operate to provide forecasts for their aircraft. It seems likely that in the future the main airline companies will combine meteorological resources to provide world-wide forecasts for their aircraft, mainly using data gathered in flight, supplemented by whatever other data are available. We should be surprised if this did not happen within the next 10 to 20 years.

As has been stated, aviation's greatest need for improved accuracy is in the short term forecasting of atmospheric conditions which, in one way or another, are sufficiently severe to endanger the aircraft or its passengers. Almost all such phenomena are very localised, with scales not much greater than 10 miles (16 km), and apart from some local knowledge of such effects as those produced by local terrain, forecasting of the exact position and time of occurrence of such phenomena would seem to be impossible without reporting stations spaced on about a 10 mile grid in the vicinity of the airlanes. Present spacing, in practically all parts of the world is about an order of magnitude greater than this. Automatic weather reporting stations already exist and a whole network of these, capable of being interrogated at will by a computer, can certainly be envisaged in the not too distant future. But their employment on a close spacing looks rather uneconomic, except perhaps in some terminal or other high trouble areas, unless a number of other significant beneficiaries besides aviation can be found. Rather, the future would seem to lie, at least for the next few decades, in relatively *ad hoc* solutions to specific aviation problems. There are a number of these in which the pressures either are, or most certainly will be, great enough for some measure of solution to be forthcoming. It is worth discussing a number of these in some detail, not only for their direct interest but because some general lessons can be drawn.

4.5 *Problems affecting take-off and landing*

First, consider the situation in relation to visibility. Twenty years ago, information on visibility required by aviation could be obtained by a process which amounted essentially to little more than somebody looking out of the airfield meteorological office window. Visibilities of several miles and cloud base heights of 500 ft (0·15 km) or more, usually do not vary much over quite large areas, and when such variations do occur, they are fairly obvious. In such visibility, the touch-down point would be visible from the control tower. When, some years ago, aircraft began to operate in one mile (1·6 km) visibility and 300 ft (90 m) cloud base, more difficult problems arose and these become more severe as minima further decrease. When such low visibilities occur, conditions can vary over quite short distances. It is no longer adequate to rely on a general visual assessment, and instead the number of lights visible at the runway is used as a measure. The visibility estimate so obtained is called 'runway visual range' (RVR). More recently, devices called transmissometers have been introduced which measure RVR remotely and allow variations to be seen as they occur. However, visibility along the runway at a height of 6 ft (1·8 m) is of interest to a pilot who has succeeded in landing, but not quite so relevant to one who is attempting to do so. What a pilot can see from a height of a few hundred feet looking downwards at an angle of 3°, as on a normal landing approach, is not always adequately described by RVR and cloud base height. Since it is not possible to mount a detector at or near the critical point on the approach path, the solution to this problem is not obvious. Some sort of backscatter measurement would seem to be the most hopeful line of attack, but a practical operational system seems some years off. It is interesting to note that, during night landings in poor visibility, the obviously better visibility of lights on the ground as seen from the cockpit, when the landing lights are off rather than on, seems rarely to be used by civil pilots. It appears that pilots prefer to have the ability, given by the landing lights, to see something of the surface texture of the runway. The greatest demand for accurate visibility measurements may come in the next decade or two, as minima are reduced and before zero-zero auto-landing becomes routine on civil passenger operations. It is interesting to note that the problems of reduced visibility can be overcome, not only by auto-land techniques, but also by dispersing the fog, and something is said of future prospects in this respect in the next section.

Another weather aspect which has become of concern to aircraft operators in recent years is the wind speed and direction given to an aircraft as it begins its landing approach. These quantities, obtained from readings of the airfield anemometer, do not always give an adequate picture of the severity of disturbances due to turbulence and wind shear which may be encountered. Some future types of aircraft, particularly large ones, are likely to be more sensitive to such disturbances than are current types. A complete solution demands a means of remotely sensing air motion and no technique is at present available which will do this. The problem can be split into two distinct cases and separate solutions to each look more plausible. The first is that of severe turbulence due to strong winds probably associated with significant effects from local terrain. Meteorological balloons

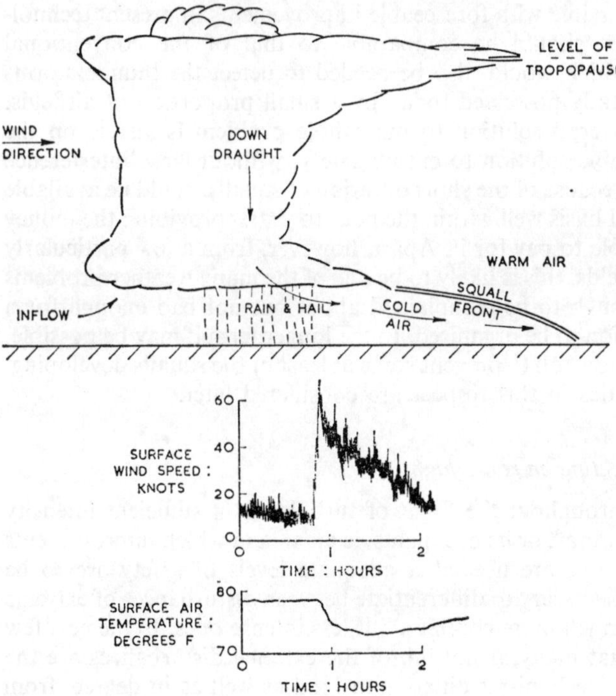

Figure 11 Illustration and time-history of passage of a thunderstorm squall

tracked by kinetheodolites, which are useful for wind variation studies would be next to useless for this problem. On the other hand, at any airfield, it should be possible to find sites at which a few anemometers could be mounted at heights not exceeding about 50 ft (15 m), the readings from which could be used to predict adequately the severity of conditions on the approach in such conditions. To do this would require extensive surveys in a variety of conditions using instrumented aircraft, such as that done in the neighbourhood of the Rock of Gibraltar by RAE in 1963, which is a spectacular but relatively simple case. Such a survey could be fairly costly but could be worthwhile at some airfields where problems are unusually frequent or severe. The second case, that of disturbances arising from squalls (*see* Fig. 11), seems to be easily the most important, as it has been the cause of the vast majority of take-off and landing accidents associated with turbulence and wind shear. This is the most difficult problem to solve without large and apparently unacceptable economic penalties; it could be solved by prohibiting all operations from an airfield when a thunderstorm was within 20 miles (32 km). The most fruitful line of attack is probably to attempt to detect remotely the temperature changes associated with the squalls, using infra-red radiometry. This would be possible with foreseeable improvements in present technology, and the cost should be comparable to that of the conventional weather radar, which would also be needed to detect the thunderstorms and which is already possessed today by a small proportion of airfields. While no easy overall solution to our whole problem is visible on the horizon, a workable solution to ensure safety, without large interference with operations because of the short duration of squalls, could be available on an operational basis well within the next 10 years, providing the money was made available to pay for it. Apart, however, from a few particularly troublesome airfields, this is likely to be one of the many weather problems which is bad enough to be complained about but not bad enough for a worthwhile solution to be organised. In the longer term it may be possible, by local weather control, to prevent some at least of the squalls developing, and the possibilities in this respect are considered later.

4.6 *Problems affecting en route flight*

The avoidance throughout the flight of turbulence of sufficient intensity to endanger the aircraft or its occupants, is an area in which improvements over current practice are needed if adequate levels of safety are to be maintained. It is necessary to differentiate between disturbances of extreme severity and the much more common but less intense ones. In the next few years, the idea that many, if not all, of the extreme disturbances are the result of physical mechanisms differing in kind as well as in degree, from those responsible for the commonly occurring disturbances, is likely to be substantiated. Extreme disturbances appear to occur either in or near thunderstorms, or in winds of at least moderate strength, associated with rather peculiar temperature stratification, blowing over mountainous or hilly terrain. So far as thunderstorms are concerned, radar techniques for their detection are well known and no fundamentally new developments are likely to be necessary in this area. It seems probable, however, that airborne weather radars operating at choice on either of two wavelengths, one of which would be around the 3 or 5 cm used at present and the other about a tenth of this, to provide scanning capability both for overall cover of the storm and accurate detection of its top, will find a place in SSTs and other vehicles which operate above the tropopause. Since the determination of storm top heights with current airborne weather radars cannot be made with sufficient accuracy for SST operation an attempt will have to be made to devise techniques which will allow the maximum height attained by storms in a particular area on a particular day, to be forecast with reasonable precision. Research in this area is showing promising results and such techniques could be in operational use within five years. The ability to forecast the other important class of severe disturbance, those which occur in mountain and lee wave conditions, will certainly improve. Fairly clear limits are placed on such improvements by the spacing of the upper air data available. The best current practice allows the existence of such disturbances to be forecast with sufficient precision somewhere in areas about 100 by 200 miles (160×320 km), but the inadequacy of the upper air network over large areas of the world prevents the application of these techniques to many affected regions. It is not clear that the techniques themselves are capable of much greater precision than is currently obtained, but in certain particular areas it might be possible to develop more accurate methods, particularly for the precision with which the times when severe disturbances occur, rather than their exact position, can be forecast. Such developments might well be costly and since the economic benefits are to be measured in terms of how many flights are unnecessarily diverted out of these areas, it is by no means certain that substantial improvements over the best current methods are likely to be applied operationally to any great extent in the next few decades. This period is, however, likely to see the production of maps showing the areas of the world where severe disturbances of orographic origin occur, and the meteorological situations giving rise to them. It should, in fact, be possible for complete, and so far as possible continuously up-dated, weather information in map display form to be provided for the pilot; knowing his route in relation to the weather ahead, the pilot should be supplied by traffic control with an appropriately revised flight plan, communicated through his display system.

4.7 *Weather modification*

In recent times, growing interest has been shown in the possibility of weather control, and ambitious programmes are being put forward in the USA[12], and the subject has also been given considerable attention in Russia[13]. The programmes of work proposed tend to be seen particularly in relation to agricultural applications, but are in fact, general enough to have wide fields of possible application, including aviation. Earlier in this chapter, some potential benefits of weather modification, on local rather than global scale, have been indicated in various obvious operational applications. Weather control is not as revolutionary an idea as it sounds, on a local scale at least, because it has been practised throughout the ages, for example, by fruit farmers lighting fires to protect their crops against frost.

Likewise in the last war, FIDO, a fog dispersal system using paraffin burners along the approach path and runway, was used to good effect, and but for the operational complexity and possible fire dangers, might have been widely used in later years. More recently the dispersal of supercooled fog by feeding from above with dry ice has been used operationally in a number of countries and is likely to be more widely used in future. Some success has also been claimed recently in dispersing warmer fogs. Automatic landing will still, however, probably be a more economic method than fog dispersal.

As has been said earlier, thunderstorms are the greatest weather hazard to aviation, and clearly one would like to see a means of thunderstorm dispersal in the future. To some, this is by no means an impossibility, in a limited sense, even in the medium term. The damage which such storms cause on the ground in some parts of the world, is probably a greater incentive to such development than aviation. However, there are roughly ten million thunderstorms in the world each year, each storm converting energy equal to a one megaton (1 Tg) bomb, so this is not likely to be easy. Nor, unfortunately, would it be an unmitigated benefit if there were no more thunderstorms. In many parts of the world, often those in which thunderstorms cause most damage, thunderstorms provide a substantial part of the annual rainfall. Without them, large areas of the world would be desert.

In considering weather modification in general, it is clear that the energy levels involved effectively prevent any kind of global attack. Although vast quantities of energy, by ordinary standards, can become available in time from nuclear sources, it is salutary experience to consider the comparative levels of natural atmospheric energies. The following are some examples[14]:

Weight of 1 cubic mile of air (4·166 km^3)$=5\times10^6$ tons (5 Tg).
Weight of earth's atmosphere$=5\times10^{15}$ tons (5×10^{18} kg).
Kinetic energy of the winds$=10^{14}$ kW hours.

Energy of evaporation or condensation of rain
$= 3 \times 10^{17}$ kW hours per year.
Cost of energy of rain at 1 penny (0·4 NP)/kW hour$=$£10^{15} per year.
Energy of a large (megaton) nuclear bomb
Energy converted in one thunderstorm $\Big\}= 10^{9}$ kW hours.
Cost of a thunderstorm at 1 penny (0·4 NP)/unit$=$£4×10^{6}.
Number of thunderstorms per year$=$10 million.

More subtle means of weather modification will have to be devised, perhaps by catching nature in some kind of state of neutral equilibrium, when relatively little energy could tip the balance in the desired way. Far too little is known about the atmosphere for it to be clear how this might be done, or whether, indeed, such states exist. Before any development is possible along these lines, it will be necessary for there to be 'models' of the atmosphere which can be used to test ideas. The possible ill-effects of irresponsible experiments in weather modification could make atmospheric nuclear testing appear relatively innocuous. If weather modification on a large scale ever does become a reality, it will provide man with a means of good or ill far greater than any he now has. Its effective use for good will demand a degree of international collaboration and agreement well beyond anything achieved to date. For this reason, therefore, it is clearly sensible to think in the first instance in terms of purely local weather modifications. However, it is evident that, even on this scale, enormous energies are involved, when one remembers that one cubic mile (4·166 km^3) or air weighs five million tons (5 Tg). Thus, a frontal attack on a major weather feature, like a depression, is impracticable. However, such features are of the nature of self-developing instability systems, and this leads one to consider whether it would be possible to control the small beginnings from which they grow, by the introduction of more reasonable amounts of heat; even in this case, however, the artificial heat required to raise the temperature by a few degrees to attempt to trigger the depression in a different area, would be the equivalent of hundreds of megaton bombs.

On a smaller scale, the triggering of local thunderstorms by artificial heat sources may be less impracticable. It might be possible to induce thunderstorms in one area rather than another, or prematurely to mature and set them off on a smaller scale. The strong electrostatic charge, which appears an essential element in thunderstorm development, may indicate other possibilities for influencing their growth, such as artificial ionisation at an early stage, or prematurely contrived discharge. If local thunderstorm control in this way were economically possible, it would clearly have attractive applications in reducing thunderstorms and associated dangerous squall conditions, at airfields.

There may be some possibilities for weather modifications in airfield vicinities, but the prospects for *en route* weather control are much less promising. There seem no real possibilities for large-scale effects, and one is led to consider the prospects for temporary local control over disturbances in the aircraft neighbourhood. The ideal would be to smooth out turbulence ahead of the aircraft by energy beams or force fields radiated from the aircraft. This seems a far-fetched idea, but it is interesting to note the recent attention given to the possibilities of influencing supersonic flow fields and sonic booms by radiated electrical fields[16]. In the scheme proposed the air molecules ahead of the aircraft would be ionised through a corona discharge, and the nose of the aircraft charged in the same electric sense. This would produce an electrostatic force field which would influence the flow of air as the aircraft approached. This could, in principle, produce some effect, but only on the non-lifting part of the flow and the consequent reductions possible in sonic boom levels would be limited. On the possibilities of smoothing turbulence ahead of the aircraft, there appear prospects that some effect might be achieved by electro-magneto-hydrodynamic means. It has been shown by Shercliff[15], that vorticity in a fluid can be suppressed by a magnetic field applied transverse to the vorticity, and impressive experimental demonstrations have been given in which vortices behind a bluff body in mercury have been dispersed in this way. Turbulence always involves vorticity in all directions, but damping the components of vorticity perpendicular to an applied magnetic field would reduce the overall level of turbulence, assisted by the natural tendency of turbulence to isotropy. How effective this type of scheme could be would depend on the magnetic field strength that could be applied, and on the power requirements to produce it. Clearly, however, the benefits of such a protective, or smoothing, force-field shield, make it a line of development well worthy of consideration for the future.

There is a converse effect to be considered, and this is that aircraft operations, particularly at high altitudes, may have some effect on weather, perhaps more in the future than in the past. The discharge of carbon dioxide and other combustion products into the upper atmosphere might perhaps have significant effects on the earth's climate. While this does not seem likely, present knowledge does not allow the possibility to be entirely dismissed.

5 COLLECTIVE PICTURE

Taking a summary view, looking into the future, we judge that there will ultimately be air transport operations largely unhampered by weather and insignificantly disturbed by atmospheric conditions such as turbulence, first by avoiding large disturbances, and second by alleviating the effect of more prevalent turbulence. Taking the high speed transport field as the one where most advanced techniques are likely to be applied, we visualise long-range hypersonic craft taking-off and landing vertically in a controlled atmospheric environment, and in a pilot-managed, high-reliability auto-control mode. In cruising flight, it should be possible to fly above the weather, but in the climb to cruise altitude, assuming that full global weather control is impracticable, it appears necessary still to consider flight through turbulent regions, and here it is necessary to look to configurations insensitive to disturbances, possibly in semi-ballistic type flight, or using auto-gust alleviation systems, both to reduce structural loading and to improve passenger comfort. These considerations will be of major importance also for the slower aircraft, which are certain to be in extensive use for shorter range work, and it is in these operational regimes that full detection of turbulence, with comprehensive weather map display for the pilot to enable major disturbances to be avoided, will be most essential. It is here, too, if we can be permitted one visionary speculation apparently beyond the expectation of current technology, that it would be most desirable to have the means to suppress or reduce widespread turbulence ahead of an aircraft, by propagated force fields, producing something in the nature of a turbulence protective, or smoothing, shield.

In this chapter we have not endeavoured to cover the development of aircraft other than in the major transport class, in relation to the atmospheric environment. A few words may be appropriate to finish with, on the future of light and personal aircraft, which are of special interest. Immense advances may be foreseen for ease of flying these aircraft, and in their operating capabilities for vertical take-off and landing. However, they can be expected to be operated in a much less controlled way than the big transports, so that it would seem necessary to provide for their operation in all kinds of weather. They could be equipped therefore, with all auto-aids, and gust detectors and alleviators, and so on, but they would still, essentially, appear a highly refined extrapolation of today's aircraft in their class. The unrestricted use of such aircraft could, of course, result in aerial chaos comparable to the city road transport situation of today, and their private use will probably be limited to uncongested air spaces, but they would still have widespread use as air-taxis, possibly on an auto-basis.

References

1 BURNHAM, J. The influence of weather on aircraft operation. AGARD Report No. 494, October 1964.

2 Atmospheric Conditions. TSS Documents, Standard No. 6. Air Registration Board. March 1966.

3 JONES, R. F. Clear air turbulence. *Science Journal.* February 1967.

4 BURNHAM, J. and SPAVINS, C. S. Gusts on the approach at Gibraltar airfield. Unpublished Ministry of Technology Report.

5 MCPHERSON, A., BURNHAM, J. and NICHOLLS, J. M. A study of mountain wave conditions over the Sierras. Unpublished Ministry of Technology Report.

6 GARNIER, R., THIBORD, L. and DUVERGE, P. Rapports présentes à la

Sous-Commission de météorologie de la Commission permanente du transport supersonique. Direction de la Météorologie Nationale, Paris, 1965.

7 BURNS, A., HARROLD, T. W., BURNHAM, J. and SPAVINS, C. S. Turbulence in clear air near thunderstorms. Unpublished Ministry of Technology Report.

8 BURNHAM, J. and LEE, J. T. A study of turbulence intensity inside thunderstorms and of its relationship to weather radar echoes, based on measurements made in Oklahoma. Unpublished Ministry of Technology Report.

9 ROBINSON, G. D. Some current projects for global meteorological observations and experiments. *Quarterly Journal of the Royal Meteorological Society*, Vol **93**, October 1967.

10 CALVERT, E. S., SPARKE, J. W., SHAYLER, J. S. and MORRALL, J. C. Safety and regularity in air transport in marginal visibilities. RAE Tech. Note Mech. 380. August 1963.

11 ZBROZEK, J. K. Atmospheric Gusts: Present state of the Art and Further Research. *Journal of the Royal Aeronautical Society*. January 1965.

12 Weather and Climate Modifications. Problems and Prospects. National Academy of Science—National Research Council, Washington. 1966.

13 POGOSYAN, KH. P. The air envelope of the earth. Leningrad 1962. NASA Translation 1965.

14 SUTCLIFFE, R. C. *Weather and Climate*. Wiedenfeld and Nicholson 1966.

15 SHERCLIFF, J. A. *A Text Book of Magnetohydrodynamics*. Pergamon Press. 1965.

16 CAHN, M. S. and ANDREW, G. M. Electroaerodynamics in Supersonic Flow. *AIAA Aerospace Sciences*. January 1968.

7 The world of the structural engineer

W. G. Molyneux

I INTRODUCTION

The pace of developments in aerospace technology is probably more rapid than in any other technological field, so that any prediction of the state of affairs for more than a few decades hence is likely to prove unreliable in the fullness of time. Nevertheless, there is some value in the exercise, particularly, as is generally the case, when such a prediction is based on an extrapolation of established trends, for the foundations are sure, providing a reliable guide to the near future and, although prediction and practice inevitably diverge the farther we go into the future, there may be something to be learned from the 'might-have-been' of the discarded path.

In surveying aircraft structural progress during the past century it is apparent that the process has been largely one of relatively slow evolution, halted or accelerated at particular times by a revolutionary development.

For example, the structure of the earliest man-carrying, heavier-than-air machines can be said to have evolved from the braced-frame structures of man-lifting kites, whose origins are lost in antiquity, but revolutionary developments were also involved in the application of cambered wing sections for aerodynamic efficiency and the internal combustion engine as a source of power, which accelerated the pace of subsequent development. In meeting the demand for ever-increasing speed the wood and wire structure of the earliest designs was soon over-taxed, to be replaced progressively[1] by the steel braced-frame biplane structure; the two-spar, fabric covered monoplane; the two-spar metal covered monoplane; and the stressed skin, light-alloy structure that is the dominant structural form today. Many revolutionary developments in aerodynamics and powerplant were associated with this progression, but from the structural engineer's viewpoint it was the revolutionary development of the stressed-skin concept that virtually halted the evolution of the two-spar, braced-frame structure and necessitated a fundamentally new design philosophy, while the associated development of the high-strength aluminium alloys

Originally published in the *Aeronautical Journal of the Royal Aeronautical Society*, July 1968.

and fabrication techniques for sheet metal working provided the means with which to fashion this new structural form without excessive growth in the ratio of aircraft structure weight to all-up weight.

There has followed more than two decades of evolutionary development of this structural form, in course of which complex structural problems in the aeroelastic field have been recognised and provided for, fatigue problems, with the attendant necessity for closer attention to detail design, have been accommodated, and cabin pressurisation problems and a host of others have all been taken into account. It is perhaps fortuitous that this same period has seen the evolution of the electronic computer from nestling to fledgling, for it is almost inconceivable that the multitudinous calculations for the design of today's aircraft could be accomplished in reasonable time without its aid.

Nevertheless, computer development itself is still in its infancy, with a course to run that will have major consequences on every sphere of human activity, not least on the refinement of aircraft design and fabrication where only the surface of the potential contribution has so far been scratched. At the same time the 'traditional' steel and aluminium alloy materials that have been the mainstays of the stressed-skin fabrication era have also evolved, the major improvements in properties that have been achieved reflecting the growth in skill and understanding of the metallurgist. Fully weldable, high-strength, aluminium alloys and ductile, ultra-high-strength steels appear to be almost within the metallurgist's grasp, while at the same time, by the exercise of a variety of skills in the physical sciences, a whole range of new materials is beginning to emerge. These materials have the potential to accommodate the more exacting environment of the second century of flight and to revolutionise aircraft structural design, fabrication and maintenance techniques.

In what follows an attempt is made to extrapolate these current trends, so as to anticipate the course of future developments insofar as they are likely to be reflected in the design of aircraft structures.

2 CONSEQUENCES OF THE COMPUTER

The second century will inevitably be marked as the era of the computer, particularly so far as aircraft structural design is concerned. In the main the existing application of computers[2] is as a more grandiose desk calculator that enables problems of greater magnitude and complexity to be attempted than would otherwise be feasible to a realistic time-scale. This activity represents the evolutionary phase of development in which the analyst has adapted to a more powerful aid performing a task with which he is essentially familiar. However, even in this task the computer is having manifold effects, particularly in relation to the evolution of analytical techniques, for of necessity attention is becoming focused on to those that most readily lend themselves to computer applications.

So far as current trends can be anticipated it appears that within the next few years finite element methods of structural analysis will become the norm, for they are well suited to the capabilities of the computer in which matrix techniques can be used to assemble and solve simultaneous equations with some thousands of unknowns. Furthermore, these methods are flexible in their applications in that a single set of programs can suffice for the whole range of problems, from the detailed three-dimensional analysis of some local stress concentration, to the major internal load and deflection analysis of a complete structure. Recent developments[3] have been towards the extension of these techniques from static problems to dynamic ones, from small deflection problems to those of large deflections, from elastic analysis to problems of plasticity, and to other non-linear problems in the structures field.

The object of design, of course, is to obtain a structure that is an optimum for the specified loading environment, but in seeking this optimum the designer has only a limited freedom of action. There are practical limitations imposed by factors such as local manufacturing capacity, minimum skin gauges, material properties, cost, ease of manufacture, maintenance and inspection and a host of others. Insofar as the designer arrives at the final design by a logical process, in which due weight is given to these various factors, and their consequences assessed in relation to some form of merit function for the structure, there is no doubt that the process can be codified and programmed for the computer. Indeed the effectiveness of a simple semi-empirical iterative technique for the redesign of an aircraft structure to obtain a least-weight design approximate has already been demonstrated[4,5] when minimum gauge thicknesses are specified for the skins; more rigorous optimisation methods[6], appropriate to any mathematically definable merit function, are now under development using mathematical programming techniques. Currently these techniques are too demanding in computer time and capacity to be economical for other than the crudest optimisation, but a breakthrough in the future seems assured.

A further field for computer applications is in reducing the tedium of the design process, and one such area that is receiving attention is that of structural idealisation and analysis interpretation. For the overall analysis of a fully schemed structure it is impracticable to attempt to analyse the true geometry in all its details. In consequence analysis is preceded by the time-consuming chores of idealisation and interpretation, in which the real structure is first represented in some idealised form that is tractable for analysis and the results are then interpreted for the real one. These tedious chores are demanding of the skill, art and experience of the stressman and cannot be relegated.

A more promising approach for the future that has been discussed by Taig[5] is to provide for the definition of the geometry of the real structure in compact, numerical terms, to define the finite element grid in relation to the basic structural surfaces, and to determine the behaviour of the real structure within regions bounded by grid lines by a substructure analysis representing all significant features of the real structure. With the development of such an approach it would be possible to obtain the local stresses at different parts of the real structure within an idealised element in a fully consistent manner, with a consequent improvement in the precision with which results are interpreted; but many other benefits would also derive from an integrated approach along these lines.

For example, the combination of a procedure that defines member sizes that are optimum for the local conditions in association with a concise numerical definition of geometry for any area of the structure, available on demand, has a potential for virtually eliminating the conventional drawing office and lofting stages of design, while also providing a more powerful design tool.

However, much depends on the versatility with which this information can be presented to the engineer, and in this respect it has been recognised that one of the most digestible forms of presentation is by graphic display. Some indication of the potential of the developing field of computer graphics is provided by the current activity with devices such as 'Sketchpad'[7]. Here, not only is a particular numerical definition of a structure displayed, with the numerically defined points linked with appropriate lines and curves to re-create the outlines of the structure, but any chosen cross section of the structure can be displayed, as a plane section or as an isometric or three-dimensional view. Furthermore, by using a 'light-pen' to which the screen for graphic display is sensitive, it is possible for the engineer to communicate geometrical information directly to the computer by way of the display screen, to modify the geometry as required while retaining a record in the computer store of what has taken place. With a sufficiently precise and detailed definition of structural geometry for the finalised design available in the computer store it is a relatively short step to program a read-out from the store in a form suitable for input to numerically-controlled machine tools, and this effectively short circuits many of the stages between the design office and production.

At the same time, with the ever increasing storage capacity of modern computers, the widening communication links from one computer to another, and with provision for a continuous flow of relevant information from other sources, the computer can take into consideration not only the local production facilities available on a day-to-day basis, but also those for a group of firms and perhaps ultimately, the national or even international capacity. Since the numerical definition of geometry is immutably enshrined in the computer store the information can readily be routed to wherever there is spare capacity, or to where the job can be done with most economy; indeed the necessary financial negotiations and contractual agreements could also well be decided automatically by computers interrogating each other!

The computer reference is again available when the dimensional accuracy of parts is to be checked on inspection, although special inspection equipment and techniques may need to be developed for the purpose. Certainly with the facility for a precise control of geometry throughout the manufacturing process, and with a common standard of reference, mismatch on assembly of components manufactured in different areas should largely disappear.

The intimate linking of design and fabrication by way of the computer will also lead to streamlining in other areas of production engineering; for example, raw stock and spares control problems, progress chasing, etc, are all amenable to 'computerisation', and will be more intimately co-ordinated with production than at present.

Material selection and the updating of materials data is another problem area for the designer to which the computer will bring relief. By the nature of things the quantity of data on materials properties is continually increasing, new alloys are continually being developed, information on costs and sources of supply are continually changing, and the designer is often not inclined to be venturesome because up-to-the-minute information is rarely to hand. We are at the beginning of the codification of information of this kind for storage in the computer and of the development of an appropriate method of communication between designer and computer that enables an evaluation of the merits of different materials to be made in relation to the design problem, so that the most suitable materials are automatically selected.

The application of the computer to the dimensional inspection of structures has already been mentioned, but the computer might also prove to have a more general application to the inspection of aircraft in service.

Most inspections in the field consist of an examination of a component in an attempt to detect some significant change in its condition, but this implies a memory of its previous condition, and human memory is far from infallible. The computer is more reliable in this respect, but the difficult problem of quantifying the features that are looked for by a skilled inspector, in a form suitable for the computer, is far from being solved. As an example of the way in which computers might come to be used in this field, the whole aircraft might be subject to X-ray photography under controlled conditions at intervals during its life, the photographs digitised in relation to density distribution over a specified area, and the history of such records maintained in the computer store. Structural deterioration from, say, corrosion might then show up as a significant change of relative density, detectable by computer analysis.

Another area of application is to the analysis of structural defects in service. The problem is to provide for the recording of defects in a form suitable for computer analysis so that rapid surveys can be made at intervals throughout the operating life to detect statistically significant defects at an early stage and initiate the necessary design modifications, or other corrective measures. The problem is more in association with minor defects than major ones, since the latter are unlikely to be ignored for want of analysis, but the contribution of the computer to the detection and elimination of minor defects will do much to eliminate a source of irritation to the operator.

3 STRUCTURAL MATERIALS

The general consideration of materials for the second century of flight is the subject of Chapter 15 of this book by Kennedy. However, it seems appropriate in this chapter to give some independent consideration to those materials which, in the author's view, will have a dominant role as materials for aircraft constructions, since design and fabrication tehniques are intimately related to the material to be fashioned.

The engineer is all too aware that every use of a material involves a compromise, in which the different properties of the various materials available are weighted appropriately in relation to the design environment, and a proper selection made by balancing one factor against another. With an increasing understanding of the parameters that determine the physical properties of materials, and an improved ability to manipulate them, there is every likelihood that it will prove possible in the future to produce materials to order with the optimum compromise of properties appropriate to the environment, but it is interesting to examine the progress along this path.

The history of the development of aluminium alloys and steels, which for some decades have been the dominant metallic materials used in aircraft structures, is a catalogue of improvement of those material properties of first concern to the designer. Ultimate strength in particular has been vastly improved over the years, both in wrought and cast materials, and a value that is one per cent of the elastic modulus is now commonplace in steels while a similar value seems within sight for light alloys. Although achievement of the theoretical maximum strength of some 10% of the modulus is still remote, at least for bulk materials, the ingenuity of the metallurgist is far from exhausted, and alloys with strengths perhaps double those of existing ones are a distinct possibility for the future. Of course, while materials with high static strength will continue to be much sought after, the environment for an aircraft structure is often such that other properties of the material are of equal or even greater concern. For example, a trend that is likely to continue in the future is towards demanding ever-increasing aircraft life, and this in turn emphasises those properties of the material that are significant from a fatigue viewpoint, or that are time-dependent, such as fracture toughness, crack propagation and stress corrosion characteristics, or creep and ageing properties. Similarly, the trends towards increasing flight speeds and associated higher temperatures at one end of the scale, and exotic fuels that require a cryogenic environment at the other, lends emphasis to the effects of temperature on material properties.

Where high temperature effects are concerned an ultimate limit for load carrying is determined by the liquidus temperature of the material, but in any given instance the limit of viability is determined not so much by this temperature as by the temperature at which a rapid reduction of strength with temperature occurs, and by the material's ability to compete with other materials with different high temperature characteristics. In these respects current aluminium alloys seem somewhat poorly placed, their limit of viability occurring at about one third of the liquidus temperature whereas for certain steel and nickel alloys the limit is nearer half the liquidus temperature.

It seems likely, however, that continuing research into aluminium alloying techniques will lead to improved properties in this respect. The conventional solid solution and precipitation hardening techniques restrict the range of alloying constituents that can be incorporated, but techniques now emerging from the laboratory such as splat cooling, powder mixing and compaction, controlled internal reactions and gas phase processes, markedly extend this range and whole classes of new alloys must result. One can only hazard a guess at their potential, but improved high temperature strength is one almost certain development and improvements in other respects, such as fatigue, creep and weldability properties can be predicted.

Suffice it to say that alloys of steel and aluminium which became the mainstays of aircraft structures in the first century of flight have far from run their course and will continue to find a place for some decades to come.

However, while recognising a continuing future for the 'conventional' aircraft structural materials, some thought must also be given to the prospect for other materials with a growth of potential such as may conceivably lead to their virtually ousting the conventional materials in airframes of the more remote future.

So far as metallic materials are concerned the most likely prospect for general structural fabrication, in both subsonic and supersonic roles, lies with the alloys of titanium. Within the space of two decades titanium has developed from the status of a relatively exotic material to one in which production is currently measured in some tens of thousands of tons per annum, and its widening use in aerospace and general engineering industries must lead to reductions in raw material costs that will widen the areas of application in which it is cost effective.

Titanium alloys already account for upwards of some 3% of structure weight on certain current civil transports, and design studies have shown that, provided full use can be made of the seemingly excellent weldability of certain of these alloys, weight savings of up to 25% of the equivalent aluminium alloy structure might be achieved. In practice, of course, a 'snowball' factor would be associated with structure weight savings of this order; for example, for the same payload it would be possible to use less wing area, less fuel and possibly less powerful engines, all or any of which would further contribute to overall weight saving and reduction in direct operating costs. Material cost, supply and problems of fabrication are the main factors currently inhibiting a more extensive use of titanium, but in all these respects the situation is improving rapidly and if this trend continues it must result in a period, beginning in the near future, in which there will be a general use of titanium alloys for major airframe fabrication.

Of course, a prognostication of this kind is scarcely an imaginative exercise, for production military aircraft of dominantly titanium structure have already been built. Furthermore, as flight speeds increase to the point where temperature considerations preclude the use of unprotected aluminium alloys and where titanium is in direct competition with steel and nickel alloys, there is a significant range of Mach number in which the case for titanium seems overwhelming, even by today's standards; as evidenced by the extensive use of titanium proposed for the Boeing supersonic transport. Since titanium alloys are at an earlier stage of development than those of either aluminium or steel, and might thus be regarded as having a bright future potential, the advantage of these materials has every likelihood of increasing as the years go by.

In stretching the imagination to other metallic materials with a future potential, one is inevitably attracted by the outstanding specific characteristic of beryllium, with the attendant promise, on theoretical grounds, of an associated high specific strength. There is already a significant use of beryllium for space applications, where the high cost of the material is more readily justified and where features such as poor fracture toughness may be of lesser significance than in general airframe applications, but attempts to develop beryllium alloys for the latter field have not yet met with much success.

The continued development of novel alloying techniques could well lead to more acceptable alloys in the future, and the consequences of such a development would be of major significance, for the weight saving potential of beryllium could well amount to some 50% of that of the equivalent aluminium alloy structure; so gross a factor on weight implies that in any extensive structural application of beryllium a different form of structural lay-out would be used as compared with that in aluminium alloy.

Mention has already been made of the strength that can theoretically be developed by a material, but the fact remains so far as bulk (i.e. largely isotropic) materials are concerned, the strengths that can be attained are an order of magnitude removed from this ideal. No doubt the future will see continuing progress towards the ideal, but it is worth considering alternative solutions to this problem.

For some decades it has been recognised that higher strength than is exhibited by the bulk material can be obtained when the material is in the form of fine wires or filaments, and is a consequence of reducing the incidence of surface damage or defects from which failure originates. Early examples of the application of this philosophy are found in steel 'piano' wire and Taylor wires, while the outstanding example of recent years is provided by continuous filament glass where filaments with a specific elastic modulus (i.e. modulus÷density) some 20% greater than that of aluminium alloy and a tensile strength that is some 4% of the modulus are commercially available. Glass is particularly noteworthy in this respect since, in general, bulk glass is regarded as having a very low tensile strength (perhaps some 0·1% of the modulus), dictated by the extreme notch sensitivity or brittleness of the material.

The future will undoubtedly see some further improvement in glass filaments, both in specific modulus and specific strength, but there are also other contenders in prospect. Beryllium wire, for example, offers prospects for a significant improvement in strength as compared with bulk beryllium, and similar results can be obtained from fine wires of most of the conventional metallic materials, i.e. steel, titanium alloys, nickel alloys. However, it must be recognised that there are many problems and difficulties in utilising filamentary materials in structural fabrication, and it transpires that if strength is regarded as the main criterion then only those materials whose filamentary strength is some two or three times the strength of more bulky forms (sheet, forgings, etc.) merit consideration. Glass filaments satisfy this criterion handsomely, and there is the further feature that a composite based on glass filaments (e.g. one with glass filaments in a resin matrix) has other properties quite different from bulk glass; for example, the notch sensitivity is generally a vast improvement on that for glass. Features of the latter kind can, of course, lend peculiar attraction to filamentary forms of materials in particular circumstances.

A continuous filament material of more recent origin than glass is carbon fibre, generally produced from the graphitisation of commercial textile yarn. It is worth noting that carbon holds a unique position among materials in that it has the highest absolute, and specific, modulus of all—a specific modulus that is some twenty times greater than that of aluminium alloy—and it has an element of uniqueness too in many other respects, for example, its theoretical strength should increase with temperature, its melting point is the highest of the elements and, in the single crystal, diamond, form it is one of the hardest materials known. There is evidence too that a measure of control can be exercised over the fibre modulus, simply by controlling graphitisation temperature, resulting in a 'family' of carbon fibres. In this material more than any other would seem to lie the prospect for the 'ultimate' material for aircraft structures of the future.

So far as bulk carbon is concerned current technology is at the very beginnings of progress along this path, but in the form of continuous filament a material with a specific modulus some nine times that of aluminium alloy and a strength of 0·5% of the modulus is already available commercially. As more is learned of carbon filament technology further major improvements in strength and modulus must ensue; within a decade or two we may anticipate that carbon filament will be available commercially with a specific modulus some fifteen times that of aluminium alloy and a strength approaching 1% of the modulus, for these values have already been achieved in the laboratory.

However, while filamentary materials commonly provide a higher strength than for similar materials in bulk, the highest strengths that have currently been achieved are for materials in the form of 'whiskers' that comprise single crystals, where the possibilities of surface defects are at a minimum; single crystal strengths of some 3% of the modulus have been obtained for a variety of materials.

Attention has accordingly been concentrated on techniques that enable the production of materials in single crystal, whisker form and it has been found that in many cases whiskers can be obtained from an environment determined by the substrate on which the whisker is grown and the atmosphere in which it grows, which must be of an appropriate chemical composition and at an appropriate temperature and pressure. Single crystal whiskers only a few microns in diameter, but millimetres or even centimetres in length, have been produced for several materials.

Perhaps the most promising of these are whiskers of silicon carbide, aluminium oxide, boron carbide and beryllium oxide, all of which have a high melting point and some of which offer the prospect for commercial production at reasonable cost. All of these also have specific moduli much higher than for aluminium although none approaching that of carbon, but their specific strengths are higher because of the far greater achieved strength in the single crystal. Against this must be set the problems of 'harvesting' and sorting a crop of whiskers grown in this way, and subsequently of utilising a material in this form in structural fabrication. Commercially viable processes are not yet in sight for these activities but there is every prospect of their development in the future.

At the same time the facility for depositing materials on to a substrate from an appropriate atmosphere by vapour phase depositing has applications outside the whisker field, and is being utilised currently to produce a boron or silicon deposit on a continuous filament of some other material with attractive properties compared with other filaments currently available commercially. There are possibilities for the development of this process to obtain bulk materials with similar properties.

Suffice it to say that so far as materials of the future are concerned the bulk forms of material with which designers are familiar will be under increasing pressure from materials in continuous filament and whisker form, and the technology of design and fabrication must accordingly evolve to keep pace with this development.

4 THE UTILISATION OF MATERIALS OF HIGH STRENGTH AND STIFFNESS

In the foregoing discussion much emphasis has been given to materials with an elastic modulus in excess of currently available materials, and to the possibilities for materials with improved strength, more closely approaching the theoretical maximum strength of 10% of the modulus, but the structural advantages of such materials are not immediately apparent.

To develop the available strength in a material a strain must be imposed, and this implies deflection of the structure under load, but the structural deflections that result when a general level of strain of 0·1 is developed are so gross that there are few practical structures where they could be accepted. This feature is readily appreciated by considering a uniform beam of length πD and depth $0{\cdot}1D$ subjected to a uniform bending moment at every section: when the strain in the outer fibres is 0·1 the beam is deflected into a complete circle, its opposite ends meeting!

Because of deflection limitations the average level of elastic strain that is developed in current structures rarely exceeds 0·005, and while it is conceivable that there will be a greater variety of structural geometries in the future where higher levels than this could be developed (as, for example, in inflatable space structures) a figure of this same order is likely to continue to apply to the bulk of future structures, at least so far as the change of strain between loaded and unloaded conditions is concerned.

Where then lies the attraction of materials with greatly improved strength and stiffness?

Obviously an improvement in strength of a given material without a corresponding improvement in stiffness, implies that the higher strength can only be realised at an increased level of strain, and limits in this respect are set by the structural deflections that can be tolerated. Furthermore, although less of the stronger material is required to support a given load so that weight saving seems feasible, to achieve this simply by reducing material thickness would result in a structure of lower stiffness, with possible accentuation of structural instability and aeroelasticity.

However, a distinction can be made between the utilisation of higher strength throughout a structure and in local regions. In practical structures strain levels vary over a wide range, from a minimum at tip regions to a maximum in areas such as wing-fuselage intersections, and in areas such as the latter a locally increased level of strain can often be accepted. There are also other ways in which such a material may be used to advantage. One of these is in relation to fatigue, for a feature of structural materials is that the smaller the ratio of working stress to ultimate stress the greater the number of cycles of loading that can be sustained before fatigue cracks develop. The fatigue life of a structure is determined by this interval, together with the interval in which fatigue cracks propagate to the point at which the structure becomes unsafe, and the latter interval should be such that, on the basis of vigilant inspection and maintenance procedures, crack initiation can be detected and remedial measures initiated before a structural failure occurs. However, an unfortunate feature of many of the materials with high specific strength is that they are brittle, with little ductility and an extreme notch sensitivity, and fatigue life following crack initiation is very short; this is particularly true of materials in whisker form, although generally less so for a whisker or fibre composite because of beneficial effects from the matrix. Hence, the main fatigue life of a component fabricated from ultra high strength material will be determined almost entirely by the interval required for the initiation of cracks. The average level of working stress for these materials is likely to be so small a fraction of the ultimate, because of practical limitations on usable strain, that an infinite life might be expected in ideal circumstances, but possibly only at the expense of great care in the fabrication and handling of the structure and in protecting it in service to maintain a high quality surface, free from any defects from which cracks might subsequently emanate.

A crack leads to failure because of the high stress concentration that it induces in the material, but less severe stress concentrations are also induced by other forms of geometry as, for example, at rapid changes of section, joints and cut-outs. In areas such as these much higher levels of stress are encountered than is the average for the structure as a whole; for example, in a multiple bolted or riveted joint in current materials the proof strength of the material may be exceeded locally, with subsequent local yielding or plastic flow of the material and indeed, the designer will often be dependent on a phenomenon of this kind to even-out the stresses in the joint so that its full strength can be developed. A similar mechanism is not available for joints in a material whose behaviour is essentially elastic up to failure, but provided the strength margin of the material is adequate many of these forms of stress concentration can be accommodated elastically, rather than plastically. However, since every local area of stress concentration is of prime concern in relation to the origination of fatigue cracks, it seems likely that for structures designed from these materials of the future, far greater attention than hitherto will need to be paid to the intimate details of the structure with a view to optimising the geometry to minimise the local stress concentrations. The powerful analytical techniques for problems of this sort when computer-aided design achieves full flower makes such an approach feasible.

It also follows that because areas of stress concentration are potential sources of fatigue cracks for structures in general, there will be a major emphasis, particularly where brittle materials are used, towards the elimination of joints, cut-outs and the like, aiming for a structure that is as near to a monolithic whole as repair and maintenance considerations will allow.

Perhaps the most prolific source of cut-outs in existing passenger-carrying aircraft is the windows, along the length of the fuselage at a spacing determined more by seating arrangements than the structural considerations, and which seem to invite the structure to 'tear along the dotted line'!

The justification for windows, even in current day aircraft, seems tenuous in the extreme. They are probably the least functional item of the aircraft structure, serving only those passengers immediately adjacent to them, and in many cases (particularly for slender wing designs, and other of the advanced planforms described by Kuchemann[8]) affording a view of nothing more exciting than a vast expanse of the upper surface of the wing. The size of the windows is apparently determined more by aesthetic considerations or airworthiness requirements than by passenger demand, and range from 370 sq in (0·24 m^2) of the Viscount windows (perhaps the largest ever fitted) to as little as 24 sq in (0·015 m^2) for the windows of some current designs.

With windows such as the latter, one is bound to ask whether or not a viewing port less than six inches (0·15 m) in diameter, offering a restricted view of an often uninteresting scene to only half the passengers, and of value generally only in daylight, is commensurate with the increased design complexity, structure weight and cost of providing it.

There appears to be no single answer to the question, 'why windows?' other than that windows are demanded perhaps as a matter of tradition, by the operator. From the customer viewpoint it may be that he likes to see the terrain over which he flies, but the point has already been made that the window is often ineffective in this respect, and the objective would probably be better served by a television screen fed from a well placed scanner on the undersurface. Indeed, since even this view might be restricted by weather conditions, the alternative of using a pre-exposed, technicolour film with close-ups of points of interest *en route* and synchronised to the flight path, as in the moving map navigational system, also seems attractive, particularly since a system for projecting films during flight is already installed in some current aircraft.

It is possible, of course, that windows are considered essential by some passengers if for no other reason than that they relieve claustrophobia, but are claustrophobics so significant a fraction of the population as to merit special accommodation and, if so, can a sufficiently complete illusion of a window be created, replacing it with a back-projection or television screen, or even a mirror?

Certainly there seems little argument for retaining passenger windows in the transport aircraft of the future, and it seems likely that even the pilot's window is ultimately destined to disappear. The current head-up display system projects an image of vital instruments on to the pilot's transparency, but could almost as readily project an image of the pilot's normal field of view on to an opaque screen by way of suitable scanners in the nose. This would also offer vision prospects outside the normal visual range, such as infra-red scanning in an approach through mist or cloud. High reliability would be required of such a system but the problems are surmountable now. Furthermore, the pilot is likely to prove the most amenable to changes in this direction as he is already accustomed to dependence on his instruments in flight (often without external vision) and instrumented vision would probably be accepted. With the continued development of automatic ground-controlled landing techniques and satellite-controlled navigation, direct vision must ultimately come to be regarded as a standby for emergency use only.

The elimination of windows would have a marked effect on the efficiency of the fuselage structure, mainly because of the simpler design resulting from the elimination of the load diffusion problems around the window cut-outs, but also since frame spacing would no longer need to be determined by seat spacing considerations; a structure of significantly lower drag, weight and first cost should result, with lower maintenance and operating costs in service and potentially of greater safety.

A particular source of stress concentration and cut-out problems on current structures is the aircraft undercarriage, and because of this it seems unlikely that the conventional form of undercarriage will be acceptable for the carriage structures of the future. Some way of distributing the undercarriage loads over a wide area of the structure, rather than concentrating them at one or two points in the neighbourhood of the undercarriage stowage cut-out, is required and one direction of possible development is indicated by current experiments on the application of the air-cushion principle for undercarriages. Also the pressure to remove airports from major centres of population, and the pressure on land for housing and agriculture, may result in a revival of reservoir, lake and ocean terminals, with the return of waterborne aircraft that distribute their landing loads over the hull.

Further areas of load concentration occur at engine and control surface pick-ups, and these may largely be eliminated by devices such as distributed propulsion (surface combustion) and the jet flap. But it would be unreasonable to suppose that all the cut-out areas and stress concentration points could be eliminated entirely. Passengers and freight, for example, cannot be spirited into the aircraft, so that some cut-outs for access doors will always be needed for these.

Suffice it to say that the aircraft of the future will have far fewer cut-outs and areas of stress concentration than those of the present, and particularly where high strength, brittle materials are used, it will be fabricated monolithically. However, a structure of the latter kind, which depends for its safety on clever design to eliminate or inhibit development of cracks, but which is short lived should a crack occur, will require the parallel development of a most sophisticated maintenance and inspection technique, including a system of operational flight recording and monitoring of the structure so that it is under surveillance for crack initiation at all times.

A further way in which a reserve in material strength might be utilised of which little advantage is taken in current structures, is by pre-stressing; this is the process by which a self-equilibrating stress pattern is imposed on either the material or the structure over and above the pattern induced by the external loading, with a view to improving the efficiency of the structure. A familiar use of pre-stressing is the pre-tensioned reinforcement in the reinforced concrete structures of the civil engineering field, while for aircraft structures shot peening, to induce a compressive stress in the surface of a material, equilibrated with a tensile stress in the interior and interference fits, are other forms of pre-stress with structural advantages in particular circumstances; a familiar self-equilibrating stress field for a complete structure is that due to thermal effects when a structure is subjected to changes of temperature.

Of particular interest in the present context is the pre-stressed condition that results when glass is chemically toughened, by a process that involves ion exchange at the surface, in which a large ion replaces a smaller one. The effect is to induce a high level of compressive stress permeating a thin layer at the surface of the glass and this is equilibrated by a much lower level of tensile stress in the interior. Provided a level of surface compression is maintained when the glass is under load, surface blemishes that do not penetrate the compressive 'skin' are inhibited from developing as cracks, enabling the glass to be used to a higher level of tensile stress than would otherwise be feasible and reducing susceptibility to failures originating from minor surface defects. Glass panels toughened in this way have already developed specific tensile strengths comparable with those of the stronger aluminium alloys, and further extensions of this process could result in a bulk form of glass with a usable strength approaching that of the material in fibre form. Toughened glass aircraft structures are thus well within the bounds of future possibilities.

Although the amorphous structure of glass has little in common with other materials, somewhat similar benefits might be anticipated could a like process be extended to some of the other materials mentioned earlier, particularly those susceptible to premature failure due to surface defects. The possibilities in this respect are indicated by the 'metalliding' process that has recently been developed by the General Electric company of the USA, in which diffusion of one material into the surface of another is achieved by electrolysis using a molten fluoride salt as electrolyte. The process is currently aimed at providing protective coatings for materials and some outstanding results in this respect have already been achieved; the levels of surface compressive stress that can be developed have not yet been investigated, but a significant level of surface compressive stress is technically feasible.

Supposing that a level of pre-stress in compression could be provided over the whole of the free surface of a structure, including free edges of plates and cut-outs, such as to inhibit tensile stress at the surface when under load, then not only would a marked improvement in fatigue properties be anticipated but, such problems as stress corrosion should also be alleviated.

It is apparent then that there is advantage to be gained from a material of increased strength with unchanged stiffness, but what of the reverse situation? Obviously an immediate advantage accrues from an increase in material stiffness in those situations where strain in the material, rather than strength, is the limiting factor; and even in strength-limited structures there are advantages of increased stiffness, particularly for stability designed regions, where scantlings can be reduced with consequent weight saving, and in relation to aeroelastic requirements. Of course, it is when both material strength and stiffness are increased that there is the maximum potential benefit.

A point to be borne in mind is that since virtually the entire aerospace environment has already been penetrated by structures fabricated from the materials of today no more severe environments are likely to be encountered by the structures of tomorrow, and the emphasis is thus likely to be on improved structural efficiency in terms of weight, life, safety and cost.

Mention has already been made of potential savings in weight and improvements in life for monolithic structures with a level of surface pre-stress, and for materials where both strength and stiffness are increased there is a further direct potential reduction in weight from the reduced cross sections needed to carry the load. In the latter respect it should be noted that there are often areas of current structures designed by what are loosely called 'minimum skin gauge considerations', which relate to factors such as the minimum gauge in which the material can be obtained and that can reasonably be expected to withstand the workshop and service environment; much thinner gauge materials than hitherto will be used in the structures of the future and fabrication procedures will need to be developed with this in mind.

5 STRUCTURAL FORM AND FABRICATION

Kuchemann[8] has already indicated that the shape of things to come may differ widely from current aircraft shapes, but to a large extent the problem that will confront the designer will be unchanged; namely, how is a specified volume to be contained, a specified envelope of loads to be supported and safety to be assured with the maximum economy in material?

The stressed-skin structural form offers the best prospect for an optimum solution to this problem, at least so far as manoeuvring atmospheric vehicles are concerned, and hence it is anticipated that this structural form will be dominant in the future with little other than evolutionary developments in design taking place. Improved ways of stabilising the structure, to enable ever higher levels of stress to be achieved in ever thinner sheets without instability failures, is one area of development, with some form of sandwich construction likely to become dominant, following the development of infallible production and service techniques of non-destructive testing and inspection.

One area that has perhaps been little exploited up to the present time is in relation to pre-stressed structures, and developments can be anticipated here both in the alleviation of stress concentrations and in the avoidance of structural instability. Mention has been made of certain aspects of pre-stressing in the previous section, and there may be other possibilities, such as in the realm of inflatable structures, where different structural forms may result.

A further possibility that arises, particularly when there are gross weight savings to be achieved by using a material in a specific rather than a general way, is in relation to the concept of structures-within-structures. As an example, we may consider a fuselage shell which must not only withstand landing, taxiing, take-off and manoeuvre loads, but also the loads due to cabin pressurisation. Furthermore, because fatigue is a dominant consideration for the pressure cabin, a lower level of working stress may need to be accepted than would otherwise be the case. However, taking the pressurisation loading condition in isolation, a filament wound structure might prove to be ideal, resulting in a highly efficient structure for this component, while at the same time removing pressurisation loads from the outer shell may also lead to improved efficiency; overall, the dual structure may prove better than the single one.

A form of dual structure that is currently being used, for aircraft that are subjected to a kinetic heating environment that is too severe to be borne directly by the load bearing structure, has an outer structure which supports the aerodynamic pressure loads and the thermal environment (by combinations of ablation, insulation and cooling), while an inner structure, protected from the thermal environment, carries the major structural loads.

Features of some structures of this type have been described by Heldenfels[10]. The problems for the outer structure are more in the realm of materials than structures, although ingenuity in design is required to minimise heat conductance paths, and to maximise external radiation and heat transfer to cooling media. Future developments here seem likely to be in evolution and optimisation of existing techniques. However, the applications of magnetohydrodynamic effects to control the external flow so as to minimise heating has been mooted from time to time, and an economic technique for achieving this objective is a possible future development. For the internal structure stressed-skin construction is currently favoured, but a geodetic, braced-frame, fibre-reinforced composite structure could ultimately prove to be a more efficient structural form where there is a separate outer shell structure supporting the pressure and thermal loading. This is simply a recognition that materials such as fibre-reinforced composites can be highly efficient when designed for a unique loading condition, but their efficiency deteriorates rapidly when a general loading condition must be provided for.

However, it is in relation to techniques of structural fabrication that future technology is likely to differ most radically from the present. A feature of many of the materials discussed is that they will be far more expensive and less tractable from a fabrication viewpoint than materials that are currently available. Hence, it behoves the designer to consider possible ways of achieving economy in materials and alternative ways of manipulating them in future fabrication.

A characteristic of many of the current techniques of fabrication is that they are profligate in the use of material, ranging from perhaps 5% material utilisation for an integrally milled skin panel, to possibly 30% utilisation for riveted, skin-stringer structure; there are relatively few areas where a higher utilisation than the latter figure is currently achieved. This situation is economically viable where material scrap value is a significant fraction of prime material cost and where the material itself can be machined easily, but these considerations will not apply to many of the materials of the future. Indeed, so far as machining is concerned, it has already been suggested[9] that by the early 1970s some 70% of the materials used in the aerospace industry will be in the 'hard to machine' category, and no doubt many of them will also be in the 'expensive' category.

However, there is plenty of evidence that the industry is girding its loins to deal with this situation, and an astonishing variety of novel techniques of metal working, shaping and forming, supplementing the conventional techniques are finding commercial application; as, for example, high energy rate forming using kinetic energy, hydrostatic or pneumatic pressure, explosives or magnetic fields; joining using electron, laser, plasma or ion beam welding, diffusion bonding and adhesive bonding; material removal using chemical, electrochemical, ultrasonic, plasma torch and spark machining techniques. Many of these techniques are comparatively recent and their full potential has yet to be developed; they will undoubtedly be taxed to the full in structural fabrication in the future.

Special problems are posed by those materials that must be utilised in a filamentary or whisker form. For these to form a viable material of any kind they must first be incorporated into a matrix of some kind, through which load is transferred by shear at the fibre-matrix interface. A variety of synthetic resins has been developed which serves this purpose, these having the advantage that while the resin is in the pre-cured state the fibre-resin composite can readily be manipulated into complex shapes, so that surfaces with multiple curvature can be fabricated readily. Furthermore, thanks to the ingenuity of the chemist, resins are currently available that can be used for extended periods of time at some 400°C and for periods of some minutes at temperatures up to 600°C; we may confidently anticipate further developments along this path in the future, with synthetic resins making progressive inroads into those high temperature regions that are traditionally the stronghold of metals and ceramics.

At the same time metals and ceramics are themselves potential matrix materials, the fibre additions serving to provide composites with enhanced properties as compared with those provided by the metal or ceramic alone, but the problems of fibre incorporation into matrices of this kind and of the manipulation of such composites in structural fabrication have still to be surmounted and this is an area to which much future research work will be directed.

There are many areas of difficulty for the designer in utilising a fibre-reinforced composite material, not the least of which is the current lack of familiarity with a process of fabrication in which the material and the structure are made at one and the same time. Furthermore, current design processes are tailored to materials that are grossly isotropic, and which are machined away or built up as required, whereas to derive the greatest benefit from a fibre-reinforced composite a design philosophy of this kind is inapplicable. In assessing their potential, it is as well to recall first a few of the elementary factors relating to strength and stiffness of fibre-reinforced composite materials. With fibres of uniform, circular cross section, all aligned in one direction, the theoretical maximum value for the volume fraction of fibres in a given space is about 0·9. Hence the theoretical maximum tensile strength and stiffness of a bundle of fibres, in terms of the total cross-sectional area, will be some 10% less than for a single fibre, although in specific terms (strength-density) values as for the single fibre will still obtain. If, now, the voids are filled with a matrix material the density of the resulting composite is increased without a proportionate increase in the overall tensile strength of stiffness, since the properties of the matrix (resin or metal) will generally be much lower than those of the fibre in these respects; hence the specific strength and stiffness of the fibre-matrix composite will be lower than for a single fibre. However, the matrix makes a major contribution to the viability of the composite in that it provides for the transmission of shear loads, it supports the fibres in a manner analagous to beams on elastic foundations so that there is strength in compression, and it also provides a measure of strength in directions transverse to the fibre direction. In the latter respect the strength is determined almost entirely by the strength of the matrix itself and, for synthetic resin matrices in particular, is inadequate except for a few specialised loading conditions.

To provide a higher transverse strength, fibres can be orientated in different directions in the composite; indeed a material with reasonably isotropic properties in a plane is provided by equal quantities of fibres in each of three directions at 120° to each other—but only at the expense of a reduction in specific strength and stiffness to about one third of that for the unidirectional case. Furthermore, by the time practical difficulties in forming composites with fibre volume fractions approaching the theoretical maximum, and of developing the full capabilities of the fibre in the composite, have been taken into account the specific strength and stiffness of an 'isotropic' composite may prove to be an even smaller fraction of that theoretically possible in a composite with unidirectional fibres.

Obviously to obtain the optimum performance from fibre-reinforced composite materials it is essential for the amount and the orientation of fibres at every part of the structure to be controlled precisely, so that strength and stiffness properties are adequate for the specified envelope of structural loads. The design process must thus provide not only for the analysis of a structure in which the material is grossly anisotropic and varies in properties from point to point, but also for prescribing what the local properties must be to achieve a design optimum. The computer-aided procedures mentioned earlier have an important part to play here. It is also apparent that the most general arrangements of fibres must lie somewhere between the unidirectional and isotropic configurations, with specific strength and stiffness some reasonable fraction of the unidirectional case, and it is for this reason that only those fibres that offer several times the strength of the bulk material are attractive for composite applications.

Many of the analytical techniques for existing types of structure are applicable to fibre composite structures, but it is apparent that the same cannot be said for the fabrication techniques. For example, the current techniques for the fabrication of fibre-reinforced composite pressure vessels and rocket motor casings, by winding a mandrel with continuous filaments, pre-impregnated with resin and subsequently cured, has nothing in common with the way in which metal pressure vessels are made.

In one sense procedures for fabrication in metal are the inverse of those required for composite fabrication. In the former case large pieces of metal in the form of plate, rod, bar, billet, etc. are moulded, machined or cut to shape and subsequently joined together to form the structure, whereas for composites a more fundamental element in the form of a filament weaves and winds its way about the structure, with many fibres here and few there in accordance with the loading condition.

So far as fibre composite structures are concerned the machines that must ultimately be developed to achieve this objective will probably owe more to the textile and paper industries than to aviation. Already there is some read across in that the techniques that are used for sorting and orientating staple fibres and for felting have application to the sorting and orientation of whisker materials; weaving procedures often require only little adaptation to accommodate high strength fibres (e.g. glass cloth); and paper making techniques can be used to produce resin pre-impregnated fibre mat suitable for moulding procedures. Similarly stitching as a bonding process will supplement the adhesive bonding, riveting, bolting and welding jointing processes of conventional structures. Perhaps in the fullness of time we may even realise the comedian's dream of a knitted aircraft structure!

The promise of these materials is such that intensive developments in the field of machines for the fabrication of composite structures may be anticipated.

Certainly fibre-reinforced composite materials seem to be well suited to the fabrication of monolithic structures (as evidence by the large pressure vessels and rocket boosters fabricated from fibre-reinforced plastic), and it is from structures of this kind that the optimum efficiency seems likely to be obtained.

It may be, of course, with continuing development of techniques of vapour phase deposition, crystal growing and the like, that it may ultimately prove possible to fabricate an entire structure by direct deposition on to a prepared former, with bulk properties everywhere approaching those of the single crystal. Such a structure in carbon would require protection from oxidation for high-speed flight, but a coating process to form a durable inert carbide at the surface could be envisaged, which might also provide a level of surface compression that would inhibit the formation of cracks. Is this the optimum strength, stiffness and fatigue life structure that represents the ultimate structure of the future?

One feature of the production shop that is likely to be more commonplace than at present is the large pressure vessel, of a size capable of containing large elements of structure or even complete aircraft, and with facilities for controlling the internal environment of chemical constituents, temperature, pressure and possibly magnetic fields. The process of electron beam welding, for example, which offers the prospect of high efficiency welded joints in a wide range of materials requires a reduced pressure environment; vapour phase deposition processes also require reduced pressure and controlled temperature, and high pressures are required for autoclave bonding work.

The surface physics of materials for varying conditions of temperature and pressure, and in the presence of different vapours, plasmas and magnetic fields is little understood at present, but continued work in this field is likely to lead to novel methods of jointing and of manipulating materials. The large pressure vessels currently available for autoclaving, electron beam welding and space simulation indicate the beginnings of this activity, but the time will come when 'space-suited' workers in the production shop are a common sight.

5.1 *Space structures*

Some mention must be made of structures for extra-terrestrial environments, since passenger carrying space vehicles for scientific, if not for business and pleasure purposes, will be a common feature of the second century.

The problems of design for the habitable parts of the structure are unlikely to differ greatly from those for the pressure cabin of high flying atmospheric aircraft, particularly if some manoeuvring capability of the spacecraft is required, but for the space environment a spherical, or possibly toroidal, pressure cabin is a more likely structural geometry than the cylinder.

In view of the high cost per unit weight or per unit volume of ferrying materials from earth to space there will be a marked emphasis on economy of material for space structures, while because of the high cost of labour and difficulties of fabrication in space, pre-fabricated components will be used as extensively as is practicable. Inflatable structures, spent rocket cases, and structures assembled from combinations of these will be typical for inhabited areas.

However, it seems unlikely that the stressed-skin structural form will play as dominant a role for spacecraft as for aircraft, and for the non-pressure loaded, general purpose space structure, structures akin to the braced frame terrestrial space structures of the civil engineer[11] are likely to hold pride of place. A terrestrial braced frame space dome of some 600 ft (180 m) diameter has already been built, and structures many miles in diameter can readily be envisaged in the extra-terrestrial environment, with design determined by inertia loading resulting from manoeuvres rather than by gravity, snow and wind loads.

The strut is a typical element of such structures, and much attention will be given to the optimum design of members of this kind. Since a pin-jointed strut carries load only in tension and compression it seems well suited to some form of unidirectionally reinforced fibre-composite material. However, advantage can be taken of the space environment to use unusual materials as matrices for such a composite. For example, in a near-earth environment the equilibrium temperature for structure with a highly reflective coating and in the solar radiation field will be in the neighbourhood of 120°K; at this temperature ice, carbon dioxide, food pastes, etc. form durable solids, usable as matrix materials provided surface sublimation is inhibited. A form of strut fabrication can thus be envisaged in which the strut is initially in the form of a roll of flattened, flexible tubing with polished exterior and aligned fibres in the interior, which is subsequently cut to length and fitted with end attachments through which a requisite quantity of water or other liquid can be injected. The water vapour pressure inflates the strut which is then spun on its axis until the water freezes. In remoter regions of the solar system where equilibrium temperatures closer to absolute zero may be obtained, even more exotic materials such as solid oxygen could be used in this way.

Hence we can envisage the astronaut utilising his life support resources to stiffen and strengthen his structure for the high acceleration phases of his flight, devouring the structure in the cruise phase and using waste products to strengthen it again for the deceleration phase!

Of course, there are other features of the space environment that will be used to advantage, and in particular the limitless vacuum will be utilised for electron beam and friction techniques of joining and for gaseous deposition of structural shells using charged particles and surfaces as an alternative to a gravitational field.

6 THE MASS PRODUCED AEROPLANE

No mention of fabrication techniques would be complete without some regard for the numbers of aeroplanes that may be envisaged in production runs of the future.

Already, the production of 'family' aeroplanes for business and leisure flying is approaching 20 per day by some companies, so that the era of mass produced aeroplanes has already begun. As the world's population booms, and leisure time increases as a consequence of increased automation, there is the prospect that the demand for personal air transport might ultimately approach that for personal ground transport in the current century, with an annual production of millions of units (and traffic control problems scaled in proportion!). Different considerations, so far as structure efficiency are concerned, must apply to production at this scale than for the relatively small production runs of the much larger bulk transports that will also be operating, with particular emphasis on ease of fabrication and cost. Furthermore, some of the more exotic materials, such as beryllium for example, may be excluded from the mass produced aeroplane purely because the material is too scarce in the earth's crust to be able to meet the demand.

For the mass produced aeroplane, as for the mass produced car, the search will be for cheap, strong, stiff materials that are readily mouldable, easily jointed and consistent in performance in the service environment. It seems certain that fibre-reinforced plastic composites will play a major part in the structure of these aircraft, and an interesting possibility is provided by a composite in which the matrix is formed from an industrial bulk glass and is reinforced with a high modulus carbon fibre. Bulk glass is one of the cheapest of potential matrix materials in current large-scale production, the raw materials for its production are virtually inexhaustible, and the technology of its manipulation extends to prehistoric times. In compression structures it is commonly superior to aluminium alloy, but only low levels of tensile strain can generally be tolerated. However, even a relatively low level of strain can result in the development of a useful strength in a carbon fibre with a modulus some fourteen times greater than that of the glass itself, while a fibre-reinforced chemically toughened glass seems particularly attractive.

Because of the basic cheapness of glass, the know-how in manipulation, and its high strength potential in the absence or inhibition of cracks, its use as a prime structural material for mass produced components merits detailed consideration. Chemically toughened, load bearing glass structures may prove as familiar to the layman of the future as is the automobile body shell to the layman of the present.

7 CONCLUDING REMARKS

The structures of the second century, although perhaps different in shape from those that are current today, are expected to show only minor differences in basic structural form; stressed-skin structure will be dominant. Major changes are anticipated in the materials of aircraft construction, materials of far higher specific strength and stiffness than hitherto becoming commonplace. However, these materials will be of negligible ductility, with a consequent need to avoid stress concentrations so far as practicable, resulting in a trend towards structures as a monolithic whole and the development of pre-stressing and allied techniques to inhibit crack formation; passenger and pilot windows are likely to disappear. Composite materials based on high strength fibres or whiskers in a resin or metal matrix, and particularly composites based on carbon fibres, are likely to provide the ultimate structural materials of the future. Novel techniques of analysis will be developed, and computer technology will grow to the point where the structure can be analysed in the intimate detail that will be necessary for those materials. Novel fabrication techniques, for composite materials in particular, will also emerge.

Mass produced aircraft for personal transport, which seem a logical outcome from the increased leisure resulting from automation, may have structures formed largely from glass, with a carbon fibre reinforcement and/or in chemically toughened form.

Acknowledgment

The author acknowledges the permission of the Ministry of Technology to publish this chapter.

References

1 HOWARD, H. B. Aircraft Structures. *Centenary Journal of the Royal Aeronautical Society*, January 1966.

2 POPE, G. G. (Editor). Symposium on the use of the digital computer in aircraft structural design and analysis RAE Tech. Rep.: 66239 July 1966. ARC CP 926.

3 *Proceedings of Conference on Matrix Methods of Structural Mechanics.* October

26–28, 1965. USAF Tech. Rep. AFFDL-TR-66-80.
4 ARGYRIS, J. H. Continua and discontinua. Paper presented at the conference cited in ref. 3.
5 TAIG, I. C. Computer applications in the development of efficient aircraft structures. *Journal of the Royal Aeronautical Society*, Vol **67**, p 706, November 1963.
6 SCHMIT, L. A. and FOX, R. L. An integrated approach to structural synthesis and analysis. *AIAA Journal*, Vol **3**, pp 1104–1112, 1965.
7 SUTHERLAND, I. E., JOHNSON, T. E. Sketchpad: A man-machine graphical communication system. *AFIPS Conference Proc*: 23, pp 329–53, 1963.
8 KÜCHEMANN, D. Chapter 2, An Aerodynamicist's Prospect of the Second Century.
9 VAUGHAN, R. L. Advanced machining technology. *Lockheed Horizons* No. 6, July 1967.
10 HELDENFELS, R. R. Structural prospects for hypersonic air vehicles. Paper presented at the Royal Aeronautical Society Centenary Congress in conjunction with the 5th Congress of the ICAS, September 1966. *Aerospace Proceedings* 1966. Macmillan 1967.
11 DAVIES, R. M. (Editor). *Space Structures*. Blackwell Scientific Publications, 1967.

8 Unconventional flight

B. S. Shenstone

INTRODUCTION

Flight that is unconventional now may not be unconventional in the future. In 1903 all flight by man was unconventional. Although the 1903 limitations in knowledge and convention have dropped away, certain kinds of flight remain unconventional for practical, if not good, reasons. Unconventional flight is flight outside the mainstreams of research and development. This implies that it includes flight modes for which little or no money is available and which, therefore, can attract only few knowledgeable and dedicated enthusiasts. Hence these modes lag behind the contemporarily directly useful aspects of flight. Unconventional flight also includes flight modes which may never attract powerful research or development effort and will remain useless to mankind.

For the purposes of this chapter, unconventional flight shows itself to be restricted to rather low Reynolds numbers, the conventional concentration of Research and Development (R and D) being at the high end of the scale, and tending to become greater as time passes, as flying equipment becomes larger and faster. The range of Reynolds number (R) involved in this chapter is therefore between 1·0 and 5,000,000. Working up the scale of R, the subjects to be considered are:

1	Insects	R: 1·0 to 25,000
2	Models	R: 1000 to 200,000
3	Birds	R: 1000 to 250,000
4	Flapping Wings	R: 1000 to 250,000 at present
5	Man-powered Aircraft	R: about 600,000
6	Sailplanes	R: 500,000 to 4,000,000
7	Sails	R: zero to 5,000,000

The Reynolds number at sea level is 68,500 VL
where V is in m/s and L is in m.

Originally published in the *Aeronautical Journal of the Royal Aeronautical Society*, August 1968.

To foresee the future even a few years ahead is so difficult in this rapidly changing age, that one hesitates to say anything because it is likely to be too obvious or too ridiculous. So without shame, this writer falls back on Kahn and Wiener, who have given some thought to the problem[1]:

'One problem of long-range speculation is that the subject curve of probabilities often seems flat. That is, no particular course of events may seem much more likely than a large range of others. To avoid the dilemma of Buridan's ass, who starved midway between two bales of hay because he could not decide which one he preferred, we must then make arbitrary choices among almost equally interesting, important or plausible possibilities. That is, if we are to explore any predictions at all, we must to some extent "make them up". Clearly, the most salient of the projections we can make is one that is "surprise-free". Nevertheless, it would be very surprising if in any 33-year period the real world did not produce many political and technological surprises.'

Again: '. . . the expert with detailed knowledge, often still feels virtually impelled to make predictions that fit responsibly into known technology and known but unsolved problems. Thus he is also virtually compelled to ignore the crucial fact that in this kind of rapidly-changing situation it is new technology or other innovation that determines the trend, since it is almost impossible for him to say much about the nature of new technology or about untested or undiscovered innovations. If he understood their nature, he would already be using or investigating them.

'So a historian will find many areas in which the expert who has specific knowledge very much in mind may actually do less well at forecasting than the "naïve" journalist or the "naïve but sophisticated" systems analyst, who assumes, perhaps implicitly or unconsciously, that innovation and creativity will occur at a certain rate, even though (or because) he does not know precisely how.'

Finally: 'For most of the projections that we are discussing, in which we are looking 20 to 30 or more years ahead, perhaps the most surprising thing that could actually happen would be an absence of surprises. Therefore, there is no implication that a surprise-free projection is likely. It may, in fact, be the most likely of the various possible projections; that is to say that when contemplating a thousand things which could happen, the surprise-free projection may have a probability of much less than one in a hundred, yet be more probable than any of the other 990 possible occurrences. It could be 'most probable' and still be 'quite improbable'.'

I INSECTS

The best summary of the insect flight problem is J. W. S. Pringle's book *Insect Flight*[2], which includes a good bibliography. As would be expected, most research has been on the insect which is most damaging to man's interest, the locust, mainly by Weiss-Fogh and Jensen (*see* ref. 2) in the

1950s. Probably the most impressive research work on the aerodynamics of animal flight was published in 1941 by von Holst and Küchemann[3], which included tests of large flying models of insects (span of 0·8 m) during which the motions of the wings were measured photographically. Later von Holst in 1943[4] devised means for measuring the characteristics of small flapping wing models in an airstream and also under zero airflow conditions. Gradually we are learning about insects, so that we can use them or prevent them using us. The flight problem is one of many, but it is vital and we still do not know enough about it. As for the smallest insects with R from one to ten, Thom and Swart[5] say that lift tends toward a variation with V rather than V^2. Most insects have hairy wings, some with short stiff hairs widely spaced, and some of the small ones with wings which are mainly hairs growing laterally from the wing beam, forming a feather-like wing. At such low values of R, the drag increases sharply with respect to lift, so that instead of L/D being more than unity, D/L becomes much more than unity. Horridge suggested in 1956[6] that insects flying at this scale probably do so by rowing themselves through the air rather than using wings as aerofoils. Here is supposition, and we do not really know, for one reason because we have not yet been able to measure the rate of such wing beats. It has been suggested that they might be as high as 1000 beats per second.

As for larger insects it is now clear that in spite of rapid flapping, the usual laws of steady aerodynamics are valid, and that the hairs are there to ensure that the insect does not suffer from laminar separation and can depend on a turbulent boundary layer.

In spite of von Holst's clever use of flying models, there should be techniques for the measurement of performance in free flight and for analysing the wing motions. It is just as important to know the range, the power output and fuel requirements for friendly as for enemy insects. At present there are far too few facts known, and over-simplifying assumptions have to be made.

The next hundred years will see these problems solved, possibly by means of complex electronic equipment which could 'sight' the insect, or by the insect carrying the equipment with him. The future needs of mankind for food are bound to elevate the flight problems of insects from unconventional to conventional, and in one hundred years we may be in a position to know intimately all insects of importance and to control them to our advantage. Insect scourges will have disappeared but only if our knowledge of insect flight abilities and the ecology linked therewith is rapidly increased by research in the near future.

2 MODELS

Free-flying and tethered models had been developed during the first half of this century by enthusiasts, but practically no research in the true sense of the word had been made available until F. W. Schmitz's *Aerodynamik des Flugmodells* in 1942[7] and a second edition in 1952. This gave facts on the law of similarity, boundary-layer conditions for models, and the results of low turbulence wind tunnel tests on sections suitable for flying models, and much other new information for low *R*. It may seem surprising that this book was never published in English; it was purely a matter of cost.

Schmitz's great contribution was to clarify problems already encountered by model aircraft enthusiasts, but not understood by them. He showed clearly why efficient flying models could not be scale models of successful aeroplanes. For the first time he dealt with the scale problem of laminar separation, which is a phenomenon not met at *R* above about 100,000 and therefore of little interest to full-scale work. Laminar separation has nothing to do with laminar-turbulent transition. It is a breakthrough of the laminar boundary layer at values of *R* of the order of 20,000, the drag coefficient becoming higher than for turbulent boundary-layer conditions, and the lift coefficient becoming less for a given incidence. These effects are so serious that the only solution is to ensure, somehow, that the boundary layer is made turbulent at very low values of *R*. This requirement led Schmitz to the investigation of a series of wing sections, with means for ensuring, somehow, that such turbulent conditions took place. The now well-known turbulator wire just in front of the wing leading edge is one of the answers.

Some of Schmitz's own conclusions were: The measured wing polars enabled for the first time, proper calculations to be made of performance and stability of flying models in the region of *R* from 20,000 to 170,000. In addition the results are useful for the design of small blowers, ventilators and the design of propellers for both flying models and small ships. The hitherto disturbing inconsistencies between the aerodynamics for full-scale aircraft and the facts of experience with models were to a considerable extent explained and the causes clarified. He showed that since models and birds are flying in the same Reynolds number region and encounter the same aerodynamic problems, model designers would do well to study birds and develop models to imitate birds rather than imitating full-sized aircraft.

Apparently Schmitz disagreed with Otto Lilienthal and would say that bird flight is *not* the basis of the art of flight, except for models.

On 4th February 1945, the Low Speed Aerodynamics Research Association (LSARA) was formed in London under the chairmanship of Dr. H. Roxbee Cox (now Lord Kings Norton). The LSARA issued a number of papers on various aspects of model flying and control and testing. Most of the papers were written by N. K. Walker (LSARA Director of Research) and R. H. W. Annenberg. Several conferences were held and Schmitz presented a paper which was severely criticised by Annenberg and M. M. Gates.

The growing professional responsibilities of members as the years passed by made it more and more difficult to spend enough time and effort on the complex model problems. Wind tunnel work takes much time, and the development of performance-measuring techniques and equipment is complex and money-consuming. The leaders, originally centred at the Royal Aircraft Establishment at Farnborough, dispersed, and some went abroad. The sad result was that the LSARA lived for about ten years and then faded away. The work has never been collected and published as a whole, and mostly lies in private files. Much information from other sources also exists either unpublished, or if published, not widely known.

It is, however, good to know that some people, such as Hacklinger, have been allowed to spend some of their working time on the development of models. His report in 1964 on the development of indoor models[8] (45 minutes duration with a rubber motor) showed that he did a proper technical analysis, first-class design, stressed the model and worked out an optimum design solution.

The fact that free-flying models are not of any real use at present need not preclude mentioning what might be accomplished during the coming century.

Models are now easily controlled from the ground and have been for some time, but the models themselves can be greatly improved in speed and duration. They must however, be completely divorced from full-scale aircraft, both aerodynamically and mechanically. The need for an entirely different source of power and fuel may not be fully realised at present, but such a unit and the aerodynamics and structure to match it is bound to come. Structures and aerodynamics will have to be analysed as carefully as for a full-scale aircraft. Such developments may not occur unless a real non-sporting need to use flying models is discovered. It might be message carrying or photography or crop-spraying, using some future compact and light chemical. Models might be used instead of weather balloons, and could bring the recordings home.

3 BIRDS

The flight of birds has recently become a subject for scientific investigation, as opposed to hundreds of years of description of most other aspects of bird life. As world population grows, much more detailed knowledge of bird flying habits will be needed if we are to preserve the useful birds and control those which might menace our existence. Present gaps in our knowledge can be illustrated by the problem of 'angels' on radar scopes which turned out to be daily movements of thousands of birds just at or before dawn, and thus normally unobserved. There is probably still a great deal of bird activity going on around us that we do not know or recognise.

We have destroyed many birds because of ignorance of their ecology which is closely linked with flying habits. The gradual take-over and destruction of natural migratory bird way-stations can destroy some of the most useful birds unless we can gain more knowledge about their flying qualities, power, range and fuel requirements. There are signs that this is being recognised, and Cone's work[9] on the soaring performance of vulture and albatross is very relevant here. Concluding his analysis of the dynamic soaring flight of the albatross (*Diomedea exulans*) he writes:

'With the aerodynamic basis of albatross flight firmly established, the ecological significance of many aspects of the bird's characteristic form and behaviour becomes clear. Not only can many qualitative deductions be made, but the quantitative nature of the aerodynamic analyses allow the actual numerical investigation and correlation of many ecological factors when adequate field data are available.

'With the availability of adequate quantitative data on specific albatross flight paths and travel patterns, combined with adequate oceanographic and meteorological information on the specific environments, it should become possible to describe the primary eco-systems of the various species of albatross to a highly quantitative degree.

'When quantitative data on the aerodynamic characteristics of several different but associated species become available, it should be possible to establish the reasons for the most significant aeroecological differences between the various species on the basis of their differences in structural form and flight capabilities.'

Comparing very different soaring birds, albatross and vulture, Cone points out that 'an interesting example of adaptation to effect the most useful balance of ecological factors is afforded by the relationship between the types of food and the relative soaring altitudes of the albatross and vulture. As mentioned in the preceding sections, the albatross is limited to flight within the shear layer, and thus remains close to the surface throughout its soaring cycle. By contrast, the vulture nearly always soars at relatively high altitudes on the order of 1000 to 5000 feet (300 to 1500 metres) where the thermal shells are well developed. The albatross feeds, in general, on live marine animals such as squid, shrimp and fish, and in order to capture its mobile prey the bird must be able to move swiftly and surely. The high-speed, low-altitude flight thus greatly enhances its success in procuring food. The vulture feeds on immobile carrion, which it can easily detect from high altitudes. From such altitudes the bird's field of vision, and consequently its scanning efficiency, is greatly increased. Thus, the food types of the two birds are precisely those which can best be located and taken in a manner perfectly compatible with the basic flight patterns required for efficient soaring flight.

'An important point of similarity exists between the vulture and the albatross in that both use the same wing-flexing technique to increase the effective wing loading. Wing-flexing is frequently used by vultures during glides into the wind, where an increase in the wing loading is necessary for an increased speed over the ground. Turkey vultures and gulls make important use of wing-flexing in gust soaring. The use of wing-flexing by the albatross in the leeward glides is exactly analogous to that of the vulture and allows the bird to maintain a high absolute velocity in the leeward glide.'

If Cone and others continue such aerodynamic studies on other birds, there is a fair chance that a great deal can be learned about the ecology of many species to our advantage and in at least some cases to the advantage of the birds concerned. Thus this aspect of flight may become less unconventional as defined in the introduction.

4 FLAPPING WINGS

The bird and the insect have inspired many people to copy them at a greatly increased scale, so far completely unsuccessfully. Is it because we do not know the 'secret' of bird and insect flight? Is it because the flapping wing is no better than a propeller or a jet? Is it because what is good on one scale is not necessarily good on another?

Present-day answers are: No, Yes and Yes. Future answers may be different, but a short sketch of the reasons for todays' answers may be useful.

There is no secret about flapping flight. We know enough about the patterns of wing incidence required and the motions necessary for the production of thrust. What we do not have is a cut-and-dried method for working out the power required for flapping flight and the efficiency of the operation. It is unlikely that thrust produced by flapping is more efficient than thrust produced by a propeller, which has efficiencies well above 80%.

However, it is often stated that a flapping wing should be more efficient than a propeller plus a fixed wing. The former uses one element, the wing, for lift and thrust; the latter two. The flapping wing avoids the profile and induced drags of the propeller. On the other hand the flapping wing has two motions (flapping and horizontal) which means a higher mean airspeed than for the fixed wing and therefore more drag. This part of the problem is not solved. But there is no mystery, only ignorance. If there were a requirement for performance calculations, the basis for them could be produced after suitable and not excessive R and D.

About 35 years ago Alexander Lippisch developed flapping wing free-flying models of about 0·6 to 0·8m span which worked very well. But experience with machinery indicated that weight is saved by operating quickly rather than slowly. Thus it was obvious that if only part of the wing were flapped more rapidly and the rest left rigid, the whole drive would be lighter and simpler. Indeed it was, with only the outboard parts of the wing flapping. But the extension of the driving mechanism from fuselage towards the wing tips was still cumbersome and it occurred to Lippisch that a more compact arrangement would be better. Why not take the outboard flapping elements away from the tips and place them close together on the fuselage? They were then removed entirely from the fixed wing and placed forward of it and worked better with less weight. The flapping wings had become the equivalent of two propeller blades which did not fully rotate. That was the end of that development.

Scale appears to be the greatest difficulty. On models, the incidence changes necessary during flapping to produce thrust are easily approximated by having flexible membranes for the after parts of the wing. Mechanically only a simple up-and-down flapping is needed around axes parallel to the direction of flight. Full-scale this is impracticable, and appropriate complex mechanical hinging to give the requisite incidence pattern is necessary. To flap the wing, requires a big and heavy reduction gearing between engine and wing, because, it the whole wing is flapped, it must move much slower than would a propeller. Even if the wing were to be elastically balanced at 1*g*, the weight which would have to be moved, stopped and started twice for every full cycle would cause quite a momentum problem because as size and power increase so do the relative weights of the flapping parts.

Such a drive would be awkward, heavy, vibrating and subject to serious structural fatigue problems. Even if marginally more efficient than propellers, this margin would be absorbed by the weight and complication of the drive.

It is perhaps worth noting that nowadays most proposals for flapping wings involve their use for man-powered aircraft. This is unwise because the available power for such aircraft is at this stage marginal, and not easily measured in flight. Flapping wings should be demonstrated full-scale by an engine-powered aircraft and precise performance measurements made. If it becomes worthwhile to flap our wings, the second century can do it.

5 MAN-POWERED AIRCRAFT

Man has apparently always wished he could fly by his own efforts, not realising the amount of effort required. This refers to the present state-of-the-art, which is primitive. It is significant that Man is larger than any animal that has flown. Wilkie has pointed out that it is probable that no animal in the history of this world which had to fly professionally ever weighed more than about 13 kg. To fly professionally is to require flight for the support of life. Domestic chickens and geese do not fly professionally. It is likely that some of the larger winged prehistoric creatures also did not. The mute swan is on the verge of not needing to fly and its power/weight ratio makes the take-off and the initial rate of climb critical.' It's not surprising that Man finds it difficult to fly in spite of his cleverness.

Although the chemical processes involved are very complex, muscle achieves an efficiency (work output/chemical energy used) of 20% to 25% under favourable conditions. The chemical energy comes ultimately from the oxidation of foodstuffs. From the long-term point of view, energy production by muscle thus depends on an adequate supply of oxygen, which must be absorbed by the lungs and transported by the blood-stream to the active muscles. Experiments have shown that the steady-state output must be limited to 0·4 to 0·54 horse-power, depending on ordinary men or champion athletes[10].

Although the ultimate source of muscular energy is oxidation, the immediate source of energy is the hydrolysis of various compounds. The rate of these hydrolytic reactions is not limited by the supply of reactants from outside the muscle, and it therefore may be very high, although the total amount of energy available is limited by the amounts of such chemicals stored in the muscles. If the following assumptions are made (giving the right answer) we have the basic pattern of the human engine:

1 There is a steady oxidative energy production of 0·5 hp falling to 0·475 hp after 25 minutes as a result of long term fatigue.

2 To this is added a lump sum of mechanical work, amounting to 0·6 hp-minutes derived from hydrolytic reactions (*See* Figs. 1 and 2).

If the man-powered aircraft were to be assisted into the air and accelerated to cruising speed, the first assumption above would supply the cruising power. If there were to be no assisted take-off, the second assumption comes into play. If take-off is difficult, this lump of power would not be enough to reach flight speed, and some of the cruising power would have to be used before take-off, thereby reducing flight duration by earlier fatigue.

It is important that the best method of using the human engine is discovered. At present it is thought to be pedalling, and the optimum pedalling speed to be about 60 rpm. Where efficiency has to be sacrified to output, the speed may be increased to 120 rpm. For cruising, the full cruising power can be obtained from leg muscles alone, but for short bursts the more muscle employed, the greater the power developed. Accordingly it has been found that simultaneous cycling and hand-cranking produces about 50% more power than cycling alone, but only for a short time. Ursinus[11] found that the advantage is small after 5 to 10 minutes when the power output becomes limited by oxygen supply rather than by muscle-mass.

If we now assume that 0·5 hp can be used for cruising a man-powered aircraft we have to allow for the efficiency of the drive and of the propeller. It is unlikely at present that the propeller efficiency would be greater than about 80%, but might be better if some design development were done for this sort of slowly-rotating large diameter type. Bicycle hub gearing reaches efficiences of between 96% and 98%[12]. Any chain or

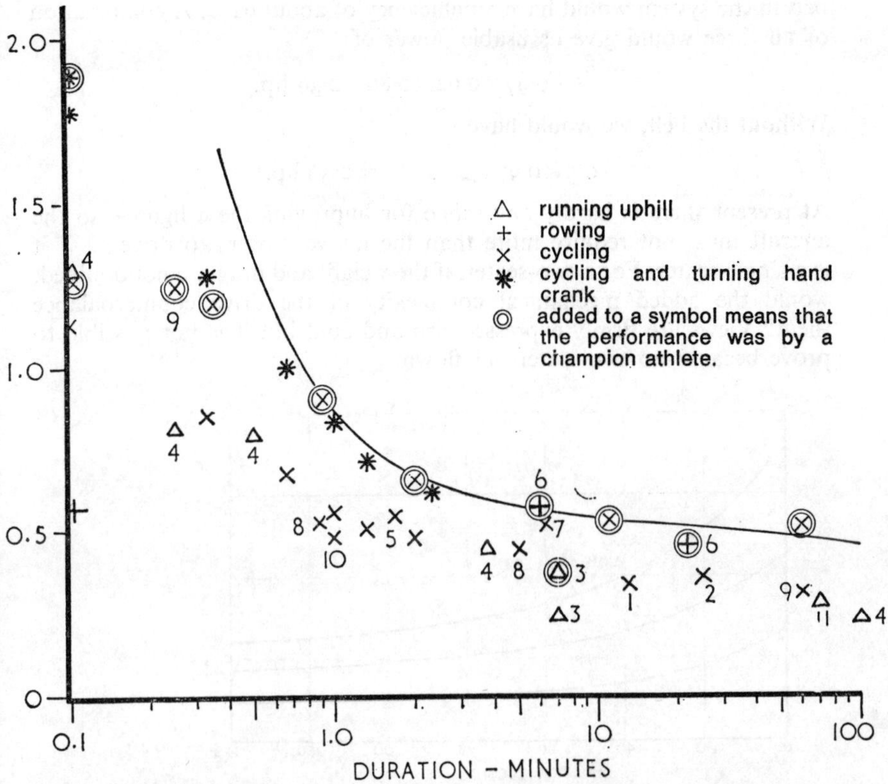

Figure 1 Maximum output of external mechanical power (hp linear scale) plotted against total duration of exercise (minutes logarithmic scale). The logarithmic scale has been used to display the experimental points clearly. The numeral indicates the source from which the experimental point was derived. Points without numerals: champion cyclists, ⊗ ref. 21; ordinary cyclists, X also ordinary cyclist performing hand-cranking in addition, ✱ ref. 29.

1, Abbott 'et al'; 2, Asmussen; 3, Bannister; 4, Bénedict 'et al'; 5, Bonjor; 6, Henderson 'et al'; 7, Karpovich 'et al'; 8, Nielsen 'et al'; 9, Raleigh; 10, Tuttle; 11, Unna. (From 'The Journal of the Royal Aeronautical Society,' p 479 August 1960)

belt in the system would have an efficiency of about 92%. A combination of all three would give us usable power of

$$0{\cdot}5 \times 0{\cdot}97 \times 0{\cdot}92 \times 0{\cdot}80 = 0{\cdot}36 \text{ hp.}$$

Without the belt, we would have

$$0{\cdot}5 \times 0{\cdot}97 \times 0{\cdot}80 \qquad = 0{\cdot}39 \text{ hp.}$$

At present there is hardly any scope for improving these figures, so the aircraft must not require more than the above powers to drive it, if it is a single-seater. For a two-seater, if the weight and drag are not doubled, would the added mechanical complexity of the drive counterbalance them? Views are firmly expressed pro and con, but it is not possible to prove because no two-seater has flown.

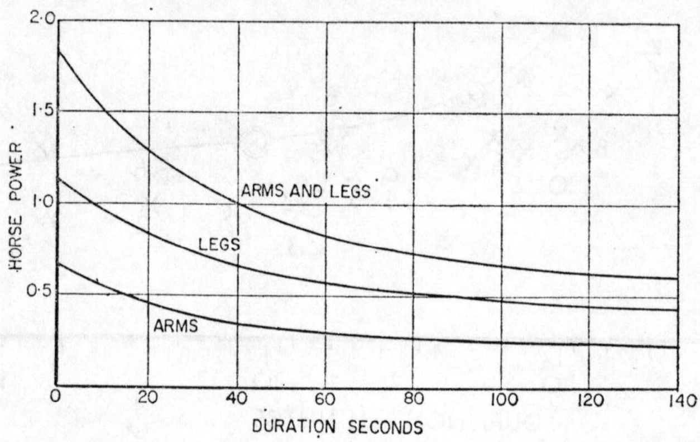

Figure 2 Comparison of horse-power produced from work done by legs alone, arms alone and arms and legs combined. Redrawn from Ursinus. Ordinary subject. (From 'The Journal of the Royal Aeronautical Society', p 48 August 1960)

In 1935 and 1936 the Haessler-Villinger (German) machine (Fig. 3) was built and flown by catapult launch and the Bossi-Bonomi (Italian) machine (Fig. 4) was also flown, normally with catapult launch, but Bossi claims that take-offs were also made without external assistance and without storage of energy. The best flights were 720 m for the Haessler-Villinger and 900 m for the Bossi-Bonomi, both with assisted starts.

The problem lay dormant except for an analysis of the problem in 1948 by Brian Worley[13] which indicated slight optimism if full advantage were to be taken of the latest developments. It was not until 1956 that any post-war effort was begun. Both T. R. F. Nonweiler and the present writer started writing and talking about it simultaneously. As a result, on 10th January 1957 a Man-Powered Aircraft Committee was formed at a meeting at the College of Aeronautics, Cranfield and the original members were:

H. B. Irving Chairman
Dr. J. R. Brown London School of Hygiene and Tropical Medicine
Robert Graham Microcell Limited
Thurstan James Editor, *The Aeroplane*
A. Newell College of Aeronautics
T. R. F. Nonweiler Queen's University of Belfast
B. S. Shenstone British European Airways

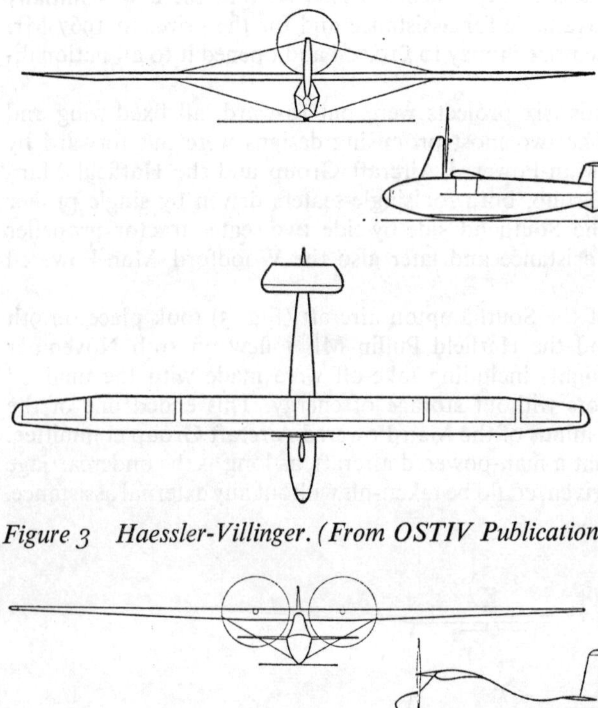

Figure 3 *Haessler-Villinger. (From OSTIV Publication VIII, 1966)*

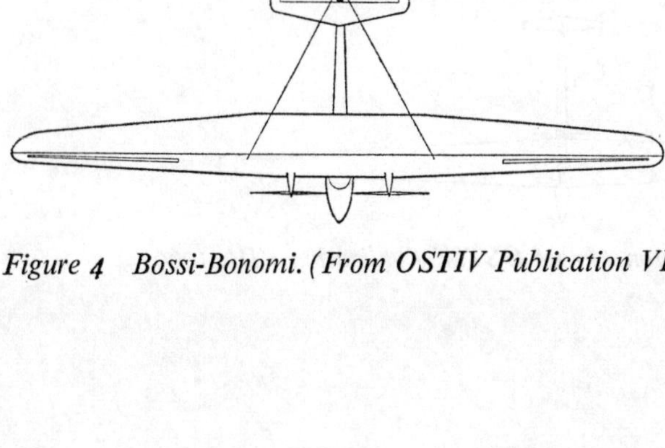

Figure 4 *Bossi-Bonomi. (From OSTIV Publication VIII, 1966)*

Initially this committee was formed *in vacuo* and had no authority outside its own members. Quite soon the committee associated itself with the Royal Aeronautical Society and a Man-Powered Aircraft Group was formed for the purpose of financially assisting promising designs. It was rightly assumed that no financially sound organisation would attempt such an unpromising project.

While money was being collected by the Group, an industrialist, Mr. Henry Kremer, in November 1959, came forward with an offer of £5000 for the first British man-powered aircraft flight around a 'figure of eight' course based on two markers, one half-mile apart[14]. The fund for assisting designers and builders finally reached £5000 so that there was initially a total of £10,000 available for assistance and for the prize. In 1967 Mr. Kremer doubled the prize money to £10,000 and opened it to all nationalities.

As a result of this, six projects were put forward, all fixed-wing and propeller driven. The two most promising designs were put forward by the Southampton Man-Powered Aircraft Group and the Hatfield Man-Powered Aircraft Group, both for single-seaters driven by single pusher propellers. Later the Southend side-by-side two-seater tractor propeller aircraft was given asistance and later also the Woodford Man-Powered Aircraft.

The first flight of the Southampton aircraft (Fig. 5) took place on 9th November 1961 and the Hatfield Puffin Mk 1 flew on 16th November 1961. Both these flights including take-off were made with the unaided strength of the pilots without storage of energy. This ended one of the main doubts in the minds of the Man-Powered Aircraft Group committee. It was now clear that a man-powered aircraft, as long as the undercarriage wheel(s) could be driven, could be taken-off without any external assistance,

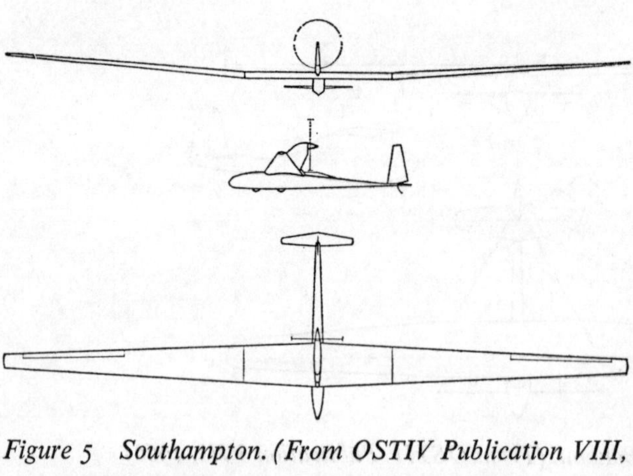

Figure 5 Southampton. (From OSTIV Publication VIII, 1966)

by the pilot. Full credit for this feature must be given to Bossi, who incorporated it in his Pedaliante in 1936.

By 1968 the only two-seater which had been completed was the Southend project, but due to mechanical difficulties in the drive, has not flown, and due to the dispersal of the interested group and the difficulty of finding a suitable flying field for the money available, has had to be given up.

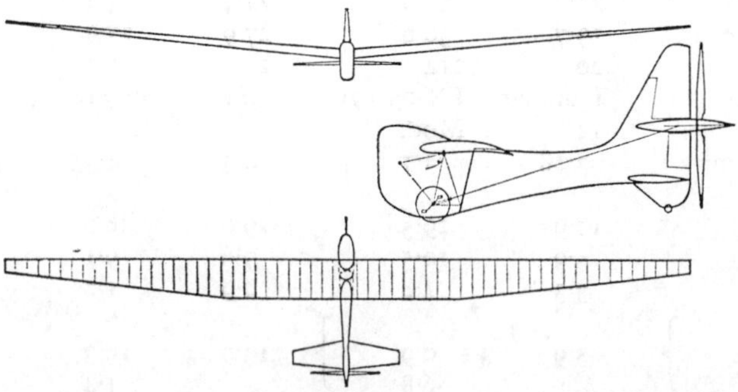

Figure 6 Hatfield Puffin II. (From OSTIV Publication VIII, 1966)

The second problem that these three aircraft solved was the argument and discussion on achievable structure weights. Original generalised estimates were based on the weight of glider wings. Detailed estimates by the designers showed much lower weights, and these were largely realised. Weights are shown on Tables 1 and 2.

TABLE 1 *Aircraft which have flown (early 1968)*

Type	*Haessler-Villinger 'Mufli'*	*Bossi-Bonomi 'Pedaliante'*	*South-ampton*	*Hatfield 'Puffin' Mk. I*	*Hatfield 'Puffin' Mk. II*
Date 1st flight	*1935*	*1936*	*1961*	*1961*	*1965*
Span, m	13·5	17	24·4	25·6	28·4
Wing area, m²	9·65	21·4	27·9	30·6	36·2
Aspect ratio	18·8	13·4	21·3	21·4	22
Wing weight, kg	24·1	45	36	29·5	38·6
Empty weight, kg	36·4	97·3	58·1	50	61·3
Flying weight, kg	111	162	121·6	113·5	124·8
Wing loading, kg/m²	11·5	7·56	4.35	3·7	3·44

TABLE 2 *Estimated and actual dimensions and weights: Hatfield and Southampton aircraft*

Type	*Hatfield Puffin I*		*Southampton*	
	Estimated	*Actual as flown*	*Estimated*	*Actual as flown*
Date of first hop	—	*Nov. 16, 1961*	—	*Nov. 9, 1961*
Span, m	24·4	25·6	24·4	24·4
Wing area, m²	29·7	30·6	27·9	27·9
Aspect ratio	20	214	21·3	21·3
Wing section root	Laminar 14%	FX-05-191 Mod.	65,818	65,818
Wing loading, kg/m²	3·28	3·7	4·3	4·35
Weights in kg				
Wing	17·9	29·5	29·0	36·0
Fuselage	5·9	10·5	7·6	9·1
Tail	3·3	1·8	2·9	1·2
Undercarriage				
Drive	5·9	0·9	13·0	10·3
Propeller		6·8		1·4
Equipment	1·1	0·5	3·0	0·1
Empty weight	34·1	50·0	55·5	58·1
Pilot	63·5	63·5	53·5	63·5
All-up-weight	97·6	113·5	119·0	121·6

The longest flight up to May 1968 was that flown by John Wimpenny in the Hatfield Puffin I, including unaided take-off. The distance between unstick and landing was 910 metres, and it took place on 2nd May 1962. It lasted about 2 minutes, and the pilot landed because of temporary exhaustion. Using Wilkie's data[10] for an ordinary cyclist (not a champion) using legs alone, Wimpenny should have tired after two minutes, had he used 0·5 hp. However, in addition he had to make the take-off effort, so he might not have been able to exert as much as 0·5 hp for his cruise. This sort of supposition illustrates the gaps in present knowledge on muscular performance and on the drag of the Puffin I at an altitude of about ten feet (3m).

The Puffin I suffered a landing accident in 1963 which destroyed the wing but did not seriously damage the pilot or the rest of the aircraft. The Hatfield group took full advantage of this opportunity by completely redesigning the wing, increasing the span from 25·6 m to 28·3 m, using the new Wortmann FX-63137 wing-section specially designed for this aircraft, and changing the structure to improve the accuracy of the external wing surface under normal flying loads. This improved machine, Puffin II, flew on 27th August 1965, and is reported to be considerably improved in performance over the Mark I. Fig. 7 shows some wing sections.

The Southampton aircraft was built mainly of spruce with some balsa. The Hatfield, Southend and Woodford machines were built mainly of balsa, spruce being used for spar flanges. The wing and, in part, the fuselage covering is the transparent plastic film Melinex (Mylar in the USA). For prototypes such as these, balsa has been shown to be lighter and more easily usable than any form of plastic or composite structure as yet available. There is clearly much progress to be made on materials for large lightly-loaded structures such as these.

These initial developments in England which have been widely reported, have revived or created interest in the man-powered aircraft in other countries. Lippisch in the USA did preliminary study and construction of a very light-weight project. At the Georgia Institute of Technology a man-powered aircraft has been built, Kimura in Japan has had the Linnet of 22 m span built and flown. Haessler in Canada has redesigned his earlier German aircraft with the much increased span of 25 m. The Man-Powered

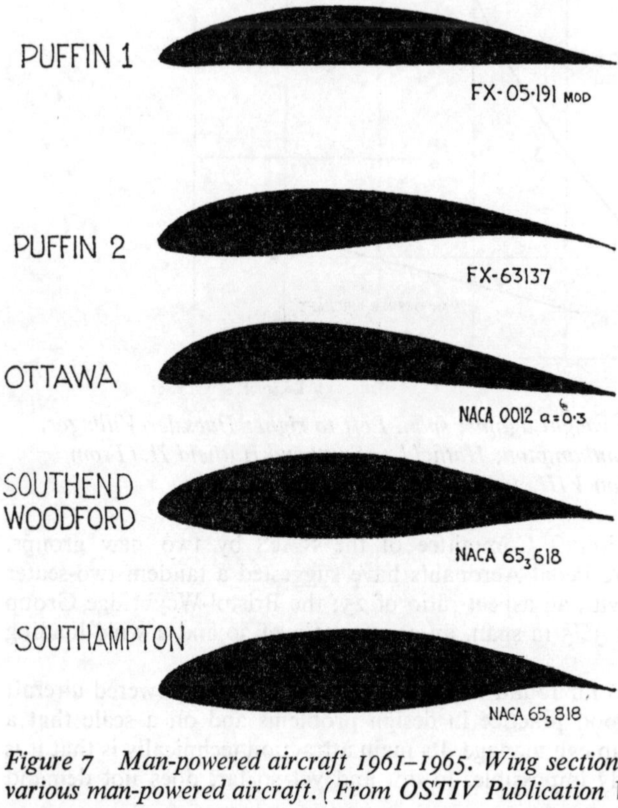

Figure 7 Man-powered aircraft 1961–1965. Wing sections used by various man-powered aircraft. (From OSTIV Publication VIII, 1966)

Aircraft Group of the Canadian Aeronautics and Space Institute is designing and building, in Ottawa, a 2-seater of 27·4 m span, using very thin large diameter light-alloy tubing for the fuselage structure[15]. Czerwinski, the designer, has developed a new method for joining such tubes, and has designed a special form of double universal to be used instead of bevel gears.

The realisation that the practical maximum span has probably not yet been designed or built, has led to two new proposals being made to the

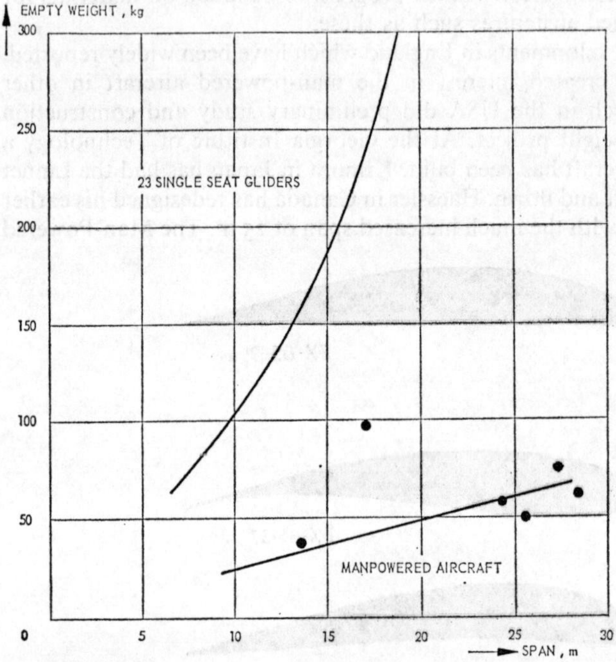

Figure 8 Empty weight against span. Left to right: Haessler-Villinger, Bossi-Bonomi, Southampton, Hatfield 1, Southend, Hatfield II. (From OSTIV Publication VIII, 1966)

Man-Powered Aircraft Committee of the RAeS by two new groups. The Hertfordshire Pedal Aeronauts have suggested a tandem two-seater of 37·5 m span with an aspect ratio of 25; the Bristol-Weybridge Group a single-seater of 36·5 m span, an aspect ratio of 30 and a wing loading of 2·55 kg/sq m.

Nobody has so far found any particular use for man-powered aircraft except as very good practice in design problems and on a scale that a young small group can manage. Its main attraction technically is that it is so difficult, nearly impossible, to do, and yet so far, does not demand sophisticated and costly materials. It may turn out to be a sport after much further development. It may fade away because it is too difficult, too marginal. But it is no more marginal today than powered aircraft were 60 years ago. They were useless then and few could see much of a future for them.

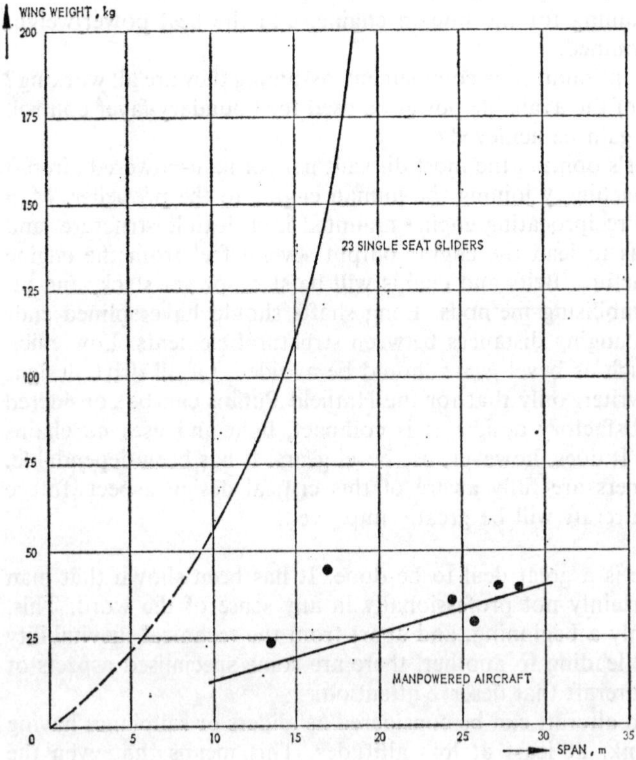

Figure 9 Wing weight against span. Left to right: Haessler-Villinger, Bossi-Bonomi, Southampton, Hatfield I, Southend, Hatfield II. (From OSTIV Publication VIII, 1966)

Before leaving man-powered aircraft in the lap of the future, here are a few items for interested people to think about in the future and answer or solve as the case may be:

1 What is the power required for level flight for past and future projects, near the ground at about, 0·1 span altitude and at heights greater than half-span?

2 What degree of laminar boundary-layer conditions can be achieved in flight on the wing top and bottom surfaces? This should include the

difficult areas, near the root and near the tip (low R and possibility of laminar separation).

3 Actual propeller efficiencies.

4 Mechanical losses in various forms of driving mechanism.

5 The energy/power problem of the unaided take-off followed by cruising should be analysed in detail.

6 The best training for the human engine, and the best power/weight should be determined.

7 What is the optimum crew complement, assuming they are all working?

8 Could part of the available power be used for boundary-layer control, and an overall gain be achieved?

9 In the writer's opinion the most difficult part of man-powered aircraft design is the machinery joining the human engine to the propeller. Man is a slow-speed reciprocating engine mounted in a flexible structure, and the designer has to lead the engine output several feet from the engine within this structure. Belts and chains will be slack or get slack, and require special stabilising methods. Long shafts should have splined ends to cope with changing distances between structural elements. Low efficiency gearing such as bevel gears should be avoided. Of all drive designs known to the writer, only that for the Hatfield Puffins can be considered a good and satisfactory design. It is compact, light and uses no chains and no belting. It does, however, use bevel gears. It has been dependable. If future designers are fully aware of this critical design aspect, future man-powered aircraft will be greatly improved.

Indeed, there is a great deal to be done. It has been shown that man can fly, but certainly not professionally in any sense of the word. This, then, can be only a beginning, and apart from the technical inevitability of one advance leading to another, there are some specialised aspects of man-powered aircraft that deserve attention.

Man-powered aircraft can be considered as gliders or sailplanes having zero rate of sink, at least at low altitudes. This means that even the slightest up-current will cause them to climb. Thermals close to the ground are weak, but no matter how weak, would cause the man-powered aircraft to climb. Thermals in general increase in power with altitude, so that if future man-powered aircraft can achieve a duration of a few minutes rather than two minutes, it should be possible for them to use favourable ground level thermal conditions to become self-contained sailplanes, needing no winching or towing, and towards the end of the flight having more flexibility than the present-day glider.

To achieve this, the future man-powered aircraft will have to be stronger and therefore heavier. Instead of an ultimate load factor of the order of 4 at low speeds and weak gust assumptions, it will have to be stronger and of the order of high performance sailplane strength. But it must not be as heavy. Comparisons between man-powered aircraft and sailplane structure and wing weights show that sailplanes are much heavier than they need be (Figs. 8 and 9.) Gliding angle (L/D) does not depend in any way on weight, and minimum sinking speed which does, is often no longer the basic criterion for glider high performance. Sailplanes could, for a given strength be lighter. It is hoped that these two kinds of unconventional and sporting aircraft may meet in this way in the future.

6 GLIDING AND SOARING (SAILPLANES)

After several centuries of ill-informed and unsuccessful gliding attempts, Orville Wright succeeded in staying aloft in an unpowered aircraft for 9 minutes and 45 seconds on 24th October 1911. It was not until 1921 that this record was broken, but in the latter part of 1921 the record was broken three times within two weeks by three different sailplanes and pilots: Klemperer in Blaue Maus, Harth in Harth-Messerschmitt and Martens in Vampyr, the records being 13, 15 and 21 minutes, but all with considerable altitude loss between take-off and landing[16].

It was soon recognised (based on Zacher[17]) that a low sinking speed was not the only requirement for a sailplane, but a good flat gliding angle was also necessary, if distance flying was to be undertaken as well as duration and altitude. After a bold experiment by Espenlaub, who was the first to build a 17 m span cantilever wing, the Konsul built by the Darmstadt Akaflieg, was an aircraft which defined more clearly the appearance of future sailplanes. It had a span of 18·7 m and an aspect ratio of 18, using Göttingen 535 as wing section. It had a long elliptical section fuselage, a large rudder and differential ailerons. Konsul was the prototype for the so-called Darmstadt School whose members, during the few following years designed a series of outstanding sailplanes. The elliptical planform wing was either directly set on the fuselage or on top of a short tower on the fuselage. This development was followed by the use of highly tapered straight wings, begun by Kakadu (Akaflieg Munich) in 1928 in which the structure was so disposed that the stiffened plywood wing skin resisted bending as well as torsion, resulting in weight saving but a complex structure. Most sailplanes of this era, and for many years thereafter, used a torsionally-stiff nose box and a separate spar for bending.

In 1929 the Wien by Lippisch appeared, which was a braced high-wing monoplane, a successful attempt to reduce weight and still use simple structures and have a good performance. During this period there was insufficient knowledge about wing flutter and aileron reversal. This resulted in some accidents, a close study of aileron design and a much improved level of detail structural design, followed by strength requirements.

Outside Germany there was also a lively activity in sailplane construction, particularly in France, England, Switzerland and Russia. The same developmental trends were to be seen. It is notable that in contrast to Germany, where in most cases, the Akaflieg and individuals were active in design, in France and England important aircraft manufacturers undertook the development of sailplanes (Dewoitine, Farman, Hanriot, Potez, Handasyde, de Havilland). In Switzerland, Spalinger came forward with his own designs and in Russia we find such names as Ilyushin, Artamonoff, Gribowski. If we name particular aircraft of this era, we should certainly mention Abrial's Vautour, Peyret's Tandem and Tscheranowski's Parabola.

During the 1920s the span and the aspect ratio increased from about 10 m and 10 to about double these values. One had recognised that by span increase the performance would be improved and in 1931 Kupper with his Austria made the experiment of increasing it to 30m. Stiffness and controllability problems were difficult to solve. As a result, thereafter, apart from a very few exceptions, sailplane spans ranged from 15 m to 20 m. In the meantime the RRG (Rhoen-Rossitten Gesellschaft) had produced Lippisch's Fafnir I. Its general features, cranked wing and fuselage and wing fairing, became typical of the Wasserkuppe school. It was followed by Fafnir II in which for the first time the pilot was completely enclosed in an aerodynamic fairing which was part of the proper fuselage shape and not an obstruction to the airflow. In Fafnir I a partial cockpit fairing led to interference, and the right answer, as above, has undergone detailed refinement ever since.

Jacobs in his various Sperber variants, the Habicht and his Reiher, and many other designers continued to use the cranked wing. Not until the end of the 1930s did the DFS (Deutsche Forschungsanstalt für Segelflug) return to the classical form of sailplane, the uncranked high wing, in Jacobs' Weihe and Olympia, one of the forms still used by designers but refined in many ways as experience grows. Apart from the Wasserkuppe School, Wolf Hirth with his Moazagotl and Minimoa with exaggerated wing cranking, was also of considerable developmental influence in the 1930s. In addition, the Akaflieg (now called FFG: Flugtechnische Fachgruppen) continued to do much original work which featured such things as camber flaps, prone pilot and much else such as the tiny light-weight but expensive Windspiel of 1933 of 12 m span and the 20 m D-30 Cirrus with an aspect ratio of 33, camber flaps, spoilers and dihedral variable in flight (1938).

In the meantime soaring was blossoming in many other countries, particularly in Poland, Czechoslovakia and Hungary, but also in Italy, Britain, Yugoslavia, France and USA. Such machines as Grzeszczyk's SG-21, Czerwinski's CW-5, PWS-101 and 103, Zlin's Z-25 Sohaj, Akaflieg Budapest's M-22, Rotter's Nemere, Musger's MG-9, Bowlus Albatross, Stanley's Nomad, the first to use the butterfly tail, and the earlier work of the Schweizer brothers, the French Avia 41P, the British Hjordis and King Kite, the Swiss Spalingers, and the Moswey and Spyr, the Russian GN-7 and Stakanovitch, the first sailplane with marked forward sweep.

The outbreak of the Second World War in 1939 put an end to what might well have been a tremendous boost for soaring. This was the specification by ISTUS (Internationale Studienkommission fuer den Motorlosen Flug) for a sailplane to take part in the 1940 Olympics. A number of prototypes were built to this requirement and competed in flight and handling tests in 1939. Jacobs' Olympia of 15 m span was chosen but the Olympic Games never took place and in later Olympics so far, soaring has not been accepted as a suitable sport.

The ten years between 1939 and 1949 saw hardly any sailplane development. Then a great forward impulse toward improved performance was given by August Raspet (USA) by his investigation of the aerodynamics of the Tiny Mite and RJ-5. He showed that by improving the quality of the wing and fuselage surfaces both in smoothness and wavyness, by improving fairings and removing gaps and leaks, a remarkable improvement in performance could be achieved. This work influenced sailplane designers throughout the world and one may safely say that such sailplanes as Pfenninger's Elfe, Kensche's HKS, Dlouhy's Spartak, the Meteor by Cijan, Obad and Mazovec, the Phoenix by Eppler and Nägele, Szuba's Zefir, Slingsby's Skylark 4 and Dart and many others have profited from Raspet's work.

Some further background to development on aerodynamic and structural aspects may be useful. Wing aerodynamics and structure are interdependent. The desire to make a wing thinner is limited by the structural requirements, the materials available and the knowledge of the structural designer. With the early lack of structural knowledge including aeroelastic problems the designer felt he had to keep the wing as thick as he could. The wing of the Vampyr had a thickness/chord ratio of 17·25% at the root. Structural developments have enabled this figure to be reduced, if so desired, to something like 12%. In the late 1920s and 30s, high cambers were used. Examples are the Vampyr with Gö 482 and the Gö 535 which was widely used on the Darmstadt types and many others. This trend, which was an effort to minimise sinking speed without using too low a wing loading, led to the ultimate in high camber, the Gö 652, used in Kakadu (1928) and Fafnir I (1930). Such wing sections were given up when it was realised that the future of soaring lay not only with duration but with range. This required speed, and low drag at low incidence, and demanded quite different aerofoil shapes. Thus we see the use of Gö 549 in the thirties and NACA 23012 just before the Second World War. Thereafter, until the present, the stress has been on wing sections which can maintain a large amount of laminar flow in the boundary layer at sailplane cruising speeds which can be more than 100 km/h with placard speeds of the order of 200 to 250 km/h, and increasing as time goes on. This requirement has led to such aerofoils as NACA 63_3-618 and some years later to Eppler sections such as EC86(-3)-914 (Fig. 10), and Wortmann sections such as the FX-61 series.

At the same time efforts were made toward the achievement of continuously variable camber. This was achieved by the HKS-1 and HS-3 in the mid-1950s, but for reasons unknown, has not been widely used. Area and camber-increasing flaps of the Fowler type have also been used, but with the small wing chords, the difficulties of structural stability and tolerances have been considerable. In addition, the difficulty of matching the wing characteristics with flaps up and flaps down has so far not been successful, the wing

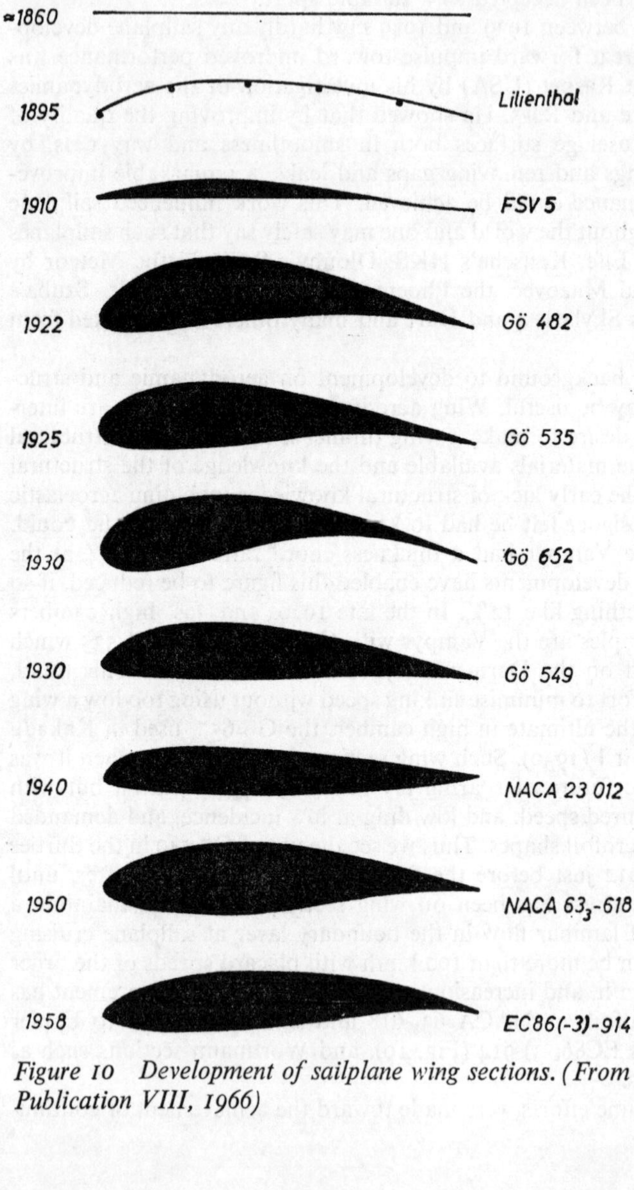

Figure 10 Development of sailplane wing sections. (From OSTIV Publication VIII, 1966)

with flaps down being improved in some resects and worsened in others. A real solution enabling the wing characteristics to be adjusted accurately and optimally to each flight condition has not yet been demonstrated.

Early monoplane sailplanes used moderate taper or elliptical planforms. In the mid-thirties there was a tendency towards sharp straight taper to save weight. However, such sharp taper led to tip stalling, and at low speeds to suspected laminar separation near the tips which added drag. The trend then changed towards moderate taper of not more than about tip/root ratio of 1/3. Later work, outlined particularly clearly by Wortmann[18], has led to a large proportion of the wing planform being rectangular (up to about half span) the outer part having a moderate straight taper with rectangular or Hoerner type tips[19]. External wing bracing has disappeared. Spanwise wing twist in the direction of washout has been widely used on tapered wings. Experience and theory both show that the greater the taper, the greater is the tendency for the outer portions of the wing to stall before the inner portions, which can lead to loss of lateral control and 'wing dropping'. This means that the aircraft may bank itself suddenly one way or the other and require considerable altitude to recover. Reducing, or washing out, incidence towards the tips would tend to delay this tip stall. This was all very well for sailplanes which were not required to fly fast efficiently. Fast flying on such aircraft as Fafnir resulted in the outer wing being negatively loaded, increasing both induced and profile drag. The present tendency, using less taper, is either to have no twist or slight aerodynamic twist which is often attained by using slightly different sections at root and tip.

The present situation is that since about 1950, a maximum gliding angle of about 1/40 has been achieved. It took 30 years to reach that from the 1920 figure of about 1/10 or alternatively the Vampyr figure of 1921 which may have been as good as 1/16. Now 1/40 is 'easy' to attain, and work is progressing towards making great improvements. However the stress is not just on best L/D but on high values at high cruising speeds, which for good designs should be better than 1/20 at say 150 km/h.

No improvements in minimum sink or in L/D can mean very much if the flying qualities are not good. For the first ten years of development controllability and stability were both inadequate. The use of short tails and balanced all-moving tails resulted in longitudinal instability, and the over-flexibility of the wings in torsion gave inadequate rolling control. As knowledge increased, tails were increased in length, and either fixed tailplanes with elevators or massed-balanced all-moving tails with anti-balance tabs were used. Wings were made torsionally stiffer, torsion and flexure axes properly located so that aileron reversal and flutter became, for a while, things of the past. However, recent efforts towards increasing performance have again resulted in flutter occurring. It shows the necessity for designers of faster and structurally more flexible designs to give more detailed attention to flutter. The requirements for the use of wing sections which could retain laminar boundary-layer conditions over a good proportion of the chord resulted in stiff wing coverings of plywood, plastic or metal which helped the problem of structural flexibility. At present there appears to be a tendency towards the all-metal sailplane including the external covering, but it appears that the most perfect surfaces are those in plastic.

However, ideal flying qualities and the highest performance are hardly achievable simultaneously because the various requirements are to some extent inconsistent. The art, today, is to find a good compromise. If it were to become a science rather than an art, it is likely that sailplane performances would improve considerably.

From 1930 until the time of writing (1968) the development of sailplanes and flying techniques have been assisted and encouraged by two international organisations: ISTUS (Internationale Studienkommission fuer den Motorlosen Flug) from 1930 to 1940 and OSTIV (Organisation Scientifique et Technique Internationale du Vol à Voile) from 1948 onwards. ISTUS was German-based and assisted, whereas OSTIV has no national leaning. For that reason OSTIV has no official or unofficial backing and must depend on the spare-time efforts of its members. Both ISTUS and OSTIV held, or hold, technical conferences at which papers

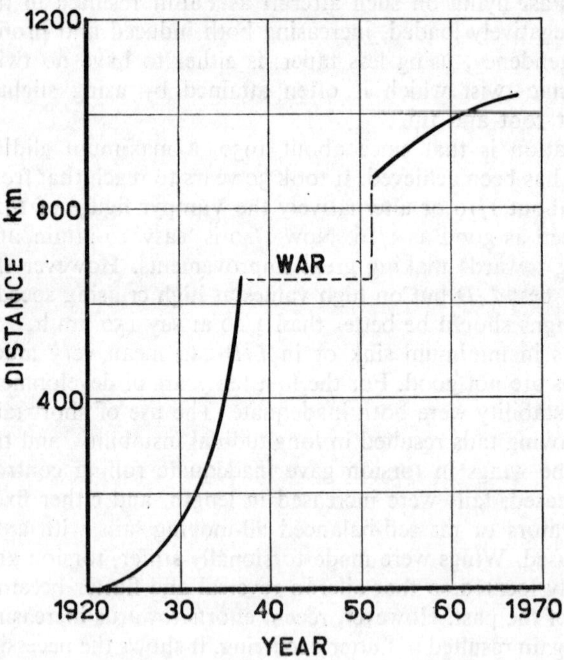

Figure 11 History of Distance Records

are presented concurrently with World Gliding Championships. ISTUS produced seven publications and OSTIV has produced nine.

It was under the auspices of OSTIV and in collaboration with the FAI (Federation Aeronautique Internationale) that the 15 m span Standard Class sailplane was introduced for international competition. This class has led to increased attention to detail design refinements, since the designer has had no scope for improving performance by merely adding span. In addition to this, OSTIV has produced two volumes of technical data on sailplanes (*World's Sailplanes, I and II*)[20], design requirements for the Standard Class, and more lately, general design requirements. These were produced and are kept up-to-date by a remarkably amicable group of international experts of many nationalities. Under OSTIV auspices special flying courses, meteorological courses and flight testing and evaluations have taken place.

The future of sailplanes may be in both of two directions: first a continuation of the present drive for increased range, speed and altitude just for the sake of the accomplishment and secondly a drive for the increased use of sailplanes for scientific investigation of the air around us.

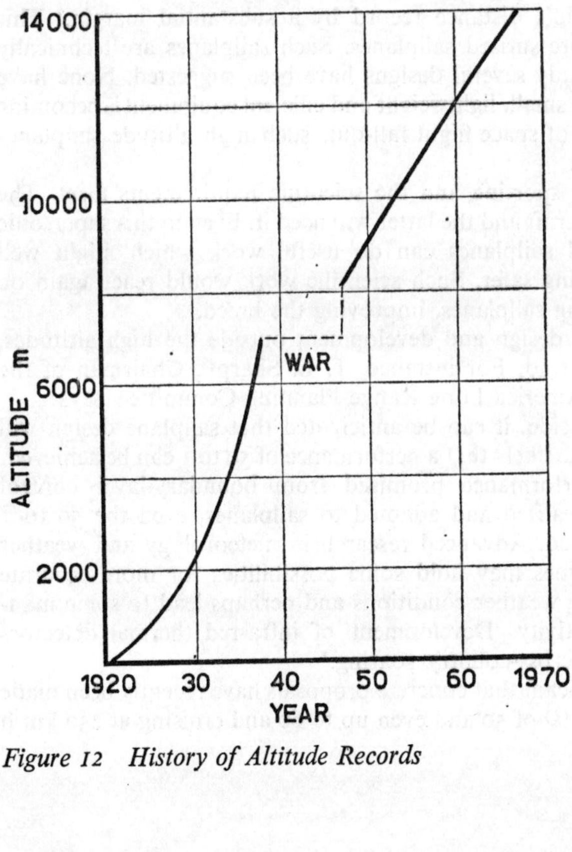

Figure 12 History of Altitude Records

The present 1967 important sailplane world records are[21] (*see also* Fig. 11 and Fig. 12):

Distance	1041·52 km
Altitude	14,102 m
Speed (100 km triangle)	136·13 km/h

Range and speed are dependent almost entirely on various forms of thermal activity which means they are limited by sun activity and are therefore daylight exercises. Altitude which is reached by using lee waves, does not depend on daylight, but there is no reason to use the night. If the lateral extent of lee waves were accurately known, it should be possible to fly far and fast along mountain ranges at heights below those which would require a pressure cabin. The records of surprisingly extensive lee wave areas shown by the Tiros satellites gives promise of such range increases which would not depend on daylight. John Aldrich, a meteorologist, sees a future for soaring in the jet-stream. He says[22]: 'The soaring pilot who is bold and skilful and lucky enough to rendezvous with the left front quadrant of a jet-stream maximum stands an excellent chance of topping the world's distance record by a substantial margin.' This implies the use of pressurised sailplanes. Such sailplanes are technically practicable and already several designs have been suggested. None have been built. Now that small, lightweight and efficient equipment is becoming available as a result of space flight fall-out, such high altitude sailplanes should be developed.

This is where the sporting and the scientific requirements meet. The former wants the aircraft and the latter will need it. Even in this supersonic age such specialised sailplanes can do useful work which might well make supersonic flying safer. Such scientific work would react again on the design of sporting sailplanes, improving the breed.

On the future for design and development outside the high altitudes, quite a lot is being said. For instance, T. E. Sharp[23], Chairman of the Soaring Society of America Long Range Planning Committee, says:

'On the technical side, it can be anticipated that sailplane design will advance and it is quite likely that a performance of 50 to 1 can be achieved. If the potential performance promised from boundary-layer control techniques can be realised and adapted to sailplanes, even the 50 to 1 ratio may be exceeded. Advanced research in meteorology and weather modification techniques may hold some possibilities for more accurate prediction of soaring weather conditions and perhaps lead to some man-induced thermal activity. Development of infra-red thermal detectors could revolutionise cross-country soaring.'

It is perhaps significant that concrete proposals have recently been made for sailplanes with L/D of 50 and even up to 60 and cruising at 250 km/h at L/D of 20.

There is no lack of optimism for the future. There are many ways ahead and so far none of them is blocked.

7 SAILS

Although sails do not fly, they pose, like wings, severe aeroelastic problems under constantly changing turbulence and air speed. In addition, sails have to be retractable, variable in area, and they are plagued with numerous interference problems.

A great deal has been written about sails, but most of it is in the category of superstition and old wives' tales.

The first proper technical approach to sails was the work by Edward P. Warner and Shatswell Ober in 1925[24] which gave the first information on pressure distributions on a sail in action on a sailboat. At about the same time, Croseck was making wind tunnel tests on sails in Germany[25]. Tanner in 1930[26], on the basis of his wind tunnel tests, gave a succinct and clear account of the basic but simplified sail aerodynamics. Further basic work was done by K. S. M. Davidson in 1936[27]. Practically all the later work has been done since 1949, and at the time of writing, the most authoritative publication is C. A. Marchaj's *Sailing Theory and Practice* (1964)[28]. This book gathers up critically what has been published with the addition of little-known work and the results of his own thoughts and experiments in Poland and in Southampton, England. It is the only outstanding contribution to the subject known to the writer. The special equipment at the University of Southampton, such as a calibrated Dragon class sailboat, a sail wind tunnel and a yacht hull testing tank has done, and is doing, a great deal to increase information about sails, the work having been inspired and directed by Thomas Tanner.

An Advisory Committee for Yacht Research has been set up by the University of Southampton, the chairmen so far having been Mr. Handel Davies, Sir Walter Cawood and Sir Robert Cockburn.

The Massachusetts Institute of Technology has also a group working on sails, but little has yet been published.

Theoretical work has been done by B. Thwaites[29] and J. N. Neilsen[30] applicable not only to sails but to flexible winged paragliders. This is one point where sails and aviation meet. Another such point is the use of wings (i.e. thick stiff structures) for sails. The replacement of the aerodynamically bad mast with something better has often been tried, but so far not with complete success, as alternatives seem to add other sources of drag or interference or reduce the 'lift', to reduce or neutralise the improvement. For instance, the Llungstrom wrap-around double sail cannot use a headsail, and thereby loses more than it gains. Work on the rigs of the Bembridge Redwings over 30 years ago, with Lord Brabazon's encouragement, illustrate the same problem, as does the work of others in the same period, such as Gidge Gandy[31].

Sail aerodynamics is confused by a series of dependent and independent variables, many of them only qualitatively realised. The shape of sails changes with the wind speed and scale and heel angle and incidence. Even if the sail profile or section shape were constant, the loading of the sail would depend on the dynamic pressure, incidence, heel, mast characteristics, boom position, twist, battening and trim.

Many early tests were done on metal sail models of fixed shape. Nowadays many of the problems of flexible sails for wind tunnel testing have been solved. However, the gap between technology and tradition is wide in this area of activity, and is made worse by rating rules which all eventually cater for design stagnation, whether like the old grotesque 'plank-on-edge' boats or the slow death of the '6 metre' because of exhaustion of rule flexibility. Sail research suffers from such rules if used to get around such rules, but only research can point out better ways of using sails and better forms of sail.

Sails are like wings, but they operate almost entirely outside the parts of the lift-drag polar used by aircraft. Tanner's 1930 oversimplification shows that sails can make no use of the polar below the point of L/D_{max}, but they use the whole of the rest of the polar, not just to 90° incidence but beyond, the stall having little significance*.

The modern sail, roughly triangular, is not an efficient shape, being too sharply tapered. No wing would be so shaped because the induced drag would be high and it would have dangerous stalling characteristics. To cure the latter, the pointed tip would require theoretically an infinite washout, or in practical terms, a lot of twist in the direction of reducing the incidence near the tip. The sail has automatically a great deal of head washout and since it operates normally beyond the stall, the wing comparison is not entirely valid. In addition the wind speed over the sail increases with altitude. For a 30 ft (9m) high sail, the air speed at the head will be about twice that at the foot, which means that the load distribution on a sail is very different from that on a wing. Even so, there is too much taper on sails, and the old Dutch compromise of a small curved gaff (or an extra-wide headboard) may well improve sail effectiveness.

Very recent work at Southampton, comparing the differences between 'thin' sails and 'thick' wing sails is very much to the point. *R* for the triangular sail varies from zero at the head to about five million at the foot of a '12 metre' class boat. At *R* of half a million the value of maximum lift coefficient for single thin sails of about 15% camber is nearly 2·0. For a thick wing-sail of section 64 A 212 of 2% camber it is only 0·8. However, the best *L*/*D* for the thin sail is only 8 and the thick wing-sail goes up to 20. That is all very well, but for the all-important close-hauled sailing, a large lift coefficient is necessary and more important than the drag. Thus it would appear that thick wing-sails are nothing like as good

*In effect, aircraft use only the first little bit of the polar and sails use all the rest.

as thin soft sails, and further, since the flow over the thin sail is unattached at both the leading and trailing edges over the whole range of incidence, existing aerodynamic theories are not applicable. Here is a very large field for future work.

One of the most difficult problems of sailing, particularly in centre-boarders, is the vicious rolling and yawing which often occurs when sailing off the wind. When running downwind with the boom hauled well in, the sail leading and trailing edges change places, as here we have incidences over 90°, possibly even 135° or more. The highly cambered part then finds itself near the trailing edge instead of near the leading edge. The Southampton group has applied the methods used in the study of dynamic stability of aircraft. For instance, if the boat is making small oscillations in roll in a smooth sea on a downwind course is the motion stable or unstable? The answer partly depends on the sign of the slope of the lift curve ($dC_L/d\alpha$) plus the drag coefficient. Drag is always positive, but after the stall the slope of the lift curve becomes negative and can exceed the value of the drag coefficient, so that there is positive aerodynamic excitation. If this exceeds the hydrodynamic (hull including keel or centre-board) damping moment, as it may well do after some critical wind speed, capsize is almost inevitable. Both the aerodynamic and hydrodynamic rolling moments induce yawing moments which may be large if one fearfully holds the tiller rigidly. Here apparently we have the reason for that unexpected broaching to in a smooth sea but there is still a great deal of work to be done. We should expect to discover in the future how to avoid these instabilities, whether by aero- or hydrodynamic means.

Sails still suffer through the relative lack of control of their shapes. It is partly because of the sailcloth, although Terylene (Dacron) has improved things, but sheeting, rigging, sail cut, the rapid change of airspeed with height and interferences caused by masts, or between sails are all involved. Recent knowledge is being applied rapidly and perhaps the most hopeful development is the trend toward much higher sailing speeds, by the increasing use of multi-hulled boats of small displacement for their hull lengths.

This trend brings with it much greater sail loadings, tending toward those met in ice yachts and results in quite different sailing techniques. The speeds of 40 knots (20 m/s) being hopefully bandied about, would certainly lead to close-hauled sailing on all courses (zig-zag down wind) and may simplify and possibly eliminate many present-day problems, but bring forth new ones such as very high handling loads. Since for a given performance the loads on inefficient sails are greater than those on efficient sails, there will be an endeavour to improve sail effectiveness just to improve handling loads.

Even higher speeds must depend on quite new hull design, but that problem is outside this chapter. The ultimate should be hulls with no more frictional drag than steel has on ice at about –5°C. A boat could then on a reach, sail much faster than the real wind speed, as catamarans already do to some extent, and ice boats even more so. Here lie problems and challenges enough to keep any sporting/research group busy for a couple of generations.

One of the obvious developments for sails is a sail arrangement which produces no heeling moment. This is not a new idea. In 1938 Rowland[32] published results of qualitative tests which were at least encouraging. Geometrically it is easy to achieve, but a workable layout seems to be difficult. However, it is bound to come, particularly if racing rules are re-cast to encourage novelty and advancement instead of the opposite.

What is needed in the future is a great deal of knowledge about known phenomena and difficulties. Such work can be planned and as sailing becomes more widespread, such research will be worthwhile because recreation is a business and money will become more easily available. In addition to what can be planned there is the likelihood that quite new ideas will come forward which would change or expand the whole outlook on sails and sailing.

References

1 KAHN, H. and WIENER, A. J. Technological Innovation and the Future of Strategic Warfare. *Astronautics and Aeronautics*, Vol **5**, No 12, 1967.

2 PRINGLE, J. W. S. *Insect Flight*, Cambridge University Press, 1957.

3 HOLST, E. VON and KÜCHEMANN, D. Biologische und Aerodynamische Probleme des Tierfluges. *Naturwissenschaften*, Vol **29**, p 348, 1941.

4 HOLST, E. VON. Untersuchungen über Flugbiophysik I. *Biologisches Zentralblatt*, Vol **63**, p 289, 1943.

5 THOM, A. and SWART, P. The Forces on an Aerofoil at very Low Speeds. *Journal of Royal Aeronautical Society*, Vol **44**, p 761, 1940.

6 HORRIDGE, A. The Flight of very small Insects. *Nature*, Vol **78**, p 1334, 1956.

7 SCHMITZ, F. W. *Aerodynamik des Flugmodells*, Carl Lange Verlage, Duisburg, 1942, 1952.

8 HACKLINGER, M. Theoretical and Experimental Investigation of Indoor Flying Models. *Journal of Royal Aeronautical Society*, Vol **68**, November 1964.

9 CONE, C. D. JR. A Mathematical Analysis of the Dynamic Soaring Flight of the Albatross with Ecological Interpretations. Special Scientific Report No 50. Virginia Institute of Marine Science, 1964.

10 WILKIE, D. R. Man as an Aero Engine. *Journal of Royal Aeronautical Society*, Vol **64**, August 1960

11 URSINUS, O. Mitteilungen des Muskelflug-Instituts 1 to 6 *Flugsport*, 1936 and 1937.

12 THOM, LUND and TODD. Efficiency of Three-Speed Bicycle Gears. *Engineering*, 20 July, 1956.

13 WORLEY, B. Can Man Fly? *Aeronautics*, February 1948.

14 SHENSTONE, B. S. Manpowered Aircraft. *OSTIV Publication VIII*, 1966.

15 CZERWINSKI, W. Structural Trends in the Development of Man-Powered Aircraft. *Journal of Royal Aeronautical Society*, Vol **71**, January 1967.

16 HIRTH, W. *Handbuch des Segelfliegens*. Franckh'sche Verlagshandlung, 1953.

17 ZACHER, H. The Shape of High Performance Sailplane Technical Development. *The World's Sailplanes*, Vol **II**, OSTIV, 1963.
18 WORTMANN, F. X. Widerstandsverminderung bei Segelflugzeugen. *OSTIV Publication VIII*, 1966.
19 HOERNER, S. F. *Fluid-Dynamic Drag*. pp 7–5 and 7–6. Published by the author 1958.
20 *The World's Sailplanes*. Vol **I** and **II**. OSTIV and *Swiss Aero Revue*, 1958, 1963.
21 *Records List*, British Gliding Association, 1967.
22 ALDRICH, J. Soaring in the Jet Stream, *Soaring*, December 1967.
23 SHARP, T. E. Preview 1966–67. *Soaring*, December 1967.
24 WARNER, E. P. and OBER, S. The Aerodynamics of Yacht Sails. *Trans Society of NAME*, 1925.
25 CROSECK, H. *Beiträge zur Theorie des Segelns*, Springer, Berlin, 1925.
26 TANNER, T. The Forces on a Yacht's Sail. *Journal of Royal Aeronautical Society*, Vol **34**, pp 880–888, October 1930.
27 DAVIDSON, K. S. M. Some Experimental Studies of the Sailing Yacht. *Trans Society of NAME*, 1936.
28 MARCHAJ, C. A. *Sailing Theory and Practice*. Adlard Coles, 1964.
29 THWAITES, B. *Proceedings Royal Society. A.* Vol **261**, 1961.
30 NIELSEN, J. N. *ASME Trans J App Mech*, September 1963.
31 GANDY, G. A proving Ground for Theories. *Yachting*, p. 44, November 1932.
32 ROWLAND, J. T. Experiments with the Kite Rig. *Yachting*, p 50, August 1938.

9 Power generation for aircraft in the Second Century

M. W. Thring

I AIMS

Power generation for civilian aircraft can only be considered in terms of what kind of flying people will want to do in the Second Century of the Royal Aeronautical Society because the weight and cost of the fuel which will be best are dependent on how much people can afford to pay, how large and how fast they want the planes to be and what the world's total aircraft fuel consumption will be in relation to the world's total supplies of various fuels.

The predictable *primary* power sources of the Second Century are the fossil chemical fuels, oil and natural gas, the nuclear fuels (fission and fusion) and solar energy.

At the present time, about 5% of the world's annual petroleum consumption goes on aircraft propulsion. I propose to base my predictions for the power generation in the next century on the following rough assumptions.

1 The world population will rise to about twice the present figure but then level off.
2 The main increase in aviation will be that arising from the desire of all mankind to have a standard of living (including travel and transport of luxury goods) about equal to that enjoyed at the present time by the average European and North American. This has two main implications.
(*a*) The number of passenger miles flown in the world will rise to 25–50 times the present figure, and the air transport of luxury goods will probably be multiplied by a somewhat larger factor.
(*b*) The main development of aircraft power generation will be in the direction of economy of fuel costs per passenger mile and not in the direction of increased speed. This fuel economy requirement will be largely the result of the limited world fuel resources.

From the point of view of the world citizen 100 years from now, the

Originally published in the *Aeronautical Journal of the Royal Aeronautical Society*, September 1968.

order of priorities in his choice of aircraft power systems will probably be as follows:

1 Safety.
2 Reliability (freedom from unscheduled delays of all kinds, especially those caused by weather).
3 Low cost (particularly low fuel cost and plane cost per passenger mile).
4 Convenience (ease of booking, avoidance of transfers, waiting, standing, queuing etc.).
5 Comfort during flight (e.g., leg room).
6 Optimum time of total journey—home to final destination. Physiological and psychological research is already indicating that even at Mach 0·8 there is considerable stress on a human making frequent flights of length >1000 km when flying east-west because of the change in meal times and night times and when flying north-south because of the temperature changes. I therefore base my predictions on the assumption that the greater part of the additional flying will remain subsonic and the basic problem is to predict how power will be generated to satisfy consideration 3 (low cost) with increased safety, reliability and comfort.

The family motor car increased its cruising speed from 20 mph to 60 mph (9–27 m/s) in the first 20 years of its development, but it has remained substantially unchanged ever since, while development has been on considerations 1–5 for the next 40 years. I predict that aviation will follow the same path with 600 mph (270 m/s) instead of 60 mph (27 m/s) but just as with cars, higher speeds (supersonic or even hypersonic) may be used in special cases. Perhaps, just as we now have a road speed limit of 70 mph (31 m/s) in Britain, so a Mach 1 speed limit will be international law at the end of the Second Century because of the sonic boom. Certainly in planning a good future for aviation, the millions of people on the ground must be considered as well as those travelling.

Under the heading of convenience, we may see a comprehensive transport system in which passengers load themselves and their luggage on to chair-carrying platforms of 6 square feet (0·56 m^2) per person which they drive on their own vehicle to a rail station where it is transferred to the train and thence on to the plane at the airport and off to another train or a hired vehicle at the other end. This would affect the design of the plane only in that the platforms would be fed in on roller belt conveyors through slots along the side or back and then clamped in place.

2 TRANSPORT POSSIBILITIES IN THE SECOND CENTURY

Table I shows various proposals for transport vehicles. A great deal has been written about the supersonic, hypersonic and ballistic vehicles but I propose to discuss the power systems which may be developed in the

TABLE I *Possibilities for transport in the Second Century*

No.	*Cargo*	*Range miles*	*Velocity mph*	*Altitude miles*	*Possible Vehicles*
1	Goods + 10,000 tons	Sea all ranges	up to 100	0	1 Air cushion vehicle (Hovercraft). 2 Belt propelled vehicle (Fig. 1).
2	Goods 200 tons	Land and sea all ranges	120	½–1	Helium filled airship[1].
3	Mass passenger and bulk goods	Inter-city < 500	200	0·01	Overhead monorail. Air or magnetic[2] bearing. Linear motor[3]. Possible use across shallow seas with towers.
4	Mass passenger and special goods	500–5,000	600	6–8	Winged aircraft. Catapult or pickaback take-off. Boundary-layer propulsion? Fully automatic pilot (flies on ground beam system).
5	1st class passenger	3,000–10,000	*M* 2–3 supersonic	10–15	Delta winged.
6	1st class passenger	6,000–12,000	*M* 5–8 hypersonic	20–35	Swing-wing[4] or 90° axis change[5].
7	1st class passenger	6,000–12,000	18,000 ballistic[6]	>50	Wingless. Launched at steep angle on returnable booster.

next 100 years for vehicles of types 1–4, as they can provide the whole of humanity with the transport needed for a life without excessive strain and rush, and without creating excessive destruction of natural beauty, noise and atmospheric pollution. The vehicle designer must serve the best interests of humanity as a whole and not look just at those problems of the greatest aerodynamic or thermodynamic interest.

I predict that the large tonnage sea vehicle of the future will be one which can go at 30–60 mph (13–27 m/s) with less power than is required to propel the present large ships. It will be capable of crawling out of the sea up a beach and its sides will let down for unloading cargo with lorries or trains. With large ships at present the bulk of the power is absorbed by skin friction[8] caused by the very high Reynolds number of the flow. The air cushion vehicle reduces this power absorption by turbulence by introducing the pressurised air between the vehicle and the water but then requires considerable power just to support the vehicle (not so much as an aircraft because of the smaller escape area around the perimeter) and it is more difficult for it to propel itself by pushing the water backwards. The ship of the future will have two flat rubber belts ribbed on the outside, running down the bows, under the flat bottom and up the

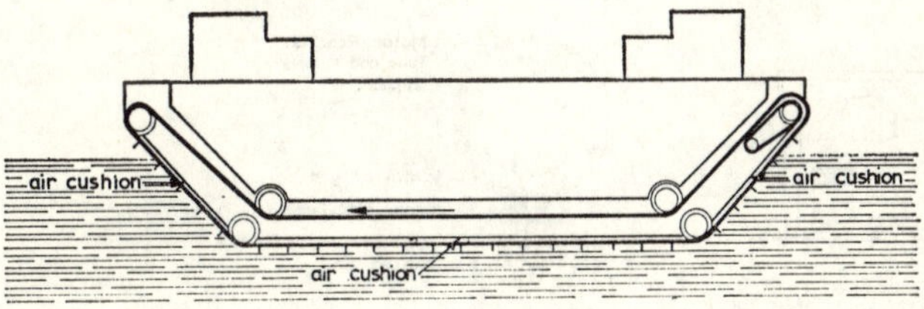

Figure 1 Air-bearing ship propulsion

stern (*see* Fig. 1). These will run backwards at the ship speed, i.e. they will be stationary in the water. The narrow space between the belts and the ship may be filled with pressurised air (the perimeter leakage gap need only be an inch (0·025 m) so the leakage is small) or water—in this case it must have a gap such that the flow between ship and belt is laminar.

Barnes Wallis[4] has proposed giant submarines filled with rectangular cargo containers for Australia, to run under the north polar ice cap. They would be powered with turbines running on combustion products from oil and liquid oxygen and the CO_2 and H_2O would be condensed. This would certainly increase the tonnage carrying capacity at a rate which would be cheap in everything but the oxygen cost, so it may well come if government development money were to be channelled in such a direction.

A helium filled airship 190 feet (58 m) in diameter, 1000 feet (305 m) long with a gross lift of 440 tons (450×10^3 kg), with 8000 hp (6 MW) diesel/gas turbine compound engines and a speed of 110 mph (50 m/s) has been proposed by L. Richards[1]. This would carry 180 tons (183 Mg) of cargo with an available cargo volume of 1,000,000 ft^3 (28,300 m^3) at a flying altitude 3000–4000 ft (900–1200 m). A flattened hull shape to give significant aerodynamic lift and the condensation of water from the exhaust to replace the fuel weight have been proposed. The problems are (*i*) the turbulent wind forces on the structure, (*ii*) cargo and crew landing and take-off.

I think it is quite possible that such a vehicle will find a place in the future for the transport of heavy engineering equipment overseas especially to inaccessible places and perhaps even for delivering cars and relieving roads within a country. Boundary layer propulsion may well reduce the power required per ton-mile still further; it is already 15 times less than that for a winged vehicle. Alternatively, it may be used to increase the speed perhaps to 200 mph (90 m/s) increasing the ton-miles per year.

One of the greatest transport needs of the developed countries is for a 150–200 mph (70–90 m/s) overhead monorail linking city-centre to city-centre, running automatically and independently of the weather. The vehicles will be supported from the rail either by a very thin air bearing inside a hollow rail (*see* Fig. 2) or by magnetic repulsion[2]. In any case they

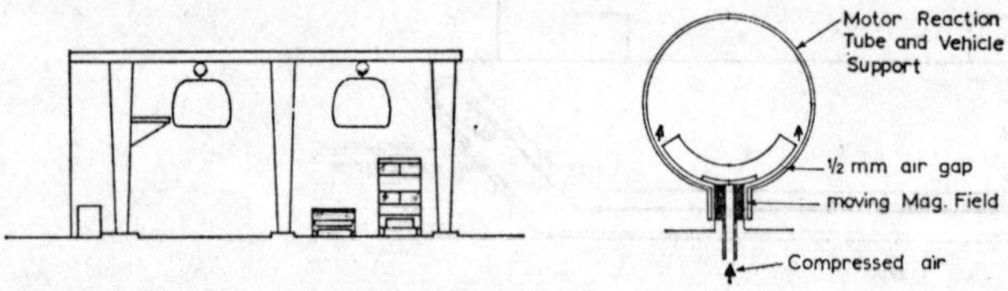

Figure 2 Air-bearing monorail vehicle

will have no wheels to cause rumble nor will they require a large volume power consuming downward airflow to support them. They will be propelled by linear motors, i.e. drawn along by a moving magnetic field in the support rails, so that the power will certainly come from a fixed central power station. At present, this power has to be picked up by brushes travelling with the vehicle but we shall find a way either of beaming microwave power along the inside of the hollow rail or of inducing a travelling magnetic field in the rail without conveying the power first to the vehicle.

For aircraft operating from city-centres there is vertical take-off or a steeply sloping catapult ramp. VTO is liable to considerable noise difficulties although it may be possible to reduce this in some way. Probably the most important contribution to be made by VTOL is for a completely flexible system for occasionally joining places which have no landing strip.

There is one possible combination of electricity carrying and guiding rail with wing support which may well enable an automatic driverless vehicle service to shuttle across the channel safely in all weather at 100–200 mph (45–90 m/s). This is a 50 ton (51 Mg) vehicle with wings which just support its weight at full speed. It runs with three-phase electric supply picked up from inside a hollow steel tube which is supported 300 ft (91 m) above the ground by suspension cables from 1000 ft (305 m) towers every few miles. It is propelled by a linear motor-moving field generated in the rail and the rail comes down to ground level in the mile near the station; here the rail is supported sufficiently to carry the full weight of the plane with small wheels or air bearing (Fig. 3).

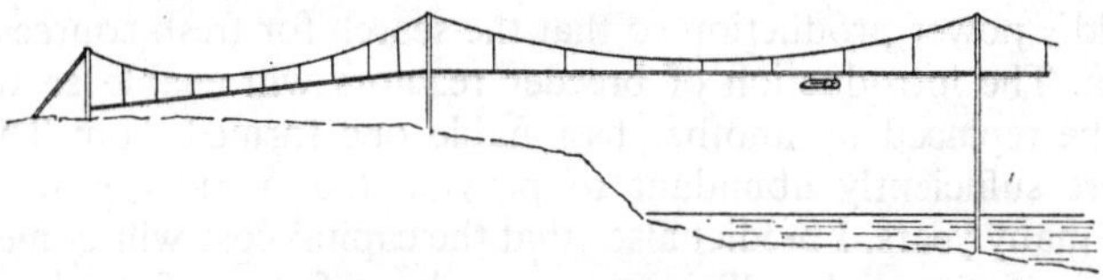

Figure 3 Winged monorail vehicle

In all the vehicles considered so far in this section the power supply can be satisfactorily obtained with existing fuels and engines or power stations, their development depends primarily on sufficient expenditure on the vehicle and system. However, when we come to type No. 4 of Table 1, the subsonic winged aircraft, the situation is quite different. The number of passenger miles flown in subsonic winged aircraft is rising at about 14% per annum, i.e. doubling every 5–6 years, and will probably continue to rise at about the same rate until it reaches the 50-fold figure around the end of the 20th century. This means that this is the field where most imaginative developments in power production should take place if technology is to serve the real needs of humanity.

3 THE WORLD FUEL AND ENERGY SUPPLIES

These can be broadly summarised as follows.

1 Proven reserves of liquid petroleum are of the order of 50 years supply at the present rate of usage. At least as much again will be found by further surveying and drilling but it will be in less accessible places and the cost of production will rise. If a method of finding out whether oil exists in a geologically suitable site (impervious dome or porous rock) without drilling a full-size well, then the search for petroleum can be rapidly and cheaply extended. This may happen within the next 100 years.

2 The Athabasca Tar Sands contain as much oil as the present known world's reserves, while the Colorado Shales contain twice as much. At present, it is considerably more expensive to obtain oil from the Tar Sands than from pumpable wells, but this difference of price will certainly be reduced as improved methods are developed for winning the tar and combining its processing with other industries which use the coke and heat.

3 Coal reserves are at least 100 times as great as oil reserves but until ways are found of making electricity from coal without conventional mining, the use of coal will continue to decline. Even when this problem is solved, coal will only be used for transport via stationary power stations.

4 The cost of electricity generation in large power stations from nuclear fission of U^{235} (the only naturally occurring fissionable isotope) will be as low as that from oil fired power stations within a decade or so. Reserves of Uranium in which U^{235} occurs as 1 in 150 are more than adequate for 10% of the world's power production so that the search for fresh sources has slowed down. The introduction of breeder reactors will enable each U^{235} nucleus to be replaced by another fissionable one formed from Th or U^{238} which are sufficiently abundant to provide the world's power requirement for many years. I predict also that the capital cost will come down by a factor of 3 and the efficiency go up by a factor of 2 when the homogeneous gaseous closed cycle MHD/vapour reactor is developed in about 50 years.

5 Finally, I will predict that nuclear fusion of two light deuterium nuclei D^2 to make an He^4 nucleus will be achieved within the century, and that an MHD method of obtaining electric power from this process without going via heat will give an unlimited and relatively cheap source of ground power generation. It may also provide a portable power pack with negligible fuel weight but it is liable to require a minimum size of the order of millions of horse-power.

I think, therefore, that portable hydrocarbon fuels will be available for all the transport needs for perhaps 50 years but that before the end of the century it will be necessary to find a method either of conveying power to moving vehicles from ground power stations or to use electricity to make a high energy fuel (e.g. electrolysis of water). I would also predict that during the first 50 years the continued expansion of air travel will make it essential to use air breathing engines using fuels cheaply derived from naturally occurring petroleum.

4 POSSIBLE DEVELOPMENTS IN CHEMICAL FUELS

The present conventional fuel for aircraft gas turbines is essentially kerosine with *C/H* ratio of 6·3, gross *CV* 11,140 kcals/kg, 8810 kcals/litre, net 10,400 kcals/kg, 8250 kcal/litre.

A special fuel has been manufactured from petroleum by a more expensive process than conventional distillation with a *C/H* ratio of 10·4 and a net *CV* of 9950 kcals/kg but 10,900 kcal/litre, while really expensive boron-based synthetic fuels have net *CV*s of 16,700 kcals/kg and 10,100 kcal/litre, but these fuels are unlikely to be used because of toxic and glassy exhaust. Another possible fuel of the future is liquefied CH_{24} (methane) which is the principal constituent of natural gas so that it is a cheap fuel. It has net *CV* 11,800 kcal/kg, 4920 kcal/litre. The possible fuel with the greatest *CV* per unit weight is liquid hydrogen with a net *CV* 28,900 kcal/kg, 2030 kcal/litre.

This can be made at present most cheaply by reacting steam with hydrocarbon vapour in an externally heated tube and removing the CO_2 but it is still a comparatively expensive prepared fuel. If electricity becomes really cheap, made by nuclear energy, pumped coal or hydrogen in hitherto inaccessible regions, it could be produced by electrolysis of water and perhaps the oxygen would be used in Barnes Wallis' submarines?

One other fuel development is important in this connection. This is a fuel[7] which is in the form of a gel in the tank but is turned into a liquid by the shear forces and heating effect of pumping and atomisation. This would substantially reduce the fire hazard in the case of a crash due to the flow of liquid fuel out of the tank.

The reason for seeking a chemical fuel with a higher *CV* per unit mass is that nearly half the all-up weight of a jet taking-off for a long flight is fuel, so a small increase in *CV*/kg can produce a large increase in payload. This is probably more important than the volume requirements of the fuel for civilian aircraft. The liquefied gases methane and hydrogen both require thermally insulated tanks and even then have a natural boil-off rate so, at present, they are only being considered for supersonic or hypersonic planes where they provide a better wing coolant than a kerosene type fuel. However, I predict that subsonic civilian transport planes will steadily increase in size perhaps up to 5000 passengers or more and that liquid methane with net *CV*/kg 13% higher than kerosene and cheaper to produce is likely to be used.

I do not think nuclear fission will become feasible to carry as a power source on an aircraft because of passenger shielding and crash hazard but possibly nuclear fusion will be developed to provide a clean safe non-thermal power source with practically zero fuel weight and cheap unlimited fuel supplies (D^2 from seawater).

5 ADVANCED POWER SYSTEMS

One possible way of combating the loss of payload due to fuel weight for a long flight is to get the aircraft to the cruise altitude and velocity with a power system it does not carry. This could be done by a tow rope, a robot piloted pick-a-back or towing launch vehicle or by beaming electromagnetic waves from the ground. All these could be developed but 600 mph (270 m/s) is only equivalent to some 12,000 ft (4 km) lift and it is necessary to take about 10 miles (16 km) to reach this speed if the passengers are not to be subjected to too severe acceleration. Thus the saving in fuel weight and increase in payload is not large and a tow system would have to occupy 10 miles (16 km). Nevertheless, I predict that a rope tow system of launching is likely to be developed because it could make a single airport able to land aeroplanes on a series of runways at right angles to a single launching line which could be despatching an aircraft every few seconds.

Then there is the possibility of obtaining a lift-drag ratio several times greater than the present figure of 18 for planes at high subsonic speeds, by boundary-layer propulsion. In the present type of plane the air behind the wings and the body has a forward velocity which is a small fraction of the plane velocity and causes the drag, and a downward velocity which causes the lift; there is a separate flow of propulsion air and combustion products which provides the thrust by its backward momentum. This thrust stream relative to co-ordinates fixed on the ground must be travelling backwards to overcome the thrust of the intake air. In boundary-layer propulsion, the engine intakes are slots along both the upper and lower sides of the wing situated near the trailing edge. They take in all the air which has been slowed down relative to the wing and eject it along the back edge of the wing with a velocity relative to the plane slightly higher than that of the plane relative to the ground. Experiments are being carried out on this system which requires a completely fresh approach to the aerodynamic shape of plane and wing and an engine which has a long narrow slit intake, no intake ram effect and a mechanical efficiency at least as good as the present engines. However, the incentive for developing the propulsion system for the airbus of the future is so great that I predict it will be successful. Perhaps we shall develop a constant volume combustion engine of entirely novel shape suited for a long narrow air intake and gas exit. One feasible solution to this problem of reducing the fuel per ton mile is illustrated in Fig. 4. This is three sets of fan blades above and below each wing, reciprocating 120° out of the phase in a direction along the axis of the wing and with an amplitude of a few inches. The blades are supported at the front edge on a vertical hinge with springs so that whichever way they are moving their angle of incidence accelerates the air backwards. Each set is driven by two free pistons one at each end, i.e. one at the end of the wing and one in the fuselage; the pistons are operated in push-pull in 2-stroke diesel cylinders with very high compression ratio to give very high engine efficiency.

There is nothing in our present knowledge of physics to suggest that the pull of gravity on the plane can be overcome by any static device other than a solid bar or buoyancy, although of course at $M=35$ in a ballistic trajectory the vehicle falls freely but maintains a constant height above the ground. The ability to switch gravity on and off without a corresponding energy introduction would of course enable one to have perpetual motion of the first kind so that the introduction of a successful antigravity device implies also a change in the laws of energy conservation. There is, however, a theoretically possible way of supporting a plane without creating downward air momentum and this is to use the pressure of electromagnetic radiation. This pressure is equal to E/c where E is the flow of radiant energy (mkg/m² s) and c is the velocity of light (m/s). This means that even with an energy flux as high as 10 MW/m² the radiation pressure is still only 1/300 kg/m², so this is ruled out.

However, it is possible that within the century we shall find a method of beaming electromagnetic radiation coherently as in a laser beam so that power can be transmitted a distance of a few tens of miles and received on the aerial constituting the underside of the aircraft with a transmission efficiency better than 50%. The wavelength will probably be in the range of 0·1–10 mm because the aircraft size must be some thousands

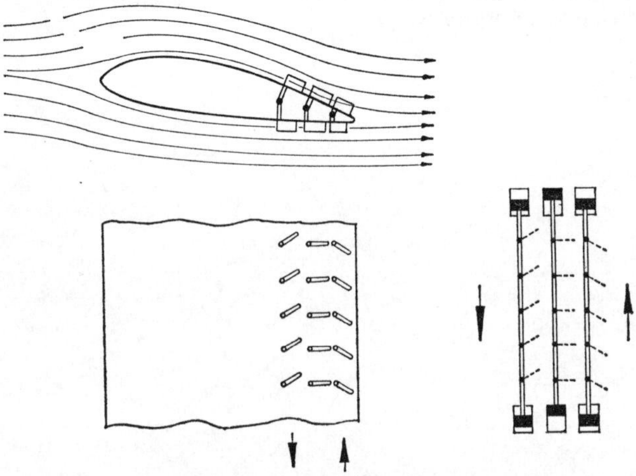

Figure 4 Direct powered boundary-layer propulsion

of wavelengths while it is perfectly possible to receive the radiation and convert it to electric power direct and not via heat. The aircraft will then have a special type of high frequency electric air-thruster, perhaps a series of motors with super-conducting magnets or more likely a mechanical device converting electric oscillation into air pressure oscillation which is mechanically rectified.

When this is developed, we can envisage a series of radiating power stations, one every 50 miles (80 km) along the air routes; those on the oceans will float on rafts. Aircraft will fly automatically along the points of intersection of beams from the successive stations, spaced therefore exactly 50 miles (80 km) apart, as if they were on a rail.

References

1 RICHARDS, L. A new Look at Airships. *The Inventor*, p 5, December 1967.

2 POLGREEN, G. R. *New Developments in Permanent Magnets*. Macdonald 1966.

3 LAITHWAITE, E. R. Propulsion Without Wheels, *Advancement of Science* 119, 1967.

4 WALLIS, B. N. The Strength of England. B.A. Section 5. *Advancement of Science*, 22, 1965/6.
5 LOMBARD, A. A. Propulsion in the New Century. RAeS Graduates' and Students' Section Centenary Symposium, *The Skyward Urge—Aviation* 1866–2016. Royal Aeronautical Society 1966.
6 GATLAND, K. W. and BONO, P. Towards Ballistic Flight. *Science Journal* p 42, December 1967.
7 Thickened fuels may reduce crash fires. *Aviation Week*, 4th July, p 36, 1966.
8 LACKENBY, H. The resistance of ships with special reference to skin friction and hull surface condition. 34th Thomas Lowe Gray Lecture Institute of Mechanical Engineers, 13th December 1961.
9 THRING, M. W. The present and future uses of machinery. (Fig. 2), *Journal of Royal Society of Arts*, p 359, April 1966.

10 Propulsion

L. G. Dawson

The rate of progress in most technical fields, and propulsion is no exception, has increased so rapidly during the past 50 years that it is becoming difficult to predict the trend of development for more than a few years ahead. On the other hand, there are fundamental limits to progress in certain directions and it is within these limits that this chapter is written. One must, of course, devote some thought to the validity of the limit. Not so long ago it was thought that the speed of sound would be a limiting speed for aircraft, for instance. One cannot go far wrong with the basic laws of physics and the intention, at least, of this chapter is not to transgress such laws.

The logical method of prediction is to extrapolate any given state of the art to the limit, then postulate an alternative solution which can again be extrapolated, and so on. In the field of propulsion this has happened again and again. Confining ourselves to mechanical propulsion, the steam engine reigned over a period of nearly a century before the internal combustion engine was refined to a sufficiently high standard to be competitive. The piston engine was accepted as the best prime mover for a period approaching half a century before the gas turbine was developed to the stage where it was ready to take over from the piston engine in aircraft. Even the turbine is outclassed at high speeds by the rocket engine.

Before trying to be logical about the future possibilities of the jet engine, it is interesting to note that the steam engine, IC piston and gas turbine engines are all heat engines which use what we now call 'fossil' fuel, that is combustible material produced in a more or less usable form by natural processes. One might even include the rocket in this general bracket—certainly the kerosene/oxygen engine—but it is stretching the point a bit with other types of fuel.

So while we think we are making a big jump forward in replacing a piston engine by a gas turbine, looked at in perspective all we are really doing is moving along the extrapolation of the basic curve of 'efficiency' of a heat engine using the familiar hydrocarbon fuel. This by no means

Originally published in the *Aeronautical Journal of the Royal Aeronautical Society*, September 1968.

detracts from the achievement of the jet engine pioneers—the gas turbine was a magnificent breakthrough in the engineering sense of production of high power for a low weight and small size, just as the steam turbine was over the steam piston engines.

The theme of this chapter is to think first about the ultimate possibilities of the heat engine, using conventional fuels and air as a working fluid. From there we probe the possibilities of alternative fuels and different methods of converting fuel energy into power.

It is impossible to imagine the way of life of the people living in the world in the next century, but it is fairly safe to assume that there will be an enormous amount of travelling about the world, and that this will be to a large extent by air, as this is a quick and comfortable way of doing it. It is very likely that there still will be a wide range of aircraft designed to fly at different speeds from slow speed short distance aircraft up to the 12,000 mile (19,000 km) range high speed giants. It is also fairly safe to assume that the engines of the future will be designed to be:

1 As reliable as possible.

2 As efficient as possible, to use as little as possible of scarce natural or man-made fuels.

3 As light in weight as possible, to allow more efficient aircraft to be made, and

4 As cheap as possible—that is to say designed to absorb little of the resources of man-hours and materials available.

Consider first the possible efficiency of future power plants.

LIMITS OF POWERPLANT EFFICIENCY

The efficiency of the propulsive powerplant depends on both the engine thermal efficiency and propulsive efficiency.

A propeller turbine and a pure jet engine could conceivably have the same thermal efficiency at, say, 200 mph aircraft speed, but the propulsive efficiency and, therefore, the overall efficiency would be very different.

Taking first the thermal efficiency of the engine, the conventional gas turbine cycle efficiency with 100% efficient components and zero pressure loss is given by a Carnot efficiency using the compressor outlet temperature as the top temperature. To get this in perspective, a figure of 900°K (627°C) compressor outlet temperature means a thermal efficiency of 76% in the stratosphere.

Accepting 900°K as a reasonable limit for subsonic engines, Fig. 1 shows in effect how the engine concept must be varied over the speed range to maintain a constant 76% thermal efficiency. While the engine efficiency stays constant at a constant compressor outlet temperature, the similar efficiency of a ramjet increases with Mach number until at $M=4$ the theoretical efficiency equals that of the jet engine. All this really shows is that at low speeds there is little temperature rise in the intake, and a high pressure ratio compressor is required, while at high speeds less and less work need be done in the compressor to achieve the respectable thermal efficiency of 76%.

The other aspect to be remembered is that of the propulsive efficiency. To keep a high propulsive efficiency the jet velocity of the engine must be matched to the aircraft speed, and the lower the aircraft speed the larger the by-pass ratio must be to bring the jet velocity of the high thermal efficiency gas producer down to the required value. In practice, this leads to lower thermal efficiency because of the losses in the by-pass turbines and fan which in large engines are transmitting and absorbing enormous

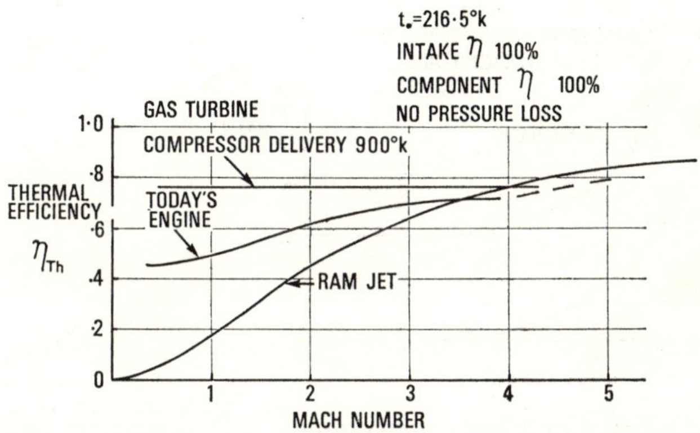

Figure 1 Theoretical gas turbine and ramjet efficiencies

horsepowers. The third curve on Fig. 1 shows typical present day figures for comparison. It is obvious because of the big gap between the theoretical and actual curves at the bottom end that we stand a good chance of making some improvement at the subsonic speeds: it is equally obvious that as the gap narrows towards the high top end we stand little chance of making a substantial improvement here.

To emphasise the point of the different engine concepts required over the speed from subsonic to supersonic Mach numbers, Fig. 2 shows how the engine design pressure ratio and by-pass ratio must vary over the speed range. With reasonable efficiency, 900°K outlet temperature is equal to an expansion ratio of 75:1, which is just about the pressure ratio of a high-performance diesel engine. In Fig. 2 the expansion ratio is kept constant at 75:1. The propulsive efficiency is assumed constant at 70% with a combustion temperature of 2000°K.

At Mach 0·8 in the stratosphere the engine is a very complex engineering job, with a pressure ratio of 50:1 and a by-pass ratio of 13; at Mach 3·5 and above, the same efficiency, or better, can be achieved theoretically with no engine compression at all—the 75:1 expansion ratio is provided by the intake compression alone.

Having established from Fig. 1 that at Mach 0·8 in the stratosphere, at present we are achieving a thermal efficiency of 47% as compared with the arbitrary theoretical limit of 76% and we can now set about estimating a practical limit.

For an engine using air as a working fluid, the limit to the combustion temperature is the stoichiometric value of about 2400°K. (There is no reason why we should not conceive temperatures higher than this using enriched air or pure oxygen—as in a rocket—but the weight of oxygen

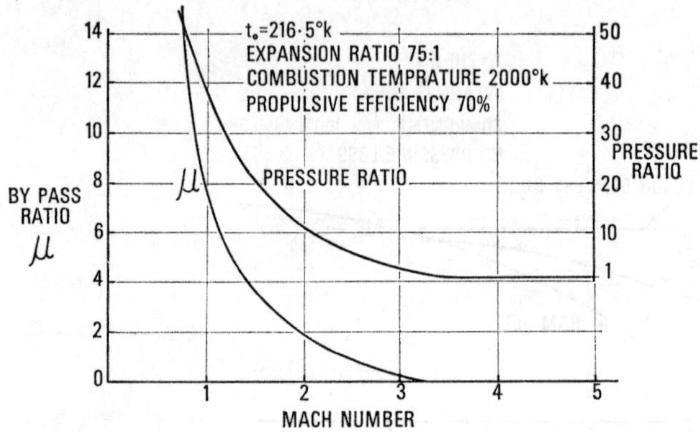

Figure 2 Engine pressure ratio and by-pass ratio

carried very quickly outweighs any advantage gained in thermal efficiency).

The limit to compressor and turbine efficiencies is obviously difficult to predict, and might justify a separate chapter, but the friction losses alone through rotating components will probably account for 3% polytropic inefficiency. With a compressor delivery temperature of 900°K the adiabatic efficiencies are 94·7% in the compressor and 97·4% in the turbine. Polytropic rather than adiabatic efficiency is chosen as a fixed parameter because this is most representative of a given standard of technology when applied to components having a range of pressure ratios. Assuming constant specific heat through the engine for simplicity, the thermal efficiency of the gas producer can be expressed approximately as:

$$\eta_{th} = \frac{T_2 - t_0}{T_2} - \left(\frac{t_0}{T_2}\right)\left(\frac{T_3}{T_3 - T_2}\right)\left[\delta \log_e\left(\frac{T_2}{T_1}\frac{T_3}{T_4}\right) + k\right]$$

where

t_0 is the ambient temperature
T_1 is the compressor inlet temperature
T_2 is the compressor outlet temperature
T_3 is the turbine entry temperature

T_4 is the turbine outlet temperature
k allows for duct losses
δ is the compressor and turbine polytropic inefficiency

At Mach 0·8 in the stratosphere, with $T_3=2400$, $T_2=900$, $\delta=0{\cdot}03$ and $k=0$, the thermal efficiency drops from 76% to 74%.

A bare minimum of, say, 5% for the pressure losses through the engine drops the efficiency another ½% to 73·5%.

The by-pass turbine and fan must also be taken into account, particularly at the high by-pass ratios we are talking about, where practically all the gas producer jet energy is transferred to the fan. Using the same polytropic efficiencies as before but with 1% pressure loss in the fan stream leads to a further 4·8% loss in thermal efficiency, or a final figure of 68·7%.

So as a result of making allowances for optimistic limiting efficiencies, the engine at Mach 0·8 has lost just over a quarter of the potential theoretical gain possible on today's figures. A gain in efficiency from 47% to 68·7% is by no means small and will justify all the hard work necessary to achieve it. Being pessimistic, if half this gain is achieved the efficiency would be 58%.

To predict limits of possible propulsive efficiency is more difficult, in that it does not depend only on thermodynamic factors which have predictable limits—propulsive efficiency is related to the quantity of air which can be accelerated by the engine, and this is as much a matter of structure as thermodynamics. The problem can, however, be put into perspective by taking a simple example.

Figure 2 was drawn on the assumption of 70% propulsive efficiency, which at Mach 0·8 in the stratosphere means a thrust per lb of air=20·6. Now today's designs of large transport engines give, say, 25 lb thrust/lb air, so there is no startling advance here. There are very good reasons for this, such as engine weight, pod drag, engine/airframe interference, ground clearance and so on, but sweeping these aside for the moment there is no fundamental reason why future engines should not use a great deal more air and drop their thrust per pound down to, say, 8·5. At a given thrust such an engine would pass nearly 2·5 times the air flow and be more than 50% greater in diameter. A figure of 8·5 for thrust per pound gives a useful gain in propulsive efficiency from 70% to 85%, which again will justify the work involved in the advances necessary in engine and aircraft design.

Perhaps the most useful picture of this can be gained from consideration of engine specific fuel consumption (sfc), which at Mach 0·8 with kerosene fuel and $t_0=216{\cdot}5$ is equal to $0{\cdot}193/\eta_{th}\eta_p$. Fig. 3 shows the complete picture.

Using the limit of $\eta_{th}=0{\cdot}76$ and the absolute limit of $\eta_p=1{\cdot}0$ we get a lower limit to sfc of 0·254.

Using the practical limit of η_{th}=0·68 and the conservative figure of η_p=0·7 gives an sfc=0·405. The futuristic figures of η_p=0·85 at η_{th}=0·68 gives the sfc=0·334, while the conservative values for both engine and propulsion gives an sfc of 0·475.

The full line in Fig. 3 shows the trend of reducing sfc with time, ending with todays designs going into service in the early 1970s. Since the rate of decrease of sfc must reduce as we approach the minimum, it is not difficult to predict the future trend, shown by the dotted extrapolation. the shaded area represents a margin of error, determined by our rate of success or failure in solving the problems involved.

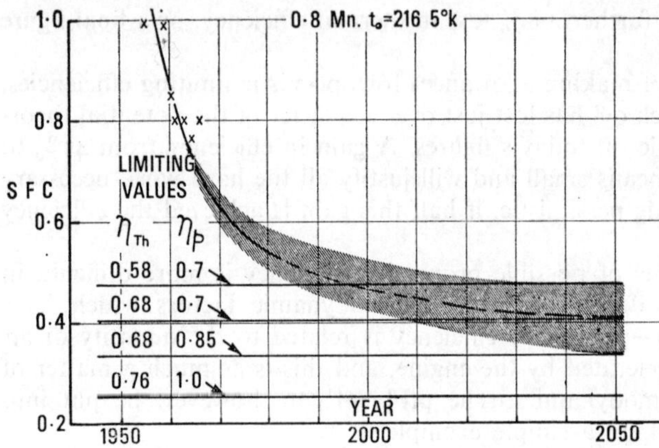

Figure 3 Possible improvement in fuel consumption

Before considering the higher Mach numbers, it is interesting to note that even at a propulsive efficiency of 85%, the arbitrary limit we have assumed for a fan engine, 15% of the available energy is wasted. What is needed here is a radically different method of imparting the jet engine power to the air flowing past the aircraft. It may well be that in the future accelerating forces can be applied to the air without having to rely on pressure gradients provided by forces on rotating blades. For instance, ionised air can be accelerated in a magnetic field—it is not inconceivable that we might do without large fans in the future. Of course, such methods would also be useful in the engine itself.

On the other hand, we might take a lesson from nature and emulate the bird. A conventional engine uses only a fraction of the air contained in the stream tube of wing tip diameter; like the bird, the wings themselves might be used to accelerate enormous quantities of air, giving rise to very large by-pass ratios indeed. The potential gains here are a 15% increase in propulsive efficiency, plus the 3% energy which was lost in the fan, less energy losses involved in the wing movement mechanism and due to the drag related to the extra weight.

The above is based on a very elementary analysis and must, of course, not be used for any serious comparisons but it does give a useful picture of the possible scope for improvement. The important lesson to be learned is that, at low aircraft speeds, the biggest problem is that of obtaining a good propulsive efficiency. While it may well be possible to produce the basic gas producer thermal efficiency of 73·5%, it is not going to be easy to avoid the large losses involved in the machinery required to give the necessarily large by-pass ratio, or to achieve the low weights which the aircraft performance demands. It also demonstrates that we are by no means at the end of the road of development of the conventional gas turbine cycle at low aircraft speeds.

At supersonic speeds, the picture changes appreciably. As shown in Figs. 1 and 2, the optimum type of engine changes towards the ramjet, and the engineering difficulties are transferred from the complicated engine to the intake and nozzle. Because any aircraft must take-off and accelerate to its supersonic design speed, a composite powerplant is necessary, and no doubt the reheated by-pass type of engine will be with us for a long time to come for Mach numbers up to say, 3·5 to 4·0.

Because of the very high intake compression ratio, the engine pressure ratio at cruise does not need to be high and the usual compromise will be reached between subsonic performance and the weight of the engine. What is needed here is a variable by-pass ratio engine, designed to match the engine over the range of flight speed. Today's compressors have a limited range of flow at a given pressure ratio, limited at low flow by surge and at high flow by choking. It may be possible in the future to deal with a very much larger flow range, and achieve a big variation in by-pass ratio.

At higher speeds of Mach 4 to 5, present day projects of by-pass type engines in which the high pressure section is shut down and the engine runs as a ramjet, are very satisfactory powerplants. As stoichiometric temperatures are approached in the combustion chamber the Mach number at which the changeover from engine to ramjet occurs will be pushed up, the engine being designed to stand the higher compressor outlet temperatures.

As shown by the curves in Fig. 1, a ramjet intake can already be made to work very efficiently and very high temperatures can be used today. It is safe to predict, therefore, that these project engines are already approaching the practical limits of the normal heat engine cycle.

ALTERNATIVE HEAT ENGINE CYCLES

The above discussion has dealt with the normal gas turbine cycle in which the heat is added at constant pressure. The question arises, can a greater efficiency be reached with other types of heat engine cycles? One or two alternative cycles are perhaps worth considering for the subsonic speeds.

The first is basically an extension of the normal cycle, that is one in which some extra compression and expansion is achieved through timed shock waves. On the assumption that the pressure ratio could be raised by a factor of 3, which would give enormous pressures in the engine, the value of T_2 increases from 900 to about 1300°K, giving a theoretical $\eta_{th}=83{\cdot}8\%$, an increase of 7·8% on the former 76%. Because of the fact that the compression and expansion processes are fundamentally inefficient, we would be lucky to see half this potential gain and it can be argued that a 4% gain in efficiency in return for the extra complexity and weight is hardly worthwhile.

The second alternative would be to use heat exchangers. The cycle which is quoted above has a high pressure ratio and is very efficient and consequently its exhaust temperature is below the compressor delivery temperature, so that a heat exchanger cannot be applied directly in this case. To make heat exchange possible, it is necessary either to intercool and reduce the compressor delivery temperature or reduce the pressure ratio which will reduce the efficiency. Thus we are faced with two heat exchange units and their dependant ducting pressure loss and weight. Unless great advances are made in the compactness and performance of heat exchangers it seems unlikely that the benefits will outweigh the penalties for aircraft use.

From an inspection of these and other cycles, we are forced to conclude that at temperatures approaching stoichiometric and with high efficiencies assumed for future components, the open constant pressure or Brayton cycle will take a lot of beating for an aircraft powerplant, where size and weight are of great importance.

The same sort of remarks apply at supersonic speeds. Consider an aircraft speed of, say, Mach 4, for which the total temperature at the engine inlet is just about the 900°K limit assumed earlier. If the limit can be raised to 1240°K limit as before, by further compression in the engine, the theoretical efficiency again rises to 82·5%. The optimistic component efficiencies assumed earlier reduce this figure by about 1·2%. The extra weight of the more complicated cycle will again tend to offset what is at best a doubtful gain of 5·3% from 76% to 81·3%. There is no doubt that the intake is a light and reasonably efficient air compressor at these speeds. Even at today's measured intake performance, the efficiency is equivalent to a compressor adiabatic efficiency of 82·9% or a polytropic efficiency of 90%. For a pressure ratio of 100 or so, this is a good performance, meaning that 95% of the compression work done could be recovered in an ideal expansion back to ambient pressure.

GENERAL CONSIDERATIONS

Having established that there is at least a limit to the trend of reduction in sfc at a given flight speed, what do we consider next?

There are really only three ways in which the engine can directly improve aircraft cruise performance:

1 Reduce the fuel load by a reduction in sfc.

2 Reduce the aircraft empty weight by a reduction in engine weight.

3 Reduce the fuel load and engine weight by a reduction in aircraft drag.

We are left, therefore, with 2 and 3.

The powerplant weight deserves a separate section and will be dealt with later.

The use of the powerplant is a theoretically powerful means of reducing the drag of the aircraft, as we know.

In a conventional aircraft the energy required (corresponding to the fuel supply rate) is given by

$$E = DU/\eta_{th}\eta_p$$

where E is the energy rate
D is the aircraft drag
U is the flight speed
$\eta_{th}\eta_p$ is the overall (propulsion × thermal) engine efficiency.

The drag can be split into three terms, body* drag, wing drag and induced drag. Suffices $_b$ and $_w$ refer to body and wing repectively; C_l is lift coefficient, and λ the induced drag factor.

Putting q for the free stream dynamic head, A for area and C for drag coefficient

$$D = (C_b A_b + C_w A_w + \lambda C_l^2 A_w)q$$

and putting $C_l = W/A_w q$ where W is the aircraft weight, yields the usual drag equation from which the minimum drag can be derived when q is varied. From this the minimum energy rate E_{min} is given by

$$\frac{E_{min}}{WU} = \frac{2}{\eta_{th}\eta_p}\sqrt{\left\{\left(C_w + \frac{A_b C_b}{A_w}\right)\right\}}$$

Splitting up the drag into its component parts like this serves to remind one that even while flying at the altitude to give minimum drag, the energy requirement could be reduced by increasing the wing area.

When boundary-layer control by suction is introduced, the drag is altered in four ways:

*Here the term 'body' may be taken to refer to all surfaces except that to which the lift coefficient refers, with a corresponding mean drag coefficient.

1 The airflow involved in the boundary layer is much reduced.
2 Most of the boundary-layer air is sucked inside the aircraft.
3 The momentum deficit in the wake (which accounted for all the drag without suction) is reduced to practically negligible proportions.
4 The profile drag due to the pressure imbalance between forward and rearward facing areas is reduced to near negligible proportions.

Most of the drag, therefore, arises from the momentum lost by the air sucked into the aircraft, and consequently a large part of the energy required to propel the aircraft will be used in pumping this air, and it can be ejected at an optimum velocity.

An expression for the total external drag can be formulated using drag coefficients with an additional suffix $_s$, whose value is chosen according to the fraction of the surface to which suction is applied. In addition to this external drag, there is, internally, a drag MU/g and a thrust MV/g for the suction airflow M ejected at velocity V relative to the aircraft. The energy which must be expended in pumping this air is $M(V^2+eU^2)/2g\eta_s$ where e is a coefficient which expresses the pressure deficit of the sucked air at the pump inlet below ambient in terms of the intake dynamic head and η_s is the thermal efficiency of the suction unit.

Accepting that the coefficient e will be the same for all sources of air, and equal to that corresponding to the wing (and this is likely to be pessimistic so far as the body is concerned) the total energy required to propel the aircraft can be written

$$E_s=\left[(C_{bs}A_b+C_{ws}A_w)q+\frac{\lambda W^2}{A_w q}-\frac{M(V-U)}{g}\right]\frac{U}{\eta_{th}\eta_p}+$$
$$+\frac{M}{2g\eta_s}(V^2+eU^2).$$

For a given surface the suction flow is, with a suction flow coefficient C_q

$$M=A\rho UC_q=2gAqC_q/U$$

where ρ is the ambient air density, and so

$$E_s=[C_{bs}A_b+C_{ws}A_w+(A_bC_{qb}+A_wC_{qw})y]q+\left.\frac{\lambda W^2}{A_w q}\right\}\frac{U}{\eta_{th}\eta_p}$$

where

$$y=\frac{\eta_{th}\eta_p}{\eta_s}\left(\frac{V^2}{U^2}+e\right)-2\left(\frac{V}{U}-1\right)$$

Again q may be varied until E is a minimum and then

$$\frac{E_{smin}}{WU}=\frac{2}{\eta_{th}\eta_q}\sqrt{\left\{C_{ws}+\frac{C_{bs}A_b}{A_w}+\left(C_{qw}+\frac{C_{qb}A_b}{A_w}\right)y\right\}\lambda}$$

y itself can be minimised by choice of V.

Experimental values have been measured for the various drag coefficients involved, and for a typical large subsonic aircraft the following drag breakdown* is appropriate:

	Area sq ft	*Drag coefficient Standard*	*Fully sucked*	*Flow coefficient* C_q
Wing	3000	0·0085	0·0002	0·0008
Body	8000	0·00242	0·0001	0·0004
Tail	1160	0·0098	0·0002	0·0008
Nacelles	1800	0·0035	0·0001	0·0004

If present day values of engine efficiency are used and the factor e is taken as 1·46 (which corresponds to a suction delivery pressure of 0·56 atm and a Mach number of 0·84) then $y=1·77$.

TABLE I

Area treated		*Energy Reduction %*	*Altitude 1000s feet*
0	None (reference)	0	39
1	Wing only	18	35
2	Wing, tail, nacelles	33·6	30·5
3	As (2)+50% body	42	26·7
4	As (2)+all body	52	23·5

Table 1 shows how the energy requirement is reduced by applying suction progressively on the aircraft.

In each case the areas remain fixed and the appropriate flight altitude is given in the last column of the table. Probably some increase in wing area would be necessary to increase the lower altitudes and this would reduce the drag still further.

The expression E_{smin} bears considerable resemblance to that for E_{min} for the conventional aircraft without suction. It can be seen by comparison that when the wing drag is reduced by suction there is an even stronger reason for wanting larger wing area than on the conventional aircraft, and this will become possible with the advent of new materials. Conversely, as new materials permit larger wing areas to be used the benefits from wing suction will become more pronounced.

In the time it takes to develop engines having the very high efficiencies referred to in Fig. 3 it is probable that suction could be applied to all surfaces and with improved structural materials will permit larger wing areas and higher aspect ratios. The aircraft body may be eliminated and

*Since the wing and tail areas are planform, not total surface areas, the coefficients, including C s are double the real surface area values.

there may well be reductions in the suction flows necessary to achieve low drag. It is possible, therefore, that energy reductions higher than the 52% quoted in Table I may be achieved, independently of any improvement in the engine performance. If a similar improvement is assumed in the pump components as in those of the basic engine then the ratio $\eta_{th}\eta_p/\eta_s$ does not change (although, of course, it is less than unity due to η_p, while $\eta_s \simeq \eta_{th}$), while with development the pressure deficit of the suction air may be reduced so that the term y may become smaller, and, with the engine improvements referred to in earlier paragraphs, it is not difficult to see a reduction of the energy requirements to under 30% of today's values.

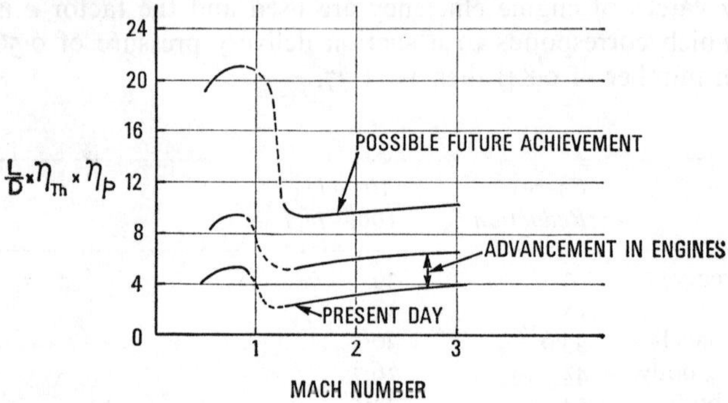

Figure 4 Aircraft performance

Figure 4 compares present day and possible future aircraft performance in terms of the most useful parameter ($\eta_{th}\eta_p L/D$). Using the well-known Breguet range equation one can easily show that this parameter is equal to

$$(R/H)\log_e(W_i/W_f)$$

where

- R range
- L/D lift/drag ratio
- H fuel heating value
- W_i, W_f aircraft initial and final weights respectively.

The parameter $\eta_{th}\eta_p L/D$ is thus a measure of the aircraft ton miles per lb of fuel used, for a given fuel, and is also equal to WU/E. The lower curve in Fig. 4 shows present day capabilities. The middle curve shows the effect of the futuristic thermal and propulsive efficiencies deduced earlier. The upper curve is a combination of the increased efficiency and the improvement which might be obtained by boundary-layer control or other aerodynamic improvements at high speeds.

The overall improvement envisaged at subsonic speeds is startling, there being over 100% potential gain from boundary-layer control and an overall improvement by a factor of 4:1. At high speeds the improvement is still large, but limited by the relatively smaller gains expected from engine performance.

There are good reasons why long-range travel in the future should be supersonic, such as:

(*a*) optimum journey time of say, three hours,

(*b*) aircraft utilisation,

but there are also good reasons why there will be large numbers of subsonic aircraft.

With the improvement to come in performance as shown in Fig. 4, we can look forward to large fleets of subsonic aircraft, which will be ideal for up to the medium range of, say, 1500 miles when carrying people and possible maximum range for freight.

ALTERNATIVE AVENUES OF IMPROVEMENT

Improved powerplants can influence the aircraft in directions other than pure cruise performance.

The whole field of vertical take-off is in its infancy, and as the powerplant is developed no doubt the VTO aircraft will come into its own. Today's helicopters are already a vast improvement on the early machines, and at the other end of the scale exceedingly light and small lift engines can already be made. No doubt in the future, the two concepts will merge, in that the lift engine concept will change towards the high by-pass powerplant because of efficiency and noise. As in the case of the subsonic cruise aircraft, we badly need a completely new method of applying a reaction force to the aircraft by accelerating the air rearwards (or downwards for VTO). Without the embarrassment of enormous rotating blades the aircraft can be accelerated to wing-borne speed by deflection of the thrust.

This is a fanciful idea, but perhaps we may be forgiven in a book of this type for posing a problem without supplying the solution—at least that is a start in obtaining a solution.

Having got used to the idea of aircraft control in VTO by forces applied by the powerplant, there is no reason why the whole control in flight should not be supplied the same way. This train of thought leads on to the ideal of a completely integrated powerplant/aircraft combination, in which the powerplant is not just a fixed means of propulsion, but can change character during each phase of the flight plan to produce the necessary forces on the aircraft during take-off, climb, cruise, descent and landing.

One is tempted here to extend the analogy with the bird, which after all must obey the same physical laws as the aircraft. While one can forsee that the pure propulsive and even thermal efficiencies of the powerplant may eventually match those of a bird, it is in the field of automatic control systems and information collection and transmission that the machinery of today really lags behind. Great advances have been made recently in powerplant and aircraft control systems but to put it bluntly, this is to be expected because, like early engines, they were so big and clumsy to start with. We still have a very long way to go before we can compete with the bird's brain and nervous system.

ALTERNATIVE ENERGY SOURCES FOR HEAT ENGINES

As stated at the beginning of this chapter, for over 100 years our heat engines have been using fossil fuel. It has been used so far for the very good reasons that it is both plentiful and cheap. The cost is purely relative, of course, but as the standard of living rises of the men who dig the coal or refine the crude oil, so does the relative price of the fuel. In years to come, conventional fuels will no doubt be relatively expensive, because of their scarcity and high real wage levels.

On the other hand, carbon and hydrogen are in plentiful supply on the surface of our planet, and in time there will be a cross-over point where it is economic to manufacture a suitable aircraft hydrocarbon fuel using solar or nuclear energy.

However, we should really be looking for an advanced fuel, synthetic or conventional, with the right combination of calorific value per pound and per cubic foot as the important criterion, together with price. Since carbon is a fairly heavy atom, one obvious choice is pure hydrogen.

The relative calorific values in CHU/lb are 10,300 for kerosene and 27,000 for liquid hydrogen. The volume of the liquid H_2 is four times that of kerosine for a given heat capacity.

Making the pessimistic assumption that aircraft weight stays the same, the advances predicted so far show that the heat energy used for a given Mach 0·8 journey will decrease to about 28% of today's values, with engine sfc, approximately 0·4 and aircraft L/D of 42.

The actual weight of hydrogen carried, therefore, will be: $10{,}300/27{,}000 \times 0{\cdot}28 = 0{\cdot}107$ of the fuel load of present day aircraft. The tankage volume will be $0{\cdot}28 \times 4{\cdot}3 = 1{\cdot}2$ times the original volume.

Furthermore, the aircraft weight will obviously decrease because of:
(*a*) the lower fuel weight and the resulting lower structure weight, and
(*b*) the reduced engine weight, which is discussed in the next section, and similar reductions in airframe weight owing to the use of advanced materials.

If hydrogen is used as fuel, therefore, we end up with the futuristic transatlantic aircraft of the next century, carrying about 5% of the fuel load of today's aircraft. This is an almost incredible figure, but it is based on a number of reasonably credible assumptions, and only demonstrates the compounding effect of such assumptions.

At present, however, there are serious drawbacks to the use of hydrogen, and these may be listed as follows:

(*a*) Cost.
(*b*) Difficult to store and transport.
(*c*) The liquid H_2 in normal aircraft tanks must be allowed to evaporate.
(*d*) Safety aspect.
(*e*) Engineering difficulties in pumps, valves and fuel system generally.

In the event of natural fuel resources becoming unavailable, and the fire danger being overcome, the manufacture of liquid hydrogen near the airfield using solar (hydroelectric) power could well make it an economical necessity for future long-range aircraft.

To put other possible fuels into perspective, Fig. 5 plots the weight per unit heat against the volume per unit heat, both relative to kerosene for, a number of exotic fuels and for a range of hydrocarbons.

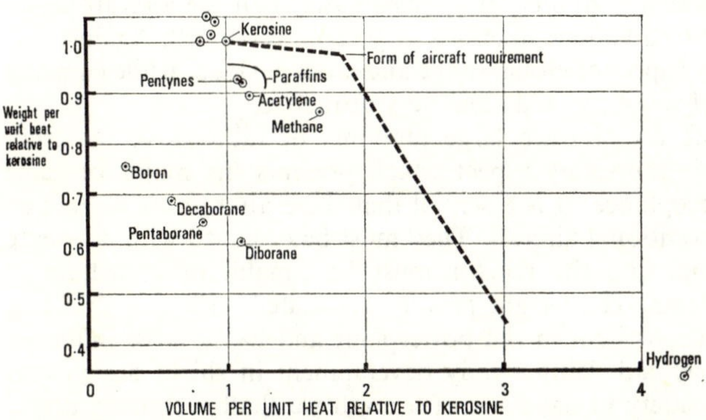

Figure 5 Relative weight and volume of possible fuels

A given aircraft can accept fuel volume up to a certain limit with very little weight penalty to cover tankage within the structural space available. Beyond this limit, the weight penalty due to the acceptance of further fuel volume is greatly increased because structural penalties are incurred. This situation is illustrated qualitatively on Fig. 5 by the dotted lines.

Aircraft and engine performance and design range requirements all affect the position and slopes of these lines, and no particular merit is to be ascribed to the example given although it is perhaps typical of a moderate range aircraft which is not volume-limited with kerosene but would be with hydrogen. For a very futuristic aircraft, as has already been shown, the steeper dotted line would embrace hydrogen, while for present day, long-range aircraft it would only just cover kerosene.

The (purely technical) merit of a given fuel can be assessed as the vertical difference between the fuel point and the dotted line. The fuels which can show some potential, apart from the boron family, are acetylene, some pure hydrocarbons, methane and hydrogen. The latter has already been discussed and acetylene is far too dangerous and costly. Cost also rules out the pure hydrocarbons leaving methane which, in the form of liquefied natural gas, is both cheap and plentiful. As well as being safer than hydrogen at least, and probably also than kerosene itself, it also has very useful properties as a coolant. Methane, therefore is a good prospective candidate for tomorrow's aircraft fuel.

There is considerable promise in the application of nuclear energy for the propulsion of very large aircraft over very long ranges. Basically this is because of the vastly larger amount of energy available per unit mass from nuclear fuels through fission or fusion than the best chemical fuels. Also the potential cost per unit of energy produced by nuclear fuel is appreciably less than that of conventional aircraft fuels.

Today's large transatlantic commercial aircraft carry some 70 tons (71 Mg) of fuel which accounts for approximately half the aircraft all-up weight and occupies a volume in excess of 3000 cu ft (80 m^3). Against this background the prospect of global range and large payload, while carrying but negligible fuel weight would seem very attractive.

Nuclear aircraft do produce large amounts of radioactive material, however, and it is the safety aspect which presents the major obstacle to their public acceptance. It is essential that these aircraft are as safe as today's large conventional aircraft. They must be designed with adequate radiation shielding, and the reactor must be capable of withstanding all possible accidents. The weight penalty associated with this shielding is the most significant item in the powerplant and has a large influence on the aircraft payload. With steady development in shield design and improvements in safety in the event of a crash we shall eventually arrive at the point where nuclear propulsion for aircraft becomes acceptable. The improvements in performance for the engine and the aircraft already mentioned can be applied to nuclear as well as to conventional fuelled aircraft to reduce the power requirements and hence the weight penalty of the nuclear engine.

Even further away in time, is the chance of transmitting energy from ground based stations to an aircraft in flight. This is an attractive idea in that no stored energy need by carried by the aircraft, like today's electric trains or Brunel's 'atmospheric railway' of the last century. The problems are vastly greater than for surface transport, and are magnified by the large horse-powers required. In addition, the large energy density needed makes the problem seem insuperable for the present. We are all used to the idea of low density energy transmission—the energy transmitted from the sun per sq ft of the earth's surface is vital to our existance and is sufficient in conjunction with our insulating atmosphere, to maintain the surface temperature roughly between +50°C and −50°C. But it is only 2% of the energy/sq ft of aircraft surface required to maintain cruising flight and to raise the energy density to the required level would presumably have lethal consequences. There are enormous problems to be overcome, but one must never be dogmatic about the ultimate outcome.

REDUCTION IN WEIGHT OF HEAT ENGINES

While thermodynamics, being amenable to theoretical analysis, has been developed to within fairly close margins of well understood fundamental limits, materials, and hence structure weight are much further off their fundamental limits, so far in fact that very little attention had been paid to defining such limits, until quite recently.

It follows, therefore, that future development will be concerned more and more with weight reduction: as this takes place the optimum engine will be able to deal with larger airflows at lower thrust/lb and will require an increase in by-pass ratio.

The big step from piston engines to jet engines was really the ability to build engines of large power for a low weight and small size. In spite of their initially high fuel consumption they provided a satisfactory alternative, and, in fact, were the only way of providing power for high-speed aircraft. The piston engine is not an ideal engine at large powers, in spite of its reasonable fuel consumption: an engine designed to run at a limiting maximum piston speed of 80 ft/sec (24 m/s) and a minimum time per $\frac{1}{2}$ rev of 3 m sec gives about 20 hp/cylinder. As the cylinder is scaled up at the same limiting piston speed the 'square-cube' law operates and the hp/cu ft and hp/lb wt decrease drastically.

The axial compressor gas turbine is in a completely different class because of its enormously greater airflow/cu ft which even at today's low combustion temperatures means much higher hp/cu ft. The 'square-cube' law operates in that the inlet air velocity is assumed to be fixed, so the power varies as l^2 and the weight theoretically as l^3. However, at a reasonable minimum size of 1 ft diameter an axial compressor will pass about 28 lb/sec (13 kg/s) of air at static sea level inlet conditions, which at today's combustion temperatures means about 5500 gas horse-power. At double the size the power will be 22,000 hp, and the theoretical hp/lb has only dropped to half the figure at the arbitrary minimum size. It is clear that at the powers necessary for large aircraft the gas turbine is much closer to its 'optimum' size than a piston engine.

The axial gas turbine has a high figure of airflow/cu ft mainly because it has a high airflow/unit frontal area: an advanced compressor today passes nearly 70% of the maximum choking flow based on compressor inlet diameter, despite the presence of rotor hub and blades, so we cannot expect any very large increase in the future. The engine length depends very much on the engine compression ratio and, therefore, on the sfc of the engine. This is obviously one of the main reasons why the modern lift engines are so very light—they are designed with a low pressure ratio without much regard for sfc because of their very short running times, resulting in a very short engine of minimum volume. Coupled with the fact that at their comparatively small sizes they are well up towards the 'optimum' as defined by the 'square-cube' law, this only means that they give very high values of hp/cu ft, and with advanced mechanical design give high values of hp/lb wt.

The subsonic transport engine, on the other hand, simply must be designed at a high pressure ratio, as determined earlier. This must lead to greater length or length/inlet diameter ratio, especially when the overall compression ratio is split up between two or three or more shaft systems. Together with a higher density than the lift engine, to give long life and reliability, increased high pressure section volume per horse-power gives appreciably lower values of hp/lb wt.

From the above reasoning, the way to reduce the weight per horse-power of the power-producing section of the engine can be broken down into logical steps such as:

1 Increase flow/unit frontal area.
2 Decrease length/diameter ratio.
3 Increase the combustion temperature to increase hp/lb/sec.
4 Reduce the density of the engine.
5 Reduce the size of the engine.

These are fairly obvious steps which one might have written down to start with, but the background is perhaps a help in determining possible limits.

Taking these in order:

1 The situation is such that only a very small increase in flow per unit frontal area, of the order of 3% can be expected.

2 The possible reduction of length of a given rotating component is dependent on the detailed aerodynamic improvements which can be made and is, therefore, difficult to predict. It is assumed that the aspect ratio of blades can eventually be doubled—this permits a reduction of length/diameter of 50%, other things being equal.

An effective way of reducing the length of an engine would be to scale it down. The anticipated advances in combustion technology and the prospect of eliminating the growth of boundary layers in an engine leaves open the opportunity to make quite small engines very efficient. With the methods of production envisaged for the next century, the increased numbers of parts which go along with a large number of small engines would not give embarrassment from the cost point of view—in fact, it would probably be cheaper to make large numbers of small parts than a few big ones and in any case small engines would find a wide market for a range of aircraft sizes.

In view of the fact discussed earlier that the advanced aeroplane will have boundary-layer control involving the pumping of boundary-layer air for its propulsion, it is likely that the high pressure section of the advanced engine will not always be called upon to drive a conventional low pressure ratio fan with a large by-pass ratio. It is often likely to have to pump boundary-layer suction air which is distributed widely over the surface of the aeroplane and again a large number of small engines would make this more feasible.

The combustion problem will be simplified from some points of view as stoichiometric temperatures are approached, because the mixing length provided in present day chambers is unnecessary. The combustion chamber will not reduce in volume as much as might be expected, however, because the primary zone must be larger to accommodate all the air required for burning.

Present day turbines already have a very high output indeed in terms of hp/cu ft but as blade speeds are pushed up even this figure must increase and it is assumed that a future turbine to drive a 50:1 pressure ratio compressor will occupy no more volume that today's turbines driving approximately half that pressure ratio.

3 The increase of combustion temperature to the stoichiometric value of some 2400°K is obviously a powerful method of increasing hp/cu ft. At 1500°K a typical engine of today gives about 300 hp/lb/sec which is increased by 50% to approximately 480. This compares very well with a good internal combustion piston engine.

The overall figure of hp/cu ft is increased by a factor of approximately 4:1. The comparison in outline of a 50:1 pressure ratio gas producer with a typical present day version of the same power is shown in Fig. 6.

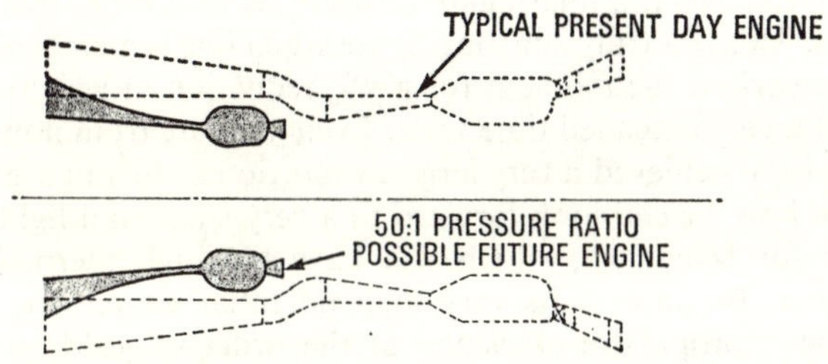

Figure 6 Comparison of present and future hp units

There is obviously an impressive difference, the volume being reduced to 25%. The by-pass section of the engine is not shown because as mentioned earlier, it will in some cases be a boundary-layer pump.

4 Having reduced the volume of the gas producer to about 25% of present day values, is it fundamental that the density must increase. In spite of the formidable difficulties involved with the very high pressures, temperatures and stresses implicit in the above assumptions, there seems to be no real reason why this should be so.

One of the most important advances in engineering materials we have known is the development of reinforced materials in which high strength fibres are embedded in a low density matrix. At the present time, the use of these materials is limited to cool components, because of the limiting properties of the matrix used—e.g., glass or carbon fibres in a resin matrix—giving very useful weight reductions for a given component, but one can look forward to the time when very high temperature composite materials will have been developed. The use of these materials may allow the high temperatures we have assumed with no more difficulty in cooling than we have at the moment for conventional materials.

Perhaps one of the most fundamental concepts which these new materials open up is that of completely new component design. The traditional method of designing separate blades for example, and attaching them to a disc in serrated grooves is very inefficient from a weight point of view. Composite materials give promise of a complete blade and disc assembly manufactured as a whole, with the blade stresses fed into the disc or drum in an orderly manner, resulting in big weight savings. The static parts of the engine, i.e., casings, bearing housings, mountings, etc., can all be treated in a similar way, reducing the number of redundant joints and flanges to the bare minimum required to build the rotating parts into the engine. The weight saving on non-rotating parts may not be so dramatic as the 50% or so possible on rotating parts, but it will still be substantial.

For these reasons, it is considered that we may assume that the density of future bare engines will be down to two-thirds of present day figures. On the assumption that the auxiliaries—i.e., the necessary fuel pumps, control system, electric supply, etc—will have kept pace with the main engine development, we arrive at a final figure of hp/lb wt of 6 times that of present day engines. This is a truly startling figure when one remembers how hard we have to work at the moment for a 5% reduction in weight, but we are thinking of a very extended time scale. Even 50 years from now we doubt that we will have achieved a very large proportion of this figure.

So far we have seen how we can look forward to a very small and light high pressure engine for Mach 0·8, which will have the high thermal efficiency quoted earlier. Because of its very high potential jet velocity, it will have a very poor propulsive efficiency of the order of 30%, so the by-pass turbines and fan or the boundary-layer control pumps are a very essential part of the powerplant.

In an up-to-date transport engine the high pressure section weighs approximately 45% of the total engine weight, or in other words the by-pass turbines and fan weigh 1·2 times as much as the gas producer.

The problem of reducing the conventional fan and turbine weight as much as for the gas producer does seem more difficult. Even at today's levels of propulsive efficiency, we can foresee a weight reduction of only, say, 50% for the rotating fan blades, and as the propulsive efficiency is raised the fan diameter and weight must increase. It is reasonable to assume also that the fan shroud and stators can also be reduced in weight to 50% of present day values. The replacement by boundary-layer pumps will alter the situation completely.

The above arguments tend to underline the assumption that the simple gas turbine cycle will be with us for a very long time: it is difficult to see how other more complicated cycles can compete with the exceedingly small and light gas producer engines predicted, since the potential gain in efficiency is small.

At supersonic Mach numbers, the engines picture is less optimistic: of the ways of improving the engine discussed above, only reductions in density and engine length offer any real hope. In present day ramjet engines, the flow/unit frontal area is already pretty well at the limit, and the combustion temperature is already approaching stoichiometric. The engine length is very short, but there may be some gain from burning at compressor outlet velocities with no diffusion. Composite materials can again give large weight reductions, but only of the order of 50%. As mentioned earlier, at the higher Mach numbers the engineering difficulty is mainly in the intake and nozzle, and it is in these components that full use of future composite materials will really pay off.

COST

In a future society the meaning of cost will presumably still relate to the number of man-hours needed for manufacture and the scarcity of the basic material, so our present basis of analysis cost will hold to a certain extent.

New design methods will cut the man-hours to a fraction of today's figures. Already the design of certain engine parts can be specified on tape, which can be fed direct to a machine. As the technique is improved and elaborated more and more of the engine parts will be manufactured in the same way. The new materials open up radically different production methods, with the possibility of whole sub-assemblies being manufactured at a time. The more complicated tools required will only be justified by a large number of items produced and one can foresee fewer different types of aircraft and powerplant than we have today in the world, but much greater numbers.

The development time of an engine at present, costly in time and money, will be very much less in the future. The more detailed and accurate design methods will mean that the lives of critical parts will be more accurately predicted and many of the unknown factors in today's design will be taken into account. The result must be that future engines will absorb appreciably less man-power in design, development and manufacture.

NOISE

No chapter on future aircraft powerplants would be complete without some references to possible noise reduction. The level of noise round our overseas airports is already high, and although subsonic aircraft can be made quieter, the advent of the supersonic aircraft will surely tend to hold or increase this level.

The trend towards increasing efficiency does however, tend to bring about noise reduction as a natural corollary. The increase of propulsion efficiency requires reduced jet velocity and ultimately, if forward propulsion is achieved by neutralisation of the aircraft wake instead of by the production of additional rearward jets, it should eliminate this cause of noise altogether. In the case of the large fan engines, the reduced by-pass pressure required for high propulsion efficiency will lead to low air velocities within the fan if the fan efficiency is to be kept high, and ultimately, when fan efficiency is improved by blade boundary-layer and wake control, to the near elimination of fan noise. Again, it has been suggested that boundary-layer suction will be by a large number of small (and very efficient) units buried in the aircraft. The noise from these will be of small magnitude and probably of a high enough frequency to render it easily absorbed before emerging from the aircraft.

The design of a quiet supersonic aircraft is a more difficult problem. As pointed out earlier, the ideal cruise powerplant tends towards the ramjet as the Mach number is increased to 4 or 5. The obvious first choice of a take-off powerplant is a by-pass engine, but it can be quickly demonstrated that the engine has to be severely oversized to pass enough air to give the conventional take-off thrust required at the low level of jet velocity needed for low noise.

There are two possible ways of improving the position, which are:

(*a*) Augment the engine air flow at take-off by installing separate fans.
(*b*) Reducing take-off thrust required.

The use of separate fans would tend to be ruled out today because of the extra weight, and because they must be retractable to reduce supersonic drag. It is just possible that with future techniques and materials they might become a possibility. Another way of looking at the problem is that the extra fans are used to give VTO, thus combining two functions, or in a launching platform which returns to base while the aircraft accelerates to supersonic speeds.

The reduction of take-off thrust required is a powerful method of noise reduction. A typical aircraft today will cruise at a lift-drag ratio of about 19, but the L/D at take-off may be as low as 4 or 5. It is likely that improvements here will prove more difficult to achieve than at cruise conditions.

Acknowledgment

The author wishes to express his thanks to all those at Rolls-Royce who have contributed towards the preparation of this chapter.

11 Man-machine hybrid: an evolving entity

G. Melvill Jones

INTRODUCTION

In Chapter 2 Dr. Küchemann writes 'The task of viewing the prospects over a whole century may be regarded as being at once impossible, presumptuous and not without a touch of the absurd', which lies close to the first response of the present author. But on reflection, the very impossibility of reliable prediction a century ahead surely stands to change the tune somewhat. Over this large time-scale, bold flights of fancy into the unknown must take equal place with cautious realism and an author may presume to feel released from some at least of his usual conditioned restraints. In turn, yielding of constraint strengthens the writer's arm, until, not without a certain sense of trepidation, one is at length tempted to plunge into the guessing pool along with a goodly company of fellow men.

And so, having taken up the pen under the present title let us first glance briefly at the broad areas for future endeavour in the field of Man-machine integration as seen from the perspective of a biologically orientated observer. Of course, this title carries with it an implicit assumption that, in one way or another, Man will remain in control of his machines. It would indeed be a sad prospect if, within but one century, we were to relinquish the prime control to our own Machines.

What, then, are the essential prerequisites for Man to be in control? First, he must be assured of adequate data acquisition, for which so far as we know today, all information must be channelled ultimately through our several sensory systems. Again, to be in control, and so far as we know today, Man's entire information outflow must be channelled through peripheral neuromuscular pathways. What fruits are to be had from the almost incredibly versatile Central Nervous System (CNS) of Man without tangible channels of input and output? Thus it seems logical to examine first Man's sensory input systems in the context of data acquisition from the machine and its environment, and second his so-called

Originally published in the *Aeronautical Journal of the Royal Aeronautical Society*, October 1968.

motor, or output, systems in the context of neuromuscular control of the machine. From this standpoint it would seem natural to examine next the whole human organism as a component in the complete Man-machine ensemble or control loop, and finally perhaps one may indulge in some speculation on the impact of this new entity, the Man-Machine-Hybrid, upon our evolving Social Structure.

SENSORY INPUT

It was natural in the first phase of flying to use our natural senses in their natural way. It seemed obvious that with clear vision one could *see* one's height, one's speed, one's attitude and change of attitude relative to the ground, just as in natural life. Why naturally too, the ears could hear the speed of airflow through the rigging. Certainly they could hear the accustomed speech of a comrade shouting aloud nearby, or later, to improve signal to noise ratio, through a speaking tube. Even the precision sensations from our inertia sensing organs of balance, now known to generate wildly misleading sensations in a wide range of flight manoeuvres, fortuitously complemented data gleaned from other senses, in the very early days of straight and level flight. But inevitably the developing machine soon extended our environment beyond its natural evolutionary limits, and then our natural senses, by themselves, proved quite inadequate. Discovery of the stall called for more precise knowledge of air speed; advent of the engine necessitated knowledge of variables quite inaccessible to direct sensation; entry into cloud, denying vision, exposed the vulnerability of our inertia-sensing organs of balance and so on. But 'Nothing venture nothing win . . .'; the very ingenuity which fathered the machine soon overcame these initial biological obstacles by creating new techniques for human data acquisition; an air speed indicator, an altimeter, a variety of direction indicators, a rate of turn indicator, an angle of slip device, a gyro-stablised artificial horizon, a fuel gauge, an rpm indicator, a pressure gauge, etc. Then gradually this sparkling stream of new inventions grew into a veritable torrent of piecemeal information packages crammed for the most part into the single sensory channel of vision. And today the point is being reached when the ability to formulate the whole picture from these many separate bits of information can break down due to rates of change becoming too fast to accommodate the necessary visual scanning and mental formulation times; and then catastrophic failure may occur[1]. Apparently we are once again brought up short by basic biological constraints, this time through asking our peripheral and central nervous systems to assimilate too many separate threads of artificially coded signals, piled in upon one another at too great a speed. In a way this may seem strange since surely a cross-country runner, twisting and turning on his little feet and always erect in an unstable state of equilibrium, must handle far more densely packed sensory information without being more than barely aware of the intrinsic difficulties involved. But of course in this outdoor exercise he is using the whole range of his genetically wired neural communication pathways in their natural 'design' modes. Perhaps, taking this clue, one can forsee rich dividends in store could we but organise our information presentation into modes which better match the sweeping range of our natural sensory aptitudes; which brings us to a point when we might profitably glance at some of these aptitudes with the central aim of guessing how advantage might be taken of them in the Second Century of Aviation.

Vision

The human eye can be likened to a small camera, in which one minute part of the retina, or photographic plate, is capable of particularly high discrimination. This leads to a roughly 2° cone of accurate vision, surrounded by a large solid angle of progressively attenuating acuity. Consequently, there is high directionality in our central vision which calls for versatile gaze-shifting ability, achieved in practice by what is probably the greatest precision of muscle control found in the body. The oculomotor system responsible for this control comprises in each eye three pairs of unusually fine and rapidly acting muscles operating differentially in three orthogonal planes relative to the skull. Two modes of muscle actuation generate on the one hand very rapid (up to 800°/sec) intermittent shifts of gaze direction at minimal intershift intervals of about 1/5 sec and on the other hand, remarkably smooth pursuit or tracking movements[2, 3] whose goal is target lock-on during target and/or head movement, at least for the time between the shifts of gaze mentioned above.

In natural life these agile physiological mechanisms subserve the overall interest of vision and no doubt the dispersed visual instruments around the cockpit of today have evolved around this remarkable physiological capability. But with too many instruments, too much time can be literally wasted during the 'jumps' from one instrument to another and outside and inside the cockpit. Evidently there is much to gain from minimising this time wastage by assimilation of data into composite displays although this matter will be discussed at greater length later. But more than this, one may guess that an accurate measure of directionality in vision could be effectively employed to drive useful components of machine response. Perhaps visual tracking of a command target spot in the instrument display could generate appropriate response sequences. Alternatively visual fixation on another aircraft might direct the necessary avoidance action, or train air-to-air weapons on an enemy target.

All the while our small 'pencil' of accurate vision is flipping from one point of interest to another, a wide angle of parafoveal or peripheral vision acquires data of a different kind. The tendency is to relegate peripheral vision to a rather subsidiary role. And yet one knows it is possible to thread one's way along a city pavement densely packed with a moving throng while reading an absorbing book. Indeed G. O. Trevelyan in his *Life and Letters of Lord Macaulay* describes this very thing as one of the favourite recreations of perhaps the most avid reader of all time.

In a more objective sense, Fig. 1 illustrates human controller describing functions obtained from four subjects by Levison and Elkind[4] in an experiment employing foveal, 30° peripheral, and 56° peripheral vision in

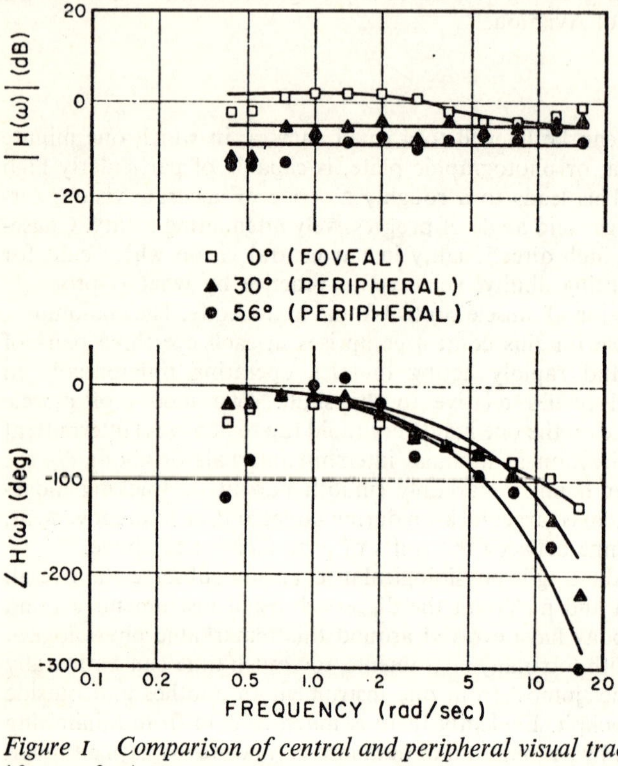

Figure 1 Comparison of central and peripheral visual tracking capability (from ref. 4)

a one-axis tracking task. It is striking to see how the shape of the describing function was essentially invariant with respect to the mode of tracking, although not unnaturally the remnant error (fraction of signal power not linearly correlated with the forcing function) and mean square error were increased with peripheral vision. Again, Hasbrook has recently demonstrated a sizeable reduction of foveal vision load using a system of flashing lights in the peripheral visual field to signal, for example, roll during an instrument approach to land[5]. Evidently when used in its natural mode, peripheral vision can provide strikingly useful information and it seems fair to predict that in the Second Century we may learn to use with more effect the subtle mechanisms of this wide-angle retinal facility.

The retina itself, apart from its enormous range of change in sensitivity, its colour vision and the almost quantal level of its ultimate threshold, is emerging in its own right as an intricate information processing mechanism. Thus, the neural signals from cells within the retina reveal features such as heightening of image contrast at dark-light borders, selective detection of linear images orientated on particular components of the retina and even selective detection by single retinal nerve cells of images moving at particular speeds in particular directions[6]. Thus we are endowed with retinal mechanisms exhibiting great complexity in both spatial and temporal pattern discrimination. Hopefully, as our understanding of retinal 'design' improves we should be able to adapt our information presentation to these intrinsic capabilities of the eye.

Even then to stop at the level of the retina must be to prejudice ultimate information utilisation. Already we are becoming aware of the kinds of processing which take place as retinal messages pass through the neural chain to our visual sensory cortex, at the back of the brain. At the mid-relay station in the brain (lateral geniculate body) further information organisation occurs, and at the cortex itself *single* cells can be found which carry, for example, the message of straight lines orientated in a particular direction and moving in a particular way located anywhere over a relatively wide retinal area[6].

To an extent, some utilisation of intrinsic retinal capability has already occurred in a somewhat haphazard way. For example, a broad-tipped rectangular pointer judged against equally broad rectangular division markers can provide improved presentation of dial information compared with the intuitively guessed fine needle tip judged against point-markers on the dial face. It seems that, albeit somewhat unwittingly, advantage has here been taken of the special pattern-recognising capabilitites of our visual system; in particular a neural affinity for processing orientation and continuity in straight-line, dark-light borders. Again the orientation of an artificial horizon line parallels a retinal equatorial zone of increased sensory unit density. Moreover maintenance of the artificial horizon bar parallel to the natural horizon might be expected to foster a natural aptitude for normal perception of a real horizon. In practice, however, the matter is apparently not so simple as this. A single bar seems too far removed from the natural three-dimensional scene to take full advantage of this potential aptitude-matching. The results of psychological experiments with moving horizon and moving aeroplane presentations has repeatedly led to uncertainty regarding the most effective mode of these two displays despite the obviously 'unnatural' attitudes which must arise with the 'moving aircraft' indicator. If, then, this simple bar horizon of yesterday proves too remote from the natural scene to manifest clear-cut advantage over the 'unnatural' moving aeroplane presentation there must surely be substantial advantage to be had from more realistic attitude display.

Seen from this perspective it seems likely that the symbolic mode of attitude display developed during the First Century will give way widely in the Second Century to increased realism presented in more or less graphic displays. Of course, research and development along these lines are already well advanced in the laboratory[7], and, at least in the military sphere, some advanced presentations have been successfully carried into the air. In some instances this display has been optically projected in a direct line of forward vision, thereby avoiding time-sharing between vision inside and outside the cockpit. An added potential advantage of the projected optical display is the capability of its stabilisation relative to the human head and eyes. During low level flight in turbulent air, for example, a pilot can be shaken too violently for visual tracking of instruments displayed on the cockpit panel, even if time permitted a glance at them. But our vestibular system achieves automatic head and eye stabilisation with higher dynamic response than is available from the visual tracking system and hence inertial stabilisation of an outside image could substantially facilitate the difficult task of quantitative data acquisition in these circumstances.

Perhaps in the Second Century of the Royal Aeronautical Society, we will see synthetic three-dimensional holographic displays, projected and stabilised in the line of forward vision, introduced as routine instrumentation which will become taken as much for granted as was the 'Standard Instrument Panel' of the Second World War.

Hearing

As with vision, great versatility obtains in the physiology of the auditory system. The absolute threshold of sensation is exceedingly low being associated with a 10^{-9} cm amplitude displacement of the ear drum (threshold power at the ear drum is about 10^{-16} watts per cm^2), the overall dynamic range is enormous extending over about 130 dB and the band width associated with normal function is broad, extending from around 30 Hz to about 15 kHz. Moreover coupled with accurate relative pitch discrimination are all manner of internal neural processing capabilities, such as, for example, heightening of contrast (reminiscent of the retina), brought about by active reflex inhibition of local regions in the cochlear or sensory membrane of the inner ear.

Indeed when one contemplates the comparative ease with which a trained musician processes the voluminous information flow associated with a full orchestral score, one's imagination leaps with flights of fancy at the potential rewards awaiting successful exploitation of this exquisite, miniaturised, sensory transducing system. Of course, we use our Man-

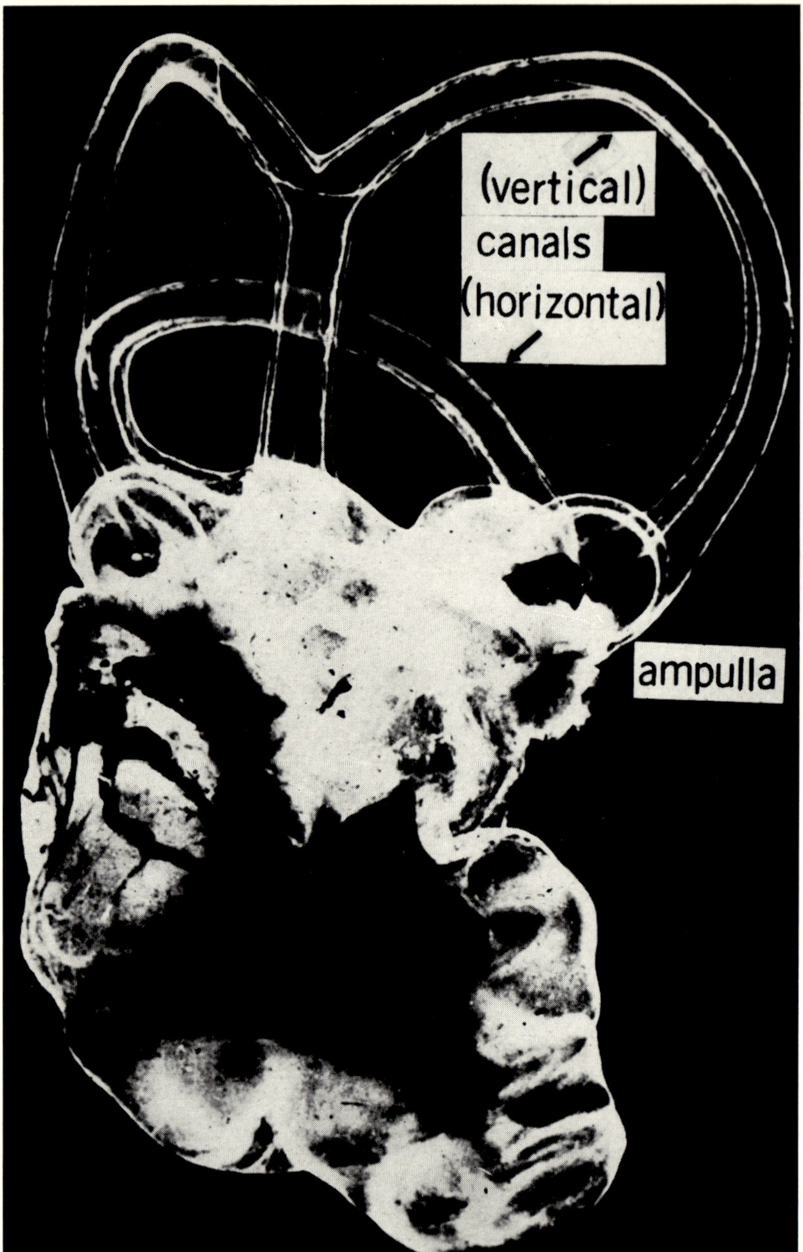

Figure 2 Decalcified preparation of the inner ear (adapted from A. A. Gray, 'The Labyrinth of Animals'; J. and A. Churchill, 1907)

made speech extensively in its conventional mode, although even here there is urgent need for technological improvement. It surely cannot be long before technologies such as pulse coded modulation become universally employed to ensure virtually perfect two-way reconstruction of original speech.

However, pioneering studies have intermittently shown that in addition to speech, the ears can successfully be used to sense continuously modulated representation of air speed or angle of incidence, with an accuracy at least equal to, and often greater than, that attained with more conventional visual displays[8]. Although the approach has not yet found universal application one can readily guess that extensive developments will occur in this sphere before we are far advanced in the Second Century. Possibly, for the more remote future, one might conjure up a vision of new continuously modulated 'languages', learned when young, and perhaps not unlike the language of music as we know it today, but carrying multiple channels of continuously graded quantitative information.

Alternatively could one imagine a 3-D aural aid to the perception of orientation, bearing in mind the remarkable direction-finding capabilities of our hearing? Not only does the binaural system carry inherent direction finding capability, but each ear has astonishing directional properties resulting from the complex accoustical structure of each pinna (outer ear and lobe)[9]; and again the current impact of stereophonic high-fidelity sound serves to emphasise the importance of auditory directional capability in natural life. Perhaps a carefully structured field of sound could be driven by inertia-sensing transducers in an aircraft in such a way as to appear to move in *any* degree of freedom relative to the skull. All the expertise associated with current visual attitude displays, such as quickening of response and integrated gravity ties, could be incorporated with the overriding advantage that the sensory system would be operating in its natural mode with attendant ease of signal interpretation and relegation of information processing to subconscious level in the CNS. But at this point it would perhaps be wise to come back to more solid ground and examine the characteristics of our own biologically evolved inertia sensing organs of orientation.

Organs of balance

It is customary to speak of the five senses of Man; sight, hearing, touch, taste and smell. However, there is a sixth special sensory organ concerned specifically with acquiring information about movement and attitude from inertial transducers in the skull. This balancing organ or, more correctly, vestibular system, is integrally associated with the inner ear organ of hearing seen at bottom right in Fig. 2. Essentially it comprises a six degree of freedom inertial guidance system. Three circular-shaped tubes arranged at right angles to one another contain fluid, the relative movement of which is accurately transduced by a water-tight 'swing door' or cupula, which deflects through an angle directly related to the relative volume flow round each fluid circuit (Fig. 3). Lying beneath these three rotation sensing tubes is a system of translational accelerometers having components of response again represented in three orthogonal planes. As will be seen the response characteristics of these devices are nicely matched to the natural patterns of head movement in everyday life. But

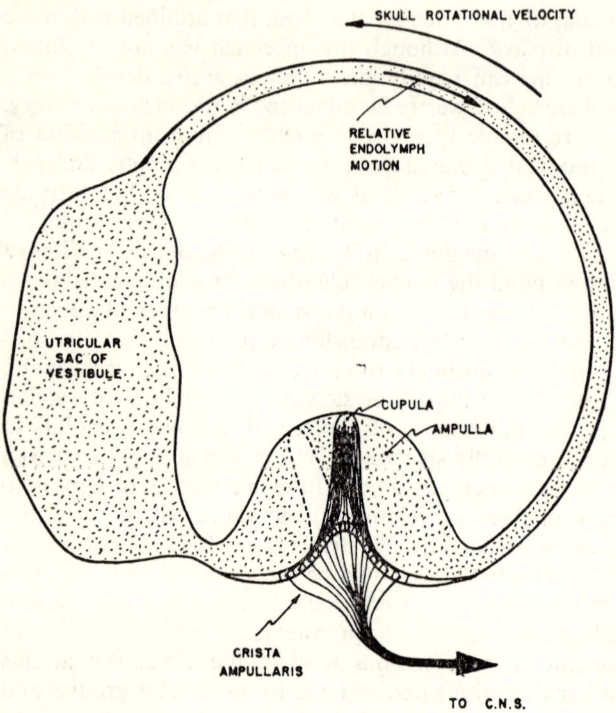

Figure 3 Diagrammatic representation of a semi-circular canal (from ref. 10)

the movement profiles of real flight extend widely outside the range of natural experience and it turns out that in these circumstances the inertia sensing systems in the body usually tend to feed wildly misleading information to the brain.

To appreciate the reasons for this it is first necessary to look briefly into the physical nature of the end organ. Each of the rotation sensing tubes, idealised in Fig. 3, comprises a very thin segment of tube (Reynolds number <1) in which viscous forces therefore predominate. Of additional significance is the fact that the cupula (the 'swing door') exhibits an elastic restoring force approximately proportional to its instantaneous angular deflection. Thus assuming parabolic flow throughout, a transfer function of the following form may be written to express the response characteristics of the contained fluid during angular motion in the plane of the canal[10].

$$\frac{\phi}{\Omega}(S)=\frac{J\alpha}{k}\,\frac{S}{(T_1S+1)\,(T_2S+1)}$$

where ϕ cupular deflection
Ω canal's angular velocity in its own plane
J polar moment of inertia of endolymph
α angle of cupula per relative angular displacement of fluid
k cupular restoring torque per relative angular flow displacement of fluid

$$T_1 \simeq b/k$$
$$T_2 \simeq J/b$$

b viscous torque per relative angular velocity of fluid flow

Figure 4 then illustrates the probable frequency response of this transfer function deduced from mixed theory and experiment. Note particularly that the response, namely cupular angle (proportional to angular volume flow of fluid in the circuit), is expressed as a ratio to the *angular velocity* of the excitation. The striking feature which emerges is a flat response with virtually zero phase over a range of frequencies extending from about 1/10 Hz to 5 Hz (asymptotic cut-offs at $1/T_1$ and $1/T_2$ radians per sec) which presumably covers the natural range of evolutionary experience. Thus at least the mechanical components of this rotation sensing system apparently transduce *angular velocity* with good fidelity over the normal range of head movement in everyday life. Furthermore, experiments have shown that it is indeed this mode of information which is preferentially carried through the nervous system into the brain[11]. Fig. 5 illustrates the pulse frequency modulated message carried by action potentials in single nerve cells in the brain during sinusoidal rotational movement of the head. In the figure the averaged excitation angular velocity is compared with the averaged frequency of action potential pulses carried in the nervous system. Evidently the neural response essentially follows the angular velocity of excitation, at least at this frequency.

However, reference to Fig. 4 will show how at lower frequencies than those experienced in normal life the velocity transducing characteristics no longer obtain. Serious errors develop in the system when it is appreciated that throughout our natural life we learn to interpret the signals in terms of angular velocity transduction in the end organ. Indeed 90° phase advancement at very low frequencies (Fig. 4) implies a sluggish response to *angular acceleration* and yet is seems clear from human and animal experiments that prolonged accelerations are indeed sensed as continued steady velocities. Obviously in the slow angular movements associated with normal flight these inertia sensing elements must feed essentially misleading information to the brain with consequent development of genuine illusions about what is happening.

Possibly in the future one may learn to employ the illusory sensations in a quantitatively useful way, although this seems somewhat unlikely. A more useful approach probably centres round, first the experimental establishment of basic system attributes permitting rational appraisal of when the system is likely to become misleading, and second the development of methods by which the misleading signals can be preferentially attenuated. Central neural components of the overall system can apparently bring about adaptive changes in the sense that, in a particular conditioning environment associated sensory signals can be selectively suppressed[12]. Much remains to be found out about the initiation of such

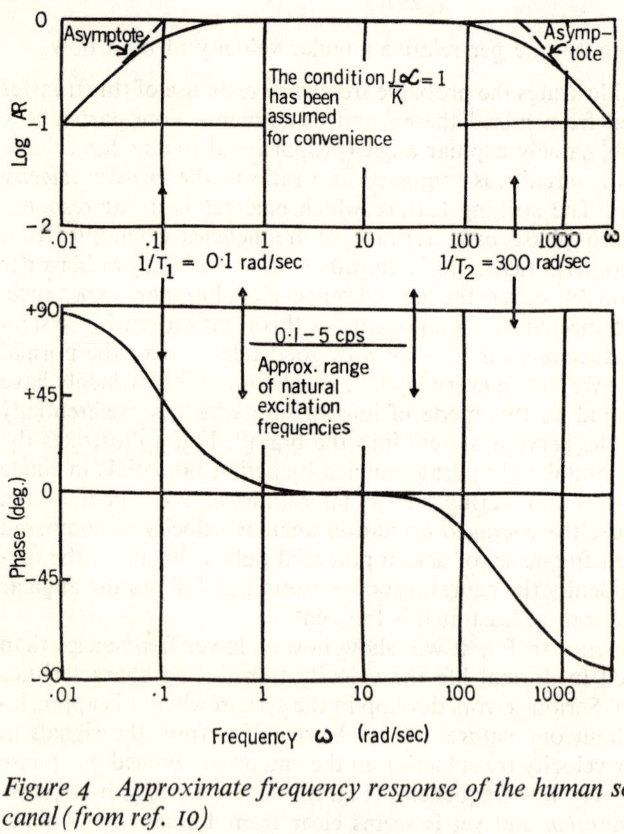

Figure 4 Approximate frequency response of the human semi-circular canal (from ref. 10)

adaptive changes, their dynamic characteristics, the duration of their retention, their transfer from simulated to real environments and the basic neurological mechanisms involved.

Acting within their normal range of frequencies, however, it must be remembered that these organs provide remarkably precise angular velocity

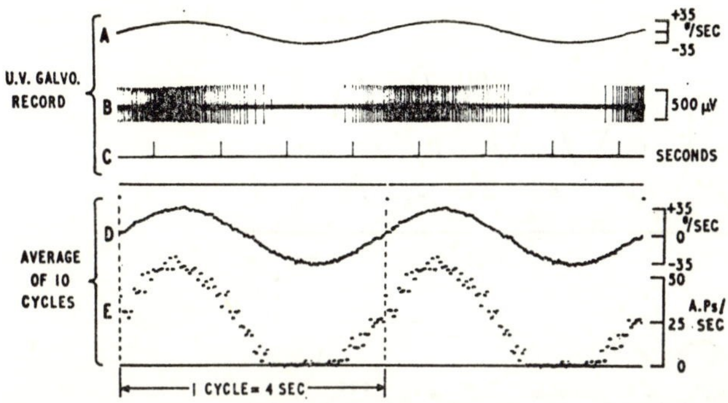

Figure 5 Response of a single canal-dependent nerve cell in the brain stem of a cat to sinusoidal rotation in the plane of that canal. a=stimulus angular velocity; b=original train of action potential; d=averaged stimulus angular velocity; e=averaged action potential frequency throughout the cycle. Averaged records are written out twice to simplify visual interpretation of the record. (From Benson, A. J., Guedrey, F. E. and Jones, G. Melvill, in press 1968)

information to the brain. Might it not be possible, therefore, to use the inertially derived angular velocity signal from our canals in a profitable way? In this connection the inherently unstable characteristics of the rotating wing configuration leads to a particularly difficult problem in instrument flying[13]. Could it be that harnessing the head to the airframe in a suitable manner would provide immediate direct access to 3-D phase-advanced information about angular movement in inertial space, upon which the pilot could base his reflex patterns of stabilising response through pathways in the central nervous system already highly adapted in evolution for this very purpose? Of course the low frequency cut-off (Fig. 4) of the human inertial sensing system would necessitate retention of DC information in some instrumental form.

A further point emerges here. One output from the system acts, through the ocular muscles, to stabilise the angular degrees of freedom of the eye relative to inertial space[14]. Thus during real head movement as when running, or during high speed flight in turbulent air, the eyes are automatically stabilised relative to space by the rapid generation of suitable compensatory reflex eye angular velocities relative to the skull. But this very fact, namely automatic ocular stabilisation, necessarily prejudices vision within the cockpit, since the induced stabilisation is relative to inertial space rather than the internal cockpit environment. Hence, in these circumstances a blurring of cockpit instrumentation can prove a serious hazard. Fig. 6 illustrates quantitatively measured discrepant angular velocities between the eye and the cockpit during and on recovery from a spin[15].

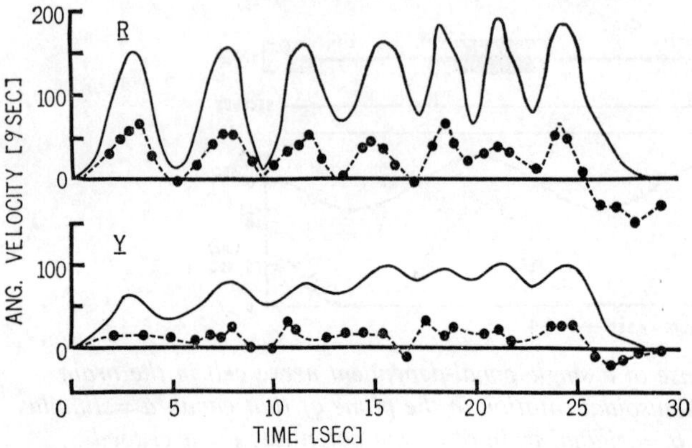

Figure 6 Roll, R, and yaw, Y, components of response during an 8-turn aerodynamic spin with the subject looking at his instrument panel inside the cockpit. Continuous lines give required eye angular velocity for visual fixation on the outside world; intermittent lines give actual eye angular velocities relative to the head. For fixation on the instrument panel the intermittent line should lie on the abscissa (from ref. 15)

Two features with possible future implications arise from this observation. First, the attendant penalty of visual blurring calls either for highly stylised visual instrument presentation in these circumstances or employment of non-visual pathways of sensory information. But second, the very fact that the cockpit framework is made to slip upon the retina in a quantitatively determined manner during transient rotational disturbing functions, presumably itself provides directly usable and suitable phase-advanced information about what is happening[16]. No doubt a pilot subconsciously learns to use such information during training. Perhaps we should consider ways of enhancing its value.

The response characteristics of our translation acceleration sensing devices are much less well known than those of our rotation sensing devices. However, if we assume three-dimensional linear acceleration, then evidently the prolonged linear accelerations associated with normal flight must sum vectorially with the apparent acceleration due to gravity to yield extensively misleading impressions of where the gravitational vertical lies. Fig. 7 illustrates pictorially the misleading impressions a pilot stands to acquire from this source of sensation during accelerated and decelerated flight in the horizontal. Flight in a straight line at constant speed (*A*) is associated with a *g* force vector indicated in the figure by the arrowed line. But on forward acceleration (*B*) a new force vector is

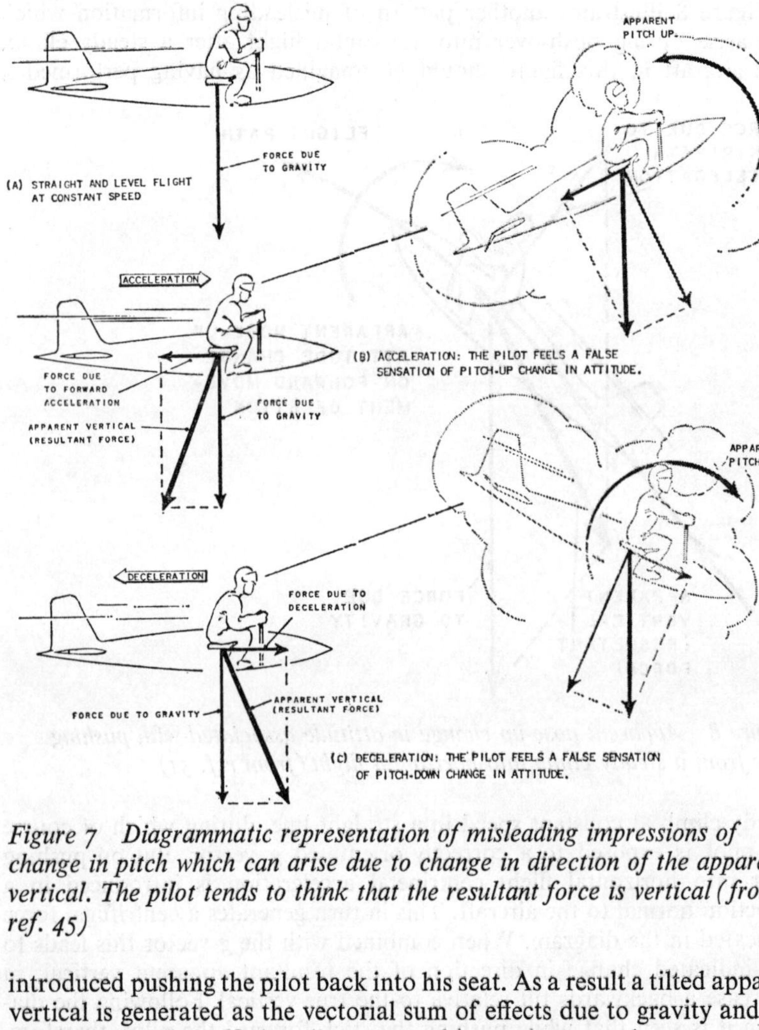

Figure 7 Diagrammatic representation of misleading impressions of change in pitch which can arise due to change in direction of the apparent vertical. The pilot tends to think that the resultant force is vertical (from ref. 45)

introduced pushing the pilot back into his seat. As a result a tilted apparent vertical is generated as the vectorial sum of effects due to gravity and forward acceleration. If the pilot senses this resultant as the true vertical, then in his mind's eye he stands to gain a false impression of nose-up change in attitude when no such change has in reality occurred. The disastrous consequence of correction for this illusory experience near the ground is obvious. In the lowest diagram of Fig. 7 we see a similar consequence of opening the air brakes at high speed. This time the pilot stands to gain false impression of nose-down attitude change. When this is imposed on the real attitude change associated with initiation of descent a pilot can feel himself to have gone through the vertical into an inverted attitude. It can be very difficult to throw off the inverted sensation once established.

Figure 8 illustrates another pattern of misleading information which can arise during push-over into horizontal flight after a steady climb. The aircraft in this figure should be imagined as having performed a

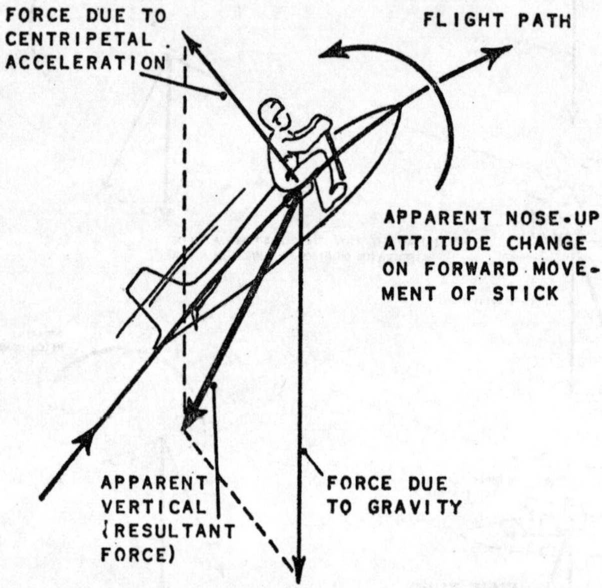

Figure 8 Apparent nose-up change in attitude associated with pushing over from a steady climb into horizontal flight (from ref. 51)

steady climb at constant speed in a straight line, during which of course the pilot is exposed to a correctly orientated *g* vector. But on pushing over into horizontal flight centripetal acceleration is introduced in a direction normal to the aircraft. This in turn generates a centrifugal force indicated in the diagram. When combined with the *g* vector this leads to the indicated change in direction of the resultant apparent vertical, in this case a backwards tilt relative to the true vertical. Following the diagram it is seen that while pushing the stick forward the pilot, therefore, stands to 'feel' a backward rotation of the resultant force vector, which in turn would generate a false impression of nose up change of attitude when in reality the pilot was initiating a diametrically opposite manoeuvre in a nose-down direction. In both cases illustrated in Figs. 7 and 8 the sensation of apparent body attitude change relative to the vertical becomes widely divorced from reality. Thus, throughout the total flight envelope the misleading impressions which can arise are multitudinous and there is an urgent need for their further investigation in the Second Century, particularly in view of the SST programme lying ahead of us.

It is important for example to define the typical situations in which troublesome illusions stand to arise. In addition we have much to learn about the means by which their conditioned suppression can be brought about by training in simulated and real flight. And then in turn we have to explore how such conditioning, once established, may be inadvertently broken down, for example, through inadequate reinforcement by real flight experience, or by sudden startle through unexpected flight emergency.

The foregoing discussion has tacitly assumed straightforward acceleration transduction at the end organ. But current experiments indicate this may be an over simplification[17, 18]. Thus, for example, with increase of excitation frequency it appears that there is rapid development of phase lag on acceleration in the central neural signal even over the normal range of head movement[18]. The implications are at present uncertain. But it is certain that in order to understand knowledgeably the signals to which a pilot is exposed during normal and abnormal flight, the next century must encourage basic experimental investigation in both the relevant peripheral transducers and the ultimate perceptual interpretation of the message received by the brain.

Touch and 'feel'

Everyone knows that in the absence of vision a hand can detect more than the presence or absence of solid articles. A handkerchief can be separated from the pocket lining, a key from a coin, a familiar key can readily be selected from many others on a key ring by blind feeling with the finger tips. Even two similarly shaped objects having slightly different surface texture can be discriminated with ease. Indeed human tactile sensation is still employed for quality control of fine surface texture in some industries where it has not yet proved possible to improve on the human capability. But fine touch is only one of the sensory modalities found in the limb surfaces, which also exhibit separate mechanisms for sensing warmth, cold, pressure, and acute and aching pain. In addition, of course, there are deep sensory mechanisms subserving such features as joint angle, muscle length, muscle tension and their rates of change. It is of more than passing interest that apparently the neural output of muscle length and tension endings does not appear to contribute substantially to conscious sensation; their cortical representation is meagre compared with that in the cerebellar and the brain stem structures. But sensations from joints, capsular ligaments, etc., comprising the so-called kinaesthetic information certainly do reach sensation. It must not be forgotten that knowledge of the message sent to muscles also contributes substantially to the overall picture of what is happening, although this particular feature will be more appropriately discussed below. Not unnaturally it now begins to emerge that as with the special senses, considerable data processing may take place at an early stage in the peripheral sensory pathways. For example, the threshold of sensation of cutaneous tissue adjacent to a local stimulated area may become substantially raised for up to about 100 milliseconds after the exciting stimulus, indicating contrast heightening akin to that in the eye and the ear and no doubt more complexity will emerge with further research.

Viewed from these perspectives, current utilisation of peripheral sensory capability seems but to have scratched the surface of what may lie in store. To be sure, constructive use has been made of shape recognition in the design of control terminals and obviously the knowledge of control locations and their force-displacement characteristics plays an important role in the general field of human engineering today. Yet surely this must represent but a small proportion of the available potential.

For the future then, one can see in a general way that continued basic research into peripheral sensory mechanisms should yield information of direct value to the human engineer. Again in a general way, one can foresee increasingly productive usage of the tactile and kinaesthetic sensory mechanisms in instrument design. To be more specific an imaginative approach has recently been made using conventional describing function criteria of success to compare tracking capability using various tactile stimuli and conventional visual presentations [19]. Fig. 9 shows the results from one subject comparing responses to purely visual and purely tactile stimuli. The tactile stimulus was in this case a pointer moving over the palm surface of one forefinger. As in Fig. 1, the phase curves are essentially indistinguishable from one another although the mean square error in tactile tracking was usually roughly twice that for visual tracking. Evidently this proves a promising approach; the more so since even performance with tactile sensation over the forehead gave similar results. In the light of these kind of data it does not seem an outsize leap of the imagination to envisage, for example, employing movement of the unloaded finger tip over a suitably designed tactile indicator for closed loop vehicle control.

Perhaps in the far-out future it might prove possible to gain direct access to single sensory neural units, along the lines already achieved in the efferent or motor pathway, described below. In this case a new realm of potential utilisation of our inherited faculties could open up so fast that it would be rash indeed to attempt further prediction.

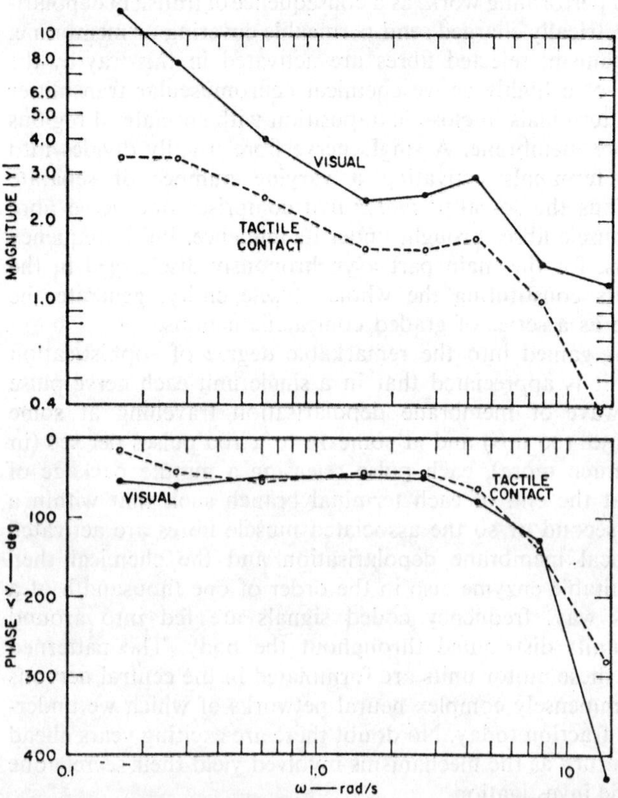

Figure 9 Describing functions for visual and tactile tracking (from ref. 19)

MOTOR OUTPUT

The neuromuscular systems of Man's skeletal structure comprise in the main a large number of nicely arranged differential actuators bringing about torques and movements at inter-skeletal joints of the limbs, trunk, head and neck. Other muscle systems exercise fine control over such functions as respiration, articulation in the tongue, change of gaze direction and so on. What then, do we know of the ways in which these multitudinous muscle systems are put to their several purposes in our everyday life?

Physiology

Essentially the muscles themselves are constructed of many separate parallel contractile elements, the muscle fibres, capable of generating active tension and performing work, as a consequence of transient depolarisation of their electrically charged semi-permeable covering or membrane. In the intact organism, selected fibres are activated in this way by the miniature release of a highly active chemical neuromuscular transmitter agent from nerve terminals in close juxtaposition with specialised regions of the muscle fibre membrane. A single nerve fibre usually divides into numerous nerve terminals activating a varying number of separate muscles fibres. Thus the so-called *motor unit* comprises one nerve fibre and the several muscle fibres brought under its influence. Pulse frequency modulated signals, for the main part asynchronously discharged in the many motor units constituting the whole muscle entity, generate the ultimate response as a series of graded contractile actions.

Some insight is gained into the remarkable degree of sophistication in 'design' when it is appreciated that in a single unit each nerve pulse is carried by a wave of membrane depolarisation travelling at some 100 to 200 mph (50–100 m/s) and at some 10 to a 100 pulses per sec (in some instances much more), each pulse releasing a minute package of active chemical at the end of each terminal branch such that within a thousandth of a second or so the associated muscle fibres are activated electrically by local membrane depolarisation and the chemical then destroyed by a suitable enzyme also in the order of one thousandth of a second[20]. In this way, frequency coded signals are fed into around $5{\cdot}0 \times 10^5$ motor units distributed throughout the body. The patterned messages feeding these motor units are formulated in the central nervous system through immensely complex neural networks of which we understand but a small fraction today. No doubt there are exciting years ahead in the Second Century as the mechanisms involved yield their secrets one by one to scientific investigation.

Very briefly, in the present state of knowledge we recognise the simplest input-output pathway in the Sherringtonian concept of a segmental spinal reflex arc[21], depicted as pathway (*A*) in Fig. 10. Even at this elementary level considerable versatility obtains. Summatory and occlusive effects can cause interaction with other incoming signals; superposition of sequential inputs may lead to long transients in the response to step input change and the segmental reflex arc (input and output in the same segment of the spinal cord) is very substantially influenced by signals from higher and also lower levels. Furthermore as illustrated diagrammatically in Fig. 10 even the final common pathway, that is the motor nerve itself, branches at an early stage to feedback negatively upon itself (pathway *C*).

It is, however, difficult today to speak of a final common pathway since it is now well established that, among the reflex components of the segmental reflex arc an elaborate internal feedback system is available from sensors of length and tension of the muscle and its tendon. This system can apparently itself be controlled after the fashion of length servo system[22] through a pathway controlling what amounts to the bias and gain of the internal sensors. Both direct open loop and closed loop control can operate simultaneously and their relative effectiveness can be changed by a wide range of influences arising from higher levels in the nervous system[23].

In the context of the present article, the foregoing must suffice to establish essential features of mechanisms located at a segmental level. But as indicated above the excitability of nerve cells and reflex systems at the segmental level in the spinal cord comes under direct and indirect influence from higher levels. Fig. 10 indicates diagrammatically long-loop reflexes (*B*) which reflect from neural centres in the lower levels of the brain (brain

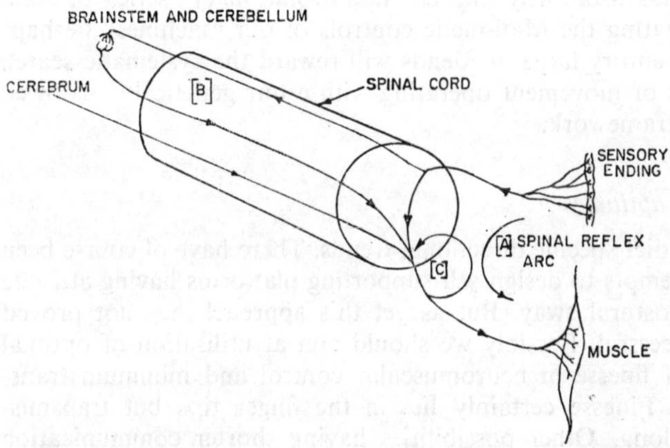

Figure 10 Some input-output elements in the control of muscle

stem) and the cerebellum. Thus any input to the segmental reflex systems generates both segmental response and long-loop influences which substantially and rapidly (50–100 ms) modify the final output or response, presumably playing an essential role in obtaining optimal shaping and organisation of the final output signal[24].

Over and above the automatic reflex processes thus far mentioned, the classical work of Pavlov defined a wide range of conditioned reflexes; that is responses which become very specifically tied to particular conditions. And finally, as we all know (or hope!) we are free to exercise voluntary control over much of our muscle systems. But the detailed mechanisms of these latter two facilities, both so obviously germain to the aviators' task, are only beginning to be understood today.

Evidently then, with some 400 muscles in the body, containing all told some 25×10^{10} muscle fibres and fed by some $5 \cdot 0 \times 10^5$ motor neuron units, the whole being backed by elaborate organising and processing machinery in the central nervous system, we are endowed with a formidable total capability of muscle control, ranging in heirarchical order from the essentially automatic, through the enormous array of conditioned responses which cater for much of our everyday life, to purposeful voluntary movement under the direction of human free will. How does all this bear on the Second Century prediction?

First, enough has been said to indicate great complexity in the 'genetic' wiring handed down ready made from generation to generation. Again as with sensory input, the overall system, having painfully evolved during the long course of evolution, must be highly matched or, adapted, to what might be called the evolutionary environment. Would it not, therefore, be naive to expect comparable performance in our muscle systems from a somewhat arbitrarily chosen 'non-evolutionary' series of limb movements actuating the Man-made controls of our machines? Perhaps in the Second Century large dividends will reward the systematic search for basic modes of movement operating within our genetically endowed neuromuscular framework.

Matching man's aptitudes

It is hard to predict specific directional trends. There have of course been rudimentary attempts to design self-supporting platforms having attitude controlled by postural sway. But as yet this approach has not proved particularly successful. Possibly we should aim at utilisation of optimal combinations of finesse in neuromuscular control and minimum transmission delays. Finesse certainly lies in the finger tips but transmission lines are long. Other possibilities having shorter communication lines with the central nervous system might rest in, say, the neck muscle systems controlling attitude of the head, the external ocular muscles turning the eyes in the skull or even the exquisitely organised muscles of the tongue. Perhaps this is becoming too fanciful: but then again perhaps not. Has not man's home-made talent for manipulation of his muscular tongue given him the speech which alone goes far to lifting him above his fellow animals? And are we not already embarking upon speech controlled programming of the man-made digital computer? From here it needs no great leap of the imagination to envisage a speech control capability which could immediately 'cash-in' on the ready made flexibility of spoken language.

Again, would it be feasible to establish three degrees of freedom in control through head rotation relative to the body? Perhaps head movements fore-aft, left-right, and side-to-side might control pitch, yaw and roll. Not only would the four limbs be released, but the natural neuromuscular systems subserving auto-stabilisation of the skull relative to space[25] might be entrained to aid short term stability of the machine. Alternatively, or in addition, could one imagine harnessing the muscles which guide the eyes with such precision for example by measuring eye angle relative to the skull. As already indicated visual lock-on during auto-stabilisation of the skull platform represents one of the finest examples of combined muscle control systems in natural life[14]; it would seem rational to employ this ready made system combination of high dynamic response for guidance of the Man-made machine.

Of course we must not lose sight of the fact that our endowed neuromuscular systems certainly have proved highly adaptable to our Man-made goals. The pen which writes these very words is sped through an endless series of sweeping curves by the finger and arm muscles driven in turn by intricate learned patterns of programmed neural signals. Again with constant practice from a young age unnatural patterns of skilled neuromuscular activity give the professional musician his unsurpassed dexterity. Presumably then we stand to gain substantially in the Second Century by imaginative investigation of the nature and limits of this kind of flexibility in our neuromuscular systems.

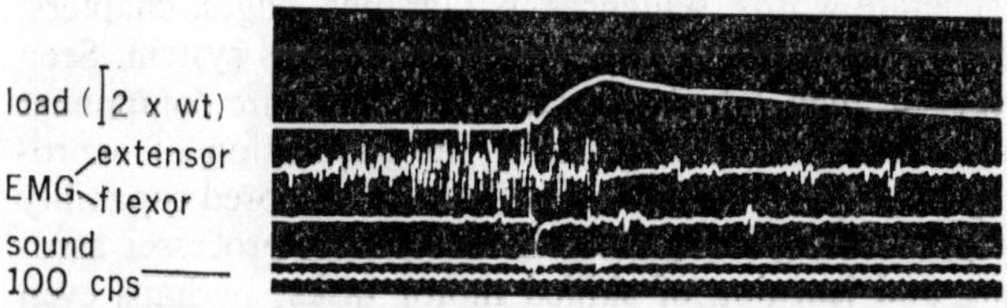

Figure 11 Electromyographic response of calf muscles associated with landing from a 15 inch jump to the ground (from ref. 26)

Training and maturation

The acquisition of skills brings us to the realm of training and learning. Today our training programmes, so far as neuromuscular control is concerned, are themselves based largely on the old evolutionary process of 'trial-and-error' and much remains to accrue from a detailed systematic approach to optimal acquisition of relevant skills. In this connection Fig. 11 gives some insight into the neuromuscular organisation of skilled performance. It illustrates time relations between the event of jumping to the ground and the associated neuromuscular activity effecting deceleration on landing. Electrical evidence of muscle activity begins some 180 milliseconds before, and is terminated about 120 ms after impact. Yet from supplementary experiments it can be shown that *effective* reflex feedback from the event of landing cannot begin to develop in the muscles concerned before about 120 ms after impact[26]. This striking observation, therefore, tells us that the entire very rapid landing process must be brought about by the early release of a complete pre-programmed sequence of highly organised and accurately timed neural activity, the course of which is inaccessible to neural feedback from the actual mechanical event of landing. By inference one can see that much of our daily muscle activity must be composed of sequential release and integration of multitudinous packaged messages, the precise code and shaping of which has been formulated and stored somewhere in the central nervous system through environmental exposure; but always ready for release on command. We can see the infant boldly feeling its way around obstacles on the floor, acquiring in great profusion the necessary basis of stored data upon which to build progressively his adult capability. The writer's intent can perhaps be grasped by invoking by way of illustrations the special neuromuscular capability of speech, the essential components of which begin to be acquired while the child searches for and stores those programmes associated with his limbs. Each syllable comprises a detailed co-ordinated neuromuscular event involving thousands of nerve fibres and motor units. But in due course these separate entities, each one of which becomes highly stereotyped (the mature adult finds it difficult, if not impossible, to change his accent in a foreign country), can be smoothly inter-related in many different ways to generate words, sentences, paragraphs, pages, chapters, books and eventually libraries of capability in the nervous system. Seen in the context of speech, then, which reflects but a miniature component in our total body musculature, one can infer the acquisition of neuromuscular 'words', 'paragraphs', 'books' and 'libraries' of stored capability as a major, and sophisticated, component in the overall processes associated with training and learning of skilled motor tasks; perhaps even skilled mental tasks too! For the future then, what kinds of sequences are most readily acquired and stored; in what ways and to what extent can already acquired data be transferred and applied in other directions; what sequences of programme release are preferred by the available system; how can optimality be approached in a given context of control?

One thing seems certain. Different kinds of response capability have optimal time gates in our maturation cycle for their acquisition. Learning a language is effortless when a child; musical dexterity must be gained when young for effective development of professional techniques. And now we know from psychological experiments that animals deprived of a particular environment over a particular maturation period can never acquire certain particular skills or behavioural patterns which comprise normal components of animal behaviour[27]. Even the physiologist is demonstrating absolute dependence of successful utilisation of genetically 'wired' neural circuits upon their timely exposure to natural stimuli within prescribed maturation time gates. Thus a kitten's eye, which at birth will generate appropriate neural responses in the visual cortex of the brain, will, if denied access to vision for the first six weeks of life, never again establish objective communications with the cortex[28]. Perhaps it is too fanciful to imagine beginning training for the task of aircraft control in infancy; and yet in a more sophisticated parallel, do we not start the installation of basic skills in our schooling system from the age of five onwards? From this perspective one can already see our contemporary computer-programmer being nurtured in the feeding grounds of our educational systems, in which he acquires mental skills in sequential series of specific exposures which are presumably roughly matched to an heirarchical series of maturation time gates. Roughly matched, that is, through endless trial and error procedures. One can dimly see a future trend, perhaps realisable in the Second Century, in which detailed knowledge of appropriate maturation time gates will permit rational organisation of exposure to the learning process and yield new levels of total capability as far removed from us today as current capability is from that of primitive man.

As to the execution of the motor task, one facet which will perhaps be brought to bear stems from recent experiments with single motor units[29]. By placing a single miniature electrode in close apposition to a single muscle fibre, it is possible to detect and record a discrete pulse of electrical activity following each pulsed neural unit drive already described. The remarkable feature has emerged that apparently it is possible to train such motor units to be activated specifically by voluntary control in the absence of activation of other adjacent motor neurones. Not only can the single motor neurone be activated in isolation. In short order, one can learn to effect controlled patterns of activity covering a wide range of different rhythmical sequences. Investigational work in this field is progressing along the lines of determining the versatility of this capability within a complete muscle. Already it seems that several different units can, with practice, be employed simultaneously in quite different ways. It does not need much imagination to generate a picture of immense output versatility if it were to prove possible to activate individually each of say 300–400 separate motor units in the body of a single muscle. Yet perhaps this conception is not entirely fanciful when presumably the emerging capability reflects nothing more than that which permits selective utilisation of sequential programmed release associated with normal neuromuscular activity in the manner already outlined. The idea becomes less fanciful again in view of the current approach to utilising specific response capability in single motor units as input to control units actuating prosthetic artificial limbs. Bearing in mind the enormously rapid increase of technological capability which would be expected in the next hundred years, one can perhaps predict a systematic utilisation of trained individual motor unit capability as the source of inputs to multiple control actuation.

For realism one must of course visualise potential capability in this sphere in the educational context of previous paragraphs. That is to say, such capability as is emerging today in the control of single motor units amounts merely to that which has been acquired by mature adults during exercises covering but a few hours or days or weeks. Imagine choosing the right maturation periods during the long course of total educational exposure and perhaps we can visualise in the hazy future an individual reclining on his couch and releasing itemised neural programmes in selected motor units for the ongoing programming of his control computer! In the long term event it might even be possible to gain access to *neural* components in the peripheral, and later the central, nervous systems, and with appropriate long term training to utilise these stage outputs for the manipulation of Man-made mechanisms. Why even today, Man-made electrical excitation of the visual cortes (brain) of Man is being successfully applied to the problem of providing 'sight' to the blind.

But of course any human capability, be it associated with direct control actuation or control manipulation through computerised assistance, must today be considered in the context of the overall control system comprising the total static and dynamic relationship between the Man and his machine. The next area for consideration, therefore, entrains the Man-in-the-Loop.

MAN-IN-THE-LOOP

During, and since, the Second World War, systems control technology, backed by developed mathematical tools and the rapidly emerging potential of analogue and digital computer capability, burst from its embryonic state into a discipline in its own right. No longer could human capability be considered out of context with the machine; no more could design and development of the machine progress effectively without pertinent dynamic knowledge of the human response. Thus, growing weight is currently being placed on the characteristic describing functions of Man-in-the-Loop.

Initially dynamic simulation of the aircraft aided the Man in learning to control the machine. Today human transfer functions are being simulated to aid the design team in matching the animate and inanimate components of their close-loop Man-Machine-Hybrid. In practice, owing to inherent variation and adaptability in the Man it still proves necessary to parallel laboratory simulation with real flight verification. And even then the experienced investigator is at pains to point out that his conclusions must be treated with caution since meaningful, mission-orientated tasks usually involve multivariable, multicontrol roles for the pilot[30]. Evidently there is plenty of scope for development of this theme in the years to come. For example, given a human describing function in a particular set of experimental circumstances, what consequences will result from superposition of multiple additional tasks? To what extent can the describing functions derived in the fixed laboratory simulator be applied to the real flight environment, where a wide range of additional sensory information becomes available to the Man? How does the human element adapt to pressure of the on-going stress? And from the results obtained in one simulator configuration to what extent is it possible to predict human dynamic response in other configurations?

Probably answers to many such questions will emerge from the growing use of *airborne* simulators. In the writer's view the part played by airborne simulation will grow rapidly in the Second Century, and not only in the context of vehicle design. A potentially serious problem is currently arising from the difficulty in obtaining adequate realistic exposure of civil aircrew to the wide range of fearsome possibilities presented by the formidable machines they are asked to fly today—and how much more so tomorrow. Perhaps the airborne simulator, emulating flight characteristics of the jet transport, yet safe to operate over an extended flight envelope, will fill the gap. In which case one can envisage the introduction of airborne simulation as an essential back up to both basic and continuation training on type.

Another more futuristic aspect of Man-in-the-Loop devolves on the design and development of modes of locomotion in such foreign environments as zero, or reduced, gravity. One can, of course, readily envisage application of conventional systems technology to the control of say, jet-propelled personal locomotion in outer space, or 'Pogo' locomotion on the moon. But, bearing in mind conclusions drawn earlier from Fig. 11, it must be evident that the matter may not be so simple as this. It seems that in our natural life on the earth's surface much of our movement is conducted through the controlled release of patterned neuromuscular activity, imprinted deep down within our neuromuscular systems during those early plastic years of developmental maturation in a 1 *g* gravitational field. Suddenly the astronaut is asked to move in zero gravity space during extra-vehicular activity and we are sufficiently naive to wonder at his early fatigue in the new tasks. In the writer's view his difficulties probably stem in the main from the fact that virtually every single component of every single one of his multitudinous on-going muscle actions must require the formulation of brand new patterns of central neural activity specific for every task, when in natural life the greater part of normal activity merely reflects programmed release of detailed patterns already stored during those early plastic years. What is the transient penalty we have to pay in our first tentative ventures into space? To what extent can the formed adult hope to make good his deficit, and how would the control problem, designed round the formed adults capability, be changed in the new generation born and raised in space? The questions are manifold and one cannot but suspect our current systems concepts will have to expand considerably in the Second Century before they can rationally be applied to matters such as these.

The adaptable man

Even today, nearer at home, we are finding that fixed describing functions are no longer adequate to describe the adaptable Man. Intrinsic adaptive capability in the biological organism can continually modify dynamic characteristics in many different ways. For example, it has recently been shown[31] that the human operator can readily move reversably from linear to non-linear modes of operation. On the whole it seems that, at least in the single-loop compensatory tracking of random inputs, the tendency is to linearise the system performance insofar as this is possible. But, for example, the imposition of non-linearity in the simplified longitudinal control system of Fig. 12 could, in these circumstances lead to untenable situations. Apparently the human controller obviates this by introducing non-linearity into his own dynamic processing, although

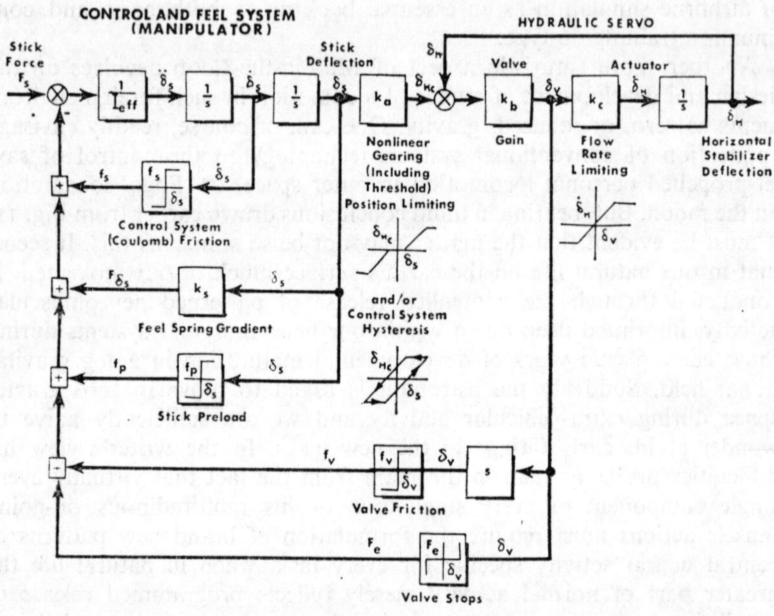

Figure 12 Simplified aircraft longitudinal control system block diagram showing common non-linearities (from ref. 52)

not, it should be noted, without some attendant performance decrement[52]. The important feature here is the ease with which he can apparently move continuously from linear to non-linear processing with consummate ease and for progress in control design we have to understand the rules which govern his performance in these circumstances.

The human operator can quickly take advantage of another signal feature, namely predictability. Figs. 13(*a*) and (*b*) illustrate the amplitude and phase data (continuous lines) describing the response of Man's visual tracking system when he attempts to follow a visual target smoothly oscillating in a horizontal plane through a peak-to-peak amplitude of approximately 10° [32]. As substantiated by the results of other investigators[33, 34], the development of significant phase lag appears to be considerably postponed relative to the phase data to be expected from a suitable damping factor of 0·6 in a second order system matching the amplitude data (intermittent line in Fig. 13(*b*)). One advantage for the visual tracking system is evident in the intermittent line of Fig. 13(*a*) which gives the mean error in reading a fine cursor line on a numerical scale attached to the oscillating target. Two seconds were allowed for reading the cursor

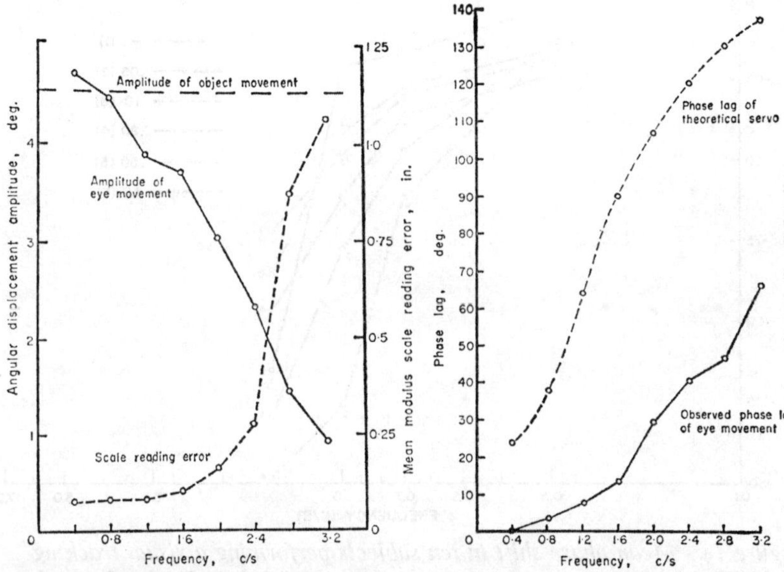

Figure 13 Visual tracking response to a sinusoidal target stimulus of amplitude 4·5° (from ref. 32)

line. It is striking to see how postponement of error increased roughly in accord with the actual observed phase data, in Fig. 13(*b*). Presumably so long as the eye and target are in phase with one another, relatively clear vision is possible near the velocity nodes whatever the loss in amplitude. In Fig. 14 it is intriguing to note how the phase curves move towards the calculated 'second order' line in Fig. 13 as the task changes progressively from a predictable to a random one[53]. Presumably just as the human controller can skip unawares from linear to non-linear modes of operation so can he change his strategy continuously to match the current level of predictability in the task. Certainly over frequencies here employed, not more than one or two cycles seem necessary for full advantage to be taken of the predicting facility.

The applied significance of a human predictability component has been examined in the control context by several authors[35, 36, 37]. As C. R. Kelley points out[38]:

'The usual engineering models of the human operator employ techniques that were first developed in control engineering to describe and model control mechanisms. If there are ways in which the human operator differs from inanimate control mechanisms that are significant for representing the human operator, the usual form of engineering model would

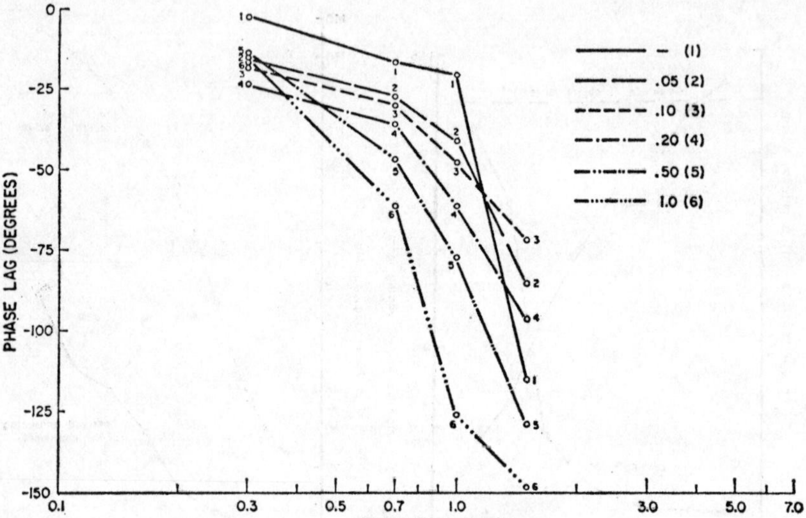

Figure 14 Mean phase shift in ten subjects performing a visual tracking task. The six curves represent responses to signals of six levels of stimulus predictability. Numbers at top right give band widths in c/s of white noise around the centre frequency (from ref. 53)

not be expected to include them. Human operators differ from control devices in that they can understand their task and its environment, can remember the past and predict and plan for the future, and can employ these abilities in control. All of these psychological aspects of the human operator are important in manual control, yet all are absent from the usual engineering models of the human operator'.

Also more realistic modelling would include:

'(1) A direct internal representation of the controlled element and its environment.

(2) Freedom of the internal representation from present time (fast-time operation) so that it can be used to extrapolate and predict controlled element behaviour.

(3) Control action based on such predictions.'

The inference is then drawn that:

'Adaption in such a model would occur automatically as a consequence of changes in the internal representation, reflecting actual external changes. The internal changes would bring about change predictions and as a result of the changed predictions, the control action taken would be changed "adaptively".'

Presumably in any given situation adaptability and practice work together in the approach towards optimisation. But it is important to appreciate that the criteria of optimisation may change; and with such change the individual must surely change his optimising procedures. For example, the success criterion may change from high precision control in low-level high-speed flight to one in which the goal is maximum

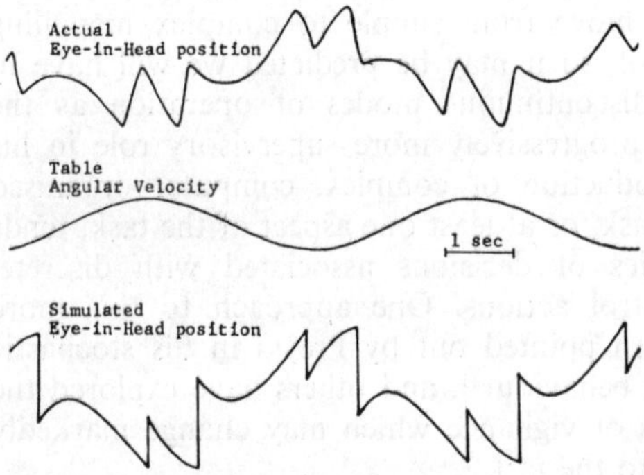

Figure 15 Comparison of actual and calculated optimised response of cat eye movement during sinusoidal rotation in the dark (from ref. 54)

safe endurance of operator control at high altitude during long-range flight. The human 'figure of merit' must constantly be changing in real life and it would be unrealistic to assume that a single criterion will elicit the full versatility of the operator as an optimising control system. That optimisation capability is deeply built into animal neuromuscular systems has recently been demonstrated by Outerbridge and Wettstein[54] in the oculomotor (eye movement control) system of the cat. In Fig. 15 the upper trace gives the eye angle relative to skull during an involuntarily generated intermittent compensatory tracking response on sinusoidal rotation of the skull in a horizontal plane. In the absence of vision, smooth pursuit eye movements, driven by the vestibular or balancing organs discussed earlier, occur in a compensatory direction, these being interspersed with quick fly-back return movements. Now it has been shown[39, 40] that each fly-back introduces an additional term in the oculomotor response which summates with the analogue drive derived from the inertia-sensing semicircular canals of the balancing organs. Using a dynamic programming technique, the above authors searched for the optimal timing and magnitude of fly-back for minimising mean square angular velocity error between eye and 'target' (i.e. outside work). The lowest trace in Fig. 15 gives the derived optimal response and it can readily be seen how closely both size and placement of fly-backs in this optimal response pattern matched the involuntary reflex pattern obtained in the live animal experiment.

Thus, it is emerging that the biological element represents a far more elaborate control system than can be modelled by simple transfer function, and one may readily predict that the Second Century will herald in extensive and productive research into the wide range of describing functions, the adaptive characteristics, predicting capability and optimising features with which we are apparently so richly endowed.

However, just as we must move from simple to complex modelling of ongoing continuous control, so it may be predicted we will have to move from continuous to discontinuous modes of operation as the human operator takes on a progressively more supervisory role in his control task. With the introduction of complex, computer organised control systems, the control task, or at least one aspect of the task, tends towards an heirarchical series of decisions associated with discrete, rather than continuous, control actions. One approach to this more difficult field has recently been pointed out by Preyss in his stochastic modelling of human learning behaviour[41], and others have explored the more psychological attributes of vigilance which may change markedly with duration of application to the task[42].

For the future then, it surely behoves us to join forces of the systems engineer and the psychologist in closer association than has yet been envisaged. Perhaps one can imagine the growing interdisciplinary faculty of Bio-Engineering being complemented by a new faculty of 'Psycho-Engineering'. Of course the engineer and psychologist have perforce been brought together from time to time in the past. But for optimal progress in new problems associated with supervisory control a new discipline must surely arise in the Second Century in which the expertise of the psychologist and engineer are each respected and understood by the other, and in which at least at the post-graduate level, teaching and research training are really integrated, as is now at last becoming the case in Bio-Engineering.

The 'human element'

At this juncture the biological investigator becomes involved. For however adept the systems engineer, the bio-engineer or the 'psycho-engineer' may be at modelling the human operator in a particular simulated closed-loop environment, unpredicted eventualities can and do lead to sudden

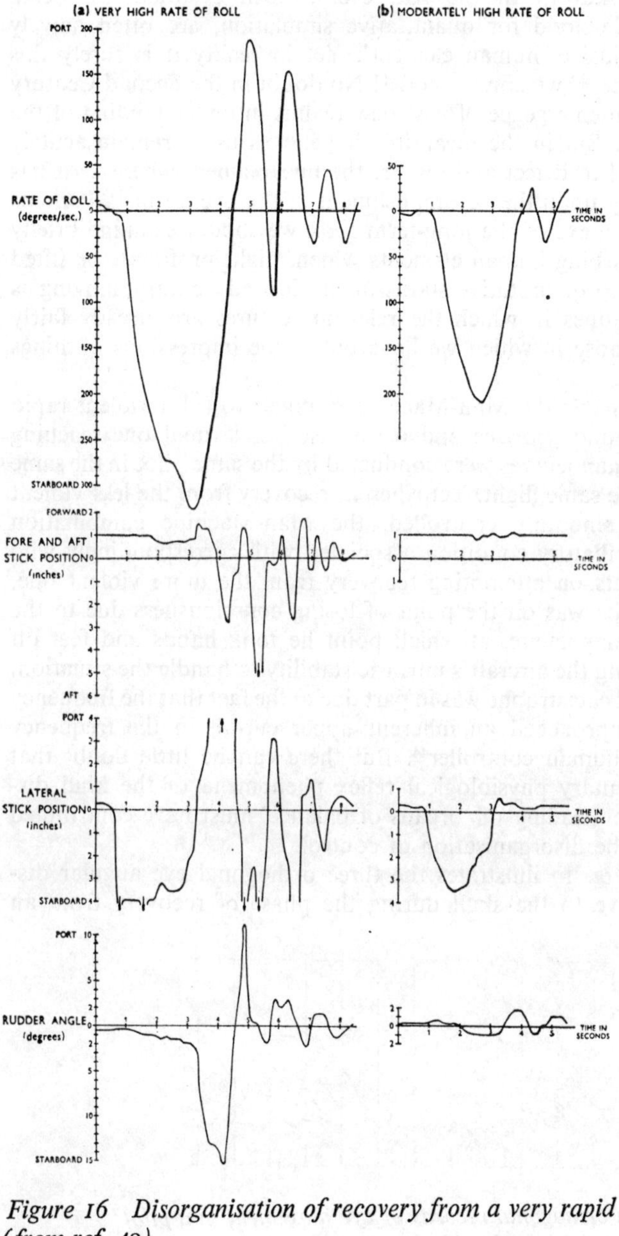

Figure 16 Disorganisation of recovery from a very rapid roll manoeuvre (from ref. 43)

catastrophic breakdown in the real event. Such eventualities, being insufficiently understood for quantitative simulation, are often naively ascribed to the elusive 'human element'. Yet in reality it is surely this very 'human element' we aim to model! No doubt in the Second Century we can expect the emergence of vast new realms in understanding of the 'human element'. But in the meantime it behoves us to remain acutely aware of the need to detect and respect the unexplained, wherever it has significant bearing upon the accomplishment of the overall task. Perhaps in this present context of the long-term look we should examine briefly some of the disturbing human elements which might profitably be lifted from qualitative to quantitative status in the Second Century, moving as we do so from zones in which the relevant features are already fairly well defined to those in which we have but vague impressions of things which matter.

Figure 16 compares the Man-Machine response to (*a*) a violent rapid roll reaching around 360°/sec and (*b*) a fast but normal one reaching 200°/sec[43]. Both manoeuvres were conducted by the same pilot in the same aircraft during the same flight. Yet whereas recovery from the less violent manoeuvre was smoothly controlled, the Man-Machine combination set up a wild oscillatory condition associated with correspondingly wild control movements on attempting recovery from the more violent one. The pilot stated he was on the point of losing consciousness due to the violence of the manoeuvre, at which point he took hands and feet off the controls leaving the aircraft's intrinsic stability to handle the situation. Probably the near catastrophe was in part due to the fact that the frequency of oscillations approached an inherent upper cut-off in the frequency response of the human controller[44]. But there can be little doubt that additional involuntary physiological reflex phenomena of the kind discussed under the heading of 'organs of balance' must have contributed substantially to the disorganisation of control.

For example Fig. 17 illustrates the three orthogonal eye angular displacements relative to the skull during the phase of recovery from an

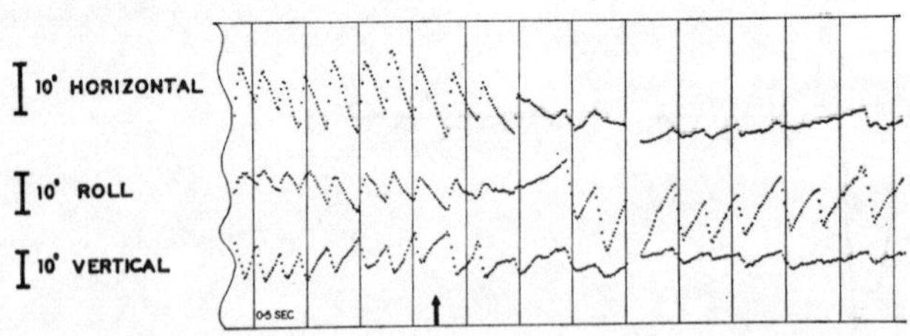

Figure 17 Three orthogonal records of eye movement in a pilot recovering from an 8 turn spin. The arrow indicates moment of intiating recovery (from ref. 15)

8-turn spin in a Vampire trainer jet aircraft. At the point of recovery, and in accordance with the transfer function given above for vestibular response, a reversed vestibular signal dramatically drove the roll component of eye movement in a 'reversed' direction, so that even if the aircraft were then stationary in roll the pilot would literally see a rotating horizon on his retina. Of equal significance in this context is the additional feature that the involuntarily driven response would even prejudice reference to the aircraft instruments, and indeed pilots have complained that in these circumstances it was impossible to read their instruments because the whole instrument panel seemed to be going round too fast to see them[45].

Figure 18 illustrates another possible situation in which involuntary reflex response could introduce hazardous conditions which would be

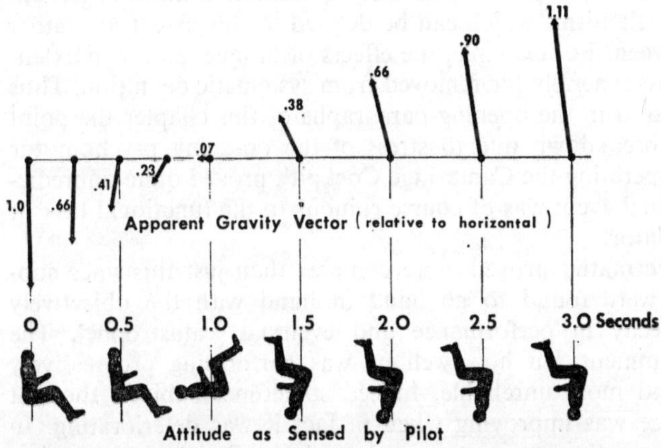

Figure 18 Possible illusion of change in attitude associated with push-over from a 6 degree climb at 250 knots such that after 10 sec there will be a 2 g negative acceleration (from ref. 46)

hard to simulate on the ground. On pushing over into horizontal flight from a steady climb as previously indicated (Fig. 8), the resultant *apparent* gravitational vector must sweep round in a 'backwards' direction during introduction of centripetal acceleration due to curved flight in a vertical plane. As we have seen there is evidence that such an event may be misinterpreted in some circumstances as rapid nose-up change in aircraft attitude[46]. If the consequent impression of nose-up change in attitude suddenly captures the pilot unawares he would be liable to respond with a corrective *forward* stick movement which of course would worsen the situation rather than improve it. The possible consequences become more significant in a physiological sense in view of recent experimental evidence that rotation of a linear acceleration vector relative to the head tends to induce compensatory reflex eye movements even in the absence of real inertial *angular* movement[47]. This raises the very real possibility of generating not only erroneous conditioned reflex control actuation through the limbs, but also involuntary reflex oculomotor response which, being compensatory for the apparent rotation, would be liable to introduce a false *visual* illusion of cockpit nose-up change in attitude fortuitously organised so as to reinforce direct sensory illusions deriving from the balancing organs[48]. A situation such as this could be particularly sinister in view of the possible adverse changes in longitudinal stability sometimes associated with introduction of negative airflow incidence[46].

The disturbing factors so far described, derive essentially from objective reflex responses to sensory signals which become misleading through mismatching with the flight environment in a manner which today can be at least in part described by relatively quantitative knowledge. But the biological mechanisms which can be defined to this extent are rather few and far between. For example, the effects of fatigue, although extensively studied, are generally far removed from systematic definition. Thus as already discussed in the opening paragraphs of this chapter the point of catastrophic breakdown due to stress of the on-going psychomotor control task in operating the Cambridge Cockpit[1], proved quite unpredictable, yet the actual event was of course ruinous to the functional task of 'flying' the simulator.

In practice, the matter proved more complex than just this since subjective changes were found to go hand in hand with the objectively demonstrated decay in performance and eventual 'catastrophe'. The 'pilot's' own comments on how well he was performing progressively became more and more unreliable. Indeed sometimes subjects thought their performance was improving when in fact it was deteriorating. In these circumstances there was a marked tendency for them to project their own failure upon imagined mechanical defects in the apparatus or to infer that outside interference was being purposely introduced. Not infrequently this situation would lead to overt anger, manifesting itself as an attempt to wreak vengeance by violence. Remarkably the severe irritability so introduced tended to persist for long periods, sometimes several days. Thus measureable changes were associated with subjective effects of quite a different kind, the influence of which is difficult if not impossible to quantify at the present time.

The latter effects observed in the Cambridge Cockpit were manifest over the relatively short time-scale of a few hours. As we are all aware however, extended working hours are also associated with changes due to prolonged wakefulness rather than the immediate effect of the ongoing task. The consequences of sleep deprivation are illusive and very hard to quantity. But Fig. 19 obtained from experimental flights in long-range maritime aircraft[49], may give some insight into at least one aspect of the covert kind of effects with which we have to contend. Two of the several variables measured in these flights were mean heading and altitude errors during manual flying periods, sampled throughout a series of 16-hour flights flown by each of four crews on four consecutive occasions with 24 hours rest between flights. To assess turbulence the output from a vertical accelerometer was recorded simultaneously with aircraft direction. The figure shows how, during the first half (8 hours or so) of a flight there was negative correlation between performance and turbulence; that is increased turbulence improved the absolute limits within which a pilot flew. He, as it were, rose to the occasion by tightening his dynamic response so that his mean excursion from intended course was reduced, rather than in-

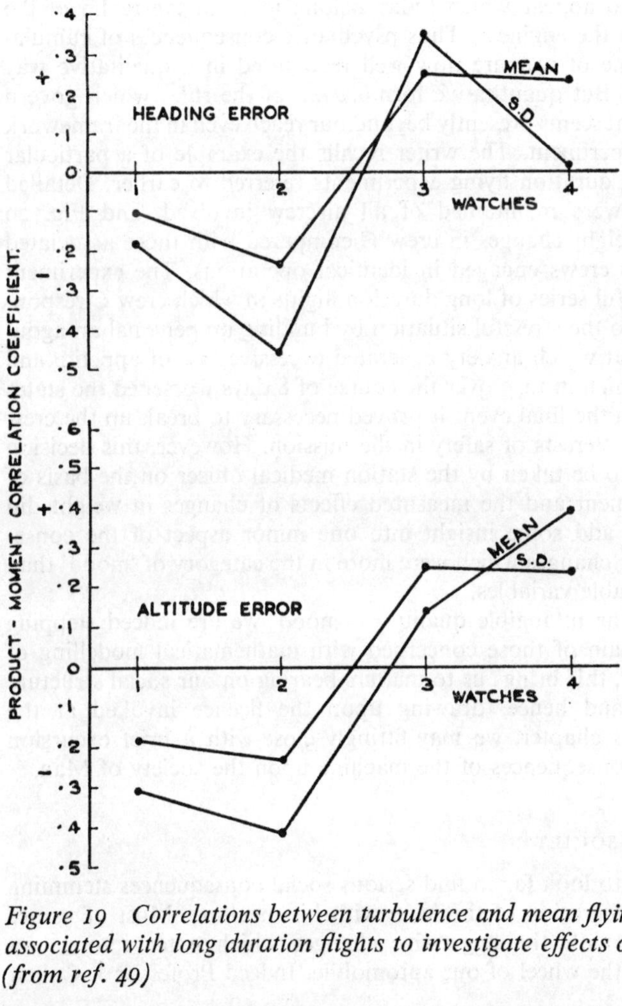

Figure 19 Correlations between turbulence and mean flying error associated with long duration flights to investigate effects of fatigue (from ref. 49)

creased by the turbulence. But this capability of rising to the occasion apparently deteriorated during the second half of these long flights, when the correlation between mean deviation from flight path and air turbulence became positive. Importantly, pilots were unaware of this effect; it seems as though the alerting effect of turbulence evidenced in the fresh pilot was in the nature of an automatic, subconscious response itself susceptible to the influence of stress. The possible consequences of loosing such an auto-adaptive response due to effects of fatigue, indicate rather subtle penalties which run the risk of being missed in laboratory and even flight experiments on fresh, well slept, human subjects.

Over a still longer period of continuous application, cumulative effects of stress begin to appear which today belong more in the realm of the psychiatrist than the engineer. Thus psychiatric consequences of cumulative stress in time of war are now well recognised in a qualitative way when they arise. But quantitative formulation of the rules which govern their development seems presently beyond our reach even in the framework of controlled experiment. The writer recalls the example of a particular crew in the long duration flying experiments referred to earlier. Detailed weight records were maintained of all aircrew involved, and Fig. 20 illustrates the weight changes in crew *C* compared with those associated with three other crews engaged in identical operations. The experiment involved a stressful series of long duration flights in which crew *C* responded as a whole to the stressful situation by building up personal antagonisms to a point at which anxiety generated excessive loss of appetite and hence weight, which in turn over the course of 8 days worsened the states of anxiety etc. In the final event it proved necessary to break up the crew structure in the interests of safety in the mission. However, this decision had necessarily to be taken by the station medical officer on the basis of qualitative judgment and the measured effects of changes in weight did little more than add some insight into one minor aspect of the consequences of major changes which were more in the category of 'mood' than objectively definable variables.

On reaching the intangible quality of 'mood' we are indeed stepping beyond the domain of those concerned with mathematical modelling of the Man. Rather, this brings us to matters bearing on our social structure and behaviour and hence, drawing upon the licence invoked at the beginning of this chapter, we may fittingly close with a brief excursion into the future consequences of the machine upon the society of Man.

MAN-MACHINE SOCIETY

We do not have to look far to find serious social consequences stemming from effective integration of the Man with his machine. Most of us are all too familiar with the aggressive changes which enter into us on climbing behind the wheel of our automobile. Indeed Professor J. Cohen

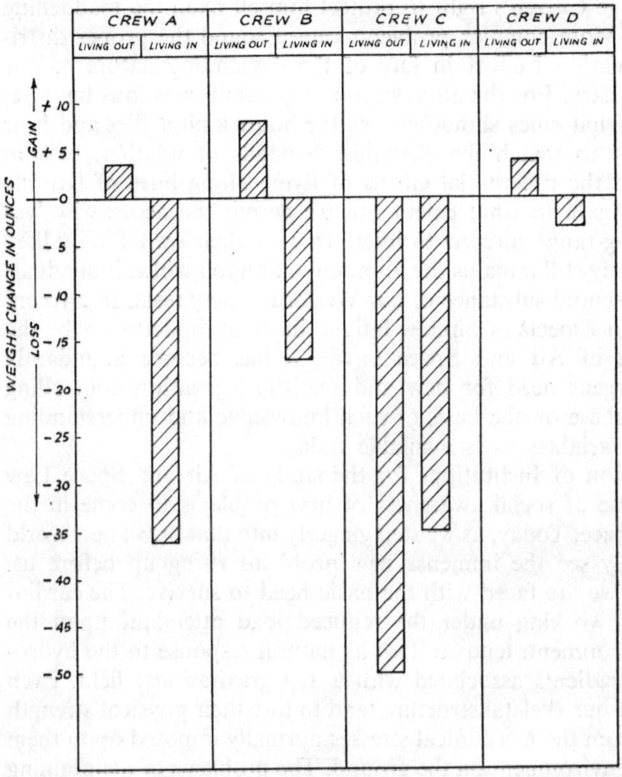

Figure 20 Weight measurements in 4 crews associated with long duration flying experiments referred to in the text. Each crew comprised 10 subjects (ref. 55)

speaking at a recent meeting of the Royal Society for the Prevention of Accidents, drew attention to the 'Centaur-like' creature which the new Man-Machine combination tends to breed; a new entity, the bio-robot, is born, part human, part machine, with the whole ensemble tending to become de-humanised to the extent that at one extreme the human half of the Centaur may almost be entirely at the mercy of his intended slave, the machine. The human society surrounding the Centaur then stands to be dangerously jeopardised, for example by the frightful carnage which follows in the Centaur's wreckless path, and the fearful cost to human health directly attributable to its pungently poisonous breath.

If the emerging Centaur on the ground is already in need of social control, what of the problems created by the new environments of flight and space? Perhaps the most obvious social problem in conventional flight centres around the Citizen's right to protect himself from the maddening effects of noise. More complex problems centre round the proper distribution of total national effort in face of the demanding nature of our aeronautical industry. For the aircrew, too, important questions have yet to be resolved. What rules should govern the hours a pilot flies and how should they vary to match the changing demands of weather, season, aircraft type and the differential effects of flying along lines of latitude and longitude. Again to what extent can we permit the necessarily 'vagrant' life of long-range aircrew to interfere with their own family life? After all, the family still remains the founding unit upon which individuals and hence the essential substance of our Western society rests. In a recent study of the legal aspects of aircrew fatigue in flight, undertaken by the McGill Institute of Air and Space Law[50], it has become abundantly clear there is urgent need for new and realistic legislation controlling matters such as these on the basis of such knowledge and understanding of the relevant variables as is available today.

The inauguration of Institutions for the study of Air and Space Law heralds in a phase of social awareness of new problems to come in the age of Man-in-Space. Today, as we step gingerly into that brave new world we can but dimly see the immense new problems rising up before us. First, of course, we are faced with the basic need to survive. The cardiovascular system, working under the reduced load attendant upon the zero gravity environment, tends to lose its natural response to the hydrostatic pressure gradients associated with a 1 *g* gravitational field. Even the very bones of our skeletal structure tend to lose their physical strength when removed from the mechanical stresses normally imposed upon them by their natural environment on the ground. The problems of maintaining life in a closed ecological environment are manifold, and there seems little doubt that marathon adjustments will be necessary in our way of life, when small nacelles of human kind are boxed together in mini-social structures while hurtling for years, even generations, through interstellar space.

Perhaps in these new circumstances we shall have to learn how to control elusive human elements such as mood and temperament. The suggestion may seem almost absurdly fanciful. And yet at a recent meeting of UNESCO, attempting to foretell the long-term consequences of Brain Research, attention was drawn to the experimental fact that at the flick of a switch an otherwise mild domestic cat can be turned into a raging animal through electrical excitation of selected locations in the hypothalamus. Conversely, in Man, electrical excitation through electrodes permanently implanted in the brain has been made to change, at will, the violent and threatening behaviour of a paranoic psychotic into that of a relaxed and smiling person. In a sense this is nothing new. For centuries it has been known that the simple hormonal act of castration can cool the ardour of the raging bull. And modern methods have shown objectively that the maternal aggression of a lactating mammal can be literally controlled by hormonal manipulation. At the UNESCO meeting Dr. K. E. Kayer even suggested 'It is not inconceivable that specific anti-hostility agents can be placed in the water supply to make a peaceful population . . .'

The prospect of control of Man's brain is formidable enough. But the 'normal' course of Mankind is threatened further still by the vision of controlled manipulation of our deep genetic structure. It is true that through countless years of selective breeding we have become accustomed to genetic control of a kind in animals upon the farm. But today, through the growing facility of genetic mechanics one can forsee a much more powerful tool, having horrific possibilities, and yet perhaps carrying a survival message in face of deteriorating genetic structure which accompanies the artificial maintenance of unhealthy life.

And so, with bold adventures into outer space, the glimmering of brain control and rapid strides into the heart of our inherited genetic structure we find Man, through his own endeavours, outpacing the slow progress of evolution by natural selection. We are entering into a new, Man-made, phase of evolution; perhaps it might be termed the 'Homological' phase of evolution, which, when superimposed upon the old biological one raises visions for the future filled with an admixture of excitement and alarm.

However, without annihilation of the Man we cannot retract from this general course of progress; rather the Man-made influence on our evolutionary progress must continue in its rapid growth. But if the bio-robot of Cohen, the Twentieth Century Centaur, is already jeopardising our coveted human qualities, themselves the product of a long biological evolutionary process, surely the Second Century must forcefully encourage intensive and imaginative research into the long term consequences likely to be introduced by the impact of Man-machine integration upon the hastening course of our evolutionary development.

References

1 DAVIS, D. R. Pilot Error. Air Ministry Publication 3139 A, London, 1947.

2 ROBINSON, D. A. The mechanics of human saccadic eye movement. *Journal of Physiology*, Vol **174**, pp 245–64, 1964.

3 ROBINSON, D. A. The mechanics of human smooth pursuit eye movement. *Journal of Physiology*, Vol **180**, pp 569–91, 1965.

4 LEVISON, W. H. and ELKIND, J. I. Two-dimensional manual control systems with separate displays. *3rd Annual NASA-University Conference on Manual Control*, Los Angeles, March 1967.

5 HASBROOK, A. H. and YOUNG, P. E. Pilot response to peripheral vision cues during instrument flying tasks. OAM Report AM-68, Oklahoma City, Oklahoma, 1968.

6 HUBEL, D. The visual cortex of the brain. *Scientific American*, November 1963.

7 HATFIELD, J. J. A programmable display synthesising system for Man-machine communications research. *Proceedings of National Symposium on Information Display*. Boston, Massachusetts, October 1966.
8 BURROWS, A. A. and FORD, H. K. Flying Personnel Research Committee Report No 966. Air Ministry, London, 1956.
9 BATTEAU, D. W. The role of the pinna in human localisation. *Proceedings of the Royal Society B*, Vol **168**, pp 158–80, 1967.
10 JONES, G. MELVILL and MILSUM, J. H. Spatial and dynamic aspects of visual fixation. *IEEE Transactions on Bio-medical Engineering*, Vol BME **12**, No 2, pp 54–62, 1965.
11 JONES, G. MELVILL and MILSUM, J. H. Neural dynamics of vestibular transductions. *Proceedings 18th Annual Conference on Engineering in Medicine and Biology*, Vol **7**, pp 38–9, Philadelphia, November 1965.
12 COLLINS, W. E. Vestibular responses from figure skaters. *Aerospace Medicine*, Vol **37**, pp 1098–104, 1966.
13 JONES, G. MELVILL. A note on some human factors in helicopter flying. Flying Personnel Research Committee, Memo 142. Air Ministry, London, 1960.
14 JONES, G. MELVILL. The vestibular contribution to stabilisation of the retinal image. *1st NASA Symposium on the Role of Vestibular Organs in the Exploration of Space*, SP–77, pp 163–72. Pensacola, January 1965.
15 JONES, G. MELVILL. Vestibulo-ocular disorganisation in the aerodynamic spin. *Aerospace Medicine*, Vol **36**, No 10, pp 976–83, October 1965.
16 JONES, G. MELVILL. Vestibular inaptitude in the environments of flight and space. *Journal of Laryngology*, Vol **53**, pp 207–21, 1966.
17 YOUNG, L. R., MEIRY, J. and YAO, T. L. Control engineering approaches to human dynamic space orientation. *2nd Symposium on the Role of Vestibular Organs in Space Exploration*, NASA SP 115, pp 217–27, Ames Research Center, California, 1966.
18 JONES, G. MELVILL and MILSUM, J. H. Neural response to adequate stimulation of the cat otolith organ. Communication to 24th International Congress of Physiological Sciences, Washington DC, August 1968.
19 BLISS, J. C. Human operator describing functions with visual and tactile displays. *3rd Annual NASA-University Conference on Manual Control*, pp 67–80, Los Angeles, March 1967.
20 KATZ, B. *Nerve muscle and synapse.* McGraw-Hill Book Company, London, 1966.
21 SHERRINGTON, C. *The integrative action of the nervous system.* Yale University Press, 1906.
22 HAMMOND, P. H., MERTON, T. A. and SUTTON, G. G. Nervous gradation of muscular contraction. *British Medical Bulletin.* Vol **12**, pp 214–18, 1956.
23 ELDRED, E., GRANIT, R. and MERTON, T. A. Supraspinal control of the muscles spindles and its significance. *Journal of Physiology*, Vol **122**, pp 498–523, 1953.
24 ECCLES, J. C. Circuits in the cerebellar control of movement. *Proceedings of National Academy of Sciences*, Vol **58**, pp 336–43, July 1967.
25 OUTERBRIDGE, J. and JONES, G. MELVILL. Vestibularly driven head movements in man. *Proceedings Aerospace Medicine Association Annual Meeting*, pp 312–13, Washington DC, 1967.
26 WATT, D. and JONES, G. MELVILL. Observations on the neuromuscular control of purposeful movement. *Proceedings Aerospace Medicine Association Annual Scientific Meeting*, Bal Harbour, Florida, pp 174–75, 1968.
27 HARLOW, H. F. Love in infant monkeys. *Scientific American*, 1959.
28 WIESEL, T. N. and HUBEL, D. H. Single cell responses in cortex of kittens deprived of vision in one eye. *Journal of Neurophysiology*, Vol **6**, pp 1003–17, 1963.

29 BASMAJIN, J. B. and SIMARD, T. Effects of distracting movements on the control of trained motor units. *American Journal of Physical Medicine*, Vol **46**, pp 1427–49, 1967.
30 NEWELL, F. D. In flight and ground simulation measurements of pilot transfer characteristics in the compensatory role tracking task. *3rd Annual NASA-University Conference on Manual Control*, pp 99–120, Los Angeles, March 1967.
31 GRAHAM, D. Research on the effects of non-linearities on tracking performance. AMRL-TR-67-9, 1967.
32 JONES, G. MELVILL and DRAZIN, D. H. Oscillatory Motion in flight, pp 134–51 in *Human problems of supersonic and hypersonic flight*, ed. Barbour and Whittingham. Pergamon Press, London, 1962.
33 STARK, L., VOSSIUS, G. and YOUNG, L. Predictive control of eye tracking movements. *IRE Transactions on human factors in electronics*, HFE-3, pp 52–7, 1962.
34 DALLOS, P. J. and JONES, R. W. Learning behaviour of the eye fixation control system. *IEEE Transactions on automatic control*, AC-8, pp 218–27, 1963.
35 ZIBOLLZ, H. and PAYNTER, H. M. Possibilities of a two-time scale computing system for control and simulation of dynamic systems. *Proceedings of National Electronics Conference*, Vol **9**, pp 215–23, February 1954.
36 SHERIDAN, T. B. On precognition and planning ahead in manual control. *IEEE 4th National Symposium on human factors in electronics.* Washington DC, May 1963.
37 YOUNG, L. R., GREEN, B. M., ELKIND, J. I. and KELLY, J. E. Adaptive dynamic response characteristics of the human operator in simple manual control. *IEEE Transactions on human factors in electronics*, Vol **5**, No 2, pp 6–13, September 1964.
38 KELLEY, C. R. A psychological approach to operator modelling in manual control. *3rd Annual NASA-University Conference on Manual Control*, pp 165–80, Los Angeles, March 1967.
39 JONES, G. MELVILL and SUGIE, N. Evidence for a new component of response in the oculo-motor system. *Proceedings 23rd International Congress of physiological sciences*, pp 419, Tokyo, Japan, September 1965.
40 SUGIE, N. and JONES, G. MELVILL. A mathematical model of the vestibulo ocular system of the cat. *Proceedings of International Congress of Medical Electronics in Biological Engineering*, Tokyo, August 1965.
41 PREYSS, A. E. A theory and model of human learning behaviour in a manual control task. ScD Thesis, MIT, February 1967.
42 FRASER, D. C. and SAMUEL, G. D. Aircrew fatigue in long range maritime reconnaissance; effects on vigilance. FPRC Report No 907, 10, Air Ministry, London, 1955.
43 JONES, G. MELVILL. Loss of aircraft control during a single rapid rolling manoeuvre. FPRC Report No 933, Air Ministry, London, 1955. Also in *Medical Aspects of flight safety*, pp 92–101, AGARDograph 30, Pergamon Press, London, 1959.
44 HICK, E. E. and BATES, J. A. V. Ministry of Supply report No 17-204, London, 1950.
45 JONES, G. MELVILL. A study of current problems associated with disorientation in man-controlled flight. Flying Personnel Research Committee report No 1021, Air Ministry, London, 1958.
46 MARTIN, J. F. and JONES, G. MELVILL. Theoretical man-machine interactions which might lead to loss of aircraft control. *Aerospace Medicine*, Vol **36**, pp 713–17, August 1965.
47 BENSON, A. J. and BODIN, M. A. Effects of orientation to the gravitational vertical on nystagmus following rotation about a horizontal axis. *Act Oto-laryng*, Vol **61**, pp 517–26, 1966.

48 JONES, G. MELVILL. Disturbance of oculo-motor control in flight. *Aerospace Medicine*, Vol **36**, pp 461–5, 1965.
49 JACKSON, K. F. Time relationships in pilot performance. *Occupational Psychology*, Vol **33**, pp 80–97.
50 SLIGHT, V. C. A study of the legal aspects of flight crew fatigue. Institute of Air and Space Law, McGill University, Montreal 1967.
51 JONES, G. MELVILL. Disorientation in flight, pp 1046–59. Vol **II**, *Advances in Aeronautical Sciences*, Ed. Th. von Kármán, Pergamon Press, 1959.
52 GRAHAM, D. Effect of some control system non-linearities on single loop compensatory tracking. *3rd Annual NASA-University Conference on Manual Control*, pp 215–21, Los Angeles, California 1967.
53 MICHAEL, J. A. and JONES, G. MELVILL. Dependance of visual tracking capability upon stimulus predictability. *Vision Research*, Vol **16**, pp 707–16, 1966.
54 OUTERBRIDGE, J. S. and WETTSTEIN, U. Optimisation in the vestibulo-ocular system. *Proceedings of 7th International Conference on Medical and Biological Engineering*, pp 291, Stockholm 1967.
55 JONES, G. MELVILL. Aircrew fatigue in long range maritime reconnaissance, FPRC Report, Air Ministry, London, 1965.

12 Hovercraft

R. Stanton-Jones

INTRODUCTION

The introduction to a book called *An Anthology of Partly Baked Ideas* states that it was compiled to provide an opportunity for the publication of ideas either half or partly baked, which might not otherwise be suitable for publication in established learned journals.

The author of this admirable anthology, I. J. Good, has defined partly baked ideas (PBIs) in some detail in the opening chapters where he gives a clear indication that reputable scientists and engineers should at very least endeavour to incubate their proposals to the point where the degree of 'bakedness' should be greater than 0·5, i.e. better than half-baked.

However, in attempting to predict a hundred years hence on the basis of the current astronomical rate of technological growth, invites science fiction, where, with few exceptions, the 'bakedness' is likely to be negative. and probably at least of the order of –2.

HOVERCRAFT HORIZONS

Forecasting the long-range future of a new technology like hovercraft, which has only emerged within the last ten years, I am at a considerable advantage, since the level of science fiction that creeps into my proposal will be harder to detect. At the same time I am at a distinct disadvantage, since the background on which these technical predictions have to be made, is extremely slender.

The scope to which the applications of the air cushion principle could be applied could lead to a somewhat vague and prodigious list in which just about everything was operated on air cushions 100 years from now.

Consequently I propose only to make the point that the scope of applications is enormous, that it can be applied either in its own right or in conjunction with other means of support, and then to concentrate or analysing, in the light of our present knowledge, what might be possible

Originally published in the *Aeronautical Journal of the Royal Aeronautical Society*, November 1968.

at the two extreme ends of the size scale of vehicle transport using this principle.

It is fairly natural for the engineer to try to determine the limits of the technological field in which he works. On the one hand, how big is it possible to make a device and on the other, how small can it be made and still be practical? Neither of these approaches takes any account of whether society really wants what the engineer is proposing. There is nothing new in this. The scientist has always forced his inventions on the world unasked, and somehow or other, the developers of that invention create a market and generate a requirement that did not exist before.

The danger of this approach is that it is likely to be based too particularly on technical considerations alone, whereas in practice, many other sociological, political and other outside influences determine the course of development of engineering art.

THE 'TRANSPORT GAP'

One of the main reasons put forward for the interest in air cushion vehicles is that there is a gap in the transport spectrum waiting to be filled by an efficient vehicle in the speed range from about 35 to 150 miles an hour (17–70 m/s) and for this gap there are three main contenders: the helicopter, the hydrofoil and the hovercraft. On the grounds of versatility, the helicopter is well established and in the next 100 years could easily win the race for medium sized vehicles. The hydrofoil has been established for a long time, but is purely sea-going, whereas the hovercraft may establish itself on its amphibious capability.

PRESENT STATE OF HOVERCRAFT ART

The current state of the hovercraft art can be summarised in a fairly simple way. Vehicles of up to 200 tons (200 Mg) can be built with a fair degree of certainty that they will perform as expected. Without any stretch of technology either to propulsion systems or structure this will apply to a size of certainly 750 tons (760 Mg) and maybe 1000 tons (1 Gg).

The current range of cushion pressures, like the wing loading of aeroplanes ranges from 25 lb/sq ft up to about 70 lb/sq ft (1·2–3·3 kN/m²). In general the smaller the craft the lower the cushion loading, while the larger craft have a higher cushion loading. However, higher cushion loading requires a greater lift power and leads to uncertainty about the pitch and roll stability. It is this effect that is limiting the use of higher cushion pressures until more is known about the dynamic behaviour of the craft.

The specific structure weight of existing vehicles lies between 10 and 20 lb/sq ft (0·5–1·0 kN/m²) and up to 300 tons (300 Mg) all-up weight it would appear practical to keep it down to 25 lb/sq ft (1·2 kN/m²). With these figures of cushion pressure and structure weight the power requirements vary for smaller craft from about 120 hp/ton (90 W/kg) for speeds of the order of 40 knots (20 m/s) down to 80 hp/ton (60 W/kg) for speeds of the order of 70 knots (36 m/s) on the larger machines.

These larger vehicles can achieve 45 to 50 knots (21–26 m/s) using only 50 to 60 hp/ton (40 W/kg).

This combination of structure weight per sq ft cushion pressure, and power loading leads to an overall structure weight of about 30% of its all-up weight, a basic dry weight of around 60% of all-up weight, leaving a disposable load approaching 40% of all-up weight.

The ratio of weight to resistance, i.e. *L/D* is about 10 for smaller vehicles, and for the larger machines of 200 tons (200 Mg) it approaches 20. The effective overall lift drag ratio, with due allowance for the power to lift the craft as well as to propel it, reduces this to about 5 for the smaller machines and about 12 for the larger ones.

A useful comparison of the state of the art of air cushion vehicles with other forms of transport can be made by looking at the SRN4.

This machine, weighing 170 tons (173 Mg), with a disposable load of 65 tons (66 Mg), has a calm water speed of 75 knots (38 m/s), although the practical cruising speed is 50 to 55 knots (26–28 m/s) under average service conditions. To carry its 60 tons (61 Mg) of payload at these speeds it requires the power of four Proteus engines with a cruising output of 13,600 hp (10 MW). This same power would propel a 70 ton (71 Mg) aeroplane with a payload of about 15 tons (15 Mg) at 350 to 400 miles an hour (156–179 m/s). A cargo ship with this same power would weigh about 12,000 tons (12 Gg) and carry some 7000 tons (7 Gg) of payload at a speed of 16 knots (8 m/s), while a hydrofoil boat would be of the order of 220 tons (223 Mg) with a payload about the same as the hovercraft, i.e. 65 tons (66 Mg), at 40 to 50 knots (21–26 m/s).

Perhaps even more illuminating is a comparison of current operating costs. The aircraft, the hovercraft and the hydrofoil lie between 15 and 25 cents a ton mile direct operating costs, but the cargo ship is of the order of $\frac{1}{3}$ cent, and a passenger ship about 5 to 10 cents.

TOWARDS THE LARGE HOVERCRAFT

If we take these current figures of weight, loading, etc, and see how they might be extrapolated in the future from the present state of the art, we might reasonably guess that the pattern will be something like that illustrated in Figs. 1, 2 and 3. This suggests that cushion pressures for hovercraft in the range of 5000 to 10,000 tons (5–10 Gg) would probably remain below 200 lb/sq ft (9·7 kN/m^2), that structure weights of 40 to 50 lb/sq ft (2–2·4 kN/m^2) should be practical and that the general relationship of structure weight, basic weight and all-up weight, would be unlikely to vary very much.

There is, however, a trend of thought that cushion pressures could be very substantially increased and Fig. 4 shows a plot of cushion pressure to length ratio against the length to breadth ratio of existing hovercraft, an area of study being undertaken by the Americans, and the way in which these compare with cargo ships. On this curve the actual pressures in lb/sq ft for a 4000 ton (4 Gg) hovercraft have been superimposed and indicate that in the American study the average cushion pressures that are being considered are some 300 lb/sq ft (14 kN/m²) which are in the order of twice those which we would obtain by extrapolation of our current knowledge.

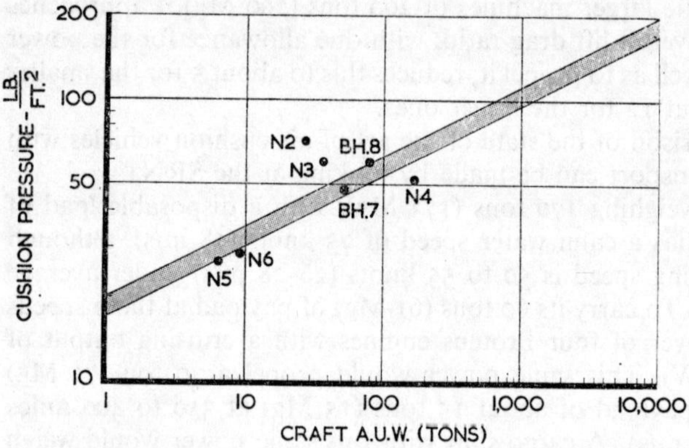

Figure 1 Trends of cushion pressure with all-up weight

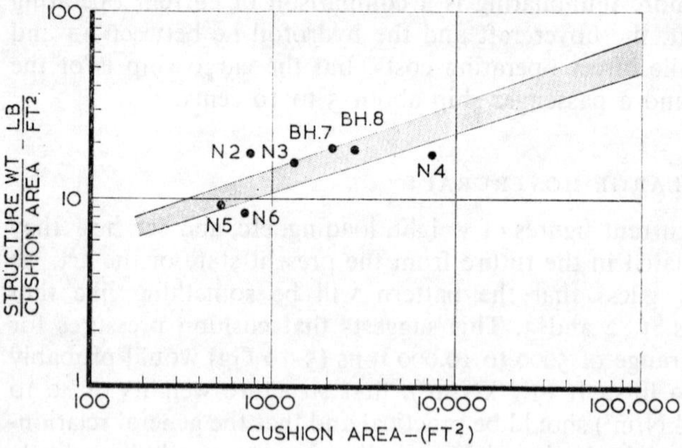

Figure 2 Trends of specific structure weight with cushion area

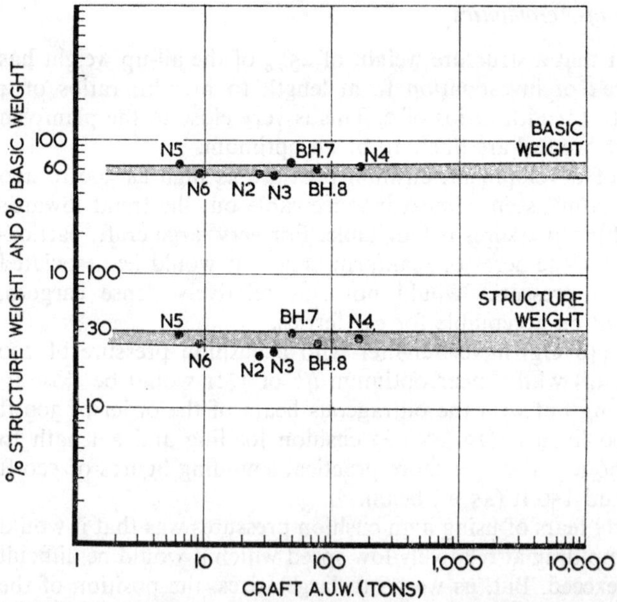

Figure 3 Relationship of % structure weight and % basic weight with craft all-up weight

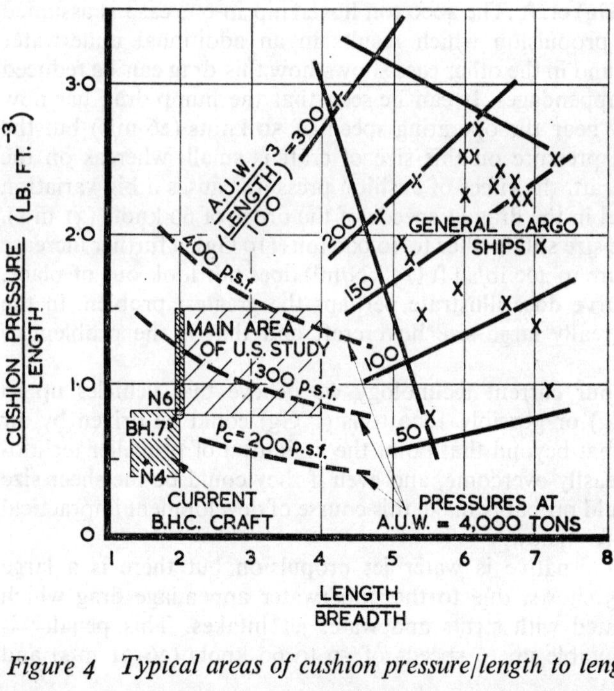

Figure 4 Typical areas of cushion pressure/length to length/breadth ratio

It is significant that a structure weight of 25% of the all-up weight has collapsed the area of investigation from length to breadth ratios of 4 down to length to breadth ratios of 2. This is very close to the planform dimensions which we feel are likely to be an optimum.

At this stage of development, cushion pressure as high as 300 to 400 lb/sq ft (14–19 kN/m²) seem almost inconceivable but the trend towards these higher cushion pressures is inevitable. For very large craft, particularly cargo carriers, the acres of planform area that would be associated with low cushion pressures would not suit relatively dense cargoes, though they might be acceptable for car ferries.

A 10,000 ton (10 Gg) hoverfreighter with a cushion pressure of 200 lb/sq ft (9·7 kN/m²) with a near optimum *l*/*b* of 1¾:1 would be close to 500 ft (152 m) long but with the outrageous beam of the order of 300 ft (91 m). At a 400 lb/sq ft (19 kN/m²) cushion loading and a length to beam ratio of about 3:1 we get more practical sounding figures of 500 ft (152 m) length and 150 ft (45 m) beam.

One of the early fears of using high cushion pressures was that it would result in high jump drag at relatively low speed which it would be difficult for the craft to exceed. But, as we go to larger sizes, the position of the hump drag becomes closer to the operating speed and its level in relation to the overall drag curve becomes much less critical. Fig. 5 shows the comparison of the drag characteristics of a 10 ton (10 Mg) craft compared to a 4000 ton (4 Gg) craft .The 4000 ton hovership in one case is assumed to have marine propulsion which results in an additional underwater appendage drag, and in the other case shows how this drag can be reduced without marine appendages. It can be seen that the hump drag has now been moved very near the operating speed of 50 knots (26 m/s) but the effect of cushion pressure on this size of craft is small, whereas on the 10 ton (10 Mg) craft, the effect of cushion pressure causes a big variation in hump drag and in the drag at speeds of the order of 60 knots (31 m/s). If we increase the size still further to 10,000 tons (10 Gg), a further increase of cushion pressure to 400 lb/sq ft (19 kN/m²) does not look out of place. However, this curve does illustrate perhaps the greatest problem in the development of really large size hovercraft and that is the problem of propulsion.

Returning to our current technology we believe that vehicles up to 750 tons (760 Mg) or possibly 1000 tons (1 Gg) could be driven by air propulsion, but that beyond that point the limitation of propeller technology will not be easily overcome, and even if they could be the sheer size and the noise would probably make this course of development impractical even in 100 years from now.

The obvious alternative is water jet propulsion but there is a large penalty as Fig. 5 shows, due to the underwater appendage drag which would be associated with struts and water jet intakes. This penalty is likely to be acceptable up to speeds of 50 to 60 knots (26–31 m/s) and though there is little doubt that the system could be developed to 100 knots (52 m/s), the resulting lift drag ratio is still likely to be too low for commercial applications.

It is unlikely that very large sea-going hovercraft need to be amphibious so there would appear to be little to stop the development of water jet intakes to absorb several hundred thousand hp. The engines are already available, the pumping machinery is likely to be within the state of the

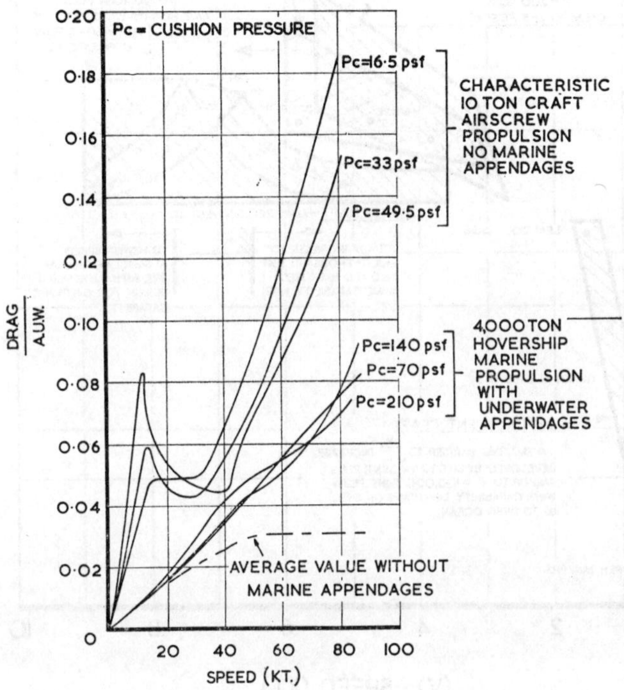

Figure 5 Variation of drag/A.U.W. with speed

art, and a nuclear power source to cope with a few hundred thousand hp should be quite possible.

But, with all that, a speed of much more than 60 knots (31 m/s) or possibly 80 knots (41 m/s), would probably be a practical limit unless we could assume a radical improvement in materials, particularly for the skirt.

In recent years, the biggest advances in hovercraft technology have been made on the skirt systems, which in a very short period of time have become comparable with motor car tyres in terms of life. Not only that, but they have resulted in something like 30% reduction in total power requirement for a given speed and sea state.

Figure 6 shows how we think the hovercraft might well develop, principally as a result of improvement in skirt technology. Being optimistic, we might get to 80 knots (41 m/s) of speed with 20 hp/ton (15 W/kg) for craft ranging from 100 to 10,000 tons (0·1 to 10 Gg).

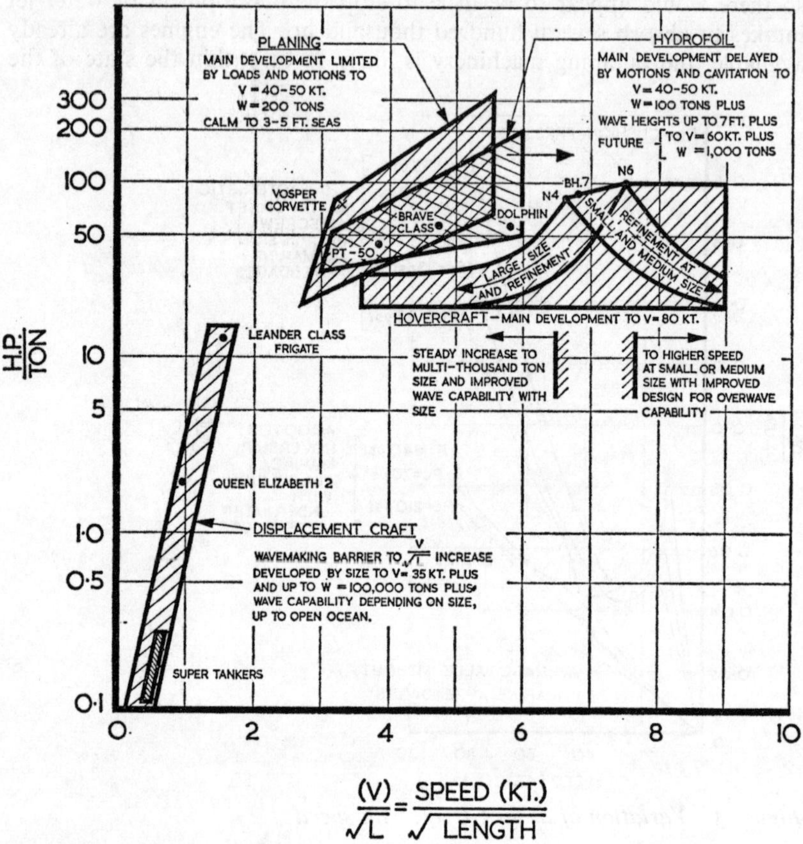

Figure 6 Trend of power variation with size

AUTO-STABILISATION

A sea-going hovercraft is a Surface Effect Ship and it follows that unless its size is large enough in comparison with the irregularities over which it travels, these irregularities will influence the motion.

The hydrofoil boat designers have tried to overcome this problem by going to fully submerged auto-stabilised foil systems operating below the surface of the sea and, hence, below the irregularities.

Hovercraft skirt systems have been designed which very successfully platform over irregularities which are equal to half the skirt depth. The difficulty has been that if the waves are greater than skirt depth, the craft, which is far from boat-shaped, will suffer severe impact. However, on a very large craft of multi-thousand tons where the skirt could easily be 20 or 30 ft (6–9 m) deep, the likelihood of impact would be reduced and the influence of the irregularities would be minimised as it is with large ships.

Nevertheless, some form of auto-stabilisation for hovercraft needs to be developed.

Some years ago Sir Christopher Cockerell proposed an actuated skirt system to achieve auto-stabilisation of a hovercraft. This might well be done by the rapid variation of air-supply to the cushion using fans with variable pitch blading.

Such a development might also substantially reduce the steady level of lift power which is required by the cushion with a consequent further improvement in economics.

In most engineering systems some 15% of power is wasted in giving the craft stability. Auto-stabilisation could well retrieve that 15% loss and should be easier to apply to a vehicle, such as a hovercraft, which would not be dangerous as a result of a failure of the system.

HYBRIDS

Since water jets are probably the answer to the propulsion problem, and since the air cushion principle lends itself to hybrid development, it is not unlikely that if intakes have to be put into the water that they will sprout hydrofoils, to provide the pitch and roll stability. This would allow at one and the same time, cushion pressures and skirt lengths to be increased, and the length of beam ratio to be extended into more boat-like shape (*see* Fig. 7).

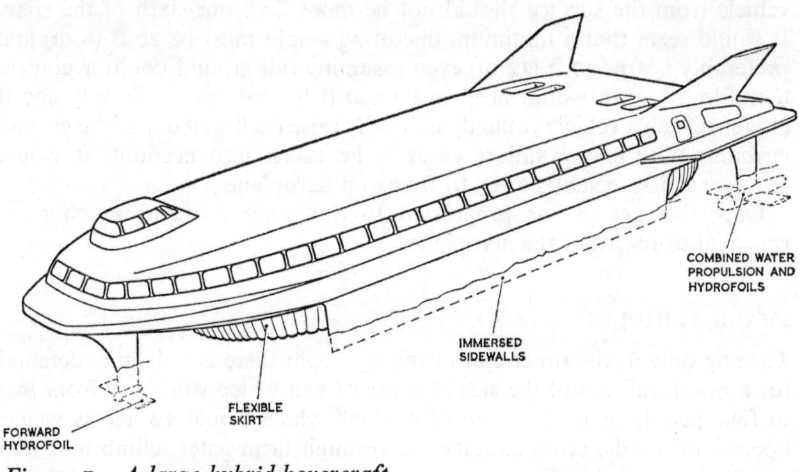

Figure 7 A large hybrid hovercraft

The combination of the air cushion with a hydrofoil could well alleviate the problems of both systems and realise the development of really large practical craft much sooner than it otherwise would have been.

RAM WINGS

One of the first ram wing air cushion vehicle designs was made by Mr. Toiro Karrio of Finland. Since then there have been many artists' impressions and proposals for ram wing hovercraft capable of travelling at speeds up to 200 knots (103 m/s) (*see* Fig. 8).

Figure 8 A possible ram wing configuration

I do not think that this is a likely development because to take any real advantage of ground cushion effect the clearance height of a ram wing vehicle from the surface should not be more than one-sixth of the span. It would seem that a minimum operating height must be 20 ft (6 m) and preferably nearer 40 ft (12 m) even assuming full ground contour control therefore its span would need to be 200 ft (61 m) and preferably 400 ft (122 m). Such a vehicle could dynamically raise itself well out of the ground cushion. With engine failure cases to be taken into account, it would clearly need sufficient power to fly as an aeroplane.

Once this design has progressed to this stage it seems much more practical to re-invent the aeroplane.

SMALL VEHICLES

Turning now to the other end of the spectrum there is a definite demand for a hovercraft about the size of a car or van which will carry from two to four people or up to a ton of payload which would go across water, operate on roads, cross country, go through farm gates, climb hills and operate in wooded territory.

This is a very similar cry to the one that followed the early development of the helicopter and in those days there were predictions that before long we would all have our personal run-about helicopter to go to work, relieving the congestion on the roads and operating with the freedom of a bird. This freedom is still a long way from being realised and I do not know whether the helicopter experts will predict that it will come about in the next 100 years.

The requirement for the small hovercraft seems less dangerous but is, technically, equally difficult. The first part of the problem may well be within reasonable grasp.

To lift a small vehicle the size of a car on skirts to a height of 4 ft (1·2 m) with moderate levels of power is possible, but having lifted it, such a craft would have a disconcerting tendency to go in any direction.

The next two major problems are, how to make it sufficiently stable and prevent it from overturning, and how to propel it.

Propelling it on reasonably level ground and across the water is straightforward enough but to make it controllable within the confines of a road and to climb hills demands the invention of a new type of wheel. It could be that the types of wheels being developed for moon vehicles may provide a solution but even then we need to invent a wheel which provides substantially more traction.

Having solved that problem we must now ensure that this wheel, in addition to giving traction on reasonably smooth ground can take large obstacles of 3 ft (0·9 m) or more as well, but we are still left with the problem of propulsion over water.

This implies two different sorts of propulsion, and with the present state of the art, such a vehicle ends up with at least 150 to 200 hp/ton (110–150 W/kg), much mechanical complexity and disappointing performance.

However, given improvements in materials and in small engine development plus the very strong demand for a vehicle with the versatility of a horse, it would appear to be a problem which requires to be solved and which ought to be solvable by the end of the twenty-first century.

13 Avionics—towards the twenty-first century

P. A. Hearne

I THE PLACE OF AVIONICS

In considering the spectrum of possibilities which advances in electronics offers during the next fifty years it is worth a short backward glance over an equal time span.

In 1918 the use of electronics was almost entirely devoted to developments in the radio frequency field, principally those of communication and sound amplification. Wireless telegraphy under pressure from naval and merchant marine requirements had reached the stage where it formed a dependable means of communication, albeit with the need for specialised human skills, for the world's modern shipping.

There had been a limited amount of aeronautical radio developments during the war, to reduce the size and weight of wireless telegraphy and the later radio telephony transmitters; much of the later stages of this work was carried out at Biggin Hill. However, these developments had scarcely any influence upon either the design and construction of aircraft or perhaps more important only a small influence on their operation. It was to be some time before the exploitation of the radio frequency spectrum on the ground and in the air contributed to increase greatly the operational capability of the aeroplane during the 1930s.

In 1968 we are in an era in which there is a rapid development of electronic technology and its application over a very wide range of fronts. The application of electronics is no longer restricted to exploitation of the radio frequency band for communication purposes but has been widened to a point at which it has become, if not a replacement for man, rather an essential element of the control and guidance of all present and future types of modern aircraft ranging from sailplanes to spacecraft. Not only has the science and application of electronics expanded to encompass a much wider field of applications than its early progenitors could have

Originally published in the *Aeronautical Journal of the Royal Aeronautical Society*, November 1968.

imagined but, during the past 50 years there has been added to it the exploitation of fundamentally new discoveries such as the principles of radar.

When it is realised that the developments in the past fifty years have sprung from the limited base of vacuum tube technology which was available to workers in the 1918 era, the problems of forecasting accurately the developments which can spring from the wide base of 1968 electronic technology can be described, and will be described, as impossible. Nevertheless, since electronics is an essential part of future forms of aerospace activity, it is necessary to attempt to chart some course of future development.

In considering the range of applications of electronics it is important to establish one dominant fact. Electronics and advanced systems are not a replacement for the unique creative ability of the human being in the control of aerospace vehicles. Rather they enable two very significant improvements to be made to a situation in which the human crew is controlling the aircraft without the benefit of artificial help from his electronic friends. First, they will enable the human crew to control aircraft in critical flight regimes with considerably less workload and stress than is presently the case and hence will improve the overall safety of operations.

Secondly, because of the introduction of failure operative control systems, they will enable the crew to operate regularly in more critical operational conditions than are currently considered acceptable for overall operational safety reasons.

The development of a truly integrated crew-system relationship instead of using systems as a crew replacement media will produce a situation where vehicles will possess a much greater operational capability than they have today and where, because of the infinitely greater reliability will, in fact, be used regularly for operational tasks and situations only encountered or ventured into rarely, and at some risk, today. In turn this greater operational capability arising from the improvements in control engineering will lead to a greater number of different species of aerospace vehicles to take advantage of these new operational potentials. Thus, one might expect to see arising from these new capabilities such developments as high-speed submersible air cushion vehicles for anti-submarine hunting and killing, genuine V/STOL airbuses capable of operating with surface transport regularity from town centres in all weathers, hypersonic strike aircraft capable of both sub-orbital and low altitude supersonic flight to say nothing of safe and reliable space transport vehicles.

2 AREAS OF ADVANCE

There would appear to be six principal areas in which technical advance in electronics and the electronic technology springing from it will influence aerospace vehicles in the future.

(*i*) Electronic devices, for example, integrated circuits whose development will dominate and control the rate of progress in every area of electronic systems.
(*ii*) Sensors, for example, inertial gyroscopes, development of which will enable the performance of some existing tasks more precisely and make possible some new tasks which cannot be done at all today, but whose development will not, with one or two marked exceptions, have the same effect as device development.
(*iii*) The field of systems which will become much more complex as device development allows truly remarkable reductions in size and weight of electronic units.

Associated with the developments of devices and systems there are three equally important advances. These are:
(*iv*) A much closer integration between the structure, aerodynamics and propulsion features of a vehicle and the control systems to the extent where significant trade-offs are made between these areas to give an overall improvement in design and operational capabilities.
(*v*) A marked reduction in cost and much more widespread adoption in all types of aircraft of systems currently thought to be advanced or 'far out'.
(*vi*) A much greater deployment of advanced electronic systems on the ground or in space or planets which will be more closely integrated with the in-flight control of vehicles than is generally the case today.

2.1 *Devices*

The development of devices may have been said to have started with the development of electro-magnetic methods of position measurement and the invention of the thermionic valve by Fleming. While improvements and refinements in electro-magnetic devices will undoubtedly take place in the field of smallness and, hopefully, reliability, it is, I think, more likely that the majority of advances of electro-mechanical components (with the exception of super-conductive devices) will be made in electrostatic devices using capacitance and other effects. These should be appreciably simpler and cheaper to manufacture than magnetic devices and offer more scope for development and replacement of mechanical by electronic techniques which is an important step towards improved reliability. Early examples of these devices are the capacitance pick-off units which already on measuring machines show the capability of detecting accurately changes of one-millionth of an inch and in gyroscopes can measure one tenth of a second of arc.

The future capabilities of these techniques are shown by the Honeywell electrically suspended gyro shown in Fig. 1. In this instrument the spinning gyro sphere is suspended in a vacuum by electro-static means and the gyro sphere's position is measured by means of optical pick-offs, another

Figure 1   Honeywell electro-static suspension gyro

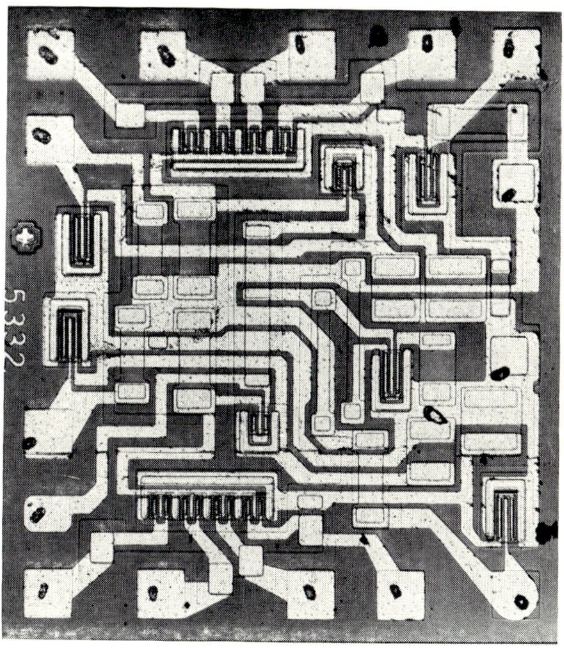

Figure 2   Micro-circuit chip. Approx. $\frac{1}{8}$ inch square

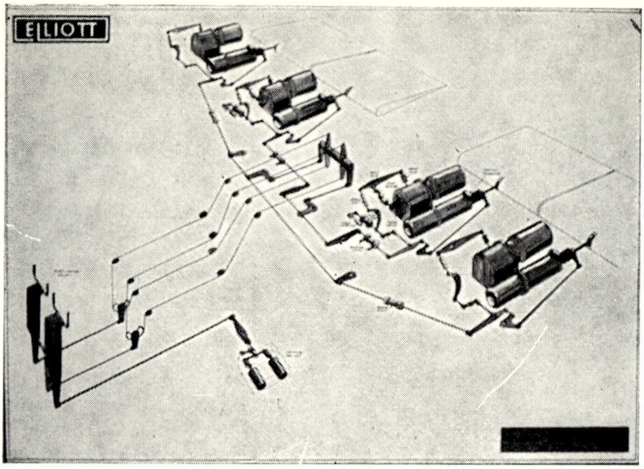

Figure 5 VC 10 flying control schematic

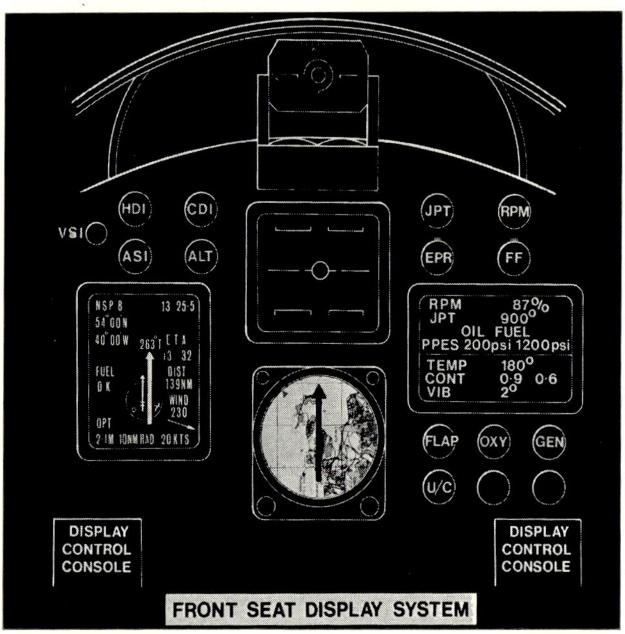

Figure 8 Integrated cockpit display system

non-magnetic device. Once set spinning in a vacuum the gyro sphere will spin for several weeks without further power and can form the central attitude reference for any vehicle from submarines to spacecraft.

However, it is in the field of what is loosely termed semi-conductors that the major advances in devices will occur. Although the early crystal wireless receivers were in fact dependent upon the type of semi-conductor effect, the real origin of semi-conductors stems from the discoveries in the Bell Laboratory in 1947.

Initially, these semi-conductor devices were costly to manufacture and this high cost together with the degree of reliability which could be achieved did not justify the adoption of the transistor for the majority of fields of application for electronics in the late 1940s.

There was, however, the one exception of aircraft electronic systems. Under the particular spur of the Korean and Cold Wars, aircraft requirements and their associated systems requirements advanced rapidly and it is perhaps noteworthy that one of the major fall-outs of aerospace programmes can rightly be regarded as the development of low-cost reliable semi-conductor devices. The introduction of semi-conductors would have come about eventually without the support of aviation development, but in a much later time-scale and over a more prolonged period. Aviation, by forcing the pace, has brought about the earlier advent of an era of automatic industrial control by means of digital computers, reliable world-wide satellite communications and the wide-spread use of computers for administrative, scientific and design processes and similar occurrences which are a necessary part of the organisation of a technologically based society. This small sidetrack illustrates the marked trigger effect which the advanced requirements of aviation can have upon the development of new technologies.

As a starting point to the next 50 years it is worth considering the present day integrated circuit which makes use of deposited metal conductive and resistive paths upon a silicon crystal substrate which forms the junctions of the transistor and diode elements of the circuit.

Hence, in one small piece of material no larger than ⅛ inch square, it is possible to generate electronic switching or shaping functions which previously would have required numerous valves or discrete transistor or resistor components occupying much greater space volume and weight (Fig. 2).

Although its smallness and inherently greater reliability resulting from elimination of individual components and reduction of external connections, gives great advantages, there are some problems. Prominent among these is the fact that the very smallness of the device allows the packaging of a large number of them in a small volume which poses difficulties in achieving an orderly interconnection solution as well as requiring great care to get good heat conductive paths to avoid having an expensive electric fire. This latter problem is being solved by development of low consumption circuits.

One attractive feature of devices of this type is that they lend themselves to manufacture by a mass production type of process in which a high degree of quality control can be achieved and in which the effects of human variations and skills can be largely eliminated.

Moreover, the manufacture of these units by means of process technology rather than by the use of human skills has enabled a massive reduction in cost to the extent that modern computers using integrated circuits are perhaps roughly half to a third of the cost of the discrete circuit computers which they have replaced (Fig. 3).

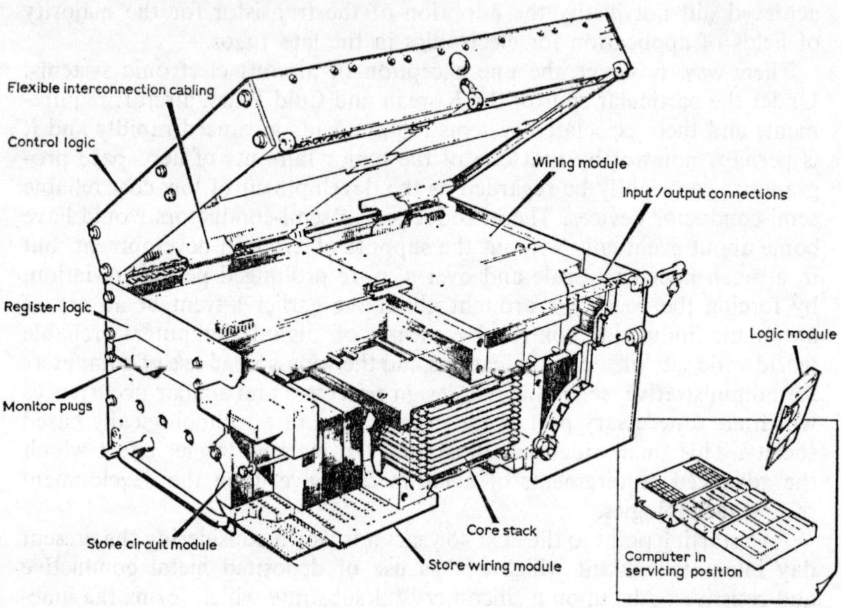

Figure 3 Elliott 920M airborne digital computer. Weight 30 lb. $\frac{3}{4}$ ATR Short case

This reduction in device cost will have a major effect upon the widespread use of computers and other electronic units which will continue at an ever increasing rate. As an example, most 1968 light aircraft except those used for pure sport, have radio equipment which is quite sophisticated by commercial airline standards of 1950. World airlines also are at the beginning of a procurement phase which includes the purchase of highly accurate inertial navigation systems, devices which did not exist up to twelve years ago and then in a much cruder form, and the installation of elaborate airborne digital computers. The reductions in cost, which will continue to occur as device development proceeds, will accelerate these trends towards the installation and use of electronic systems with a high degree of operational capability and it is this feature of cost reduction which will be one of the major features of electronic technology during the next 50 years.

While the integrated circuit will form the basis for most production avionic equipment for the next five to ten years, as the manufacturing technology advances, a still further reduction in size with ultimate advantage in reliability and cost will be achieved by the introduction on a massive scale of what is currently termed large-scale integration.

Basically, transistors and integrated circuits are manufactured by a mixture of chemical processes, high temperature diffusion and photographic methods and it is presently usual to deposit a large number of such devices on a single slice of silicon before cutting them all ultrasonically into the smaller single elements, discarding faulty devices.

With the improvement of manufacturing methods it has become possible to deposit circuit layouts which contain a considerably greater number of active junctions than the 7 or 8 junctions usually found in integrated circuits. Provided satisfactory yields can be obtained there seems no reason why the cost of large-scale integration (LSI) circuit devices should not be reduced to a point comparable with present day micro-circuits, which in digital form are selling for between 10s. and 15s. The use of a small number of LSI devices for a computer, for example, significantly reduces the cost of assembly and testing processes involved in its manufacture. Costs will be further reduced by use of semi-conductor memories in which the memory element, instead of using a core plane, can be composed of a large number of transistor or diode junctions deposited in an LSI slice.

A variant of this technology is to manufacture a standard LSI unit on which the interconnections are made by a computer controlled electron beam welder. By using a variety of programmes a large number of different functional units can be built up from a standard production LSI element.

The initial benefits from this type of technology will be in digital equipment. Furthermore, digital computers lend themselves to a much greater standardisation of circuit elements. However, the expansion of this technique to analogue computers will not be long delayed, particularly if means are found, such as the computer controlled welding technique referred to above, of producing specially designed analogue devices from basically standard mass production elements.

It does not seem unreasonable to expect that the type of such devices will result in quite powerful airborne digital computers occupying a volume no greater than a beer can, becoming available at prices less than £1000 (1968 currency values) towards the end of the era. Table 1 (p. 300) shows the cost reduction so far achieved in digital computers and an extrapolation over the next 20 years indicates the high degree of confidence which can be placed upon this reduction in price being achieved.

The availability of elements of this type will enable the wide-spread use of sophisticated electronics in advanced commercial and military vehicles and will, of course, allow not only the introduction of multiple fail operative systems as a standard design practice (*see* section 2.3) but will also permit the use of electrical signalling systems which can usefully substitute for mechanical and aerodynamic controls with significant weight and cost savings.

Although digital techniques will undoubtedly become more widespread, another device, as yet imperfectly understood, promises further development possibly along lines as yet undefineable. This is the molecular electronic function generator, or moletronic device as it is sometimes known, which produces an output as a function of the input because of the shape and distribution of the crystal material of the device. For example, one could pass a voltage through the input and obtain the sin/cos function of the voltage out. One can imagine some totally different types of electronic unit possibly manufactured by chemical and ultrasonic milling from a solid block of silicon, which could become say, a standard adaptive autopilot computer for a future family of aircraft.

Three other device techniques should be briefly identified:

Fluidics A great deal of work is going into digital and analogue devices which use air or liquid as a computing medium rather than electrons. Several successful examples have been built including engine control and autopilot computers. In quantity they should be even cheaper than future electronic systems and have the advantage of requiring little or no power supplies, ram air or fuel being quite suitable. In fact, someone has suggested, probably facetiously, that the weightless, spaceless zero power computer could be achieved by using the honeycomb skin of an aircraft as the basic computer element. It is, however, probable that they will be restricted to the less complex control tasks because of the difficulty in organising large fluidic computers.

Tunnel Diodes This is mentioned merely to show what is possible from a combination of techniques. Currently a computer using a combination of microwave and computer techniques has run at a clock rate of 150 m/cs, i.e. some 50 times faster than most existing computers. One is currently being designed to run at 750 m/cs. At these speeds the computers have to be small because the speed of light or electro-magnetic propagation intervenes if the components are more than a few inches apart from each other. The objective of these high-speed machines is aimed partly at reducing the size of a machine for a given number of calculations rather than speed for its own sake.

Superconductivity As temperature is reduced to within a few degrees of absolute zero electrical resistance of a metal disappears and quite improbable computing speeds and data transmission can be achieved for very little power. Another feature is the capability of producing very large electromagnetic forces for very small amounts of power and the potential of producing a direct d.c. transformer. There is no doubt that the problems of producing the requisite low temperatures in aircraft will be overcome and that extensive use will be made of these effects in future aircraft both in the device area and for sensors.

It does not seem unreasonable to conjecture that the exploration of the effects of super-conductivity will allow the exploitation of detection systems with very low signal strengths and signal to noise ratios. If this is so the way is open to the development of new forms of sensor units (*see* section 2.2) both in terms of the existing radiation spectra and the new exploitation of hitherto unusable or unknown phenomena.

A further field which suggests itself is that in which electro-magnetic forces may be used either in terms of a power generation system, possibly in a spacecraft orbiting through the earth's magnetic lines of force or conceivably in some type of yet-to-be-conceived anti-gravitational system. This latter suggestion is offered in the belief that physics neither started with Newton nor terminated with Einstein, and the realisation of Universal Fields beloved by science fiction writers may well come about as the result of the exploitation of the properties of super-conductivity.

2.2 *Sensors*

In the field of sensors, one sees first the development of more refined methods of measurements which we are currently making by cruder means, in such items as air pressure, aircraft attitude and rate of change of attitude etc. These refined methods will take the form of elimination of mechanical moving parts, or alternatively if cheaper, the reduction of these parts to an absolute minimum by means of improved manufacturing techniques which permit a smaller number of very accurately made parts to replace a complex assembly using compensatory methods to eliminate manufacturing deficiencies. For example, in the field of inertial gyros it is probable that laser beam gyros or a neutron spin device will replace the present rotating wheel. Alternatively, ways may be found to manufacture, say, the gyro sphere previously described much more cheaply.

A further example is the development of single silicon crystal diaphragms to replace a pressure sensor for air data systems and replacing the present complex mechanical capsule assembly used in 1968 systems.

Secondly, there will be an intensive effort to discover and exploit new 'windows' in the atmosphere around the aircraft to improve long-range measurements, detection and identification for military and civil applications. Some of this will involve the exploration and development of existing techniques such as radar, using advances in computers and new types of amplifiers for detection and reconstruction of very low signals in high noise environments and to compensate synthetically for other physical shortcomings such as aerial size. Additionally, one will expect other attempts to defeat signal attenuation from long ranges either by massive increases in radiated power, possibly using super-conductive devices, or more elegantly by discovering signal characteristics which do not suffer from the same debilitating effects of attenuation as signal amplitude. This increase in range is particularly important as the current radar frequencies which give the most accurate measurement and definition are the ones which suffer from the most serious atmospheric attenuation.

The net effects of the work on radiated signals should be to allow a very close representation of visual conditions to be displayed to pilot or ground controller in all types of weather conditions at ranges up to several tens of miles. In turn, this should lead to the development of strike aircraft systems which have the true capability of hitting any type or size of target in any type of weather with accuracies comparable to those achieved in fully visual conditions. These new types of RF measurement will also prove useful in the exploitation of automatic landing and collision avoidance systems in the civil aircraft field, by the development of easily interpreted near visual monitoring displays.

Beyond the extension of the use of existing types of techniques there is the development and use of radiated phenomena which are presently either imperfectly understood or as yet undiscovered. As an example of the former, the present work on the detection of clear air turbulence by measurement in the infra-red radiation group may be cited.

Physical measurements have shown the association of the phenomenon of clear air turbulence with changes in water vapour and CO_2 in the atmosphere and this fact has enabled systems designers to apply the latest techniques of infra-red sensor head and circuit design to the measurement of these changes.

Currently, detection ranges of up to 50 miles (80 km) seems possible with the development of existing techniques and refinements should ultimately allow a range of 100 miles (160 km) or more to be achieved in the upper atmospheric levels.

Throughout the next half century one can imagine that the association of particular physical measurements and changes with hitherto unexplained or arbitrary events, will lead to a significant increase in operational safety, e.g. detection of high altitude hail, detection of squall turbulence lines.

Since undiscovered phenomenon are by definition not definable, it is not yet possible to be specific as to the areas in which advances may take place, but it seems likely that in three or four areas physical and engineering breakthroughs may take place.

First, there will be the development of a sensor which is complementary to radar, but which exploits a different radiation window for the purposes of detection and ranging. It seems probable that some form of the laser will form part of this development if only because of the ability to make use of high emitted powers to overcome the signal attenuation at extreme ranges.

In a different spectrum the 1968 use of infra-red reconnaissance and battle-field sensors is proof of the power of these techniques. Used in association with the synthesis and signal correlation techniques made available by the use of small powerful airborne digital computers, the infra red system might well replace the radar system for use in areas with poor reflected signal characteristics, say in an ECM environment.

Secondly, it seems also certain that there will be a considerable emphasis on the development of sensors for underwater exploration and warfare which will undoubtedly stimulate research into other methods of detection and discrimination rather than the present crude method of ultrasonic pulse ranging.

Thirdly, developments in high magnetic field phenomena such as super-conductivity and magneto-hydrodynamics may well stimulate research into an area, hitherto derided, of anti-gravitation. The existence of universal fields and the relationship of electro-magnetic force to gravitational force are ones which may theoretically be possible and the technology of the next 50 years could well begin to make it usable by the designers of tomorrow's flying machines. Whether or not this work results in any practical outcome cannot be forecast at this time, but the streams of theoretical work and electronic technology development would seem to be sufficient in the next century to enable at least a crude laboratory demonstration to be performed if the fundamental basis for an anti-gravitational force exists at all in our universe.

The fourth sensor area in which a complete departure from present day concepts may take place is in the field of body or neuron current detection and possibly the understanding and use of a radiation spectrum now loosely known as telepathy.

At its simplest, this technique would involve the detection of the signal and feedback currents passing to and from a pilot's muscle from the brain which would be amplified and passed to an actuator thus obviating the need for physical movement. As progress is made the currents could be picked up at the brain centre itself and ultimately use some form of thought transmission which would be amplified and used in some co-operative computer or control system. Lest this be thought, or rather believed, too far-fetched, it should be pointed out that computer systems are already being organised in the pattern of human organisations and computers are being built which can be trained along Pavlovian lines to learn from punishments and rewards.

A useful start has already been made on the human end of the man-machine interface as it is so soullessly called. Considerable work has already been done in the development of computer displays which can be interrogated by a human merely placing his finger on the piece of information which he wants processed. Work is starting on a more personalised computer access system which may, in fact, be a number of electrodes which are permanently implanted in the human body, possibly at the time of the eleven plus or the AFRAeS examination. With such connections the engineer or pilot of the future can connect himself to and use any computer control system without the need for elaborate peripherals. In the same way as man can couple himself to a power tool by his own limbs to perform tasks beyond his physical capabilities, so in the future we will achieve this very close coupling with a power brain to perform tasks beyond man's unaided intellectual and perceptive capabilities.

2.3 *Systems*

Having dwelt at some length on the devices and sensors which will form the basis of future systems let me turn to trying to forecast the shape of some of the systems themselves.

The first thing is that with the reductions in size, weight and cost it will become possible to build electronic control systems which are as reliable as the main structure and propulsion units of the aircraft. This will come about partly from the increased reliability associated with micro-electronics, but principally because multiple systems with adequate redundancy can be contained within reasonable weight, price and volume limits.

Figure 4 shows the very marked increase in reliability which can be achieved by multiplexing system channels calculated on the basis of both

MEAN TIME BETWEEN FAILURE OF ONE CHANNEL/HOUR	NUMBER OF CHANNELS 1	2	3	4	5
200	5×10^{-3}	5×10^{-5}	7.5×10^{-7}	1.5×10^{-9}	1.5×10^{-10}
1000	10^{-3}	2×10^{-6}	6×10^{-9}	2.4×10^{-11}	1.2×10^{-13}
5000	2×10^{-4}	8×10^{-8}	4.8×10^{-11}	3.8×10^{-14}	3.8×10^{-17}

M.T.B.F./HOUR OF MULTI CHANNEL SYSTEMS

Figure 4 Mean time between failure of multi-channel systems

present day and future predicted failure rates for single systems. Since the mean time between failures of all engines or main structure is not higher than 10^{-9}/hour, a four channel system at present day reliabilities or a three channel system in future will not reduce the overall safety level of the aircraft below present day safety levels. Indeed, there is strong reason to hope that the widespread use of multi-channel electronic systems of the reliability indicated in Fig. 4 could make very significant improvements to existing safety standards. An early example of this is seen in the work currently being undertaken on the Trident and VC 10 Automatic Landing Systems. Currently these systems must achieve a failure probability of less than 1 in 10^7 hours and this failure probability must be maintained over the last 30 seconds of flight immediately before touchdown.

With the advent of reliable multi-channel electronic control systems it becomes possible to consider the widespread replacement of many of the mechanical control links which have become increasingly heavy, complex and demanding of maintenance in modern aircraft. A prime example of this is the future replacement of mechanical flying control runs with electrical signalling, routing electrical signals from stick position pick-offs via electrical conductive paths to the electrical power amplifiers and thence to servo-motors mounted on the main control actuators. Readers who are aware of the complications of the mechanical control runs on existing large aircraft (Fig. 5) will realise the enormous simplification which the substitution of electrical signalling will bring. Indeed, the autopilots of the Buccaneer and VC 10 already rely upon this technique for injecting autopilot control demands into the main flying control actuators and considerable flight experience has already been gained. Lest designers hold their hands up in horror at this approach it should be pointed out that mechanical failures in conventional flight control systems are by no means unknown.

The advent of electrically signalled flying controls will greatly aid the ease with which auto-stabilisation and automatic flight control demands can be applied to the aircraft. Once all control demand signals are in electrical form, signal processing and combination can be carried out far more easily than with today's cumbersome electrical/mechanical/hydraulic means. Stick/control surface gearing changes and elevon/rudder 'mixing' variation with speed will also be achieved in a much lighter and more reliable manner than the very awkward mechanical means used today (Fig. 6).

One interesting feature of an electrically signalled control system is that it enables the human pilot to modify the automatic control inputs if he wishes to change the flight path slightly; this is generally called stick steering. A particular form of stick steering which will be increasingly used in future aircraft of the new aerodynamic shapes described in Dr. Küchemann's chapter is called 'Manoeuvre Command'. In this system the auto-stabiliser/autopilot is used to modify the response characteristics of the aircraft to give consistently good handling characteristics throughout the speed range such as a constant rate of roll for a given stick deflection or a given angle of pitch for a given stick position instead of the acceleration or rate type control of the natural 'untampered with' aerodynamic surfaces. This is particularly important for these new shapes where the aircraft's or space-vehicle's natural stability and control effectiveness in the roll and pitch axes may be so widely different and vary so much through the speed range as to cause extreme difficulty to the pilot. This form of automatic stabilisation will almost certainly be required to give acceptable harmonised handling qualities at all stages of flight.

Electrically signalled controls will not be restricted to flying control systems. Many of today's systems which involve cranks, torque tubes etc, to transmit control demands or even control power will be replaced by redundant multi-channel electronic signalling systems with considerable savings in weight and size.

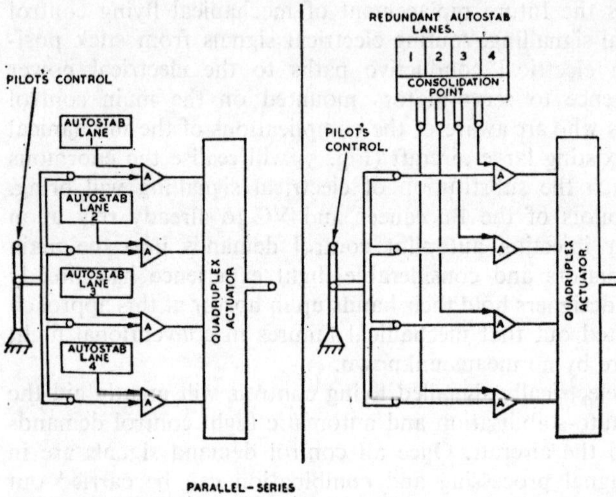

Figure 6 Possible integration of auto-stabilisation in electrically signalled flying control system

In the field of automatic flight control, auto-stabilisation and auto-pilots will become even more an integral part of any aircraft design even in the lowest weight ranges. It is possible, however, that these will be markedly different from the simple displacement and rate controllers used today. First, there will almost certainly be a widespread adoption of digital techniques probably using very small and powerful digital computers of the LSI variety, which will enable us to discard the present day analogue systems with their limited capabilities and drift and tolerance problems, and their often complex electro-mechanical elements. Digital autopilots will be accompanied by the use of adaptive techniques possibly of the learning machine type organised along the conditioned reflex lines already referred to. The combination of the two techniques will make automatically controlled flight of the future a much more flexible under-taking than the present day when autopilot capabilities restrict automatic control to a limited number of carefully prescribed manoeuvres. This flexibility of automatic control will extend even further if the man/machine coupling, either implanted or by telepathy, used in conjunction with a learning computer organised on bionic or neuron model lines becomes a reality in this time-scale. Man will be able to perform a wide range of flight manoeuvres at will using the intelligence of the human being and the perception of the automatic control system in a way presently undreamed of. Automatic control systems of this sort will enable aircraft to be manoeuvred freely, accurately and safely throughout their entire manoeuvre envelope with minimum demand on the human pilot. As an example, an aircraft making an automatic landing today must turn on to the final approach heading ten miles (16 km) from the runway for a long drag-in and suffer itself to be coupled to the ILS beam through an automatic pilot whose control engineering mathematics might just scrape an 'A' level. The whole process is tedious and time consuming, but the problems of today's autopilots demand that the task they are called upon to perform should be strictly defined and as limited as possible and strictly linear. The aircraft of the future will, I imagine, be able to approach the airfield in a close tight pattern re-entering into atmosphere with rapidly changing height, airspeed and configuration and make an approach and landing with possibly a curved final approach using the human/automatic bionic system described here.

As well as being much lighter than existing autopilots the future systems will be sufficiently powerful, and reliable, by virtue of their multi-channel redundancy, to enable designers to use them to provide a high degree of synthetic stability thus allowing the size of fixed aerodynamic stabilising surfaces to be reduced. Use of the automatics this way to reduce tail surface drag and weight has already been used successfully on one high-speed British aircraft. The trend will certainly continue and will have major favourable advantages on the payload ratios in the aircraft described by Dr. Küchemann.

Digital computers will find a widespread application to many different techniques in the aircraft. Their use in inertial navigation is well known and in the not too distant future this will be extended to cover all forms of navigation both horizontal and vertical. In its expanded form the navigation computer will correlate all the position inputs from various aids and height sensors, and relate the aircraft's position in three dimensions to the flight plan information fed in via a tape before take-off, or ATC information and clearance fed in via data link in flight. From this information the computer can generate steering demands to the automatic and human pilots. If the aircraft/engine performance characteristics are also stored in the computer and engine fuel flow continuously fed in, the computer can calculate range available at any given point in flight, indicate stand-off, endurance and diversion range available etc, at different airports and generally act as a storehouse of flight and cruise control information.

As an extension of this latter technique, a separate computer whose store was programmed with the overall aircraft and system performance characteristics will be used to monitor the functioning of the aircraft and its systems in flight and relieve the human crew from the very demanding task of continuously assessing the operation of the aircraft and its, then greatly increased, numbers of various types of systems. This latter variety of computer will be used in conjunction with an airborne recording system or possibly with a ground-based recording system with which it is in continuous touch by data link. This computer plus recording system will enable immediate diagnosis of trends and faults in flight with a print-out of the fault diagnosis for use by the ground crew at the next maintenance stop. It should be noted that with the advent of redundant digital systems the very high degree of self-check inherent in digital systems will often enable aircraft to continue for several flights without the need to replace faulty units immediately.

Small size digital computers will also be used in the control and organisation of such systems as communications, cabin conditioning etc. Most important they will certainly be used for engine control. Multispool engines, particularly when used with variable inlets and nozzles already pose very severe control system design problems. Digital computers, either electronic of fluidic using engine air or fuel, offer the most promising method of providing the engine control systems of the future. Not only will they be able to cater for the widely different control requirements for the wide range of flight regimes of a variable mode type engine, say from take-off as a turbo jet at 200 kt (100 m/s) to hypersonic cruise in the scram-jet mode, but by virture of their tighter control loops they will allow these engines to be operated in an optimised condition much nearer their limiting conditions such as surge lines, with considerable improvements in actual propulsive efficiency achieved.

Before leaving computers two points should be emphasised. The first relates to the organisation of computers and computing systems in aircraft. Although many people have advocated the all-singing-all-can-do massive computer this seems to be a foolhardy approach. It seems much more likely that the parallel of computers and human organisation will be repeated in the systems field. In any complex organisation staffed by human beings one does not rely upon one central omnipotent person but instead delegates authority in a hierarchical manner. The same approach to airborne computing systems using a number of smaller co-operative computers in a hierarchy or federated organisation with a full flow of information offers important safety and reliability advantages (Fig. 7). If computers fail or go sick, their work can be shared among the remaining units; if computers become overloaded, similarly. The whole concept of the system is that of a sane division of responsibilities among a number of able individual units, which seems much more attractive than the facist supercomp proposed by the big computer men. These types of computer systems will be greatly aided by the developments in multiplexed data transmission which should allow digital information to be piped around aircraft using a very small number of wires or probably, by 2016, no wires at all, developments which will considerably reduce system weight.

The other developments associated with computers will be the development of electronic displays with much greater flexibility or scene shifting capability than is possessed by existing mechanical instruments. The very

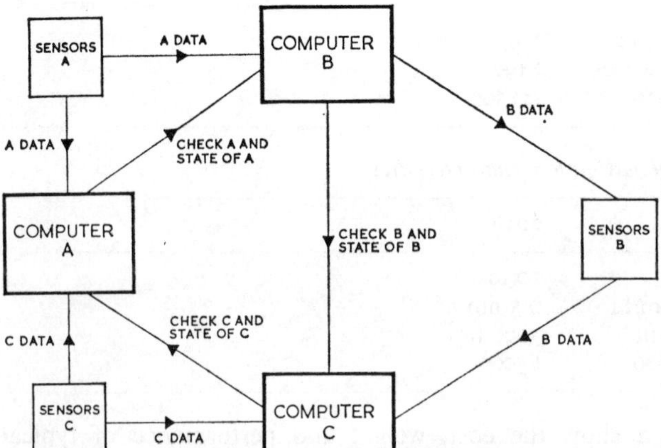

Computer A carries out a specific computation with data coming from Sensors A. In the event of Computer A failing, Computer B is informed of this by the check A facility. B is kept informed of the stage reached in the computations by A and has sufficient spare capacity to take over A's calculations. To this end the data from sensor unit A is also fed to B. Similar reasoning applies to Computers B and C.

Figure 7 Federated computing system for integrated avionics

much greater amount of information produced by computing systems compounded with the much greater number of such systems in the aircraft will raise the premium on instrument panel space even higher than today. Developments of electronic instruments in which large numbers of alternative data displays can be selected for various modes of flight are already under way and though initial designs will use cathode ray tubes, the later displays will use solid state devices such as electro-luminescent mosaics. These displays will enable an optimum match of information symbols, from the control and sensor systems, with what will almost certainly be an accurate representation of the immediate outside world derived from radar or similar sensors. This combination, which is similar to that used in today's head-up displays, will enable pilots to control the aircraft using the familiarity of direct vision cues backed up in the form of symbols by the derived cues which are not perceived easily, or at all, by human vision (Fig. 8).

Before leaving airborne equipment I would like to refer to my earlier claim that cost and weight reductions will result in a much more wide-spread use of automatic control devices in fifty years time.

TABLE 1 *Airborne digital computer characteristics*

	1966	2016
Weight	35 lb	<10 lb
Speed	1 us access	1 us access
Storage	8192 words	8192
Cost	£15,000	<£500

TABLE 2 *Inertial navigation characteristics*

	1966	2016
Weight	50 lb	10 lb
Accuracy	1·5 nm/hr	0·5 nm/hr
MTBF	1500 hr	5000 hr
Cost	£30,000	£5000

Tables 1 and 2 show the cost, weight and performance of typical present day airborne digital computers and inertial navigation systems respectively and my estimates for their cost in 2016. It will be seen that there is a cost reduction of between a factor of 6 for the IN system and of 30 for the computer. Weights are reduced by a factor of 5 while performance is significantly increased in both cases. I see no reason why the executive aircraft of the future should not use these types of systems as freely as the Küchemann Kollossal Waverider or the Morien Mach 1 Metro Bus.

Turning finally to the developments which will take place in ground or space fixed aids used in the overall aviation operation infrastructure. Satellite communication will obviously be used universally on long-haul and probably short-haul routes and indeed some airlines are already beginning to experiment with it.

As in aircraft, the increasing complexity of ground organisation will require the wide-spread deployment of computer based systems to match the speed and volume of air traflc with correspondingly effective ground operations. A start has already been made in this field with the use of computer aids to air traffic controllers in which the routine logging of aircraft movements is carried out by the computer and digested information presented to the controller. In later systems, such as the Mediator system now under development in the UK, the computer will be used to process raw radar data and to do such tasks as conflicting traffic detection and generally relieve the controller from the technical procedural, decision making routines and allow him to concentrate his human intelligence upon strategic decisions (Fig. 9).

In this way the efficient use of computers in air traffic control systems would enable a much better utilisation of air space to be achieved than with the present systems, which are based upon the limits of memory capability and decision making tasks of unaided human operators.

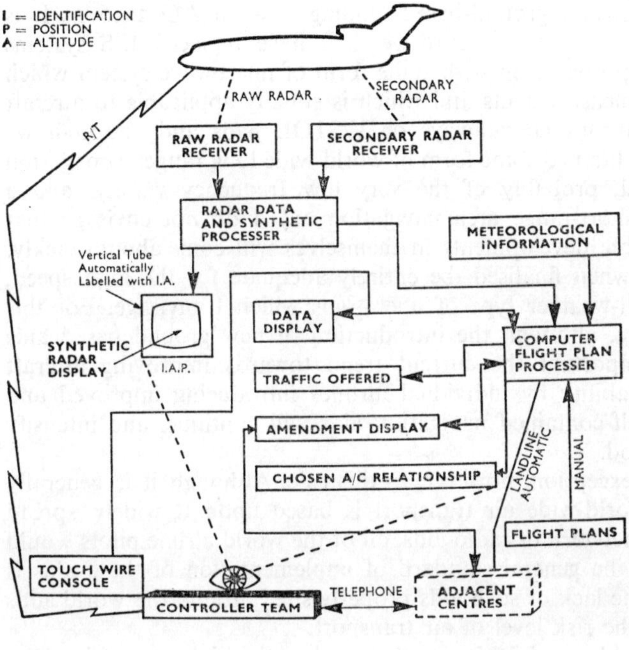

Figure 9 Mediator computer based Air Traffic Control system

Provided the problem can safely be reduced to strict mathematical terms the implementation of a system of area type control is envisaged which would allow aircraft to fly under controlled conditions at any point on, and any path within, a given area as opposed to the present linear control which constrains aircraft to follow fixed airways. This type of air control would prove increasingly important as the volume of air traffic increases and as the growth of V/STOL inter-city traffic takes place.

The computers used in air traffic control will inter-communicate in any communication network with the computer based navigation system in the aircraft thus allowing more rapid and efficient air-ground communication than currently conducted by VHF voice. Ground ATC computers will also inter-communicate with computer based airline operation centres, computer controlled freight warehouses, computer controlled flight planning centres and computer recorded passenger check-in, so that the general preparation of the aircraft flight plan, make-up of the load sheet, air traffic control clearance etc, will be in a parallel operation at high speed through electronic means as opposed to the somewhat tedious human hand, paper and pencil system still often found in 1966.

It will, however, certainly continue to prove as difficult to obtain international agreement on the standardisation of such items as ground-based navigation and ground-based landing aids in AD 2016 as it is today. Although by the year 2016 we shall have replaced ILS systems based on VHF propagation with some form of microwave system which includes range measurements and which is equally applicable to aircraft landing on conventional runways or V/STOL pads and although we shall also have obtained some form of world-wide long range propagation navigational aid, probably of the very low frequency variety, and a corresponding short-range area navigation aid, I do not envisage that any of these three developments in themselves will come about quickly. Nor will they, when finalised, be entirely adequate for the high speed, high volume, all-weather type of operations which I envisage. For this reason, therefore, although the introduction of new ground based aids will be very important the current trend towards improving aircraft operational capability by individual airlines introducing improved and new types of self-contained navigation aids will continue, and intensify during this period.

There is one exception, however, to this rule. Although it is generally believed that world-wide air transport is based upon a widely spread highly efficient network of radio aids, all of the world airline pilots would assure you that the general standard of implementation of these aids is quite low and the lack of such aids properly sited around the world adds appreciably to the risk level of air transport.

One of the problems of deploying these aids is the high cost of installation and maintenance. During the next 50 years there will be significant developments along the lines associated with aircraft equipment in reducing the cost and improving reliability and operational availability of ground-based short-range navigation and landing aids to the point where their adaption, installation and continued use by even the poorest nation will not cause any embarrassment.

3 TIME SPAN

In attempting a forecast of the future it is usual to put forward a scenario or model of the pattern of development of the most likely time at which the discovery or major implementation of the new techniques will be achieved. Such a pattern has been attempted in Fig. 10. In this diagram

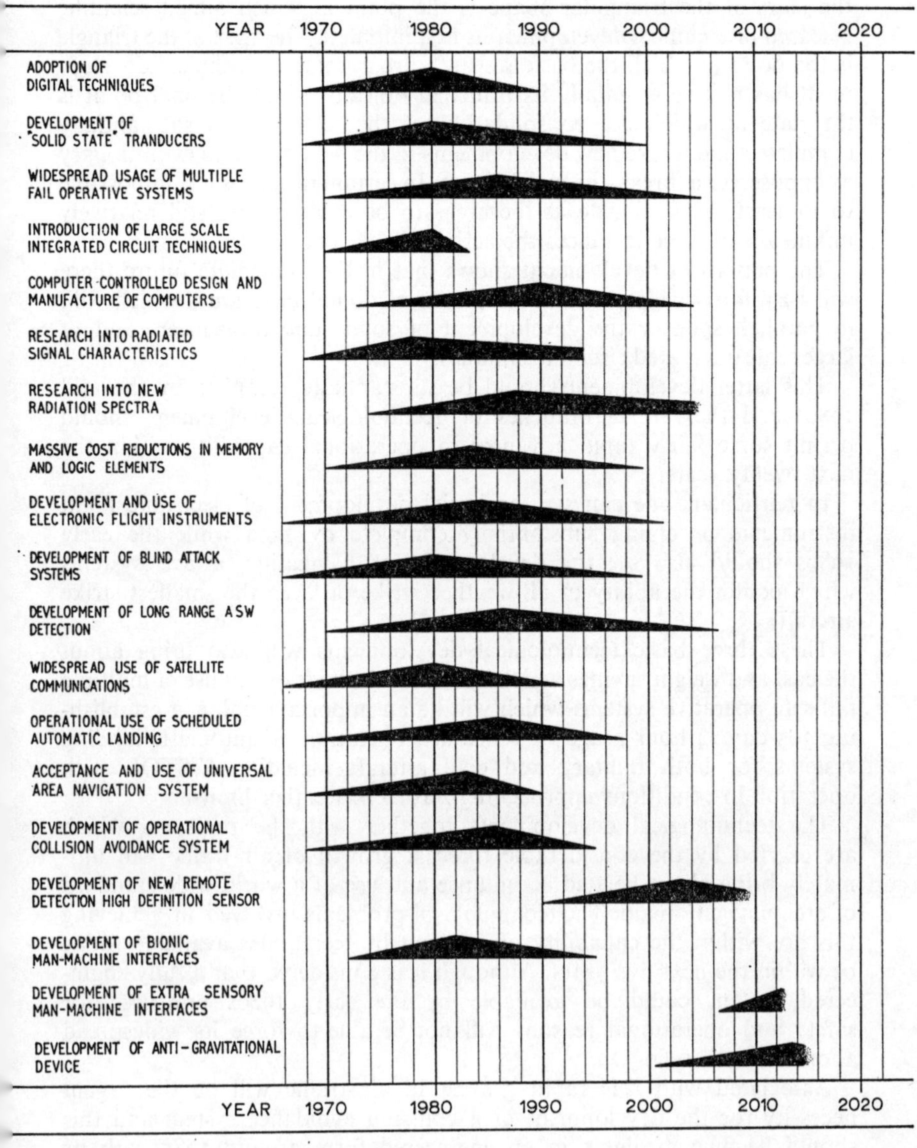

Figure 10 Avionic scenario 1968–2018

the start of the triangular shape is the point at which aimed scientific research or technical development is first initiated. The apex of the triangle is the point at which the basic research programme or technical development has met substantially its principal objectives and the final point is the date at which the consolidation of the technique is substantially completed and where new developments in the field are of an evolutionary as opposed to a 'break-through' nature. In certain instances it is impossible to forecast the final date as the areas to be explored are still relatively unknown. In these instances the activity is left open-ended.

The pattern of development shows that in the immediate future there will be a fairly widespread development of digital techniques which will, in parallel, sponsor the development of solid state transducers and of large scale integrated circuit techniques.

This latter development should be substantially complete by the mid 1980s and these three branches of technological development should permit some fairly rapid advances in operational capability within the next twenty years.

In particular, one can see the initial development of electronic flight instruments becoming substantially complete by 1990 while the early 1980s should also see the development of navigation attack systems which confer the ability of all-weather strike on even the smallest strike aircraft.

These three basic technological developments will also bring about the cost and weight savings which will allow the widespread use of multiple fail-safe operative systems which will be an important point in establishing 365 day/24 hour usage of scheduled operation of automatic landing systems for both military and civil aircraft including V/STOL with operation in conditions approaching zero/zero weather limits.

The technological developments together with the pressures which are exerted by the considerable forecast growth of air traffic will ultimately bring about belated acceptance and use of a world-wide standard of area navigation aid. The technological problems involved in achieving this are within the capability of solution by techniques available today or within the next five years. Although it is considered that a fully engineered system could be available by the early 1980s pressures for safety and operational reasons will not be able to force its widespread adoption until 1990.

Associated with this rapid growth in air traffic will be the urgent necessity for the development of a collision avoidance system and this should become available in an engineered form around 1985 with its adoption within the majority of civil aircraft taking place over the next 10–15 years.

Within the radio frequency spectra one can see the rapid implementation of satellite communication systems as the principal means of aeronautical fixed and mobile telecommunication services and this work will stem from the successful results presently being achieved with relatively low cost ground and airborne equipment via the ATS fixed station satellite.

The work on the aspects of radiated signals within known radiation spectra is already under way and it is to be expected that by about 1980 sufficient progress will have been made for the design of long-range ASW detection systems.

This work on radiated signals will also contribute greatly to the development of the collision avoidance systems previously mentioned.

Within the next decade the implication of computer controlled design and manufacture of airborne and groundborne computers should increase greatly the amount of scientific effort and the application of advanced systems thinking which can be brought to bear upon a solution to the problems currently only hinted at.

The early application of these techniques will allow the development of biomedical or self-adaptive man/machine interfaces which will allow human beings to use computing systems with much greater facility than with today's old-fashioned method of discrete manual switching and control operation.

The more widespread availability of computers and the ability to apply computers more readily to design and research processes, even to the extent of using specially designed computers, will allow research work into new radiation spectra to proceed on a scientific basis much more rapidly than in its present rather slow conjectural path.

In turn, this should permit the realisation of a first engineered model of a new high definition sensor around the turn of the century, such a device giving the equivalent of a direct vision viewing at close range from signals detected by a sensor head some 50–100 miles (80–160 km) distance.

Work in this new radiation spectra should also enable the development of new types of communication links between human beings directly or between human beings and machines in an area which is now loosely described as extra-sensory or telepathy.

Work in this area should begin at approximately the year 2000 and simple laboratory experiments demonstrating the ability of human beings to control computers by thought processes should be achieved by the year 2020.

Finally, as a challenge to the science of aeronautics itself, the widespread use of computers and the new electronic technology including superconductivity should enable a much greater amount of work to be applied to the basic physics of force and energy transmission.

If this work shows that there is no absolutely fundamental reasons which can negate the universal field theory it is possible that anti-gravitational machine could be demonstrated at laboratory level by the year 2020.

4 CONCLUSION

Let me now conclude this survey with what I regard to be two most important points.

First, that I firmly believe in Aviation. Aircraft, Space Vehicles or whatever follows them will remain one of the most important communication and defensive systems available to the peoples of this planet, and possibly other planets for a long time to come. The ignorami, the ungifted amateurs, the non-specialists of encyclopaedic ignorance who preach and practice abdication from aviation, and from the frontiers of challenge and achievement, will disappear and be forgotten long before aviation and its descendant sciences lose their importance to mankind. If we have faith, and if we fight, I am sure that members of the Royal Aeronautical Society will establish aviation once more in its rightful position as a leader industry.

The second point is that neither I nor any of my colleagues in the electronic industry believe that the multifarious electronic-trickery which I have described in this chapter will ever supersede the unique creative intelligence of man in aviation.

The control systems and computers which I have described are not a replacement of man. Rather they are an extension of his capabilities to enable him to create the technology and environment of the 21st century and to perform these tasks more accurately under trying circumstances. Computers even of the bionic type cannot think. They are now and will always be slaves, intelligent and willing hopefully, but only a means which we may use to off-load the chores and routine tasks associated with any operation to enable man to devote his unique capabilities to the areas of decision in which only he can perform.

Figure 1 *Sikorsky S64 Skycrane helicopter*

Figure 2 *Bristol 173 helicopter, with stub wings*

Chapter 14 *The future of rotorcraft*

Figure 11   *Duck during landing*

14 The future of rotorcraft

R. Hafner

'Che Sarà Sarà' (*The Duke of Bedford's family motto*)

INTRODUCTION

Having been delivered about three years ago of a paper for the Royal Aeronautical Society's Centenary Journal on the subject of rotorcraft in retrospect[1]—without, apparently, undue birth pains—I was led to think some months ago, when I was asked to write a similar paper—this time on the prospect for rotorcraft in the next 100 years—that the process would perhaps be equally painless. However, unwittingly, I had made one grave logical mistake: I did not realise that this time I was involved with an altogether different bairnie!

In looking backwards one is concerned with collecting and collating factual matter from various sources, comparing present views with the views held at the time, and with such hindsight arriving at fairly positive conclusions, a more or less routine logical process, the quality of which is mainly measured by the care in considering all important factors in sufficient detail to ensure valid conclusions.

In looking forward, however, one is involved in a rather different process. Here one has no clearly visible facts that can be assembled by routine processes into a precise pattern. Rather does one seek, with a feeling of uncertainty and vacillation, to get a limited perception of the shape of things to come—very much in the manner of the blind—from just the feel of a few touches. There are the illegible signposts which only give a sense of direction without telling where it leads eventually, and similar chips of information, disjointed, ambiguous and often apparently mutually contradictory, which certainly do not assemble into 'The Picture of the Future' but perhaps, with luck, into some limited picture of a possible future.

All this adds up to the one certain conclusion, that future telling is rather a hazardous activity. The near future can perhaps be seen with more clearness, because most of the pertinent factors already exist, so that there is only the matter of their assembly into a picture, as well as

Originally published in the *Aeronautical Journal of the Royal Aeronautical Society*, December 1968.

the times involved—very much like the short-term weather forecast—by extrapolating from the present positions and trends with the aid of the many available scientific procedures[2], such as historical analogy, scenario writing, the 'Delphi' technique, morphological research, the 'PATTERN' technique, etc. However, as we venture further into the future, these scientific methods become increasingly unreliable, not because of the logic of their deductions, but because of the shakiness of some of the premises. Whereas, in the short term, technological advance consists of predictable elements, in the long term it is composed mainly of things unpredictable and, therefore, the far future is not logically foreseeable. In the distant view details get lost in the mists of time and the *vision of imaginative thinking* must take over, where the *sight of scientific reasoning* begins to fail. In this context it has been said that 'too great a burden of knowledge can clog the wheels of imagination'. Knowledge tends to give a picture of the future only in terms of what we know already. It is a limited view, indeed a distorted view, inasmuch as it omits the effects of the knowledge still to come. There is a salient point in a sentence by Arthur Clarke[3]: 'When a distinguished but elderly scientist states that something is possible, he is almost certainly right, but if he states that something is impossible, he is very probably wrong'. This statement is no reflection on the eminent scientist, nor indeed on the value of scientific disciplines, but it does indicate the limits of such disciplines. Immediately beyond these limits does not lie the endless space of sheer fiction, as the pure disciplined scientist would assume, but a broad border area of imaginative logic, which is intellectually very satisfying without conforming strictly to scientific rules.

With these general observations on the problems of forecasting I proceed to the main theme of the paper, the future of rotorcraft. For the purpose of this paper a rotorcraft is defined as a VTOL aircraft, which, at least in the hover mode, is supported by rotors or propellers which rotate in open air in a substantially horizontal plane. This includes autogyros, helicopters, tilt-rotor and tilt-wing aircraft and the like. It does not include deflected slip stream STOL aircraft, jet lift aircraft and fan lift aircraft. Rotorcraft technology is thus a limited field which I do not expect to survive a further whole century. Rotorcraft will play a significant role in the next 'three scores and a half' years, but then they will, I think, be superseded by more advanced forms of aircraft. This paper is concerned with two major aspects: the probable development of the rotorcraft within the environment of the technical evolution as a whole and vice versa, the influence that this development may exert on other technologies.

FURTHER DEVELOPMENT OF THE ROTORCRAFT

The helicopter is a lifting device with a hover performance that is unlikely to be surpassed. In forward flight however, it is severely handicapped and in spite of some scope for improvement—certain forms of so called 'rigid rotors' are capable of operation at high advance ratios and thus somewhat higher forward speeds—its speed and L/D ratio potentials will always be very much inferior to those of the fixed wing. Thus the pure helicopter has secured a place in roles where hover time is significant, such as anti-submarine warfare, observation, ultra-short-haul, cranes, etc, and this is likely to be so in the future.

The size of helicopters is an interesting subject. It has been said that the time will come when everybody, taking-off from the roof of his house, will fly his own little helicopter. I am afraid I cannot subscribe to this idea, not for reasons of economy—given sufficient demand, helicopters could be mass produced like motor cars relatively cheaply—but because of operational difficulties. The thought of tens of thousands of amateur pilots in a limited air space in IFR conditions frightens me. Whereas an inexperienced or irresponsible yachtsman in collision with an ocean liner, at worst, will kill himself and his few companions, an amateur pilot in a similar situation will also destroy the large air liner with all souls aboard. For this reason, with the intensifying air traffic, it will become increasingly difficult to obtain a pilot's licence and thus the number of personal aircraft will be severely limited.

I think, however, that at the other end of the scale, the very large helicopter in the form of a specialised aerial crane has a future. Some years ago Fitzwilliams[4] discussed the virtues of giant helicopters, even though he did not consider special helicopter cranes. The Americans and the Russians have since produced such monsters (although not quite giants) capable of carrying loads up to 15 t (15 Mg) (Fig. 1). I believe that still larger helicopters, of low disc loading and relatively low speed, capable of carrying 60 t (60 Mg) or more, will be used in substantial numbers. By radically changing the methods of construction in civil engineering, such that large units are manufactured in factories and from there are directly carried by aerial cranes to the building sites and assembled quickly to a complete structure, substantial savings in time and money can be made. Moreover, the traffic problem on the ground is eased. Light-weight, mobile houses would be very popular. The military potential of such a large helicopter is also obvious,

The modest speed of the helicopter can be increased somewhat by compounding the rotor with other aerodynamic devices. By adding small stub wings (Fig. 2) which also act as undercarriage structure, a modest speed gain is achieved at little structural cost. More sophisticated forms of compounding have been proposed, but these have been found to be less cost-effective. We may, perhaps, therefore, see in the future some such simple forms of compound helicopters, but to achieve a substantial speed gain, radically different configurations are required which may be termed advanced rotorcraft.

All rotorcraft will benefit from improved materials, especially from the new fibre technology[5], with substantial weight savings in rotor blades, hubs, and transmission parts. The safety will be increased by advanced forms of fault detection[6], such as the acoustic analysis of gear noise[7], the spectroscopic analysis of oil[8], hot bearing detectors, and so on.

The advanced rotorcraft has been a popular field for inventors and the number of expert and inexpert contributions that have been made in a relatively short space of time is legion. A year ago, in 1967, I wrote a paper[9] on this subject entitled 'Quo Vadimus?' an illustration from which is given in Fig. 3. The message in this picture is: Behold this diversity of ideas, which, surely, cannot all be right! Hence, this question, to which —alas—there is not a straight answer, as a great number of diverse factors are involved. I have studied VTOL technology over some years and in the course of time have made a number of fundamental observations:

1 VTOL thinking—until recently—has been associated predominantly with military operations. Leaving aside helicopters and fan-lift-powered VTOL aircraft with a supersonic cabability, I am on record, since 1954, with the view that the domain of the VTOL aircraft is mainly in the civil transport field[10, 18, 19].

2 There are emerging a number of factors for VTOL transport and some against it.

The world is confronted with a population explosion[11] which is given in Table 1.

TABLE 1 *The population explosion*

Year	*Developed Nations (Millions)*	*% of 1968*	*The Whole World (Millions)*	*% of 1968*
1900	800	73	2000	67
1968	1100	100	3000	100
2000	1600	145	7000	233

The consumption of raw materials in the developed nations is today about 20 t (20 Mg) per head per annum. These materials travel an average distance of 600 miles (1000 km) in the various stages of production before reaching the consumer. This consumption has risen from negligible quantities a century ago to this figure and is expected to reach double the present value at the end of the century. Also the average distance of goods travel will increase owing to a number of social and technological factors. Thus we may expect in the developed nations the annual ton-miles of goods traffic at the end of the century to be more than three times that of today, and considering also the increased consumption of the large population in other parts of the world, this factor may well rise above five.

Regarding passenger traffic, from historical analogy, we may also expect a substantial increase in commuter travel (the average distance between the place of residence and the place of work increases), and also increased weekend travel and increased travel at annual holidays (some of the overflow of an affluent society). Widely differing figures have been quoted for these increases. I have assumed an increase of 100% per head, which with the increase in population gives a rise by the year 2000 in annual passenger miles of over 3 times that of today.

I have not included business travel in the above assessment of passenger traffic, because business activities contain two distinct elements: transport and communication. It has been said that communication is the enemy of transport and I quote J. R. Pierce of the Communications Science Division of the Bell Telephone Laboratories[13, 14]; 'Increasingly we will confer by electrical communications rather than by travelling.' There is no doubt that the means of communication will grow rapidly to levels inconceivable today. With microelectronics, lasers and hundreds of communication satellites it will be possible by the simple process of dialling to make phono- and video-contact with single persons or parties, at any point in the world. There will, therefore, only be a need to travel physically for special reasons such as to get to know personally somebody or something, for intimate discussion, for the exercise of skills such as a surgeon performing an operation or a mechanic carrying out repairs. I am inclined to think, therefore, that the need for business travel will not increase commensurately[12] with the growth of business activity.

Of the total traffic today (ton-miles), by far the biggest share is carried on roads and on water, a lesser share by rail and only a small share by air. There is still room on and in the sea, but the traffic on roads and rails is already reaching saturation point. From Fig. 4 it is evident that the road transport system cannot be increased economically by factors of three to five to meet the demand of the year 2000.

Some of this traffic will go underground (piping liquids, etc) but some also must be carried by air. A transfer of surface traffic to air even as small as 10% would increase the air traffic by more than one order. However, air space also is already congested and considering the large space that ground facilities occupy in relation to the towns they serve, one wonders whether the air transport system of today is capable at all of undertaking a large share of the total world traffic. Fig 5 shows the number and size of the airfields located in the London area and it is obvious that these lands cannot be multiplied by a factor of say 10 to meet this increased traffic.

I believe that the introduction of a VTOL transport system can materially ease the situation. It has been shown at Los Angeles and New York International airports that, if properly planned, VTOL traffic and CTOL traffic can operate within the same air space without mutual interference. VTOL ports are much less space consuming than airports, especially if they are on an elevated level, and a number of such V-ports can be set up, therefore, in towns without undue economic penalties. I quote[13] A. S. Boyd, US Secretary of Transportation: 'We must find a way to use our resources systematically—and I use that word literally—to relieve the massive congestion of air traffic that now seems inescapable. Clearly, the V/STOL industry will have no short fall of opportunity in the air

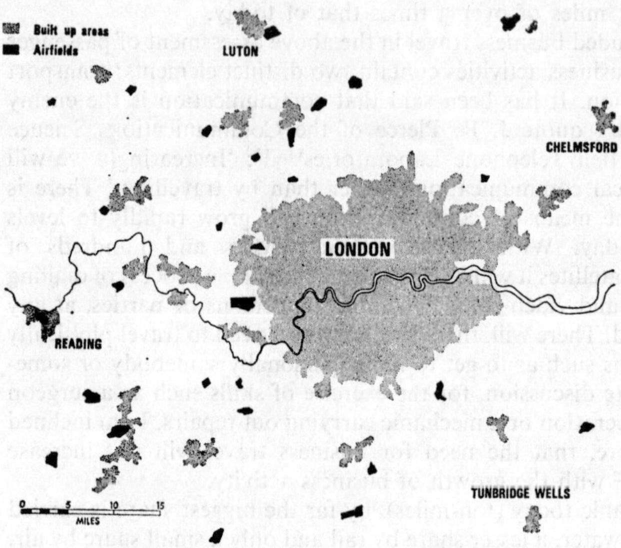

Figure 5 Airfields around London

world of tomorrow' and B. N. E. Halaby, President, Pan American World Airways: 'The need for V/STOL aircraft is already established. Such a machine looms as a key to opening wider the rich short-haul, inter-city air travel market. Effective V/STOL operations will also serve to help relieve painful airport, airways and ground congestion that is now choking major air transportation centres.'

3 A Civil VTOL transport system will have to face the competition of at least two existing transport systems:

(*a*) The railways, which will be substantially faster than at present[12], and will offer fares a little higher than at present but still lower than air fares.

(*b*) The airbus which will be a little faster and cheaper than present day aircraft. Unless a fundamental discovery is made in acoustics to harness random noise, the SST will not compete in the short-haul transport field. To keep the sonic boom within tolerable limits the aircraft is forced to fly very high which is not feasible at short stage-lengths.

4 A viable VTOL transport system, therefore, must offer:
(*a*) Fares not higher than the present air fares.
(*b*) Block speeds not less than those of present transport aircraft.
(*c*) V-ports that are reasonably compact and accessible somewhat like railway stations and much more than airports[15].
(*d*) A noise level and a flight pattern such that only a small area around the ports is seriously affected by the noise of VTOL traffic.
(*e*) An improved method of ATC which fully makes use of the properties of VTOL aircraft. There must be quicker terminal manoeuvres which are independent of wind direction and which will permit a movement rate of closer to one per minute[16].
(*f*) A safety standard like that of other public transport.
(*g*) Efficient transport over stage lengths between 100 and 1000 miles (160–1600 km). Outside these limits other forms of transport will dominate.
(*h*) Economic aircraft in various sizes up to about 200 passengers.
(*i*) The simultaneous operation of a number of suitably located V-ports in a large town.
(*j*) Simpler and cheaper processing of passengers and goods and shorter turn-round times than in present aircraft operations.

5 To achieve relatively low fares the VTOL aircraft must have:
(*a*) An empty weight to AUW ratio and a cruise *L/D* ratio comparable with that of the fixed wing aircraft.
(*b*) A high propulsion efficiency.
(*c*) A fuel consumption in hover and vertical climb no higher than that of a conventional aircraft at take-off.
(*d*) The highest possible subsonic cruise speed.
(*e*) Low first costs. The cost of the VTOL aircraft per passenger seat is bound to be somewhat higher than that of the comparable fixed-wing aircraft, but providing it is not excessive, it can be absorbed, if the other required properties of the aircraft are achieved.

6 With the advent of the jet engine and the steady growth of aircraft and of traffic, has come the growing nuisance from noise and air pollution which now has reached such a state, that the general public has become hostile to any further encroachment of air traffic on populated areas. There has also been an unfavourable public reaction to the increased hazard in the street that has arisen from the operation of helicopters into the centre of Manhattan (Pan-Am building). These signs portend difficulties to an introduction of a VTOL transport system operating around the clock from the very heart of cities. Unless the general public can be convinced that the benefits that society may reap from such a system will outweigh the nuisance and the hazards it creates, it will be rejected.

7 The terminal facilities and procedures of CTOL transport today are extremely cumbersome and expensive. Considering, for instance, Heath-

row Airport[17], one finds that about 4½ sq miles (12 km²) of surface space, a massive array of buildings including air termini in the city, the whole of London's air space and a staff of airport and airline personnel which is counted by the 10,000's are required to deal with about 12½M passenger movements per annum.

The cost of this system is about £25M per annum (airport income £15M and additional airline outlays including the air termini £10M, all of which, in the end, has to be borne by the air passenger), which works out at £2 per passenger movement or £4 per passenger flight. It is clear that with such terminal costs widespread short-haul air traffic is unthinkable. Oxford Circus tube station, by contrast, occupying a space comparable with that of the access road tunnel to Heathrow airport and with a staff of about 25, handles nearly 32M passengers movements per annum, i.e. nearly three times that of Heathrow Airport. I am not suggesting that these two stations can be compared by such a simple yard stick but I wish to highlight the colossal difference in the procedures in these two stations and to show that in this respect air transport has a long way to go to come into alignment with *any other form of mass transport*. I am also convinced that a VTOL transport system will lend itself much more to an economical solution of these terminal procedures. In the first instance elevated V-ports like tube stations, make only small demands on surface space, they combine the functions of the airport and the air terminus, booking is generally unnecessary, and the processing of passengers and goods, thanks to the compactness of the layout, can be made appreciably simpler and cheaper[12, 15].

To sum up, there is a growing pressure to take a substantial share of traffic into the air and it would appear that a VTOL system rather than the CTOL system is capable of carrying such loads economically.

8 VTOL aircraft are essentially hybrids inasmuch as they are sustained in two heterogeneous modes of flight, i.e. hover by means of one or more actuator discs, where the disc loading is the critical parameter, and cruise by means of wings were the span loading is the critical parameter[18, 19].

These two modes of flight differ from one another not only in the means of support but also in control, stability, etc, and VTOL aircraft, therefore, must be fitted with means adaptable to suit both modes of flight, or such means must be supplied in two versions, one for each mode of flight and performance penalties, therefore, are necessarily incurred. It is clearly advisable to utilise efficiently as many common parts as possible in both modes of flight, which is the concept of the well balanced VTOL aircraft[18, 19].

There are many, sometimes conflicting, aspects of balancing and here judicial compromise is the secret of good design. One such aspect is installed power which should be used efficiently in both hover and cruise. Power balanced solutions are obtained in a wide range of disc loadings from the slow helicopter to the supersonic jet lift VTOL aircraft (Fig. 6). If the design is *not* power balanced, as, for instance, is a jet-lift VTOL aircraft with only subsonic cruise performance, then parasitic weight is inevitably carried.

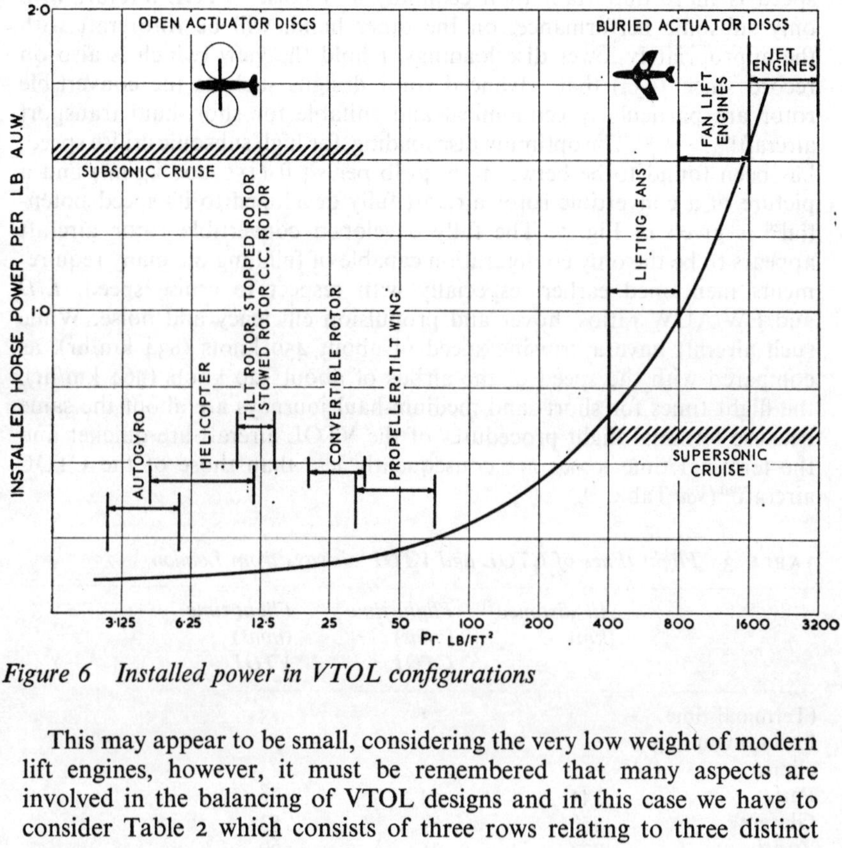

Figure 6 Installed power in VTOL configurations

This may appear to be small, considering the very low weight of modern lift engines, however, it must be remembered that many aspects are involved in the balancing of VTOL designs and in this case we have to consider Table 2 which consists of three rows relating to three distinct

TABLE 2 *The penalties of lift engines*

	Weight	*Volume*	*Cost*
Bare engine	1,1	1,2	1,3
Installation	2,1	2,2	2,3
Fuel for terminal manoeuvres and reserves	3,1	3,2	3,3

items and three columns relating to three distinct properties in which the bare engine weight (1,1) is the only one of nine elements in the array.

Rather more serious are the elements, 2,1, 3,1, 1,2, 2,2 and 1,3. I believe, therefore, that all VTOL aircraft must be substantially power balanced and the high disc loading jet-lift variants will all have a supersonic-capability. They are potential bomber and fighter aircraft, where speed is more important than economy and noise. VTOL aircraft with only subsonic performance, on the other hand, will be rotorcraft with the appropriately lower disc loadings. I hold the view (which is also on record since 1954) that advanced rotor designs such as the convertible rotor are particularly economical and suitable for short-haul transport aircraft[10, 18, 19, 20]. The optimum disc loading, for high subsonic cruise speed, has been found to be between 30–40 lb per sq ft (150–200 kg/m²) and a picture of a convertible rotor aircraft fully developed to its speed potential[20] is given in Fig. 7. The fully developed convertible rotor aircraft appears to be the only configuration capable of fulfilling the many requirements mentioned earlier, especially with respect to cruise speed, *L/D* and EW/AUW ratios, hover and propulsion efficiency and noise. While such aircraft have a cruising speed of about 450 knots (834 km/hr), as compared with the speed of the airbus of about 520 knots (964 km/hr), the flight times for short- and medium-haul journeys are about the same because terminal flight procedures of the VTOL aircraft are quicker and the terminal time losses are consequently less than those of the CTOL aircraft[20] (*see* Table 3).

TABLE 3 *Flight times of CTOL and VTOL airways from London*

	Air distance (km)	*Flight time (mins) CTOL*	*Flight time (mins) VTOL*
(Terminal time losses, including climb)		23	13
Paris	341	44	38
Glasgow	556	58	53
Zurich	773	71	69
Rome	1451	113	117

The passenger, however, is offered a substantial time saving in the VTOL aircraft because the flights terminate at convenient places near the city centre instead of at the aerodrome. This type of aircraft has, therefore, an excellent chance to establish itself in the short-haul market. It is, *ipso facto*, a very useful military transport. There is also a small version of this type which is eminently suited for a fast executive aircraft, as it will offer similar flight times for direct 'roof to roof' flights between factories or offices. It is the fastest means of personal transport possible

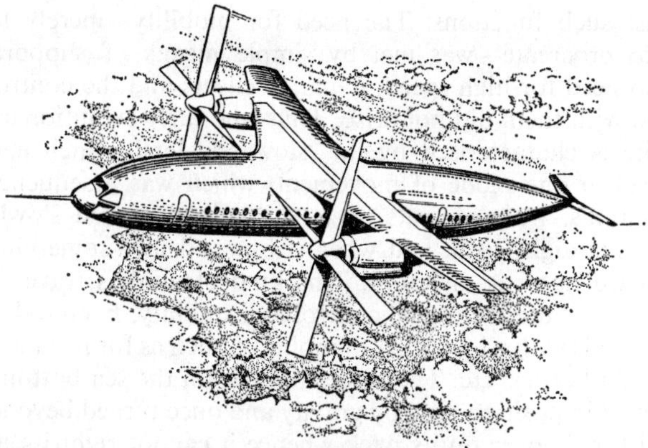

Figure 7 Convertible rotor aircraft in cruise

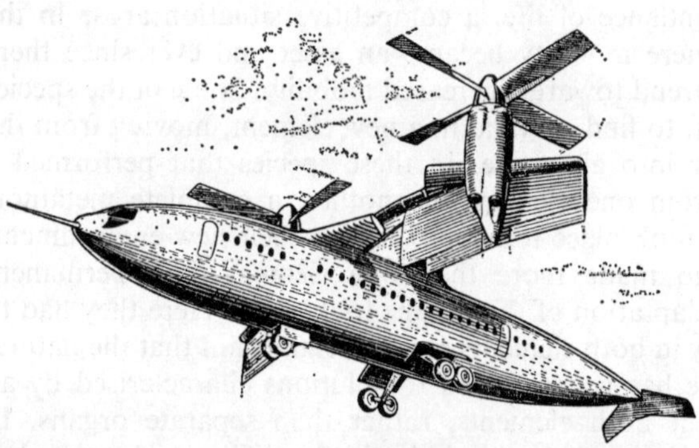

Figure 8 Convertible rotor aircraft at take-off

with the present technology and I expect a substantial number of such aircraft to be employed in military and civil duties.

TRENDS TOWARDS HIGHER ORDERS OF MOBILITY

Evolution trends towards increased mobility are a common phenomenon in nature and are a consequence of the natural order of the survival of the fittest. As in many other aspects, man has tried to copy nature in this and has achieved some measure of success. It should be noted that the bicycle and the helicopter are early examples of this trend and in some advanced forms of rotorcraft—*inter alia*—we shall see it developed to a rather higher degree of perfection. However, let us first consider the natural evolution of mobility in the animal world. The early primitive animals, whose lives comprised only a few simple functions, had primitive organs to fulfil just such functions. The need for mobility—merely to gather food and to procreate—was met by simple means of support, and as there was no need for high speed, dynamic effects and the control problems connected with them did not arise. Thus, the early evolutionary phase in animal life is characterised by the 'slow crawlers', which had only means to travel in one mode of movement, which was a sequence of quasi-static situations. Following its nose, the animal moved slowly along a more or less straight path, and when it bumped into something it did not like, it tried another direction. This kind of aimless travel is devoid of anything that could be termed control and stability. Even today we find primitive animals with a low order of mobility, such as for instance, mussels, snails, certain beetles, etc. The oyster lies flat on the sea bottom, and the tortoise is held in position only by gravity and once turned beyond a certain angle will topple over on its back whence it cannot right itself.

This early period of life was in a sense a very peaceful age with no ambitions and no competition between the species. However later, in an ever growing abundance of life, a competitive situation arose in the fight for survival, where mobility became an asset and ever since there has been the natural trend towards increased mobility. Some of the species changed their habitat, to find survival in a new element, moving from the sea to *terra firma* or into air space. In these species that performed a permanent change from one element to another, a complete metamorphosis of the organs took place to adapt them to their new environment. In other species who made more than one element their permanent habitat, a different adaptation of the organs took place. Here they had to function satisfactorily in both elements and it is significant that the natural evolution in this case has generally led to solutions characterised by an organ adapted to suit both elements, rather than separate organs. In particular, it will be observed that mobility in the different elements has been achieved by the same organ adaptable to different environments, rather than by separate organs for each element, which become parasitic when not in use. Examples are the webbed feet of water birds and the extremities of penguins and seals. In technical terminology, this means that natural evolution would not lead for instance, to an amphibious motorcar using wheels on land, as well as propellers in the water, but rather to flaps grown on to the wheels to act as pedals, or alternatively, to propelling screws shaped such that they can propel the vehicle also on the ground in the fashion of certain tractors. In short, the natural solution is generally a common means for both modes of movement.

There has also been another, even more important natural development. The primitive behaviour of the early living things was hardly in advance of the simple cause and effect behaviour of inanimate objects. The truly 'vital' factor in the fight for survival turned out to be the advent of 'sensors' together with a 'brain with a memory'. The principle of a simple statically stable support was superseded by a new one of a basically unstable supporting mechanism, fitted with sensors and a computing brain, which, together with the executive organs, acted in a 'closed loop' circuit. As the speed of movements increased, dynamic effects became dominant in the operation, which required considerable refinement in these 'sensors' as well as the 'black box', and in a long evolutionary process the advanced animal has eventually achieved this and with it a mobility, the degree of sophistication of which is quite beyond comprehension. Man has learned and, indeed, has come to like, to walk upright with a supporting mechanism of almost as many degrees of freedom as he has bones in his body, all basically unstable mobility, controlled and stabilised entirely in nervous closed loop circuits. At every step he makes he loses and recovers his balance and so achieves a new form of stable walk. But he has achieved far more than this, he has achieved a spring-board to an altogether higher order of mobility. Not only does he walk, but he can also 'learn' to run, jump, swim, ride a bicycle, fly an unstable helicopter, or perform acrobatic feats. Significant here is the basically unstable support, because the feedback mechanism is exercised continuously thereby and so developed to higher performance and there is a consistent awareness of what goes on. Man has also become more agile and versatile, because quick movements are more readily achieved in the absence of primitive stabilising forces, which tend to resist such movement.

This important principle is also fully developed in bird flight [21, 22]. In Figs. 9 and 10 a gannet is shown during a landing manoeuvre and in cruise flight. The tail of the bird, obviously, is not an elementary stabilising device like the tailplane in an aircraft, but a sophisticated variable-geometry control organ. In landing manoeuvres it is fully spread to produce large control moments and in cruise it is closed to form the streamlined tail end of the body. Evidently, the bird prefers in cruise the basically unstable minimum drag configuration together with 'black box' stabilisation through the wings, to the stable but aerodynamically inferior solution with the spread tail. Fig. 11 highlights a different aspect of the profound sophistication in the control of bird flight. The three pictures, which are consecutive frames of a film with one 16th of a second time intervals between frames, show the rapid control manoeuvres of a duck during an approach to a precision landing[22]. It will be seen that in this integrated control system, not merely isolated surfaces, but the whole body, wings, tail, legs and head are involved in a very comprehensive manoeuvre. By comparison, the classic aircraft with its large fixed vanes in the tail, appears as primitive in its movements as the tortoise. Modern developments have brought refinements in the control of aircraft (autopilots and other aids) but these are more in the nature of additions to the classic configuration (aerodynamically stable shape) rather than a radical change. Only recently—mainly as a consequence of the widely varying environment of fast aircraft (M range 0·2–6·0, q range from 2–2000 lb per sq ft (10–10,000 kg/m^2))—has thought been given to radically new flight control systems to meet these widely different conditions and finally to the idea of 'fly by wire' controls[23].

It is noteworthy that some of these apparently novel problems were already known to the designers of the early helicopters. Here also we have a wide range of environmental conditions from hover to cruise, and control methods had to be found to cover this range. However, to achieve stability in hover proved too difficult in the early days and, therefore, the helicopter pilot had to learn to fly basically unstable aircraft with the aid of the excellent 'black box' and 'sensors' given him by nature. The unstable hover mode of flight, having thus been mastered, the need in VFR conditions for stability in the cruise mode did not appear to be of primary importance, and so the helicopter, at first, did not grow the familiar empennage of the fixed-wing aircraft. Later, for flights in IFR, stabilising surfaces were added but it was also discovered that it was advantageous to augment the stability of the helicopter by means of a closed loop mechanism involving angular movement sensors and electronic circuitry (SAS). The concept, developed to the full, is embodied in the convertible rotor VTOL aircraft[20] shown in Figs. 7 and 8, where control and stabilisation in hover and in cruise are obtained only through the rotors and the wing-aileron-flaps with the aid of a stability augmentation system. Here by means of cyclic pitch variation of the rotor blades, pitching moments are produced and by differential collective pitch of the rotors, rolling moments are produced in the hover—and yawing moments in the cruise mode. Differential movement of the ailerons produce yawing moments in the hover—and rolling moments in the cruise mode. It should be noted that the control moments produced by the rotors are independent of overall rotor thrust, and similarly, the moments produced by the aileron flaps are independent of overall wing lift and these control organs are thus effective throughout the flight. Hence—apart from a small trimming surface—the conventional empennage is omitted, with considerable saving of drag and weight.

The automatic flight control system, a typical block diagram of which is given in Fig. 12, is characterised by a control and stabilisation loop containing linear and angular movement sensors, a central controller comprising integrator, variable gain, filters, cross couplings, etc, control actuation commands and also—in certain control channels (wing-aileron-flaps)—by an adaptive controller loop containing control parameter adaptation means which, over a wide range of flight conditions, maintain invariability in control. Such a flight control system not only provides stability and control in hover as well as cruise, but is capable of dealing also with gusts, blind landings, collisions avoidance, etc. Electronic or fluidic circuitry can be used in this control system. It will be seen that the primary organ in this system is the control and stabilisation loop, and the pilot with his manoeuvre commands is in the position of an outsider, merely biasing the system in one way or the other, to effect a change in flight direction.

Such an arrangement may raise doubts in the minds of flying people about its reliability. Pilots are used to flying aircraft directly (although modern aircraft are already removed somewhat from this primitive situation) and do not accept easily the idea of merely monitoring the control proceedings 'from the end of a telephone line'. However by suitable 'fail-safe' devices, the hazards of malfunctioning of such control systems can be effectively eliminated, and I am sure it is only a matter of time before there will be confidence in this form of control, and it will be used in carying degrees in all sophisticated aircraft.

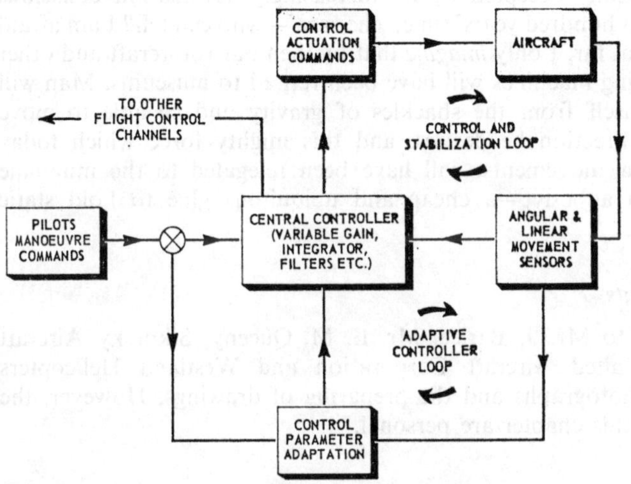

Figure 12 Automatic flight control system

In the Boeing 747 aircraft, the fin rises to the height of a five storey building and the tailplane area compares with that of a ballroom floor! The wetted surface of the impennage of transport aircraft works out at about 12 sq ft (1·1 m²) per passenger. Stability is bought rather dearly with such clumsy means! It has been shown that the aerodynamic performance of the convertible rotor aircraft in Fig. 7, with its radically different flight control system, is significanly superior to that of the comparable fixed-wing aircraft[20].

Such a rotorcraft could be available in eight years' time. However, there is a great deal to be done besides the aircraft. A new transport system has to be established which, operating from built-up areas, needs to be specified in many more aspects than the present air transport. It is in many ways more akin to a railway system. The size of stations, the size and weight of 'trains', indeed the 'gauge of the rails', and the location of the 'buffers', will have to be agreed internationally! This international collaboration, of course, will be bedevilled by national aspirations, but I hope, the great need for such a transport system will be a force, sufficiently powerful, to overcome these obstacles and we may have, in a few years' time, an agreed solution.

I expect the rotorcraft, operating mainly from V-ports in cities, to play a dominant role in short- and medium-haul air transport for a number of decades, leaving the long haul to the advanced CTOL aircraft[24] and its 15,000 acre (61 km^2) airports[25] well out of town. Looking further afield, with miniaturisation and mass production methods, sophisticated semi-automatic controls will spread from aviation to other technologies, and will be readily accepted by an increasingly automation conscious society. But in a hundred years' time, 'che sarà'—who can tell? I am afraid I cannot *see* that far, I only *imagine* that by then our rotorcraft and other magnificent flying machines will have been retired to museums. Man will have freed himself from the shackles of gravity and be able to move freely in any direction he desires, and this mighty force which today impedes all our movements, will have been relegated to the mundane function of an adhesive—a cheap and ubiquitous glue to hold static things in place.

Acknowledgments

I am indebted to Mr. J. Barlee, Mr. E. M. Queeny, Sikorsky Aircraft Division of United Aircraft Corporation and Westland Helicopters Limited, for photographs and the preparing of drawings. However, the views given in this chapter are personal.

References

1 HAFNER, R. A Century of British Aeronautics. British Rotorcraft. *Journal of the Royal Aeronautical Society*, Vol **70**, pp 235–42, January 1966.

2 JANTSCH, E. *Technological Forecasting in Perspective*. OECD, Paris 1967.

3 CLARKE, A. C. *Profiles of the Future*. Victor Gollancz, 1962.

4 FITZWILLIAMS, O. L. L. The Giant Helicopter. *The Journal of the Helicopter Association of Great Britain*, 1952.

5 HAM, A. C. Carbon Fibre Composites and their Potentialities for Aero-space Structures. October 1968. (Unpublished.)

6 HAFNER, R. Safe Mechanisms. *Journal of the Royal Aeronautical Society*, Vol **63**, pp 273–92, May 1959.

7 IVES, A. I. R. and ROSE, P. M. Detection of Gear and Bearing Faults by Narrowband Analysis of Noise/Vibration, Westland Helicopters Limited, June 1967.

8 JENVEY, T. M. H. The Application of Spectrometric Oil Analysis for the Detection of Wear and Incipient Failure in Helicopter Transmission Systems. Westland Helicopters Limited, February 1968.

9 HAFNER, R. Quo Vadimus? AGARD Symposium. Göttingen, September 1967.

10 HAFNER, R. The Domain of the Helicopter, 7th Louis Blériot Lecture. *Journal of the Royal Aeronautical Society*, Vol **58**, pp 661–702, October 1954.

11 REVELLE, R. Population. *Science Journal*, October 1967.

12 SCHRIEVER B. A. and SEIFERT, W. W. Air Transportation 1975 and Beyond. MIT Press, 1968.

13 *Vertical World*, p 10, 19th August 1968.
14 PIERCE, J. R. Communication. *Science Journal*, October 1967.
15 HAFNER, R. A Proposal for a VTOL Port. ARC 29,303, July 1967.
16 HAFNER, R. Typical Flight Profiles and Procedures for Low Disc Loading VTOL Aircraft. ARC 29,031, April 1967.
17 British Airports Authority Annual Report and Accounts. HMSO, 1967–8.
18 HAFNER, R. The Helicopter, The First of the VTOL Aircraft. *Journal of the Royal Aeronautical Society*, Vol **65**, pp 815–29, December 1961.
19 HAFNER, R. The Domain of the Convertible Rotor. 9th Anglo-American Conference, AIAA, 1964.
20 HAFNER, R. The Performance Potential of the Convertible Rotor. ARC 30,003, March 1968.
21 BARLEE, J. *Birds on the Wing*. Collins, 1947.
22 QUEENY, E. M. *Prairie Wings*. J. B. Lippincot Co., 1947.
23 SUTHERLAND, J. P. Fly by Wire, Flight Control Systems. AGARD Symposium, Oslo, September 1968.
24 KÜCHEMANN, D. An Aerodynamicist's Prospect of the Second Century. Chapter 2.
25 MASEFIELD, P. G. Airports, ATC and Air Traffic, 7th Convention of the Guild of Air Traffic Control Officers on 'Air Traffic Control and the Airport'. October 1968.

15 The prospect for materials

A. J. Kennedy

I INTRODUCTION

Modern developments in materials, and the possibilities they offer in present and future engineering, make up a subject on which there has been no shortage of publications in recent years. It would not be possible, in a single article, to review the likely future developments in all the materials of interest to aeronautics; in any case some detailed references have already been made to this subject in other chapters, notably that by Molyneux (Chapter 7) on structural engineering. The point of view adopted here is rather more long-ranging, and certainly more speculative.

Aeronautics and its offspring, space technology, are the most challenging and inspiring of all the advanced technologies, each step in their advance being beset by critical judgments and by increasingly complex decisions. Materials, as a subject, is caught up in this, and the thinking that has motivated much of the research and development on new materials would certainly not have arisen at the time it did without the pressures of aeronautical advance. In the field of more fundamental materials studies, progress has closely followed the evolution of instruments capable of examining the detailed structure of materials: the optical microscope, in the first place, then the use of X-rays, followed by electron microscopy, and a whole series of instruments and techniques—scanning electron microscopy, X-ray fluorescence analysis, and many others—which in their separate ways have contributed to our detailed knowledge of just what materials are like on the micro-scale; how their atoms are arranged, how ordered in their arrangements they are, whether there are faults or defects in the structure, what form these faults take, how different atoms are mixed locally, what happens when the material deforms, or fractures, or corrodes . . . and so on. Part of all this activity was inspired by the development of theoretical ideas about how crystalline materials, like metals, could behave in the way they did; why they were usually plastic for example, rather than brittle, and if they were brittle, why this was so.

Originally published in the *Aeronautical Journal of the Royal Aeronautical Society*, January 1969.

The very broad front on which basic materials knowledge now advances still has its main roots in classical crystallography and chemical kinetics (unrecognisable though these may be at times) together with some which have sprung up on the way, and perhaps most notably, dislocation theory in crystalline materials on the one hand and the mechanics of long-chain polymer behaviour on the other. These subjects, and the models evolved from their study, are generally the product of considerations on a fine scale—that is, involving the behaviour of relatively small groups of atoms. The ideas are sound, and quantitative experimental confirmation can often be achieved, so long as the questions asked are limited, and the material fairly simple in structure, or at least well-defined. Working from this basis has not, unfortunately for engineering as a whole, enabled a new framework for the description of the properties of materials to be established and this remains a major issue. The greater part of modern elasticity and plasticity theory is an evolution from the classical continuum descriptions. It is true that in places, and for certain conditions, links can be identified with, say, fundamental structural metallurgy, but these are at present too weak to bear the weight of serious design. The consequence is that the considerable body of enlightened and rewarding research on the development of new materials, and on the improvement of those that already exist, has contributed in the main by offering to engineering better materials for exploitation, without at the same time contributing any radical change to the ways in which engineering takes account of materials properties—or even suggesting (and there are some exceptions here, fracture mechanics for example) new ways of assessing the capabilities of materials for advanced application in complex operating conditions. It could be said in defence of this situation that what is being done by those working on materials is all that can be expected, bearing in mind their immense range and variety, the catalogue of questions that might be asked, and the great difficulties, and expense in time and effort, inherent in developing, testing and fabricating a new material. We should perhaps, on the occasion of a review of this kind, pay some passing tribute to the insight, persistence and overall achievement of those who have worked in materials, in support of aeronautics, and recognise the profound contribution made by people in this country to the variety of subjects and materials involved.

It may be useful here to indicate the structure of the present chapter. Some general fundamental considerations are first reviewed, and then the materials issues in structures and in propulsion systems examined separately. In each of these sections, the engineering factors are looked at first, before taking up specific classes of material, although it is not easy to separate the two questions completely. There follows a section on 'new' material forms and processes, although just what 'new' means here is debatable, and finally there are some thoughts on the very long-term outlook.

2 SOME FUNDAMENTAL CONSIDERATIONS

If we are to look ahead and attempt some kind of judgment on the state of materials in, say, fifty years' time, we might ask first what basic possibilities exist for creating materials radically different from those which now exist, accepting, as we must, that we have the same range of elements at our disposal. Materials derive their useful properties from the way their constituent atoms are bonded together, from their arrangement spacially, from the way different atoms are mixed—whether this occurs fairly uniformly, whether the mixing is ordered or not, whether separate phases exist, and so on—from the kinds and the distribution of faults they contain (and particularly highly localised crystalline faults), and from a series of structural features such as the size and shape of the individual grains in metals, or the extent to which such grains are isotropic, or, as with polymeric materials, the amounts of crystalline and amorphous material present. It would be out of place in an article of this kind to pursue a detailed analysis of the many basic questions involved, but it helps, in looking ahead to the future, to appreciate that the primary factors that can be varied are not unlimited, but are surprisingly small in number. The actual possibilities, when all the ranges of possible compositions and material states are included, and the effects such variations may have, are immensely large, certainly far larger than it would be feasible to examine on an empirical basis. There have to be guide lines, and whatever the basic theories of the solid state may have failed to do in helping engineering design as such, they have contributed very valuably to the way in which promising materials have been sought out and developed.

Although, as has been said, there are a strictly limited number of structural variables that can be scanned in the search for new materials, we must recognise two important issues in extrapolating into the future. First, the *range* of any individual variable may be very significantly extended in the future, and second, the technologies involved in the conversion of materials into useful forms may (indeed, they certainly will) bring existing, but so far unexploited, materials into engineering and in new and improved assemblies, as well as making the exploitation of new and otherwise 'difficult' materials a possibility. As an example of the first case, considerable developments have taken place in recent years in the control of grain size in metals. Without entering here into all the technical detail, there are benefits in having a very fine grain size (say less than 1 micron) in gaining ductility, or improving fracture toughness, while at the other extreme, when high temperature creep is a limiting factor in an application, benefits can be conferred by having very large grains; gas turbine blades with grains running along their whole length, and also blades made up of one single grain, have been made in nickel alloys and tested with some success. Extremes of grain size, and in particular the possibility of breaking down the crystal structure to very small units by, for example, very extreme rates of quenching, and by very fine dispersions, open up the likelihood of a family of materials which have structures unlike any available today. One known, and little exploited, property of such materials is superplasticity: materials which possess this can exhibit extreme compliance to plastic deformation, under appropriate conditions, and can, for example, be readily formed in this state.

It is worth devoting a little special consideration to the question of superplasticity for it exemplifies the very great theoretical margins which exist for exploitation so far as the mechanics of deformation are concerned. The underlying theories have held out such possibilities for a considerable time: indeed, one of the major theoretical difficulties has been that of explaining why plasticity is usually so limited in metals considering the potential offered by the extensive dislocation movements which are possible. (There is, incidentally, no equivalent theoretical encouragement in the case of high-temperature reactivity, although even here there is still a great deal of potential still to be exploited.)

A large number of alloy systems have been, and are being, studied with a view to obtaining some practical advantage from superplasticity. Among these are chromium-cobalt, chromium-ruthenium, chromium-iron, chromium-rhenium (any of which would have applications in air-breathing engines, and the first two of which exhibit superplasticity), low-alloy steels, aluminium-zinc, aluminium-copper, copper-nickel, copper-zinc, tin-bismuth, lead-tin, and many others. Materials in the superplasticity condition typically show elongations of several hundred per cent and figures near 1000 per cent have been reported. There seem to be two broad classes. The first embraces deformation of a material under conditions in which a phase transformation, or a structural change such as recrystallisation, occurs; the second depends upon diffusion-controlled deformation mechanisms which are dependent on temperature, strain rate and diffusion path length. In the latter case, very large effects arise when the path lengths are made extremely small by, say, fine dispersions, or by developing a very fine grain size. Conditions are usually fairly critical, as the results on an aluminium-zinc alloy, shown in Fig. 1 demonstrate.

The exploitation of such a property could prove of great value in the shaping and forming of strong and useful materials with a conventionally low ductility. Whether the achievement of a potentially superplastic structure will adversely reduce the properties of such materials under service conditions is a question which will need to be answered for specific cases. If the conditions are very different from those imposed during the fabrication operations, little difficulty may arise, or alternatively it may be possible in some cases to fabricate in the superplastic state and then thermally and mechanically treat the material to convert the structure into a form more suitable for service.

Allowing our thoughts to run beyond the extreme case of fine grain we might obtain metals which are predominantly characterised throughout by what metallurgists recognise as a grain boundary structure, with, say, very little long-range crystallinity: another way of looking at such material is to imagine it over-crowded with faults (dislocations) to such an extent

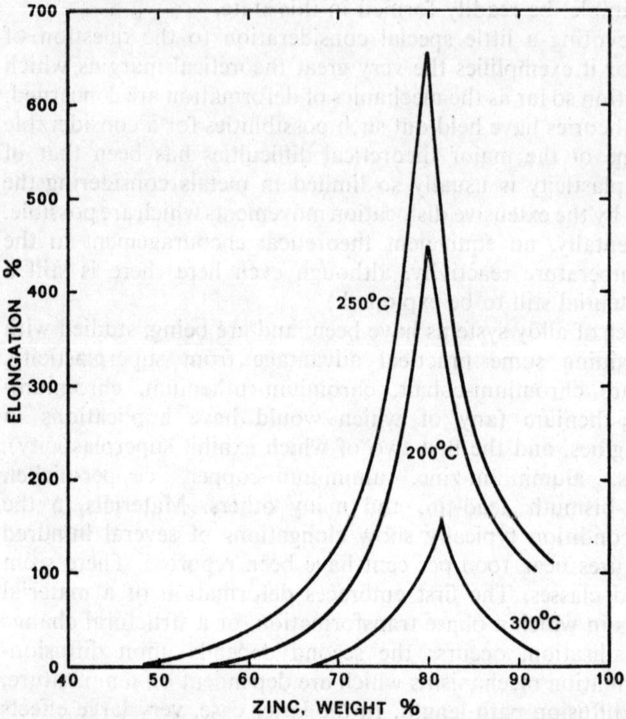

Figure 1 An illustration of the very great increase in elongation of test-pieces of Al-Zn alloys obtainable for the optimum combination of temperature, composition and strain rate. These particular alloys were tested after quenching from 375°C. The results are due to Presuyakov and Chervyakova, and are quoted in the review by Underwood[11]

that the crystalline order is severely broken down. The problem, of course, is to retain this as a stable form. It is possible that routes to such structures might be found through very rapid cooling, or rapid deformation processes, or combinations of both. Metals compacted from rapidly-quenched particles (cooled at rates as high as 10^7–10^8 deg C per sec) have been produced. The structure of the alloys achieved (that is how atoms are mixed) by such variations is different from those which exist now, and in turn, their ageing processes lead to different final structures, which may well have new, advantageous properties.

The second important factor, the availability of new technologies, is generally well appreciated. There are many examples of the influence such technologies can have: the production of fibre-reinforced composite components, for example, or the economic shaping of brittle materials. We shall return to this issue later in the review.

3 STRUCTURAL MATERIALS

3.1 *The outlook from the point of view of the structural requirements*

Fatigue and stress-corrosion have remained the structural engineer's most aggravating material concerns. Designing for creep in the structure has become a consideration with the supersonic transport aircraft, and clearly if future SSTs are to fly economically then they must operate in a regime where some high temperature deformation occurs at highly-stressed places in the structure. Handling the complex creep-fatigue situation, with thermal cycling superimposed as well, demands judgments, and forms of data handling, which pass well beyond the range of present materials knowledge. To start with, the total effects of all the possible mechanical and thermal histories are just not known metallurgically, although the worst possibilities, in the case of the aluminium alloy for Concorde, are soundly established on a straightforward empirical basis. We must, however, ask ourselves if this is the way aeronautics is to progress; the selection of materials for complex stress and thermal conditions, and the prediction of their behaviour is a subject which will certainly cause aeronautical design much heart-searching in the future, and one which merits greater fundamental study than it now receives.

Will we see any great advances in the evolution of materials with significantly greater fatigue resistance than those which exist today? The stress level at which a fatigue crack starts and spreads does depend on the alloy, and on its state (degree of heat treatment, and so on) but attempts to raise these levels have been disheartening compared, for example, with the rate of advance of creep-resistance in the high temperature alloys —or even in the static strength of the lightweight structural alloys. Why is this? Basically because fatigue develops initially on a very localised scale—in zones perhaps only a few thousand atoms in linear dimension—and the primary initiation of a crack in a crystal is dominated by local crystal features which are not easily changed if the basic element (aluminium, say) remains the same. Certain alloying additions do affect the nature and the behaviour of the crystalline faults which generate the primary micro-crack, and they can inhibit or delay, to some degree, its formation. Unfortunately for engineering the extent to which this interaction can be developed has proved disappointing. Preventing the spread of the crack by introducing crack-stopping dispersions into the matrix, or by alloying developments which result in the crack propagating more slowly, have had a rather happier history, and this is one of the features of fibre-reinforced composites that contribute to their structural attractiveness. The possibility of finding more effective ways of preventing the defects in metals from starting a fatigue crack in the first place cannot be altogether dismissed. Whatever disagreements there may be about the exact nature of the fundamental mechanism, it must involve the to-and-from movement of relatively large groups of dislocations, and it is clear that such a movement could be stopped by strong enough interactions with other atoms, or particles, or by the development of particular kinds of dislocation networks. The environment also plays an important part in the rate of crack development, and improvements in methods of surface protection would also contribute beneficially. In restricting the motion of dislocations, the difficulty is to retain in the material enough of the other necessary engineering properties—toughness under impact, for example. So far as static strength, and creep, are concerned, we can have higher strengths tomorrow if a degradation in other properties can be accepted. Strength as such is, in fact, one of the few properties that is not exploited at the present time, and one for which prediction is relatively straightforward. Fine diameter metal wires, filaments such as boron and graphite, and the so-called 'whiskers' of crystalline materials, metallic or non-metallic, have strengths and elastic moduli significantly greater than materials of conventional form. Graphite fibres with moduli up to 90×10^6 psi (620 GN/m²) and graphite whiskers with moduli greater than 100×10^6 psi (690 GN/m²) have been achieved (Fig. 2). In fine whiskers,

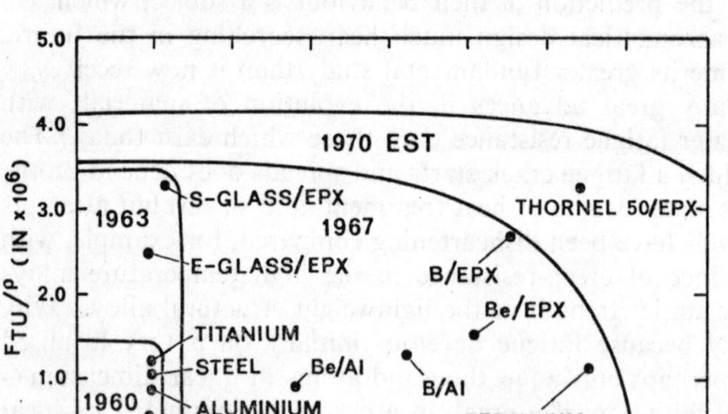

Figure 2 A plot of strength/density ratio against modulus/density ratio for various metals and composites, showing the boundaries of development for various years. Thornel is a graphite fibre. (From Korb, Gatzek and Olsen[16])

there is a considerable variation in their individual strengths, whatever the material, but the best come near to the theoretical strength of the relevant inter-atomic bonds, so one is approaching the ultimate possibility in these cases. There is no doubt at all that a whole generation of materials which exploit the reinforcement available from continuous wire or filament and from discontinuous (dispersed) whiskers will come into being, the matrix conferring the necessary plastic compliance and transferring the external imposed loads to the stronger internal fibres. A great deal of fundamental research, design, and production development has yet to be done before a clear winner emerges however. In high temperature technology the composite requirements are rather different and will be discussed in another section. Tables 1 and 2 show some of the properties of fibre and whisker materials.

TABLE 1 *Properties of continuous reinforcing materials (from ref. 17)*

	Density		*Modulus*	*Specific Modulus*	*Tensile Strength*	*Specific Strength*
	g/cm³	lb/in³	10^6 lb/in²	10^7 in	10^3 lb/in²	10^6 in
Boron	2·35	0·085	55	64·8	400	4·7
Carbon/graphite	1·5	0·054	60	83·0	350	6·5
Silicon carbide	3·20	0·115	70	60·8	400	3·5
Boron carbide	2·52	0·091	70	82·4	330	3·9
Beryllium	1·83	0·065	35	53·0	200	2·8
Titanium diboride	4·18	0·162	84	51·8	150	0·93
E glass	2·55	0·092	10·5	11·4	500	5·4
S glass	2·50	0·090	12·6	14·0	650	7·2
4H-1	2·66	0·096	14·5	15·1	730	7·6
SiO_2	2·19	0·079	10·5	13·3	850	10·8

TABLE 2 *Properties of whisker fibres (from ref. 17)*

Fibre	*Density*		*Modulus*	*Specific Modulus*	*Tensile Strength*	*Specific Strength*
	g/cm³	lb/in³	10^6 lb/in²	10^7 in	10^3lb/in²	10^6 in
Al_2O_3	3·96	0·143	62	43·4	3000	21·2
Al_2O_3	3·60	0·130	50	38·5	500–2,000	3·85–15·4
SiC	3·20	0·115	70	60·8	3000	26·1
B_4C	2·52	0·091	70	76·9	2000	21·9
Si_3N_4	3·18	0·115	55	47·8	2000	17·4
Graphite	1·66	0·060	102	170	2845	47·4
BeO	2·85	0·103	50	48·5	1900	18·4

Stress corrosion has a bearing on fatigue behaviour, but is a problem in its own right. The situation, in broad terms, is that an optimum judgment has to be made between a number of competing properties, including stress corrosion, and that in most cases, with a little reduction in the strengths demanded, the stress corrosion behaviour would be acceptable. But in the future new alloys will appear which will bring their own problems—new titanium alloys for example—and no overall generalisations are possible, for chemical behaviour is sensitively dependent on composition and microstructure, and particularly on states of precipitation in alloys. For radical improvements we must look in part to more intensive and successful methods of surface protection, and this in turn is another aspect of composite construction.

The mechanical failure of aircraft components, by fatigue or by any other mechanism, is the primary structural hazard, and means of monitoring the quality of such components, both before and in service, is a continuing concern. For aircraft operating in the future under even more severe conditions than those encountered at present, the need for a good assessment of possible internal damage increases, and we may expect considerable developments in the general field embraced by what is known as non-destructive testing. In principle, much more information is contained in the wave-forms of reflected ultrasonic signals than is (or can be) used at the present time. It is possible that quite new, if much more elaborate, techniques will be able to detect initial defects which would not now be recognised as such, and in addition monitor the changing state of the material in a component *in situ* on the aircraft, and enable damage to be recognised at an early stage. Whatever the success or otherwise of this particular line of research, one must foresee a much greater effort in such testing procedures when the fate of expensive machines (and passengers) is at stake. The application of computers to the task of sorting through a mass of complex test data, and detecting changes which may arise from a significant change in the material state of a component (and not necessarily associated with a crack) is a question which has had little serious consideration so far.

Materials capable of absorbing energy without internal damage—noise energy in particular—are already a major interest. There are two particular questions of long term concern: first, whether materials with good structural properties can be developed with a high capacity for energy absorption (good damping capacity) in the frequency ranges of interest, and second, just what special damage is introduced by noise compared with that associated with the more conventional fatigue stresses of much lower frequency.

The state of fatigue developed is certainly different at high and low frequencies (and different, of course, at high and low stresses) but just how materials may be developed with high frequency resistance in mind is a subject that has hardly been touched on. It is, however, clear that the noise components can have marked effects on other processes which involve stress relaxation—creep for example—and also on the processes of ageing in precipitation-hardening materials, and new complex possibilities of failure arise which will certainly develop into subjects of their own in the years to come. Again one must expect some of the quantitative assessments to be made by the use of computers.

3.2 *The outlook from the point of view of the available materials*

The possibilities of a better combination of properties in aluminium alloys, in the main by increasing corrosion resistance through small additions of silver, have been pursued for some time. An alloy with 0·3 per cent silver and 0·15 per cent zirconium, for example, with low impurities and with slightly higher zinc and magnesium contents than the 7075 alloy is being actively tested out. There are, however, problems with pitting corrosion with these silver-bearing alloys. As with many other materials—nickel and beryllium for example—the exploitation of some kind of powder process with aluminium has been of interest ever since SAP (sintered aluminium powder) demonstrated the benefits of an oxide dispersion in aluminium so far as high temperature strength is concerned, even if this was achieved at the expense of room temperature properties. Whether the major advance with aluminium will come through filamentary reinforcement is an open question. High strengths resulting from the addition of 50 per cent boron filaments (200,000 psi (1·4 GN/m^2)) are reported: improved fatigue strengths have been observed with steel wires in aluminium, and programmes are in hand on beryllium-reinforced aluminium, but overall the all-round properties of test pieces in such materials have not so far been acceptable. The production of specially-reinforced components for critically stressed places in the structure seems, however, an inevitable development. Beryllium itself remains unpromising largely because its brittleness at room temperature appears an inherent property which is difficult to modify metallurgically—very fine grain-size materials may again offer some margin of benefit in this connection. Another fairly remote possibility is the retention of the very high temperature form of beryllium at much lower temperatures. It is likely that this would not show the brittleness of the equilibrium form of beryllium, but so far no metallurgical means of retaining the more ductile form has been found, or has even looked likely.

At prices of about £30 per lb (£66 per kg) of bar and ingot, and about £100–£300 per lb (£220–660 per kg) of sheet, it is hardly attractive economically, although larger markets would probably halve these figures. Nevertheless, beryllium sheet has been used in aeronautical structures (a prototype rudder for the McDonnell-Douglas F-4, for example) and some of its ductility difficulties can be designed around. The Lockalloy material (62 Be–38 Al) is now available in a wide variety of forms and shapes and offers some compromise between the best that beryllium can offer. Of all the various possibilities the use of very high strength beryllium wires embedded in a suitable matrix would appear to be the most promising for aeronautics, and it seems difficult to believe that a material with such outstanding structural properties (density, modulus, strength) will not be used for certain components in the years to come.

Titanium is certain to be taken up more widely in aeronautical structures and particularly in the next stage of development of SSTs (*see* Fig. 3). Concorde has already been an influence in this direction, with 8 in (0·2 m) sections of the alloy Ti-11Sn-2·25 Al-4Mo-0·2Si being used in the structure (and a considerable amount of other alloys in the engines; 6Al-4V for compressor blades, 4Al-4Mo and 6Al-5Zr for compressor discs). Without reviewing the considerable range of alloys now being developed, it is clear that very high room temperature yield strengths are likely to be achieved in the future, possibly around 250,000 psi (1·7 GN/m²). The alloy containing 6Al-6V-2Sn already approaches this

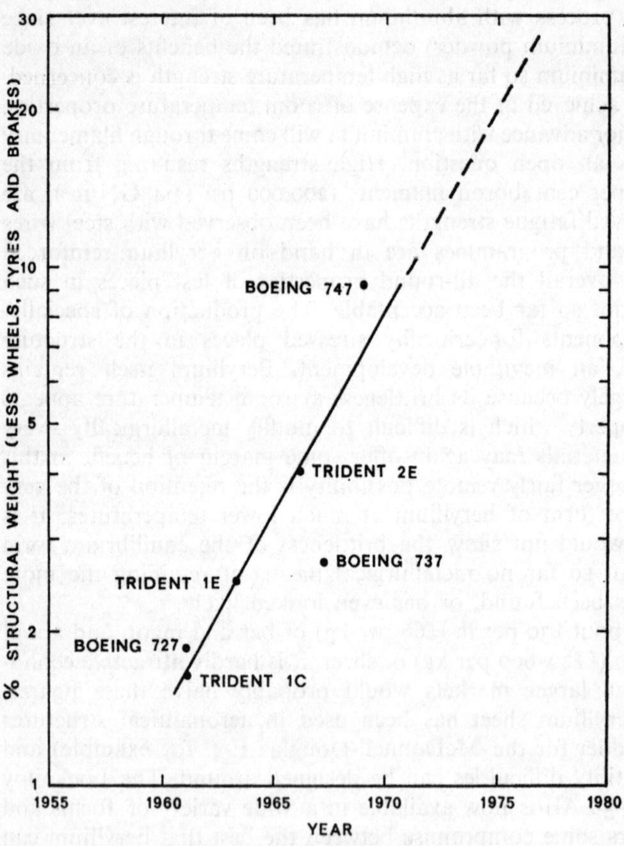

Figure 3 Growth in the use of titanium in aircraft in recent years

figure. The development of greater creep resistance in titanium alloys taken together with operations such as high temperature forging (so-called beta-forging) will extend the properties significantly, and apart from engine applications we shall see a steady increase in such components as titanium wheels, panels, stiffeners and fasteners. The all-titanium electron beam welded titanium rotor of the Lockheed Cheyenne helicopter is also indicative of future applications.

Inevitably in long-range thinking about aerospace structures, the composite forms predominate. Very large specific modulus gains are within reach with, for example, boron, graphite, or beryllium filaments in an eopxy resin as compared with E-glass or S-glass in epoxy—factors of 3 or 4. A whole range of possibilities for structural composites exists, magnesium-lithium alloys reinforced by high strength steel; silicon carbide, graphite, boron, beryllium, or refractory metals in structural and high temperature alloys.

The filaments are almost inevitably pre-coated or pre-treated in some way, and this coating can become the matrix material in some cases—for example, graphite-reinforced nickel has been produced from nickel-coated fibres. Alternatively the matrix can be introduced as powder, or as sheet, by liquid infiltration, or even in the form of wires, the latter being a useful way of conforming to some desired geometry. High energy-rate techniques, including explosive compaction have also been investigated. Certainly the newer reinforced plastics will gain ground at the expense of light alloys wherever they are not critically load bearing—that is, in the secondary structure. The potential of really strong filaments can only be realised if the very difficult problems of compatability with the matrix can be resolved. A good, but not too good, bonding is required, and no reactivity, plus a reasonable match of certain physical properties so that, for example, large thermal stresses are not developed. Radically new design methods are also needed, but once these are accepted, and the capabilities of computer solutions fully exploited, much more efficient structural components will be achieved.

4 PROPULSION MATERIALS

4.1 *The outlook from the point of view of the engineering requirements*

The most general feature of engine development over the past twenty or thirty years has been the steadily increasing operating temperatures. These have been made possible in the first instance through the evolution of successful high temperature alloys, and later by the introduction of cooling techniques. The major part of the advance has been due to the possibilities offered by the alloying of nickel and by the combination of mechanical and chemical properties exhibited by the nickel-chromium series. Other materials, such as titanium, have played a notable part, but the spearhead has certainly been the material for use in the hottest parts of the engine. There are a fair number of materials with a higher melting point than nickel, and even though most of these are much denser, it would seem worth bringing them into use if a significant margin in operating temperature can be achieved, and as a 10°C improvement a year in alloy development has characterised the advance of recent years, not much is needed to make such a change well worthwhile. Unfortunately, it is the chemical behaviour of the so-called refractory metals, molybdenum, niobium, tantalum and tungsten, that (in general) makes them difficult and in spite of an immense research effort over the past fifteen years or so, chemical protection by alloying, or by coating, has proved far more difficult to achieve than improvements in physical properties, particularly with those which are otherwise the most attractive —molybdenum and niobium. Chromium, an intermediate possibility, has advantages in oxidation resistance, but is embrittled to an unacceptable degree by certain gases—nitrogen, for example—and its optimum application may have to be finely judged, at least in the present state of alloy development. For the future, in spite of all the negative results, a more optimistic outlook is justified: chromium alloys will certainly be brought into use, and some of the refractory metal alloys have sufficiently promising properties, so far as the development of self-healing coatings is concerned (the essential need) that we would be foolish to underrate the chances of success. Although a number of cooling systems have been, and are being, worked on, the possibilities of cooling by a continuous fine diffusion through the solid component, with evaporation at the surface, have certainly not been exhausted. A great deal of metallurgy remains to be done, but we can be confident that in fifty years' time engines (and structures) using refractory metals will be flying.

4.2 *The outlook from the point of view of the available materials*

Taking up what has already been referred to in section 4.1, the immediate outlook for new materials in propulsion systems lies with carbon fibre reinforced plastics for low temperature components, a greater exploitation of titanium generally, improved methods of cooling (such as transpiration cooling) for existing, or slightly improved alloys of the nickel-chromium-cobalt-iron combinations, the introduction of chromium alloys for particular components, and the possible use in limited applications of the more brittle materials, such as silicon carbide or silicon nitride, or reinforced ceramics. In the longer term the initiative must lie with the refractory metals if significantly higher temperatures are to be resisted for acceptable working lives, as will be the case with ramjets. The key question is that of their protection from chemical reactions with the environment, plus the lesser problem (or, at least, the one that seems nearer to solution) of an acceptable mechanical toughness. In spite of very large research programmes, particularly in the USA, progress has not been encouraging so far, and, in general, acceptable chemical properties have been more difficult to develop in materials than the physical ones. What is needed is a successful coating which has the property of being self-healing, and which will survive thousands of hours up to 1400°C (say) under thermal cycling, being at the same time resistant to the corrosive and accidental impact conditions which arise in turbojets and ramjets. A successful protected material would, of course, also have applications in lift and control surfaces for hypersonic cruise vehicles: for short term applications in this connection higher temperatures (say 1600°C) would need to be endured. For rocket propulsion systems, and for ramjet cruise, protection for periods of tens of hours up to 2000°C must be aimed at: very short times at even higher temperatures are now possible (with graphite and some plastics for example) but the case of intermediate times and intermediate temperatures is proving a particular obstinate regime to deal with.

At the present time no materials, coated or otherwise, exist which will perform satisfactorily under the projected conditions, but there are some promising possibilities. The silicides and aluminides of the group of refractory metals have been intensively studied, and while many of these are good in themselves, it is their reaction (interdiffusion) with the substrate material which is often the limiting factor. A great deal of metallurgy remains to be done on ways of controlling such reactions, in particular by additions of other elements. The coatings can also be easily damaged, which is a disadvantage. So far as the mechanical properties are concerned, the development of carbide precipitation is the most effective immediate method of conferring high temperature strength, and a great deal more will certainly be done in optimising the structures developed, as well as in achieving basically higher strengths in the solid solutions by appropriate additions. Particle additions of various forms can certainly raise the high temperature strengths. Some remarkable effects can be achieved in a number of directions by quite small additions of other elements: molybdenum and tungsten have their low temperature ductility greatly enhanced by, say, 3 to 5 per cent additions of rhenium, and this improvement continues as the addition increases. Rhenium is too expensive to be considered seriously for large additions, but other platinum-group metals, possibly in much smaller quantities, or even metals in other groups altogether, might exert similar effects as well as contribute to the chemical protection of the alloys. Little is known about many of the basic mechanisms involved and there is at the present time little substitute for persistent fundamental research into these topics, and continued work in alloying generally.

A final comment here on titanium. One third of the weight of an average jet engine is titanium, and in 1967 nearly 90% of the titanium used in Britain went into aeronautical applications of various kinds. One factor that is holding up the cost of the metal is the present lack of a scrap re-cycling system. Metallurgically, it is difficult to see the high temperature fall-off in properties being pushed much above 500°C (and this is a disappointing temperature compared with the melting point) without a sacrifice of toughness and other properties, but a large number of alloys are still being worked on. Titanium has poor wear and galling resistance but even here there is good reason to expect improved coatings for titanium (as for other metals). We must look optimistically to the future and expect some at least of the difficulties which arise with the lubrication and wear of materials to be significantly reduced as a result of the researches now going on into the basic mechanisms involved and into the development of new lubricants and coatings.

5 NEW MATERIAL FORMS AND PROCESSES

One of the developments which needs to be assessed in the context of future aeronautics is that of cellular materials. These are produced by the addition of compounds which evolve gas in the melt, or by foaming an organic material containing a dispersion of fine metal powder: the product can then be fired without melting the metal. The introduction of hollow spheres into the melt is another method that has been tried out. Large pieces of foamed magnesium, zinc, aluminium, nickel or copper are now available, as well as some of the refractory metals in somewhat smaller pieces, but thicknesses of up to 4–6 in (0·1–0·15 m) are typical. In some cases the process can introduce a strengthening dispersion—as it does when oxide is produced in foamed lead. Whether foamed metals will have the structural reliability of other forms is not yet established, but they are generally good under impact conditions, suffering mainly from adverse corrosion properties, having a large surface area, the majority of which is difficult to protect with any confidence. But the use of slabs bonded to external, protecting, sheets may provide the combination of stiffness and lightness which seems most worth exploiting at this stage. Foamed ceramics have also been produced. Prices of about £1 per lb (£2 per kg) for the more common metals are typical at this stage—the refractory metals running very much higher. Apart from special energy-absorbing applications, and use in space structures, materials of this type may have advantages as electrodes in fuel cells, as thermal insulation and for a variety of other special purposes.

The need for large forgings (for example, in titanium) will bring larger presses into being; electrochemical machining will gain ground—it is often more economic to produce special material in bar form and machine to component shape, than to use some kind of casting technique and machine finish, apart from the improvement in properties which some other route may develop; laser drilling, hydrostatic extrusion, high energy-rate forming, thermomechanical processing (ausforming and related processes), electron beam welding, (used, for example, to join space structures in vanadium-niobium alloys) are all varied aspects of some of the changes that are coming about in materials technology. Each of these subjects is a study in its own right, with quite different applications (and difficulties) with different materials, and generalisations can be extremely misleading. But many of these processes do open the door to the more economic exploitation of materials which would previously have been regarded as too difficult to fabricate.

A general problem which confronts all composite construction, whether this be by simply cladding or by an internal reinforcement, is that of joining dissimilar materials. There is probably no single materials field which dominates the future of aeronautical technology so strongly as this. Many of the present difficulties can be traced back to the inadequacies of bonding, or to the deficiencies, of one kind or another, in the interfacial properties of multi-component systems: the spalling of coatings on refractory metals, for example, or the chemical degradation of an adhesive or the difficulties of wetting a particular material with a suitable metallic matrix. The chemistry and physics of interfaces are subjects which have fortunately taken on a new life in recent years, and there is much to be learned from the contributions they have made to understanding. The practical production of many of the interesting forms is, however, still a major difficulty. Testing complex components, made up of different materials also raises a new set of difficulties: a lack of bonding over a small area may have serious consequences, but be difficult to detect by present-day techniques. The specification and the quality control of such products are subjects which will need a great deal more attention before such materials can be confidently brought into use. A greater exploitation of electrodeposition and electroplating processes also looks likely, not merely to apply simple surface coatings, but to deposit alloys, particularly with graded composition to match boundary requirements and in combinations with other materials: in other words, an exploitation of electrodeposition within the general scope of composition production has considerable potential, bearing in mind that final, shaped components might be produced by such a means. It happens that considerable developments are taking place in this field at the present time, and always provided that the processes turn out to be reasonably economic, there is no reason why new and specially designed composite materials should not be produced by such means.

Finally, we should note the potentialities of new fabric forms of materials, for example, those woven from high temperature wires, or from graphite, both as components for composites, and for special purpose protection.

6 THE LONG-TERM OUTLOOK

The foregoing is not a comprehensive review of all the developments taking place at the present time in materials. Many classes of material have not been referred to at all, and in this final section the opportunity will be taken to look at some of these in assessing the more important trends.

Steels are important to aeronautics, both structurally and in the form of engine components, and for supersonic aircraft beyond Mach 3 they are important as possible skin materials, so that developments have a considerable bearing on what is possible in the future. As with other materials, the strengths available with steels have risen markedly in recent years, and there is little doubt that over the next fifty years steels with very high strengths indeed, exceeding 300,000 psi (2·1 GN/m^2), will be available. The important classes are the low alloy ausformed steels (which exemplify the possibilities of thermo-mechanical treatments) and the high nickel maraging steels. These can readily be formed, and they are weldable, but are expensive and have low corrosion resistance. Stainless steels have important possibilities as skin materials, and have already been used in experimental aircraft for this purpose. The question is whether, once the capabilities of titanium alloys have been fully taken up, it may not be worthwhile to jump direct to the nickel-chromium series which offer a much larger margin of temperature.

The promise of higher temperature polymers has been with us for some time: so far there are few commercially available products that will withstand temperatures very much in excess of 200°C for any useful engineering time, but clearly this situation will change, and materials like the polyphenylene oxides and polysulphones will operate at much higher temperatures. The promise of good materials based on fluorine or boron also remains. The development of successful high-temperature elastomeric materials, resistant to the fluids they are likely to encounter such as oils and greases, is one of the very critical aspects of high-speed aeronautics, and a field which underlines the immense influence that aeronautics has had in pushing up the capabilities of materials.

Aeronautics will progress over the next century by bringing into use a much wider range of materials, many of which will appear in composite forms. Aluminium alloys will remain for many applications, improved in their stress corrosion behaviour, but replaced in part on the one hand by reinforced plastics and on the other by fibre-reinforced aluminium. Some parts will be taken over by titanium and its alloys, which will in any case go ahead very fast in the next generation of transport aircraft, both structurally and in engines. There is still plenty to do in titanium alloy development in finding optimum alloys for particular applications. Refractory metals, starting with chromium alloys in engines, and extending to protected niobium and molybdenum, will advance into hot structures (in hypersonic vehicles, for example), into propulsion systems, and design will need to take account of the time-dependence of their behaviour. Parts of structures and propulsion units will need to be fabricated in terms of components and inserts which can be readily and cheaply replaced. For the extreme temperatures, and the thermal cycling conditions, which the leading edges, and the nose, of a very high speed, or re-entrant vehicle will experience, it seems unlikely in the present state of knowledge that any material will meet the requirements without a quite different approach to maintenance. Cooling systems will, of course, help.

For subsonic aircraft, the critical factors are economics, reliability and safety. Ways of continuously monitoring critical components, and of assessing the information derived, have a direct bearing on the detection of early fatigue, or stress-corrosion, damage, and it is possibly in this area, coupled with even more successful ways of reducing fatigue incidence by both metallurgical and engineering means, that a real advance could be won. Ideally, also, the structure should absorb as much energy as possible under the impact conditions, and real improvements in this direction are likely to be gained by re-design and by combining different material forms.

If aeronautics over the next century is not to defeat its own ends by making development impossibly expensive, there must be a better reconciliation between the activities of materials development as such, the procedures of material testing and evaluation, materials processing and fabrication, and aeronautical design. The subject is too complex and detailed to review in a tailpiece to an article of this kind, but in brief, advances are needed in the ways in which the important engineering properties of materials are characterised. These are inadequate as they stand in dealing with much of the complex judgments which have to be made; very large (and expensive) test programmes have to be undertaken to provide the basic data, and such data are often quite unconnected with any of the more basic properties with which fundamental research is largely concerned.

Materials engineering has not yet found its proper form and while much of its empiricism is admirable we must expect in the future considerable changes in the way in which the area between materials research and development, and aeronatuical design, is filled in.

Bibliography

In a subject as broad as the one under review the published work to which reference could be made is extremely extensive. The following provide some introduction to the various topics and, in turn to associated references.

1 *Selection of Materials and Design.* Institution of Metallurgists, Iliffe Books, 1967.
2 *Composite Materials.* Institution of Metallurgists, Iliffe Books, 1966.
3 LUI, K., GUIDOTTI, R. and KLEIN, M. Research and Development of magnesium/aluminium electroforming process for solar concentrators. NASA CR-66427, 1967.

4 WIMBER, R. T. and STETSON, A. R. Development of coatings for tantalum alloy nozzle vanes. NASA CR-54529, 1957.
5 NEGRIN, M. and BLOCK, A. Protective coatings for chromium alloys. NASA CR-54538, 1967.
6 REICHENECKER, W. J. and HENSCHKEL, J. NASA Contributions to metals joining, NASA SP-5064, 1967.
7 AVERY, D. H. and BACKOFEN, W. A. Structural basis for superplasticity. *Transactions Quarterly, American Society of Metals*, Vol **58**, p 551, 1965.
8 HOLT, D. L. and BACKOFEN, W. A. Superplasticity in the Al-Cu eutectic alloy. *Transactions Quarterly, American Society of Metals*, Vol **59**, p 755, 1966.
9 MORRISON, W. B. Superplasticity of low alloy steels. *Transactions Quarterly, American Society of Metals*, Vol **61**, p 423, 1968.
10 STEPHENS, J. R. and KLOPP, W. D. Ductility mechanisms and superplasticity in chromium alloys. NASA TN D-4346, January 1968.
11 UNDERWOOD, E. E. A review of superplasticity and related phenomena. *Journal of Metals*, Vol **14**, p 914, 1962.
12 COHEN, L. A., POWER, W. H. and FABEL, D. A. New metal foams. *Materials Engineering*, April 1968.
13 Cellular metals. *Product Engineering*, 9th October, 1967.
14 FLINT, E. P. Lightweight aerospace materials. *Materials Science Research*: Vol 2 (Editors Otte and Locke) Plenum Press, 1965.
15 ALEXANDER, J. A., WITHERS, J. C. and MACKLIN, B. A. Investigation of three classes of composite material for space vehicle application. NASA CR-785, May 1967.
16 KORB, L. J., GATZEK, L. E. and OLSEN, R. P. Advanced Spacecraft Structures; a forecast. ASM Westec Conference, Los Angeles, March 1968.
17 WHITE, R. C. New fibers for structural reinforcement. ASM. National Metal Congress, October 1967.
18 Technology Forecast 1968. *Metal Progress*, January 1968.
19 KENNEDY, A. J. Technological forecasting for metals. University of Bradford. National Conference on Technological Forecasting, Harrogate, July 1968.
20 Trends and Forecasts for 1968 and far beyond. *Product Engineering*, 28th January, 1968.
21 MCHENRY, H. I. and KEY, R. E. The F-111 logic. *Metal Progress*, March 1968.
22 HARRISON, J. R. Welding the critical spacecraft structures. *Metal Progress*, March 1968.
23 DENKE, P. H. Materials for the supersonic transport. *Materials Science and Technology for Advanced Applications*, Prentice-Hall, 1962.
24 EMERSON, H. C. Welding and brazing of space age materials. *Materials Science and Technology for Advanced Applications*, Prentice-Hall, 1962.
25 MANG, W. G. High energy-rate forming of space age materials. *Materials Science and Technology for Advanced Applications*, Prentice-Hall, 1962.
26 MASH, D. R. Plasma-arc spraying of space age materials. *Materials Science and Technology for Advanced Applications*, Prentice-Hall, 1962.
27 HOFF, N. J. Materials problems areas peculiar to supersonic aircraft. *Materials Science and Technology for Advanced Applications*, Prentice-Hall, 1962.
28 SCHIFF, D. Pyrolytic materials for re-entry applications. *Materials Science and Technology for Advanced Applications*, Prentice-Hall, 1962.
29 FREUDENTAL, A. M. Aspects of reliability under conditions of elevated temperature creep and fatigue. *High Temperature Materials and Structures*, Pergamon Press, 1964.
30 SCALA, E. Materials aspects of the re-entry problem. *High Temperature Materials and Structures*, Pergamon Press, 1964.
31 DIFFENDORF, R. J. Refractory pyrolytic materials. *High Temperature Technology*, Butterworths, 1964.

32 FULRATH, R. M. Controlled microstructure of refractory bodies. *High Temperature Materials and Structures,* Pergamon Press, 1964.
33 HESSING, H. Materials requirements for space propulsion. *High Temperature Materials and Structures*, Pergamon Press, 1964.
34 SRP, O. O. and GEYER, N. M. Protective coating requirements and recent developments. *High Temperature Materials and Structures*, Pergamon Press, 1964.
35 SCHWARTZ, H. S. New materials and physical constructions for ablative use. *High Temperature Materials and Structures*, Pergamon Press, 1964.
36 KENNEDY, A. J. *The materials background to space technology*. George Newnes, 1964.
37 ZACKAY, V. F. (Editor). High strength materials. *Proceedings of 2nd Berkeley International Materials Conference*, John Wiley, 1965.
38 OTTE, M. O. and LOCKE, S. R. *Materials science research*, Vol **2** (*Proceedings of Conference on Advances in Aerospace Materials*, Orlando, 1964). Plenium Press, New York, 1965.

16 Education and training in the Second Century

Joseph Black

I INTRODUCTION

Why include the topic of education and training, it could be asked, in a book devoted to predicting ahead to the Second Century of the Royal Aeronautical Society. It is obvious that in the other fields covered there is bound to be great technological progress but men, it might be thought, will teach, and learn, and study much as they do today, and as they did yesterday. Thus we could advise our student engineer that:

'the acquirements which are necessary to enable the individual to distinguish himself, or even to practise his profession with a moderate chance of success are partly abstract and theoretical, and partly experimental or practical. An engineer should, in addition to the knowledge required to fit him as well as others for the active duties of life, have such a knowledge of mathematics as will enable him to investigate as well as to apply the rules laid down by writers on those branches of the mixed sciences to which his attention will most frequently be drawn. He should be well acquainted with the principle of mechanics, hydraulics, and indeed with all the branches of natural philosophy: he should be able to draw and understand the principle of engineering drawing. Having acquired the requisite amount of theoretical information the next step is to gain that practical knowledge which is essential to the proper application of this information . . .

'the best mode to gain experience is to enter the employment of some eminent man in the profession, in whose office there will be every opportunity offered to the young engineer of witnessing the mode in which work is carried out. He will be employed first as a draughtsman: as he becomes more acquainted with details he will have more responsibility thrust upon him . . . for works as will require all his theoretical knowledge . . . he should cultivate a habit of observation taking ample notes and sketches . . . he may work his way up in a profession where merit alone can lead to distinction.'

—advice which is as relevant today, as it was when written in 1849[1].

Originally published in the *Aeronautical Journal of the Royal Aeronautical Society*, February 1969.

It is worth recalling that the Centenary of the Royal Aeronautical Society coincides roughly with that of the provision of public education in England (the Education Act 1870): even the most ardent supporters of that Bill would not have predicted an annual budget on schools, further education, colleges and universities of over £2000M within a century.

It could be argued that many fundamental ideas and methods will remain unchanged over the next hundred years. Applied mechanics based on Newtonian concepts will still serve engineers well—and they could still be learnt using traditional methods of lectures, classroom tuition and tutorials. Some concessions to course structure and curricula changes, to cover topics raised in previous chapters by, for example, A. V. Cleaver (chapter 5), D. Küchemann (chapter 2) and W. G. Molyneux (chapter 7), would suffice to keep our education up to date. In support of this argument that, fundamentally, education changes much more slowly than the science and technology which it helps to produce, we can recollect that Mach published his shock-wave schlieren photographs in 1898: his work only appeared significantly in gasdynamics in the late 'thirties, and did not come into aeronautical engineering courses until post-war, nearly sixty years after publication.

But to take the view that in spite of great technological advances in all fields there need only be modest changes in educational content would be looking back with nostalgia to the past, rather than forward with excitement to the future. There is still some justification for the taunt that education today is the last stronghold of the 'cottage industry'.

We have, therefore, as predictors to examine the scale of education and its methods. We must question whether we can begin by extrapolation from our traditional structure involving school, college, university, research-institute, training department, a structure which itself has undergone startling changes even within the past two decades. The boundaries between these institutions are likely to be much less rigidly defined in the future. To maintain an *open-ended* approach in our discussion we must use terms such as *teacher*, *learner*, and *institution*, since *university*, *polytechnic*, *apprentice-school*, *training department* may become much less significant: the student post-2000 AD will likely spend periods of education and training in all of these and in new organisations we cannot yet envisage.

2 THE AIMS OF PREDICTION

One approach to predicting the future in education is to extrapolate fairly from the present situation, taking into account the predicted aerospace developments. Legg did this well in 1966[2], basing his estimates of student numbers, cost and organisation in AD 2016 on economic factors such as population growth, defence expenditure as a percentage of gross national product, traffic growth, and on such social factors as trends in higher education, numbers of post-graduate students required and related requirements.

Such an approach, however, while a valid one for educational planners concerned with the next few decades, is based on an assumption that the present system and techniques will continue virtually unchanged into the future.

It would thus be a limited prediction, since it would ignore the much bigger question of human intellectual developments, the impact of new technologies on behaviour and response and the sociological conditions. It would be as if in examining the stability of an aircraft we still today centred the analysis on the human behaviour and strength of the pilot, and made the aircraft respond in a manner to suit him, rather than considering the pilot-plus-aircraft as a complete system with power, feed-back, automatic stability, and possibly self-adaptive control, with the human being now merely part of a 'loop'. Such an analysis does not detract in any way from the dignity, ability or judgment of the pilot as a human being, but relates him more accurately to the demands of the particular situation.

As forecasters trying to look much further into the dim reaches of the future we must concern ourselves more with an enquiry into the nature of education and training itself, and how it will be affected by the technological advances which, in large measure, it creates through the efforts of the people which are produced by the self-same education. We must regard education and training as a special form of communication from generation to generation of the vast and ever-growing knowledge and wisdom starting from the distant past—from the loom of language and the first use of tools.

As Ashby pointed out[3] there have been four intellectual revolutions in this communication:

education within the family;
the adoption of the written word (vehemently opposed by Socrates*);
the advent of widespread printing;

and now the fourth,

the technological, based on electronic recording, television and the computer.

Thus, before we get immersed in the detail of course content and structure we have to attempt to predict the *learning* situation in which the student of post-AD 2000 will find himself. Socrates could not have predicted what the effect of writing would be: Gutenberg could not have predicted what effects his invention would have on education, with libraries as vast store-houses of knowledge. We have the advantage as

*'For this invention of yours will produce forgetfulness . . . . neglected memory . . . . with written discourses you could fancy they speak as though they were possessed of sense, but they will ever repeat the self-same story.'

forecasters that already, technologically, we have achieved enough to enable us to perceive what changes may come in the next great step forward. As Cleaver observed[4], Cayley would find no difficulty in understanding the modern jet aircraft and its means of propulsion, but he would be utterly baffled by its radar.

By analogy we are at the 'Cayley' stage in education, and if taken by Wells' Time Machine to AD 2066 we might recognise some basic activities going on in what were still recognisable as the *old* classrooms. But we would also visit strangely laid-out buildings with a functional organisation that we could not appreciate; we would also have no comprehension of how the students were learning and communicating since they would be using the future equivalent of technological advances comparable to that of radar in our own time.

Before we consider the specialised aspect of future aerospace education (if, in fact, such a concept will still exist) we must bear in mind that engineering education cannot be divorced from overall educational evolution. This must include the schools where already new mathematics, technology projects, computer work, and a heuristic approach to science learning are under way. By AD 2000 the school-leavers will be the products of a technologically based system: they will not find acceptable a further education which is traditional, includes much unnecessary drudgery (disguised as good mental training) and does not sustain the excitement in the pursuit of knowledge and achievement of purpose which the schools have initiated. For example, a school syllabus for the 'Man-made World' includes such material as 'decision-making', 'optimisation' and 'other theories and techniques which contribute to our technological civilisation'[5].

3 EDUCATIONAL TECHNOLOGY

Teachers have used technical aids to enhance their teaching since time immemorial, from fingers in the sand, to the slate, the blackboard and chalk, and the familiar slide projection of stills, and silent and sound films.

More recently, on the visual side, the overhead projector, with its large illuminated surface and convenience of display both for transparencies and models has come into common use. Another recent addition is the closed-loop 'concept' film in a permanent cassette easily inserted in a back-projector, particularly convenient for student self-instruction.

The post-war advent of magnetic tape recording was first exploited educationally in the language laboratory, although it soon became useful in other applications. For example, inexpensive combination of laboratory instruction recorded on tape, and played back together with a coupled automatic slide projector, operated by pulses recorded on the tape, can improve the quality of instruction and free the teacher of a wasteful repetitive role[7].

A decade ago came the much more powerful and expensive closed-circuit television (CCTV) system which offered the opportunity of the overflow lecture, and gave a whole class the chance to look at a magnified image, to look down a microscope or through a telescope, to get close, and to penetrate into the regions of hazard or danger, and to see and hear the teacher as if he were talking directly to the student.

In addition to its complexity and expense, however, a drawback to the widespread use of CCTV has been the limitation of only being able to transmit live with no possibility of repetition. In this respect it could not compare with the ease of repetition of the film. However, with the successful development of video-tape recording, and particularly with the most recent achievement of making a range of recorders which are compatible, i.e. tapes made on one can be played back on another, a whole new world of educational opportunity is opened up, since television programmes can be recorded, played back and exchanged.

Thus, for example, lectures by distinguished experts can be recorded for play-back to lecture audiences elsewhere: we might anticipate that outstanding events like the Wright Brothers Memorial Lectures, both in the UK and the USA, would be recorded and played back to the Branches of the Royal Aeronautical Society.

Even as recently as 1965 the first definitive report on these new developments described them as 'audio-visual aids'[6]. As such, the teacher selects the material to be shown or heard to enhance his programme of teaching.

It has become clear, however, since the introduction of these devices that they are going to be much more important as learning aids rather than for teaching, in that it is the student who now has the power of recall of recorded material which can help him to learn—whether it be a mental process, a language, a skill, a craft, or a sport.

To exploit this potent new resource for learning demands that the whole educational process should be reassessed (irrespective of the level) and possibly built around these new powers of communication. Ashby pointed out that 'The characteristics of technologies is that they begin as useful arts without science (the art of healing, the art which built temples and churches, the art which built roads and bridges) and it is only later in their history that scientific principles have been so vigorously applied to them. The art of teaching is one of the last of the useful arts to become technological'[3].

But educational technology is not a mechanistic concept of teaching in which the timid introduction of some ill-assorted pieces of projection equipment used with cursorily selected and ill-co-ordinated visual material wholly replaces the teacher, and deprives the student of the intimate reaction of a guiding and encouraging mind[8].

It is rather an application of systems engineering into education, which means:

(*a*) the systematic analysis of the appropriate *learning* situations,

(*b*) the definition of objectives and the aims of the task being undertaken,
(*c*) the selection of appropriate means and the testing of these methods to evaluate the degree to which the end results aimed at are achieved.

From this approach two systems can arise:

(1) open systems which take the best model available based on experience with a pilot sample of learners, and assume that they will suit a majority of learners in similar situations (this is what good teachers have done intuitively since the beginning of education).

(2) closed systems which include devices whereby the learner modifies the system according to his progress.

The systems approach can produce new methods: its stress on objective definition, evaluation, and pragmatic adjustments balances those older ideas which are based on philosophic, psychological, or social arguments, or even simply the 'hunch' of the inspired teacher.

4 PROGRAMMED LEARNING

This emphasis on learning rather than teaching and on systems leads us to think more about the individual rather than the group. It encourages us to define our objectives much more clearly, and having done that, to plan a route which will lead the learner to their attainment.

There can be two types of route: a linear one, i.e. a step-by-step for all students following the particular programme, each proceeding at his own rate of success, and a branching route, in which individuals break off and follow branch routes at those points where they encounter their individual difficulties, and then return to the main route. In both routes the number of new ideas, or concepts is carefully thought about in the preparation of the programme, and its division into frames: at appropriate stages in the allocation of concepts there is evaluation and correction before the student proceeds to the next stage.

Programmed learning is coming into extensive use, particularly in industrial training, and much research, often associated with psychological research, is now devoted to it[9]. All learning activities, whether they be to learn engineering science subjects, to trace a fault in electronic equipment, or the complex task of pilot conversion to large jets, can benefit from this disciplined approach.

There are limitations of course, until we understand how the human brain codes information. There is the danger that 'it only rewards those responses which are in agreement with the programme . . . the doubter, the dissenter, the questioner . . . anyone with an original mind can get no stimulus or satisfaction'[3].

In spite of these reservations, however, it is certain that with the amount of research being devoted to programmed learning in all its aspects, the teaching material of our Second Century students will be presented to them in a very much better organised manner, and the step-by-step assessment will have made our present formal examination systems redundant, as we shall discuss again later.

5 IMPACT OF THE COMPUTER

The greatest single influence on education and indeed on every aspect of life in the Second Century of the Royal Aeronautical Society will undoubtedly be the electronic digital computer. Its present power, and the vastly greater power which it will possess in the future will force new attitudes and will have a profound impact on the way man thinks. The possibility of designing machines which are more intelligent than ourselves is already being discussed. According to Sutherland[10] there are all sorts of biological limitations on our own intellectual capacity ranging from the limited number of computing elements in our craniums to the limited span of life and the slow rate at which incoming data can be accepted. There is no reason to suppose that such stringent limitations will apply to computers of the future. For example if 10^9 bits (i.e. binary units) is a reasonable estimate of the amount of information stored in a human lifetime, it would take only about 20 minutes to transfer a lifetime's experience from the memory of one computer to another, using existing techniques.

The demands of advanced engineering analysis, particularly in aeronautical and space endeavours, lead to remarkably quick acceptance of the computer, but as a rapid desk calculator—an attitude damned by Argyris, 'which dooms its holder to computer solutions no more imaginative than those obtained by such machines'. Babbage and his associates had a more imaginative perception of the hoped-for use of their computing engine over 100 years ago!

Argyris has demonstrated clearly how necessary it is to utilise the computer-aided engineering analysis and computer-aided design[11]. His view is that the latest developments in man-and-computer technology will lead us to deal with computers as 'research colleagues rather than machines'. He points out that if engineering consequently becomes more of a science than an art, due to the rigorous discipline imposed by our new colleague then we will have to teach engineering in a different way—'the educator will have to become more involved with the computer . . . This can bridge the gulf between specialisations, since the same sub-routine will solve many problems. Some of the old specialised training might be retained to give historical continuity and teach intuition'.

But if the computer can have an impact on the complex processes of engineering science then will it not also be capable of converting the art of teaching to much more of a science? The role of the computer in the educational and training scene of post-AD 2000 will go, I think, beyond Argyris' engineering colleague. It will be

(*a*) an integral part of an adaptive learning system,

(*b*) a testing and monitoring device in conjunction with programmed learning packages,
(*c*) a tool for simulating scientific experiments and modelling operational exercises,
(*d*) a basis for an information dissemination and retrieval system,
(*e*) an administrative and organisational tool for the whole educational process from admission through assessment to completion.

Already, in one way or another some of these aspects are in operation: What we now see is that a computer-based learning system is needed to bring all these together; while the concept is far-reaching the technology is under way. In the USA government agencies, foundations and industry already spend over $100M a year; in the UK following a working-party report[12] a large-scale feasibility study is being made under the aegis of the National Council for Educational Technology, with the helpful co-operation of the Ministry of Technology. (The NCET was the body set up by the Department of Education and Science following the Brynmor Jones Report[6]).

Computer-based learning, or computer-aided instruction, is, in its infant state, limited in scope, and very expensive, but the advent of multi-access computers, the vast increase in storage capacity, and the remarkable reduction in cost of a unit of operation with the new generation of computers enables us to predict with some confidence that it will be an economic proposition in the future. It is also safe to assume that there will be a widespread availability of computer resources throughout our technical and social organisations.

It is appropriate that aerospace engineers should be interested in developing this new technology, since much of the engineering innovation required is coming, and will increasingly come, from the advanced demands of new aircraft, missiles and space vehicles. We have already been the pioneers in the concept that where a learning activity in real circumstances is very expensive, it pays to simulate the reality with machine simulation, even though the latter also appears to be an expensive item in itself. This is the situation with the flight simulator, which in many ways embodies much of (*a*), (*b*) and (*c*).

As recently as December 1968, the Director and Chief Training Advisor of the Civil Air Transport Industry Training Board on a visit to the USA reported that airlines there, worried about the high costs of retraining for the Boeing 747 are getting together to share ideas and are moving to a much greater use of simulators. Retraining a pilot of a contemporary jet to fly a 747 could cost £25,000, and about half that figure annually to keep the man in touch.

Some idea of how our students post-AD 2000 will learn can be gained from looking at one of the most developed systems so far available, the PLATO system from the University of Illinois[13]. Each terminal provides the student with an individual keyset, like an electric typewriter, and a television screen. The keyset allows the student to send information to the computer, and the screen presents the information prescribed by the computer program (Fig. 1 (*a*)). The programs are based on a complete set of rules, referred to as a teaching logic: over twenty different sets of logics have been written so far, the types most often used being the tutorial and the enquiry. Fig. 1 (*b*) shows a flow diagram for a tutorial logic.

In a typical enquiry logic on electrical circuit analysis the keys are divided into three groups; for entering constructed responses to questions; student control-keys to answer particular problems; and those which he

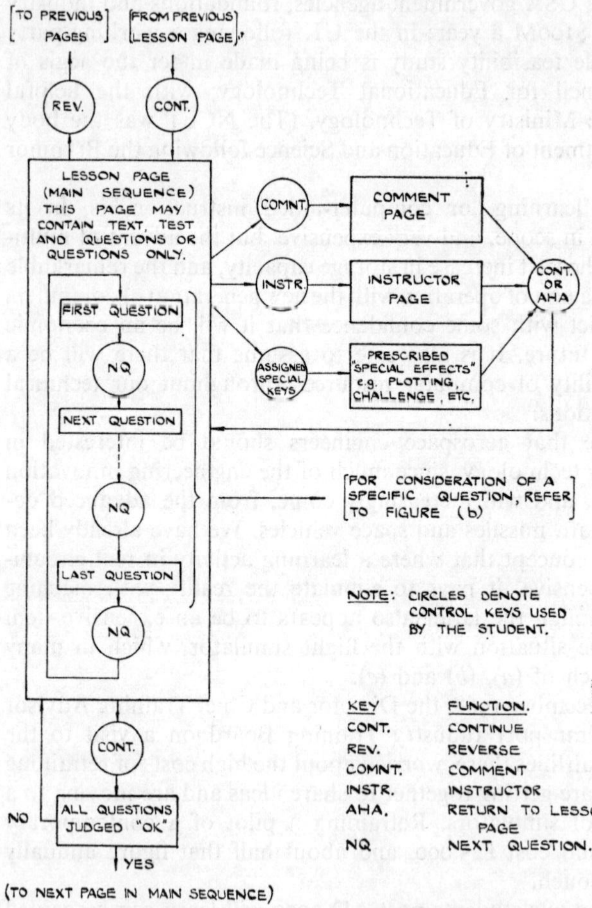

Figure 1 Computer-based Learning Systems. Flow Diagrams for a tutorial logic. (a) (above) with main sequence and special branches, (b) (right) with help branches, for the PLATO system (ref. [13].)

uses to conduct his search for needed information. There are many other systems under development, but all of them call for advances in the man-machine interface, with new types of terminal equipment, visual and graphical display and data selection. There are hints that flat panel displays which exhibit memory and costing much less than conventional cathode ray tubes, based on gas discharge and plasma techniques are on the way. We can expect startling new electro-chemical devices in the next fifty years which will make enormous contributions to this link.

The computers already coming along will cope with many hundreds of terminals so that with the new terminals fully developed, we can foresee

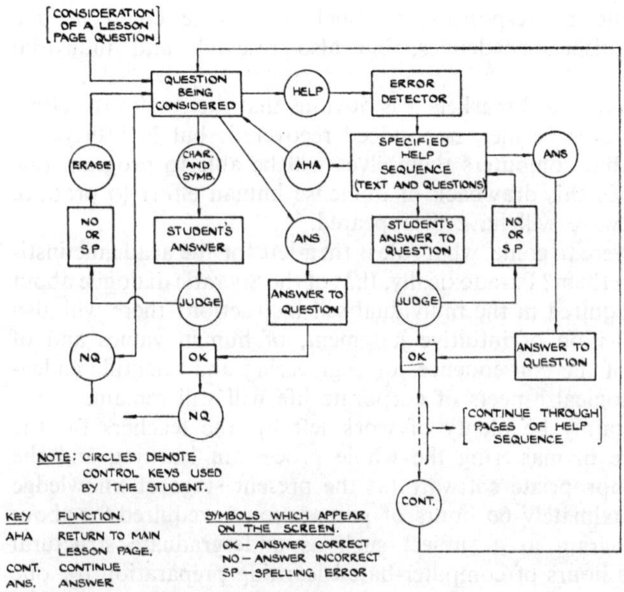

that it will be possible for an individual to dial into contact with highly flexible systems for self-instruction in classrooms, workshop, library, home and hostel. Such systems could have a profound effect on the physical facilities which we have up till now considered necessary for education, e.g., the large lecture hall, with its low utilisation, may disappear in favour of suites of study centres.

It is already clear that from the cost point of view such computer based systems depend greatly on the number of users of any given program: thus at school level the potential utilisation could be massive (for example, the Inner London Education Authority Television Service may eventually be available to almost ½ million children). Can we see, therefore, a big enough demand even in the Second Century for aerospace engineering courses, with such small groups in our teaching institutions? This situation may pose the most difficult policy problem in education in the next few decades: whether the great advantages of a fully worked out system prepared by the world's authorities in any particular subject should be used at the price of a surrender of some of the much-prized autonomy of individual academic institutions. For example, we could visualise a course in aeroelasticity, or orbital trajectories, or structural analysis which would be available as a package deal of computer tape, video-tape recording of experiments, and lectures, flight tests, satellite observations, and related topics: it would also include programmed learning textbooks, complete with tests and assessment schemes. The team which would be responsible for such a package could include not only distinguished academics, but also research and industrial engineers.

Unless there were world markets it is obvious that the cost of the effort required would never be met, nor indeed recovered: but in fifty years' time it is likely that computers themselves will be able to program raw material and hence this drawback of immense human effort to produce the requisite software will have disappeared.

If such means were to come, what role is then left for the academic institution as we know them? Paradoxically, that of the Socratic dialogue about the knowledge acquired in the individual self-instruction: there will also remain the acquisition of intuitive judgment, of human values and of an appreciation of the consequences of engineering and scientific endeavours: the sociological aspects of corporate life will still remain.

There will certainly be plenty of work left for the teachers for the foreseeable future in mastering the whole process of learning and the preparation of appropriate software (at the present stage of knowledge and ability approximately 60 hours of preparation is required for about one hour of program in a subject such as undergraduate structural analysis, and 100 hours of computer-based learning preparation for one hour of learning: hence the need for computer help in programming the material).

One important consequence of the power of the computer to monitor performance, and to provide continuous measurement and correction for the individual student will be the eventual disappearance of the conventional time/test examination. Our successors in the Second Century of the Royal Aeronautical Society will probably look back at our way of determining the success or failure of students as we look back at trial-by-ordeal as a measure of justice!

Information storage and retrieval

The information explosion which we are witnessing will carry on, leading to an ever-increasing rate-of-growth of knowledge, which already shows

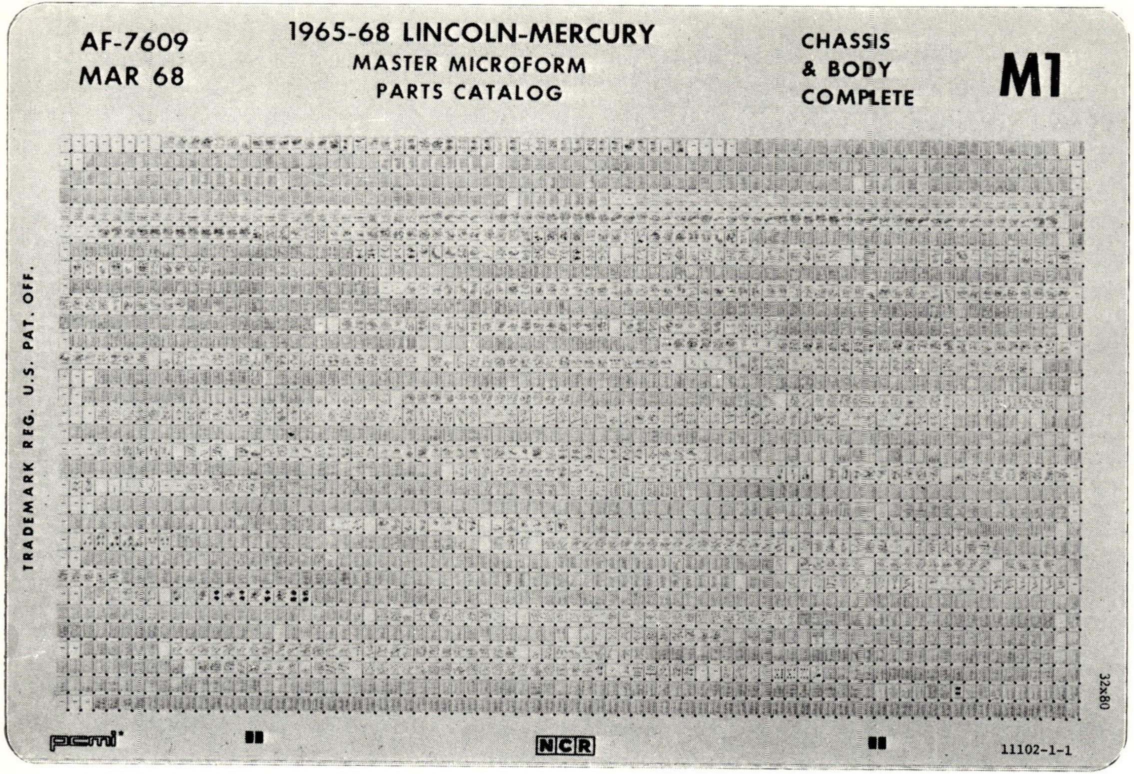

Figure 2 (left). A standard card distributing information contained on over 3000 pages. (above) Location of desired page and reproduction on a bright screen; copies can be taken off any page

signs of overwhelming the library services and resources of educational institutions. There are two problems, the storage of the mass of new material, and the retrieval of all the related information from different sources. It will only be by processing the information to come through computers that the opportunity will be presented for adaptive search techniques, and the organisation of personal files of information for ready access.

The need for subject matter to be presented in textbook form, with its rapid obsolescence, will disappear, as indeed may the technical journals in their present form. A distinction will have to be made between technical information and data, which can be codified, stored in the computer, and printed out on demand, and fundamental concepts and abstract thoughts. The textbook and journal will be devoted to the contemplative, philosophical, conjectural and predictive aspects of science and technology.

If this seems far-fetched, we have already the MEDLARS project under way, with its information storage/retrieval/automatic print-out of medical knowledge[14], and a similar approach is being made to legal knowledge. In the chemical field the Chemistry Society Unit provides a current awareness service based on the computer matching of user-interest profiles with words used in the titles of articles and papers. To the Royal Aeronautical Society, which pioneered the condensing of information on to Data Sheets, this new development should be a welcome challenge.

The means of storing and retrieving information will be considerably improved by technological advances, some already in sight, and others still unknown. For example, with the discovery of photo-chromic dyes a reduction of 200:1 is commercially available. The startling effect of this can be judged from the dimensions needed for the complete part stores catalogue and service drawings for one model of a car, on one 6×4 inch ($0{\cdot}15 \times 0{\cdot}10$ m) card (Fig. 2).

In addition to the saving in storage space, such methods lend themselves to automated information retrieval. Such a service is already available concerned with material selection data*, so that by AD 2000 we can be confident that students and professional engineers will have at their finger-tips a wealth of information, which would require half a dozen personal librarians working for one today.

It is conceivable, therefore, that a joint body of engineering teachers, industrial and research engineers, and members of the technical branch of the RAeS could collaborate to produce on, say, two 6×4 inch transparencies a complete assembly of all the basic equations, test results, material properties, design data, etc., involved in aircraft or space-vehicle engineering. Every one of our Second Century students could then have their own copies, which would eliminate hours of note-taking, and allow the teacher/student encounter to become again an opportunity for discussion, criticism, and inspiration. The lecture-room could become a

*Technical Information on Microfilm (TIM Ltd).

work-room with cubicles containing the necessary reading projectors, and the personal contacts would take place in tutorial conditions in smaller rooms.

It is perhaps worth digressing here to examine a little more what the future role of the teacher will be. Adelson, as a scientist working on a systems approach to future learning, sets out the multifarious roles of the teacher . . . 'he may be an interpreter, a source of meaning for experience and information, . . . a benign critic of performance, . . . a challenger . . . an appreciator of originality . . . a catalyst of thinking.' He considers that only when the teacher is freed from mastery of all the necessary subject matter and 'the dependence of learning on what the teacher knows is minimised' will the teacher be able to perform properly all the valuable roles outlined[15]. With the knowledge and information transfer delegated to the computer and automatic storage, the teacher will be free to concentrate on his real functions, and to establish the necessary personal relationships.

6 SOME OTHER TECHNOLOGICAL ADVANCES

We have already discussed how the widespread adoption of closed-circuit television and video-tape recording will make it easy in the future to have lecture material presented by any outstanding teacher or world authority from academic, research or industrial fields to any university or college audience at times which suit that audience. The full impact of this new availabilty is only being realised by educationalists thinking about the future. At present some modest exchanging of taped material takes place, rather like the borrowing of films, but looking further ahead, the implications are much greater for the traditional autonomous structure of academic institutions. There will come a time when departments, particularly those concerned with technological subjects, will be unable to present to their students from their own resources of staff and facilities all the subject matter deemed necessary and relevant. In these circumstances they will be able to draw on complete recorded courses.

Obviously they would prefer to use courses which they have helped to plan, and in which they have been involved in the preparation. This is bound to lead to co-operative planning preparation, joint use of scarce facilities, and clear definitions of course objectives, which may for the first time force universities and colleges to re-examine their relation to each other, to the industries which absorb their graduates, and the communities which sustain them.

The individually controlled computer learning programs already discussed will of course draw heavily on this world store of video-taped subject matter.

We should not forget, too, that in the world we are predicting there are unlikely to be economic or technical limitations on the number of broadcast channels available: we may expect, therefore, that much educational material of value to our student aerospace engineers will be available to them on national networks: if the timing should not happen to fit the local time-tables there will be no difficulty in recording for convenient play-back (assuming that appropriate copyright arrangements will have been negotiated by then). Already the success of 'O' and 'A' level scientific subjects, of technical college subjects, and advanced technology, particularly in production techniques, broadcast by BBC and ITV, have indicated the promise of longer and more penetrating programmes.

6.1 *Electronic-video recording (EVR)*

The example of the success of the long-playing record for the gramophone suggested the attraction of being able to insert a video-record in a machine which would send a signal direct into a domestic television set.

The demand for the development of a suitable technology came with the advent of the first moon orbiting satellites. The problem there was the transmission of pictures from the back of the moon, when telemetry of signals was inoperative. A combination of television camera, optical recording and electron-beam scanning was successfully developed.

Following intensive research and development, a system of optical-type film reels with miniaturised images, is now established commercially*: the reels pass automatically through a teleplayer in which an electron-beam scans the images, and an audio-visual signal is sent out, comparable to a broadcast one. A reel 213 m long, with tape width 8·75 mm contains 200,000 frames, each of which can be viewed singly if desired.

Hundreds of copies of reels can be printed from program material originally prepared on standard 35 mm or 16 mm optical film, or video-tape. It will thus be possible for institutions to maintain libraries of recorded material available for re-play by individuals or groups.

While the process at this stage will only be economic when the demand for copies of particular programs amounts to over a hundred, we can certainly anticipate that this new ability to record and play-back audio-visual material in combination with standard television sets will have a great impact on future learning systems.

6.2 *Laser-beam communication*

With the massive research and development effort which is being devoted to lasers, it is not too far fetched to assume that beyond AD 2000 it will be possible to have an individual laser-beam link to communication satellites in fixed orbit. These will carry stores of information, and will not simply

*Columbia in the USA (EVR Partnership in UK).

be relay stations: they will be permanent universities available to all who have the appropriate ground equipment*. We might, therefore, anticipate that a class of students in AD 2060 will consist of a number of students in small-group or even individual study carrels (originally a term for a small chamber in a monastery reserved for individual study and contemplation), working on a console with visual display and computer-controlled instruction, transmitted by laser-beam from the appropriate satellite.

The science of holography and associated pattern recognition is bound to have technological applications which will contribute to learning. Just as in cybernetics, where our new knowledge about mechanical systems and their mathematical analysis revealed new knowledge about animal and organic behaviour, when applied to living systems, so we may find that in establishing holographic pattern recognition we will find out how the human brain recognises familiar material so rapidly. Once we knew this, we could then present new material to be assimilated in a more easily recognisable pattern, or system of organisation.

At a more modest level, laser-beam (and microwave) links will allow for permanent connection between regional universities and other institutions, thus overcoming the prohibitively expensive cable linking.

A combination of national microwave communication links with television-telephones incorporating large screens will also facilitate post-graduate seminars, conferences, and perhaps even more important, joint academic-industry collaboration, without the physical necessity of transfer of people and equipment.

Further developments of the newly-introduced electro-writer, a device which transmits by way of telephone lines, writing and drawing, will also lead to much more long-distance lecturing and instruction, without the necessity of expensive television transmission. Such transmission of written material, when combined with future advances in xerography, could obviate the need for note-taking, since as the lecture or discussion proceeds each listener will be provided with continuous print-outs of the material the lecturer is writing on his blackboard—which is now an electronic pad in front of him.

*To demonstrate how rapidly these predictions are becoming realities, since these were written, a paper was presented to the International Broadcasting Convention by W. T. Brandon in September 1968, *Satellite Educational Television Systems for Africa and S. America.* In S. America it had been estimated that the capital cost of a single synchronous satellite broadcasting direct to schools would be between $125M and $180M. Transmission would be in the UHF band and schools would pick up transmissions on a 10 ft (3 m/s) dish aerial system costing about $200 each.

In a recent paper to the AIAA, P. A. Rubin gave an estimate of £5M per annum for the proposed educational satellite for India (one shilling per student), while corresponding figures for Mexico worked out at just over £1·5M per annum (10d per student).

6.3 *Biological advances*

With Biology taking over from Physics as the most rapidly advancing science of the next fifty years there will be a great surge forward in our knowledge of how the human brain and the nervous system work. With such knowledge we should be able to stimulate much mental effort, which in education and training is achieved by our present indirect teaching. We should also be able to communicate more directly with the immense human memory store, which in the majority of individuals is considerably under-utilised.

Already we can see in the development of artificial limbs how new discoveries about myo-electricity, i.e. the minute electrical signals transmitted by the nervous system, are enabling us to couple up the actuation of the artificial device with the thought of the desired movement, rather than a physical act to motivate the device.

Other advances are equally significant: eyesight deficiencies can be detected now by measurement of appropriate brain electrical signals responding to prescribed light pattern flashes, transmitted through surface electrodes.

The reverse process of feeding-in signals through similar electrodes cannot be ignored as a possible method of education, or at least transmission of data into human memory store, within a time-span of the next 100 years. Although indeed a frightening prospect, when the implications for good or evil are considered, it cannot be dismissed when we consider how rapidly in our own time, heart and other organ transplants were carried out, once the scientific knowledge, and technical skill were available, with scant regard for moral issues or social consequences being debated and agreed.

There would be immediate benefits in the man-computer interaction. If the link to the computer were to be directly from human electrical thought-impulses, then truly the computer would be the amplifier of man's intelligence.

The association of engineering with biology and psychology will be one of the great expansion areas of human activity within the next fifty years. Because of its particular nature much of this effort will be appropriate to the aeronautically biased engineer. It is significant that the name Bionics was coined as long ago as 1958 by Steele, of the Aerospace Division of the USAF, to describe a science of systems whose function is based on living systems, or which have characteristics of, or resemble, living systems. Bionics could be concerned with, e.g. the study of turbulence on the surface of dolphins, the emission of cold light by organisms, or the comparison of transducers, neuristors and computers to the human nervous system[16].

The subject is now taken seriously by AGARD, following conferences in the past few years, and there is now a Bionics panel allotted the task of 'the study of living and life-like systems with the goal of new principles, techniques and processes to be applied in man-made technology'.

Cybernetics, the science of control and communication in the animal and machine, even though it dates from Norbert Wiener's classic work of 1949, is scarcely yet established as an undergraduate study. However, much of the contribution of cybernetics to control theory comes into engineering studies as control engineering. In the future, we must expect the engineers to be much more directly involved in what might be termed cybernetic technology, making more use of the scientific advances.

7 AEROSPACE EDUCATION AND TRAINING

Now we can look in more detail at our own field of interest, putting it in the context of the developments outlined above. Engineering education has moved historically from a general course providing a basic scientific and mathematical training, illustrated by engineering applications, to a division into many specialised courses where the particular applications dominate the choice of topics to be studied right from the inception of the course. There have been many variations lying between these two extremes, and the debate as to which is best, and which provides the knowledge and abilities needed at any selected stage of technological progress is a doctrinal one.

In many ways, the dilemma of the advantages of a broad-based generalised approach, or the meeting of real needs which a highly specialised course provides, is met again organisationally. Thus we have many specialised institutions within engineering, the Royal Aeronautical Society being not only one of the oldest, but probably the first to be concerned with such limited activity, but now with the swing of the pendulum back to a desire for a generalised view we have the coming together of the diverse institutions into the Council of Engineering Institutions; this seeks to establish common educational ground. But at the same time to meet the needs of the qualified professional members of the RAeS itself even more specialised groups within the Society must be set up. Thus, the future educational system in aerospace engineering will have to strive to reconcile these apparently conflicting views.

The traditional professional course for an aeronautical engineer has been based on aerodynamics and structural analysis, but already we are experiencing the wider needs of such engineers, and we can predict a whole new range of subjects, and involvement, in the activities which will occupy them in the near, and not-too-distant future. It is impossible to predict the course, or even approximate curriculum, post AD 2000, but we can usefully adopt Ashby's view on a related topic: 'it is possible to make some useful predictions about the *heredity* of universities in the year 2000. It is not possible to predict about the social environment in which they are likely to operate.'

If we add to social environment the unknown state of technology as it will exist, and its acceptance or otherwise by the outside world, it becomes even more impossible. But we can attempt to predict the inheritance which the next four or five decades will present to the engineers of post-2000. We can assume a large-scale space endeavour, released from prestige feats, routine mass-based air transport, well-established fast surface transport, and integrated transport systems.

Thus the future students will still need to acquire a basic knowledge of aero- and gas-dynamics, hypersonics, structural analysis, and dynamics and control, but in the light of the educational technology discussed earlier the treatment of these subjects will have left behind our present classical approach. Because of the computer-based approach, much of the subject matter will have a unified treatment. It is likely, indeed, that new mathematical methods will produce a universal engineering. One attempt at this, called Diakoptics by its author, G. Kron[17], is based on algebraic, numerical, and topological solutions by network and computer: though originally applied to electrical engineering there does not appear to be any limitation to its application to aerodynamic, mechanical or structural problems.

They will also have to learn about relativity, celestial and orbital mechanics, adaptive control, man-machine interaction, bio-mechanics, cybernetics and bionics. Another field of great importance to our students fifty years hence, in view of the expected large scale of space-exploration programmes, and airline passenger numbers comparable to present-day train numbers, will be a professional study of administration, management, and finance.

Design as a highly organised academic discipline will also bring much greater emphasis on synthesis in place of traditional analysis. With the increased knowledge coming from psychological research and cybernetics the methodology of design will be much better understood, and there will follow from it much more teaching of problem-solving, and the encouragement of innovation. The computer will make a massive contribution, with light-pen, touch-wire and as yet unforeseen means of communication between designer and computer established as standard teaching equipment.

The TV, microwave, and laser-beam links will make joint industry/academic design projects really feasible. This link may in fact lead to the growth of a completely new physical organisation in which the activities will have merged to such an extent that a university or college department as such will not be distinguishable from a research or design out-station.

Production methods will demand engineers highly educated in scientific principles on the one hand to cope with the 'astonishing variety of novel techniques of metal working, shaping and forming . . . high energy rate forming . . . joining using electron, laser, plasma or ion beam welding . . . material removal using chemical, electro-chemical, ultrasonic, plasma-torch and spark erosion techniques' as Molyneux[18] pointed out. The sheer range and high cost of the new techniques to come will demand production engineers with extensive computer and operational research knowledge, in addition to financial and administrative skills.

Anglo-French projects such as Concorde have demonstrated the advantage of having professional engineers who possess at least a working knowledge of a foreign language; this has stimulated the innovation of one or two courses already in which engineering studies are combined with the study of at least one foreign language.

In addition to its vocational use, such a language study has a broadening effect on the intellectual development of the engineer, since through the language he learns more about the country and its people.

With a growth in collaborative projects, and the development of new techniques for learning foreign languages there will be a great expansion in the language element of the engineer's education. Quite considerable periods of a student's training will be spent abroad, and indeed we can hope that the aerospace engineer post-AD 2000 may be the equivalent of the present-day international civil servant, able to work equally well in any member state of a United Europe or America.

The penetration of aeronautical techniques into other fields such as gas-turbine trains, linear motor drives, hover devices, and all forms of automotive transport and oceanography may paradoxically lead to the disappearance of the concept of an aerospace engineer, and the courses of study may be distinguished simply under headings such as energy conversion, propulsion and systems.

Another likely new subject will be the science of forecasting the future; what we have been doing in this book on an intuitive basis, will have become, under the stimulation of the development of mathematical modelling, simulation and gaming, a discipline in its own right. Since by AD 2000 any large-scale technological endeavour is likely to absorb a large proportion of national income and effort, considerable project forecasting will be demanded before it is embarked on. A start in this direction has been made in the Ministry of Technology, who have set up a Programme Analysis Unit, whose task is to survey ahead potential developments in various technologies over the next few decades, e.g. oceanography, or computer-based learning systems (as described above). Some US universities have set up study-groups to predict, on a scientific collection of available evidence, the likely technological, scientific, and social conditions in 2000 AD.

8 THE EVOLVING FORM OF COURSES

Some hint of the form of courses comes from the latest analysis of career structure and needs in the Swann Report[19]. This shows that specialisation easily leads to fossilisation, and the faster the pace of change the sooner the specialist's skill is out of date; it follows that the aerospace engineer will need, and must be prepared for, several spells of retraining during his career if he is to remain a useful member of the profession. The logic of this dictates that first degree or diploma courses must be broad-based, so that irrevocable choice is postponed. Many short courses will be required subsequently to acquire particular skills at intervals during any one career.

While many such short courses have come in within the past decade, they have been a minor activity mainly of interest to young graduates. In the long-term future we can expect such courses to be an important function of universities and colleges, comparable in scale to their teaching of undergraduates. These institutions will no longer be the exclusive preserves of the young; they will be closely linked to the community, providing many services both for professional, vocational and leisure needs.

9 RESEARCH

The universities have been deeply involved in research and innovation since their inception, and we can safely assume that their role in this respect will continue unchanged. But what about the scale of research in the next hundred years? We can clearly see the growth of 'big' science, to a size where already the size of a facility is so great that nationally there can only be support for one or two, based on selected institutions (e.g. radio-telescopes for radio-astronomy at Jodrell Bank, Manchester and at Cambridge), or where two or three institutions must share one, e.g. the reactor at Daresbury shared mainly by the Universities of Manchester and Liverpool. The ultimate in size demands international collaboration, as with the accelerator at CERN in Geneva, involving the UK, France, Germany, Italy, and Switzerland.

In the computer field the only means to provide adequate-sized machines has already led to the establishment of the National Computer Board, with large regional installations shared by five or six universities and colleges. The evolution of this forced collaboration must be in the direction of other shared facilities, either in the form of large specialised institutes attached to a parent institution, or the creation of new centres detached from them all. In parallel, existing research institutions will provide shared facilities for post-graduate research. Some hint of this can be seen today in the use of the Atomic Energy Authority generator at Culham with different universities utilising various allocated particle tracks, and in the use of facilities at the College of Aeronautics at Cranfield for both undergraduate and post-graduate work by many university and college departments. Because of the new visual and computer links we anticipate, physical separation will be comparatively unimportant in the future.

In space research, too, up to six institutions already co-operate to mount experiments on a capsule—a British one to be launched by an

American rocket. We can expect much greater international co-operation which will affect university autonomous research.

If we examine much of present aerospace engineering research, such as gas dynamics, hypersonics, cryogenics, cybernetics, electronics, or materials science devoted to exotic materials either for construction, or for fuel, it is striking to note how much of it would have quite properly been considered the domain of classical physics before the advent of nuclear and solid-state physics drew the majority of the physicists away. In a like manner we can assume that much of present day space physics will have become the normal R and D activity of aerospace engineers.

The whole nature of experimental work at university and college level is likely to change in the next fifty years. Much of the effort now devoted to it is routine and uneconomic in terms of results achieved, and new methods particularly of simulation may teach techniques to students more logically and rapidly than experiments on the real apparatus. The ever-increasing power of the computer to deal with non-linear mathematics, to present visual display, and to extrapolate over a wide range of parameters once given a basic solution, should obviate much of the extensive testing at present absorbing a great part of university research effort, and allow for more innovation and exploration of new ideas as the role of university research.

The consequence may be that the large undergraduate teaching laboratories, housing a wide range of equipment, will dwindle in favour of simpler project laboratories very well equipped with computer and data logging and processing equipment. This will be accentuated by the growth of interdisciplinary research projects which will involve a wider range of techniques and disciplines.

Thus much of post-graduate aerospace engineering research will not only be extra-mural to the universities, but will be extra-terrestrial. Individual university laboratories will survive for the innovator and initiator of new techniques, but the rate of communication of knowledge will be so rapid that any promising hint will be picked up and processed on such a large scale that little will be left to do in the small laboratory. On the other hand the university researcher will be at no disadvantage, since the multi-access computer links, the information retrieval, and knowledge links, will make him independent of his own university's individual resources, and he will be in a position to use large shared facilities.

Inter-university space research satellites will be joint ventures of world universities, manned by teams of researchers and students. These collaborative exercises may be the models for greater political collaboration.

10 POST-GRADUATE EDUCATION

There are two aspects of post-graduate work which will change in the future. The first is the conventional one, in which young graduates continue either to study in more depth, or pursue a narrow research topic. The changes we can expect are:

1 Much less concentration on a two-to-three year period in one place, and much more joint research programmes with the student spending considerable time using extra-mural facilities, either in industry or establishments: this has started, but in two or three decades it could be the standard pattern.

2 Post-graduate study courses shared between centres of excellence so that students enrol on a nationally-based course, and not with a particular institution. Soon these courses could be internationally-based, as transport costs come down. From an investment point of view it is already cheaper to move people to expensive facilities, rather than duplicating the facilities.

The other aspect is the newer one, not concerned with narrow research topics but concentrating on product technology[20]. These courses would channel the generalised knowledge of the new graduate to all aspects of a particular device or product. For example, in the UK we already have centres for machine-tools, semi-conductor devices and fluid power. With the growth in technological products and the demands in aerospace engineering, it is likely that product technology centres will become even more important and will be major educational activities. These will be joint academic/industrial units based either in educational institutions, or located in the industry itself.

Such centres will also provide for the other great educational need of the future we have already mentioned, that of the continuing requirement for the professional engineer to acquire new knowledge and skills throughout his career, as science and technology progress so rapidly.

II INDUSTRIAL TRAINING

The general acceptance of sandwich courses, and other forms of industrial training, together with the establishment of the training boards, such as the Engineering Industry Training Board and the Civil Air Transport Industry Training Board, indicate that training of the professional engineer is now seen to be an equally important part of his career formative years as his academic training.

By the year 2000 and certainly beyond that, the academic and industrial environments may well have merged so much in certain fields, such as design and research, that it is the department head in industry who could be the teacher in certain circumstances, and conversely the academic member who may be directing the project analysis in industry.

The new techniques of educational technology will be equally powerful in giving the young engineer very rapidly a detailed knowledge of industrial processes and organisation. The same techniques of computer-console, visual display and programmed learning, will thus be in use in both academic and industrial contexts.

12 THE HUMAN ASPECT

There is an implicit assumption in all these speculations about the future that aerospace and engineering technology will continue to be an exciting and stimulating activity, to which the most talented of our youth will be attracted, by its intellectual demands and challenges.

But we in the UK and even possibly in the USA may enter the Second Century with the possibility of a serious decline in numbers and standards of those wishing to study technology[21]. There is widespread doubt about the attractiveness of careers in these fields. (It is sobering to reflect that today's 15-year-old will be the responsible engineer or manager of 45 in the year 2000 AD.)

One of the tasks, therefore, for continuing generations of aerospace engineers will be to ensure that technology presents a worthy challenge, and a rewarding career. Society must learn to hold technology in high esteem; technologists must earn this esteem.

13 EDUCATION FOR LEISURE

This is a far-reaching topic but we must allow in our future planning for a combination of more leisure time for the majority of the population, and an ever-increasing desire to study and learn for its own reward—not for vocational or professional advance.

One facet of this desire may be a dramatic expansion of the facilities offered by, for example, such a project as the Open University, operating through broadcast television. One of its major offerings, once it is well-established, could be study courses of social aspects, or historical aspects of technology: in particular, with the fantastic advances in space achievements there will be great public interest, and a thirst for knowledge about the projects, and endeavours.

Thus a new role will be allocated to the academic and industrial teachers, that of mass exposition of aerospace science and technology, as part of a whole involvement in education for leisure.

CONCLUSION

Education at all levels has always been in a state of dynamic growth, and a subject of controversy. To be of any value, prediction in this field must attempt, no matter how poorly, to range widely. Only by aiming in all directions may we succeed in even hitting one target. Should a teacher in AD 2069 bother to search through the strange relics of printed journals, he may be able to see that at least one of their ways of learning was anticipated.

In a book devoted to a historical survey of future societies, Armytage[22] covers the hundreds of man's predictions. He shows in this book how such visions can and do refresh minds for renewed grappling with the present, by arming them with ideas for man's future needs: he indicates that, out of an apparent welter of futuristic fantasies, a constructive debate about tomorrow is emerging, providing us with operational models of what tomorrow could be.

References

1 WEALE, J. *Dictionary of Terms used in Architecture, Engineering, Mining, etc.* London, 1849 (*in possession of the writer*).

2 LEGG, K. L. C. Aerospace Technology and Higher Education 2016 AD. *The Skyward Urge—Aviation* 1866–2016. *RAeS Graduates' and Students' Centenary Symposium.* July 1966.

3 ASHBY, E. Reflections on Technology in Education. *Graduate Journal*, Technicon, Haifa, Spring 1967.

4 CLEAVER, A. V. Astronautics. Chapter 5.

5 The Man-made World. Commission on Engineering Education, Washington DC, 1967.

6 BRYNMOR JONES (*Chairman*). *Audio-Visual Aids in Higher Scientific Education*: UGC and DES. HMSO, 1965.

7 BLACK, J. Use of the Tape-Recorder in Laboratory Instruction. *Chartered Mechanical Engineer*, Vol **12**, No 8, August 1965.

8 DUNCAN, C. J. Towards a Technology of Education. *New Education*, April, May, June 1968.

9 KAY, H., DOBB, B. and SIMES, M. *Teaching Machines and Programmed Learning.* Penguin Educational Special, 1968.

10 SUTHERLAND, N. S. Machines like Men. *Science Journal*, October 1968.

11 ARGYRIS, J. H. and PATTON, P. C. Digital Computers and Engineering—A Look into the Future. *Journal of the Royal Aeronautical Society*, Vol **71**, pp 244–51, April 1967.

12 BLACK, J. (*Chairman*). *Computer-Based Learning Systems.* Report of a Working Party, National Council for Educational Technology. Working Paper No 1, 1969.

13 BITZER, D. L., HICKS, B. L., JOHNSON, R. L. and LYMAN, E. R. The PLATO System: Current Research and Developments. *IEEE Transactions on Human Factors in Electronics*, Vol HPE **8**, No 2, June 1967.

14 SOMERFIELD, G. A. *Computer Based Information Retrieval Systems.* Edited by B. Houghton, Clive Bingley, London, 1968.

15 ADELSON, M. Educational Ends and Innovational Means. System Development Corporation, Santa Monica. California, Rep SP-2293, October 1966.

17 KRON, G. *Diakoptics.* Macdonald, 1963.

18 MOLYNEUX, W. G. The World of the Structural Engineer, Chapter 7.

19 SWANN, M. (*Chairman*). *Flow into Employment of Scientists, Engineers and Technologists.* Council for Scientific Policy, HMSO, 1968.

20 BOSWORTH, G. (*Chairman*). *Education and Training Requirements for the Electrical and Manufacturing Industries.* Ministry of Technology and Department of Education and Science, HMSO, 1966.

21 DAINTON, F. S. (*Chairman*). *Enquiry into the Flow of Candidates in Science and Technology into Higher Education.* Council for Scientific Policy, HMSO, 1968.

22 ARMYTAGE, W. H. G. *Yesterday's Tomorrows.* Routledge & Kegan Paul, 1968.

17 Management of design

J. T. Stamper

INTRODUCTION

If Oscar Wilde had been concerned with this chapter I think he might well have said: 'To be asked to attempt an essay on the development of an engineering specialisation over the next hundred years may be regarded as a misfortune. To attempt to predict the form that control of a complex creative process will take looks like carelessness'. Having undertaken the task, perhaps somewhat carelessly, my first attempts to assemble some thoughts led to the conclusion that I was not only careless but also foolish and as such unqualified to present any views I might have. This chapter is the result of a determination to break out of this unproductive frame of mind and to set down my personal views for what they are worth. I can only hope it has not turned out to be 'A Trivial Comedy for Serious People'.

Before launching into a discussion of the subject it is necessary to define 'Management of Design'. By this I mean the motivation, organisation and effective control of design—or, expanded further, the motivation, organisation and effective control of the relation of knowledge to specific needs in a creative way, in order to achieve a desired objective. In the context of engineering the desired objective will be to provide the community with what it wants, when it wants it and at a price that it can afford, noting that what the community needs is related to what can be provided. In this sense advances in engineering themselves create needs.

The fact that we are concerned with management of design implies problems associated with large groups of people and here I am dealing with complex designs involving the development of a creative environment and the integration of many specialised disciplines into a coherent design activity. Although my thoughts are based on experience in the aerospace industry and are concerned with the future problems of aerospace activities, I believe that most of the conclusions on design are relevant to any complex engineering design and most of the thoughts on management apply equally well to any complex management problem.

Originally published in the *Aeronautical Journal of the Royal Aeronautical Society*, March 1969.

DESIGN METHODOLOGY

First it is necessary to look at the fundamental nature of the design method and ask 'to what extent is this likely to change?' I have previously defined design as being the relation of knowledge to specific needs in a creative way in order to achieve a desired objective. This may be represented by the block diagram given in Fig. 1 which illustrates that what is needed is first related to what is known by a creative individual or group

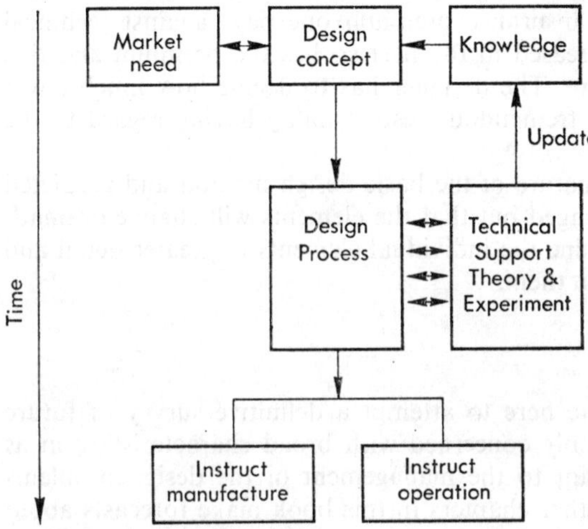

Figure 1 The design method

who develop the design concept. This is followed by an iterative process during which consequences are worked out in increasing detail, in parallel with theoretical and experimental work to fill in gaps in existing knowledge until finally the design is defined. This definition in practice takes the form of giving instructions to manufacture and describing how the design is intended to work.

It is perhaps worth remarking that the 'What do we want?—How can we get it?' approach to creative thinking is not the only one and that 'What might we have?'—How might it be useful?' has often been more effective, a point to which I will return later.

The essential elements, therefore, in the design method are as follows:

(*a*) An appreciation of the market need
(*b*) A fund of knowledge
(*c*) A creative approach to relating (*a*) and (*b*)
(*d*) An ability to work out in detail the engineering consequences of the design concept
(*e*) An ability to extend knowledge by calculation and test
(*f*) An ability to formulate instructions to manufacture and operate the design.

To the question, 'Is fundamental design methodology going to change in the next hundred years?' I am sure the answer is 'No'. I am equally certain, however, that the balance and nature of the elements of the method are going to change dramatically. Indeed, they have changed considerably in the past and the pace is accelerating. One good example is the ratio of technical support to drawing office effort which is employed today compared with typical practice in design teams before the Second World War. Any new engineering involves risk and in commercial terms technical support is the insurance premium one pays against technical risk. As the risk has increased in the past so has the premium and this will continue in the future. The designer has to decide how much cover he requires and this is a tremendous responsibility having regard to the high premium rates.

Having examined the nature of the basic design method and predicted that it will remain unchanged but that the elements will change dramatically, we must now examine the individual elements in greater detail and hazard some guesses about them.

THE MARKET NEED

It would be inappropriate here to attempt a definitive survey of future needs and indeed I am only concerned with broad characteristics, in as much as these are relevant to the management of the design problems that will be presented. Other chapters in this book make forecasts about the future on a broad front and the concensus of opinion seems to leave the designer in no doubt but that he will be called upon to undertake projects of increasing complexity.

It might be argued, however, that designs to meet specific human needs will be refined to the point that further development will become unnecessary. The knife and fork are very adequate pieces of equipment, so nowadays is the motor car, so in the future will be aircraft, nuclear power stations, etc. If this is the case it can be argued that sooner or later a plateau will be reached when everything required has been designed. Then automated factories will produce standard designs and the level of human involvement will be that required to maintain and supply the replication machinery. Presumably the rest of the time would be spent lying in the sun, playing Bingo or in other pursuits appropriate to individual tastes. While individual design concepts seem to follow this pattern, I am sure it is fallacious to extend the principle, because to do so ignores the continually changing pattern of human needs.

The fundamental need of the human race has always been survival, basically involving food, defence and sex. Once achieved, survival more comfortably by progressive control of environment has been the aim. This has led from shelter, fire and conquest to transport, communications, education, power, medicine, etc—the list is endless. The ability to survive comfortably brings the human animal to the point of facing the stark question, 'What is it all about?' and in order to avoid the difficulty of thinking about this, we have more sophisticated concepts of work, amusement and pleasure, which in turn have produced their own host of needs to be satisfied.

Survival and its consequences will in future continue to present a changing pattern of need. We are immediately involved in problems of keeping peace in the world, of population, of food, of environmental pollution and many others. In the longer term man may have to defend himself against extra-terrestrial life or, because of some foreseen natural catastrophe, have finally to face leaving the earth for ever.

For these reasons I believe that for the practical purposes of extrapolation over the next hundred years human needs are inexhaustible and the future will hold for the design engineer a continuous challenge to his ingenuity. The problem for the design manager will be to bring to bear on projects of increasing complexity the wide and expanding variety of knowledge and skill that will be essential for success.

If this is to be the future pattern then what will our approach be to the discovery of these needs? In the past there have been two approaches. The first has been the direct approach by asking the potential customer and this has led to the elaborate market surveys which are common today. The second has been to consider what might be possible and to relate these possibilities in a speculative way to the future pattern of human society. Most of the significant innovations of the past, for which there has often been no immediate demand, have been the result of this latter 'visionary' process and yet, oddly, it is currently regarded in some quarters as not quite respectable. The current vogue is to base market predictions on what are regarded rather academically as 'facts' and to dismiss other factors as mere opinion. I believe that in the coming years the weaknesses of present market research and cost-effectiveness studies will be seen more clearly and it will be recognised that the so-called visionary was, in fact, often processing a vast amount of data in arriving at his vision and including social, economic, psychological, political and many other factors in his projection.

The innovator of the future will still be a man capable of seeing further than his colleagues but his window on the future will be opened by a wide range of unusual computer studies involving the kind of factors presently regarded as inadmissible. In this way the 'visionary' approach will be formalised and made respectable and combining the data of the market study will completely supersede market research as currently understood in establishing market needs.

KNOWLEDGE

In the past the knowledge that could be brought to bear on a given design problem was usually of a personal kind and the way in which the problem was tackled was very dependent on the background or taste of individual engineers. This personal knowledge was supplemented by a limited time available for reading and searching for information in libraries. Access to libraries was sometimes difficult, identifying the works containing required information equally difficult and finally reading volumes in order to find essential information tedious in the extreme.

More recently the tremendous proliferation in written material, much of it unoriginal and repetitive, has placed the professional man in an impossible position in keeping abreast of his subject. Attempts have been made to ameliorate the situation by the production of information digests. These, however, are of limited use because they are usually straight précis or summaries of the work and lack any element of judgment. Badly needed are technical reviews which evaluate and place individual papers in context within the structure of the subject as a whole.

Because of the highly personal nature of the use of knowledge it has always been extremely difficult to reconcile effective continuity of design teams with the steady turnover of personnel which inevitably takes place.

This situation is now beginning to change for two major reasons. First, the ability to find information quickly is being revolutionised by modern techniques such as the use of microfilm and computer sorting. The improvements in access will allow much more accurate and useful classification of information and, as a result of bringing much more relevant material together, it will be easier to form judgments as to the usefulness of particular information. Already information services based on the new techniques are being offered and this development is certain to continue. Even in the comparatively short term I believe this process alone will have a marked effect in accelerating scientific and technical progress.

The second major factor in changing the situation is concerned with the increasing use of high-speed computers for calculations of all kinds. On the one hand, the enormous potential of these machines is radically altering our approach to the kinds of calculations that we wish to perform. Dr. Thwaites in chapter 4, 'The future role of computers with special reference to aerodynamic design', suggested that almost instantaneous solution of any deterministic problem will be possible within the century and, against this background and the resulting revolution in our approach to mathematics, many classical methods will sink to the level of triviality. On the other hand, the amount of effort put into programming methods is increasing rapidly and already, because of the effort involved, engineers are beginning to use other people's programs. Today we have the concept of program libraries, tomorrow these will be national or international libraries. Because of the magnitude of the programming task, this process will inevitably stimulate the active seeking of the best methods and validation of these methods against test data. This in turn will lead to standardisation of methods and clearer statements of applicability. The growing investment in programs will produce a reluctance to program new methods unless these are demonstrably superior. Perhaps most important of all, the process involves the active rejection of inferior, less accurate or less comprehensive methods and information.

For all these reasons, therefore, we are moving into a period when a long overdue and much needed distillation of scientific and technical knowledge will take place, disciplined by the old fashioned criterion of cost, in terms of human effort involved in rewriting all the selected knowledge in computer language. Whether or not all knowledge will finally be contained in one vast cryogenic computer available to all mankind is perhaps best left to science fiction but I believe that in retrospect this process of distillation will be seen as one of the major impacts of the computer on science and engineering.

In the future, then, I see a very large increase in the knowledge that will be brought to bear on design problems. Part of this will be the steady increase in what is known, but perhaps more important will be the ability to retrieve relevant information quickly, to perform calculations virtually instantaneously and to know with more certainty the weight that can be given to results.

The effect of this on the designer will be that he is released from a good deal of tedious and detail uncertainty and he will be free and have the tools at hand to consider more deeply the fitness of his design for its purpose. One significant effect on the design team as a whole will be that much of its detail 'know-how' will be depersonalised and at this level effective continuity will be maintained in a library of computer software and data storage files.

THE DESIGN PROCESS

Having looked briefly at the market need and knowledge, which are the raw materials of the design method, we must now consider the design process involved in relating these two elements.

Phases in the design process

I think it is most instructive to look first at the microstructure of the process which may be regarded as consisting of three essentially different phases. The first of these is a creative phase during which knowledge is related to the problem in a speculative and creative way and the end result of which is an idea. The second phase is one in which the consequences of the idea are worked out in detail, perhaps by a drawing layout or perhaps by calculation. The essential feature of the second phase is that it is deterministic, the input data are fixed and the output follows as the logical result. The third phase is that of assessment of the results, during which judgments are made and decisions taken. The end of this cycle may be regarded as setting boundary conditions for further problems which are tackled in the same manner.

Alternatively this basic sub-routine of the design process can be expressed in the following statements:

Phase 1 I wonder how we can achieve *X*?
Suppose we do *Y*,
Phase 2 Given these facts
That means the following will result,
Phase 3 Will the result achieve *X*?
Yes/No then we will/will not do *Y*.

The design of a complex project involves innumerable iterations of this general form each of which is complete in itself.

It may seem obvious that design problems, large and small, should be approached in the same way and that this process is basically the same as the macrostructure of the design method described earlier. The purpose, however, of looking at the fine structure is to draw the further conclusion that for practical purposes the three phases are of two different kinds. Phases 1 and 3 involve ill-defined boundary conditions and criteria and may be regarded as probabilistic and Phase 2 is deterministic.

One way of looking at a large design team is to regard it as an explosion of the simple case of one designer working at one drawing board. The resulting enormous sheet of drawing paper has, however, been cut into hundreds of pieces and fixed on to hundreds of drawing boards. In a similar way the thought processes, which in the simple case are contained within the head of one man, are now scattered throughout hundreds of designers who communicate or fail to communicate in the normal human way. To enable a large team to function, the overall problem has to be broken down into many smaller ones which can be tackled by individual designers. Each designer then applies the design method and in the process begins to outline the design on paper and to produce a host of further problems to be solved. Throughout the design, consistency, compatibility and adequacy can only be ensured by human communication, much of which is of a homely kind. For instance, 'You can't put your unprintable pipe through the spar there, that's where my valve is mounted', is the kind of communication that is familiar to all designers. Communication, therefore, is the *sine qua non* of a design team and gradually as the design iterations are performed the inadequacies, inconsistencies and incompatibilities are resolved until finally the design is defined.

In essence, then, in the large design team, one man at a drawing board is replaced by many men working on a very large drawing board with all the attendant difficulties of communication. In this situation, drawing paper is the basic clay which is moulded by innumerable design iterations to produce the final definition of the design and, at any stage, drawings provide the data base upon which the design process proceeds.

Computers in the design process

We are presently at the point of realising the enormous impact that computers can have on the deterministic phase in the design process. For some years we have been using increasingly powerful programs for calculation purposes and we are beginning to see the application of visual display and automated drawing techniques to the geometrical problems of the drawing board. Design is beginning to be seen as an intimate dialogue between the human individual who can have ideas, judge results and take decisions and the high-speed computer which can store, sort and retrieve information, perform logical operations on the data and display results. Current developments in computers such as multiprogramming, random access and visual display will lead in the coming years to a number of significant changes.

First, let us consider the question of design calculations. For some years there has been a feeling that although computers had released us from the trap of the early 1950s, in which there was only time and money to look at a problem one way by manual methods, e.g. flutter calculations, increasingly they had led to a situation in which a feel for the work was being lost. Data were fed in and results obtained quickly, but a feel for the sensitivity of the results to assumptions and the method was more difficult to obtain. The advent of visual display techniques makes it possible by suitable design of software to have the ability to look quickly at intermediate stages of a calculation. In this way a much more intimate feel for the working of the computer will be developed. In addition, multiprogramming, the development of control from remote consoles and developments in program languages will lead to open shop programming and control. Thus we are already in sight of a situation in which individual engineers can be engaged in intimate and continual contact with a very large computer, without even being aware of the fact that colleagues are similarly engaged with the same computer.

Secondly, the storage and manipulation of geometrical data is already well advanced and it is clear that, in the next few years, this kind of information will be available instantaneously to individual designers as a result of similar developments to those that have been mentioned.

Thus the next phase of development of the design process will be characterised by recognition of the power of computers in the deterministic phase and a growing intimacy in the dialogue between individual designers and the computer. At this stage the computer will still be used essentially to amplify individual abilities. Communication for the purposes of design will still be carried on outside the computer and drawing paper will remain for most purposes the basic clay being moulded in the design process.

However, for each of the programs run by individual engineers, basic design data will be fed to the computer. It is clearly then only a short step to the concept of basic design data held within the computer which can be referred to for individual programs. This data bank will be updated as the design proceeds and in this way all subsequent work will be based on the latest data. No one who has managed a large design team will under-rate the significance of this becoming an automatic process. It is also clear that it will then be possible, following the change to any data, for the computer to sort automatically through the routines affected and to signal this fact to the control stations of designers whose work had been affected. At first this might take the form of a warning light at the designer's visual display unit, allowing him then to interrogate the computer as to the nature of the change and to decide the appropriate action to take. Further in the future the computer will automatically re-run affected routines and will only ask for a decision from the appropriate designer if significant results are altered outside pre-determined limits. This work will be performed when the computer is not heavily loaded, for instance, overnight. Each morning designers will then return to their task with an analysis of what has been affected by the previous day's work and by precisely how much. This step in development will, of course, be a very significant one because the computer store will now become the basic clay being moulded in the design and by way of links between data files and programs within the computer some of the present essential communication between individuals will become automatic and take place within the computer.

That this kind of development in the design process will take place I have no doubt. However, it will involve the systematic analysis and programming of design methods and big advances in computer power and software development, particularly for information retrieval. For these reasons I think that the rate at which this development takes place will be limited by the human effort (cost) involved in its implementation rather than by human ingenuity. Nevertheless, in one hundred years' time I believe that drawing paper will have been superseded by the computer store as the basic clay moulded in the design process. Designers will start with clean reels of tape or their equivalent and, by way of the processes I have described, mould these into a complete description of the design. Much of the communication involved in ensuring consistency and compatibility of the design will be done automatically within the computer and the designer, relieved of this detail responsibility, will be free to work in a more creative way, with more regard to adequacy of the design and to the social implications of his work. Within this time-scale I also believe that the ability to use computers in a heuristic way will continue to develop and will have a strong influence on the development of the design process.

In suggesting that drawings will be superseded by the computer store as the basic definition of the design, the question is raised 'Will drawings be required for any purpose?' and this leads to an examination of the functions of drawings. Other than design definition the most obvious function is the instruction of manufacture and the method of manufacture is, therefore, the key to answering the question. Inasmuch as the manufacturing processes are automated, it will be convenient to give instructions in the form held within the computer. For this reason instruction tapes for automated machine tools will be generated within the computer. Similar tapes for the manufacture of tools used in other processes will be generated in the same way. Where it is required to instruct human operators it will be necessary to use the most efficient methods of communicating with human individuals. These will include spoken and written words and methods of visualisation such as drawings, models, etc.

The other important function of the drawing is to provide the designer with a vehicle for the continuous visualisation of his design throughout the design process. In the future this function will be enhanced by direct visual display of computer-held information with the ability to retain hard copy of this if required. In addition, models will be made from easily formed materials, using control tapes based on computer-held numerical data and automatic drafting equipment will produce drawings from similar data. Developments in holography will lead to the practical use of three-dimensional imagery in the design process. Nevertheless, because of the essential simplicity and power of the drawing I am sure that it will continue to be used as one of the weapons in the armoury of visualisation, but will no longer be the medium for recording the complete definition of the design.

Communication with computers

It is clear that the increasing importance of the computer in design, as well as in many other applications, will focus more and more attention on the problem of human communication with computers. Already much work is being done on simplified programming languages, and in specialised applications many are already in use. In the future the work being done on character recognition will, for some purposes, cut down the effort of information transcription into written languages digestible by computers. Inevitably, because the spoken word is the commonest, most rapid and most intimate form of human communication, audio communication with computers will be a sought-after ideal. At present the amount of information, in electronic terms, that has to be handled to make this possible looks discouraging but already special purpose computers that will respond to simple commands are under development. I believe that, in a world becoming more and more automated and reliant on automatic machinery, we will see the development of a simple, unambiguous spoken language for the purpose of instructing machines of all kinds and it is interesting to speculate as to whether this might become the ultimate international language for basic communication purposes. Whether or not this proves to be the case I am sure that in a hundred years' time communication with computers will be by the spoken word.

Before leaving the subject of communication with computers, I should mention the possibility of advances in neurology and electronics leading to the tapping of electrical signals in the brain which could be fed directly to the computer. In this way one might imagine the specialised abilities of a large number of human brains co-ordinated and controlled by a computer in such a way that a complex design could be created by simply 'thinking' about it. I believe that, if they occur, developments of this kind are well outside the next hundred years and would involve modification of the relatively undisciplined thought processes of the human individual as well as advances in computer technology completely outside our experience.

THE INDUSTRIAL ENVIRONMENT

Before discussing questions of design management I think it is appropriate to consider the environment in which the design processes envisaged will be carried out.

Historically in the manufacturing industries, as the means of cheap production have involved more expensive plant, so the larger organisations able to afford the equipment have swallowed their smaller rivals. Even the largest organisations are now involved in mergers in order to widen markets, further reduce the costs of production and alleviate the burden of research and development, sales, product support, etc. These results have followed the enhancement by machines of man's physical abilities. Now we have started down the road of enhancing man's mental abilities and the situation I have envisaged in a hundred years is one in which a design team's productivity will be enormously increased in parallel with a huge investment in equipment. Such design teams will only be able to be supported by the largest organisations and will be so powerful that small rivals will be unable to compete. Thus I predict a situation developing in design which exactly parallels that which has already occurred in production. As far as companies producing complex designs are concerned this will simply accelerate the merger process which is already well under way. In the time-scale we are considering it seems clear that only a few very powerful organisations will remain selling their products in international markets.

In the merger process the usual pattern has been centralisation in the first instance in order to understand and control the situation and to reap the obvious benefits of the merger. This has been accompanied and followed by a bureaucratic development of the central control. This development is the result of a natural drive to know what is happening, leading to more and more data being requested. In practice, however, understanding of the situation is often very little improved, the centre preoccupied with analysing data gets out of touch, policies become unreal and control is lost. In extremes, the negative principle that any dynamic situation can be controlled by stopping it, is invoked. Classically, realisation of this state of affairs has been followed by a policy of decentralisation which if carried to extremes, largely negates the purpose of the original merger.

Obviously neither extreme of centralisation nor decentralisation makes any sense and in the future this kind of simple-minded neurotic oscillation will not take place. Wider acceptance of cybernetic principles applied to management will lead to the natural homeostasis of large organisations, within which levels and depths of control will be settled by the functional requirements of the control mechanism.

Turning now to the question of company motivation it is interesting to speculate as to what this might be in a hundred years' time. Again historically the original motive was often personal gain. Personal wealth and ideas were combined with individual labour to produce a product from which a personal profit was made. In order to maximise this profit costs were held down and wages were low. This led to the ideas of exploitation of the worker and the evil of profit, which caused revolutionary changes in some parts of the world and from which we are still suffering today. The growth of companies involved more capital and public shareholding and this was paralleled by the formation of labour unions for the protection of workers. Today the scale of very complex engineering projects has created a situation in which government financial support is necessary and monolithic trade unions negotiate wages at a national level. In this way the worker is involved indirectly by taxation in financing the project, the professional managers of the large company no longer benefit directly from the company's profitability and the shareholder is increasingly regarded as a source for borrowing money at a fair market rate. Who, then, in this environment stands to gain if the company is successful? The answer must surely be everyone.

In the future the giant companies, with large labour forces and handling projects involving national finance will increasingly be identified with the affairs of state. Profit previously regarded as bad because it was personal will become respectable and the standard of living regarded now as a function of national wealth will become identified with the profitability and success of large business organisations. The old 'establishment' will give way to a new elite controlling the major industrial organisations and the 'cabinet' will become the state board of directors. Anyone who objects to this view on the grounds that there is more to government than running a business, should reflect on the amount of time that any manager in a large organisation spends on human problems and how much attention large organisations give to education and training, welfare, sports and social activities, not to mention planting gardens around factories and offices.

Already amateurism in government and civil service is being attacked and in the next ten or twenty years radical changes in governmental machinery are inevitable if we are to continue to prosper.

In political terms, the change that I foresee is not the nationalisation dogma associated with socialism in which control is taken over by the State for the greater good of the people; nor is it a kind of Marxism in which each has according to his needs. Indeed, it is not democracy as it is currently understood, as no business could run on the basis that everyone had an equal say on every topic as they clearly do not have equal competence. In the future each will have equal opportunity, will say according to their recognised competence and have according to their efforts.

Politics have been defined as 'the essence and art of government' which in turn has been defined as 'the conduct of affairs in the best interests of the people'. In a more enlightened world one hundred years' hence, I believe this will be synonymous with a new concept of business management.

The environment, therefore, in which I foresee complex designs being created in the future will be that of very large scientifically managed companies closely identified with the policies, objectives and finance of the state and whose prosperity will be regarded as vital to the national economy and hence to the general standard of living.

MANAGEMENT

The question is sometimes asked, is management an art or a science? My answer to the question is that it is an art and it will be a science. Science itself has its roots in the magic and mystery that surrounded alchemy, astrology and religion. The structure of systematic knowledge that today we call science was begun when man made his first attempts to measure and to codify his experience of natural events. In the same way today we have started to make measurements in management and to assemble the experience man has of managing situations into a rational structure.

That we have not yet progressed very far is exemplified by the kind of criteria which are still sometimes used in judging a man's potential as a manager. No doubt five thousand years ago an impressive appearance and a loud voice were excellent qualifications for a high priest of the Gods of the Nile. Today one would look for very different qualities in a man one trusted to predict the behaviour of a river. Until quite recently points were being scored on personnel assessment forms for answers to such questions as: Is he of good appearance? Can he communicate well? Is his timekeeping good? Now we are told a comparison of a man's performance against objectives is what matters and management by objectives is being widely preached.

The comparatively new science of cybernetics, which is concerned with communication and the control of systems, has given rise to the concept of industry as an exceedingly complex probabilistic system, the behaviour and properties of which are amenable to mathematical analysis. This is already the subject of study by a few and I believe will soon be widely accepted.

These are but two examples of new thinking which is being applied in management and which in the next one hundred years will firmly establish management as a science. However, the subject of management is extremely wide and I do not propose to attempt any kind of review in the space available. I began by defining management as motivation, organisation and effective control and it is under these three headings that I will restrict my remarks to those aspects which I believe are central to management of the design process.

Motivation

If we assume that the highest level of design management is motivated by the overall company policy and objectives, then we must ask the question: What is it that motivates the individual members of the team? The cynic may quickly answer, pay; but a moment's reflection will demonstrate that this is wholly inadequate for the purpose of design. Let us, for example, compare the performance of men in two different situations. The first is interested in his work, clearly understands what is required, has the environment, information and facilities to do the job and feels that his work is important and that he is fairly treated. The second is disinterested, uncertain what he should do, cannot obtain necessary information, is inadequately equipped, does not feel his work is important and is nursing some grievance concerning his personal treatment. It is obvious that, if one is looking for a high level of directed creative effort, the output of the second man will be negligible compared with the first. It is a primary function, therefore, of design management to provide leadership and to develop a creative environment in which individuals feel they can contribute to the design as a whole. In this way an *esprit de corps* is established, morale is high and the best work is produced.

In the past, when engineering projects were less complex and teams were smaller, the Chief Designer was not only clear about his aim but also had a detailed knowledge of the means by which his objective could be achieved. This situation allowed an authoritarian kind of design management in which detail instruction was given direct from the highest level and the team provided the necessary hands to work their master's will. The Chief Designer was able to maintain personal contact with all the members of his team and had little difficulty in generating the necessary team spirit. This kind of management, however, is clearly limited by the ability of one man to cope with every aspect of the project.

Today we have large teams working on very complex projects. It is no longer possible for the chief designer to have a detailed knowledge of all the means and he has had to concentrate his attention on formulating design objectives and to rely on his specialists to provide the ways and means. In this situation responsibilities must be delegated, which in turn demands that communication must be good. Neither of these conditions can be met in the absence of organisation. A natural hangover from the past is the situation in which a chief designer fails to delegate and continues to concern himself with detail instruction. This leads to an over-loaded chief designer and growing uncertainty among the staff, as detail instruction from the top fails to cover every aspect. The chief designer thinks he needs and recruits more people and the confusion in the team grows worse as the chief designer becomes more isolated from his team in trying to cope with all the problems. This kind of vicious circle can result in a large inefficient design team badly led by a hopelessly overworked design management. A further danger of high level detail direction is that in practice it can often be wrong and the resulting neurosis at working level between authority and what is right, usually results in nothing at all being done.

In the future, I foresee this situation becoming steadily more difficult. As knowledge and specialisation in various branches of knowledge continue to increase it will become more and more difficult for one man to formulate forward-looking but practical design objectives. Thus, not only will he have lost his grasp of the means but also of sensible ends. It will then be vital to have a creative atmosphere in which many minds have the freedom to exchange and try ideas so that realistic objectives may be formulated and clearly understood. In the past the discipline that is necessary for the execution of any design was a natural extension of the authoritarian leadership. In future, discipline will be just as vital for success but the slavish carrying out of instructions will be replaced by a responsible approach to the work, combined with an understanding of its relevance and importance to the whole design. In this design team of the future the high level of specialist knowledge required will only be brought to bear by clear delegation of responsibility and the full potential of the team will only be realised if mutual understanding is of a high order. I believe it was Confucius who wrote: 'If language is not correct, then what is said is not what is meant: if what is said is not what is meant, then what ought to be done remains undone'. I am certain that in a hundred years' time the importance of clear statement and first-class communication will be greater than it has ever been since that was written over two thousand years ago.

Summarising then, the motivation of the future design team will depend on the development of a creative environment in which:

1 Individuals have a clear understanding of the relationship of their work to the whole design and of what is expected of them.
2 There is freedom to generate ideas.

3 Individuals are interested in their work and feel that it is necessary and important.
4 Access to information is easy.
5 Necessary equipment is available.
6 Individuals feel they are fairly treated.
7 There is clear delegation of responsibility to enable specialist skills to be brought to bear.
8 First-class communications exist at all levels.

These conditions will only be realised if the organisation is rational and easily understandable and it is to this that we must now turn our attention.

Organisation

> Find out, if you can
> Who's master, who's man.
> *My Lady's Lamentations*—SWIFT

Organisation is fundamentally the giving of an orderly and understandable structure to related parts of a whole function and is essential when more than one individual is concerned with a single objective. The start of the problem, therefore, goes back to the single individual and it is here that we should begin in seeking to predict the organisational patterns of the future.

The development of design organisation. Referring to Fig. 2 the earliest engineers can be represented as having been equally concerned with all aspects of the problem. If there was a need to raise money or lobby a case in the House of Commons this was just as important as analysing a structure or making a part. Their success was dependent on the ability to recognise and deal with all the necessary functions of the total job. Obviously the scale of task was severely limited by the ability of one man and in order to undertake more difficult tasks more men had to be involved.

At this point it is fundamental that help was required with some aspect or part of the work and what was later to become the acute problem of specialisation was born. In practice, assistance was usually required first in manufacture, then in design and soon after that in financial matters. In this way the basic structure of most present day engineering organisations represented in Fig. 3 was set up in the first stage of organisational development.

It is important to note that already the trade-off between design, manufacture and costs is in the hands of the manager, who alone can be responsible for the engineering compromise and that the true role of specialist departments is to assist the manager to meet his objective. This relationship existed while the original engineer was still the manager; the situation became distorted when he was replaced by a manager neces-

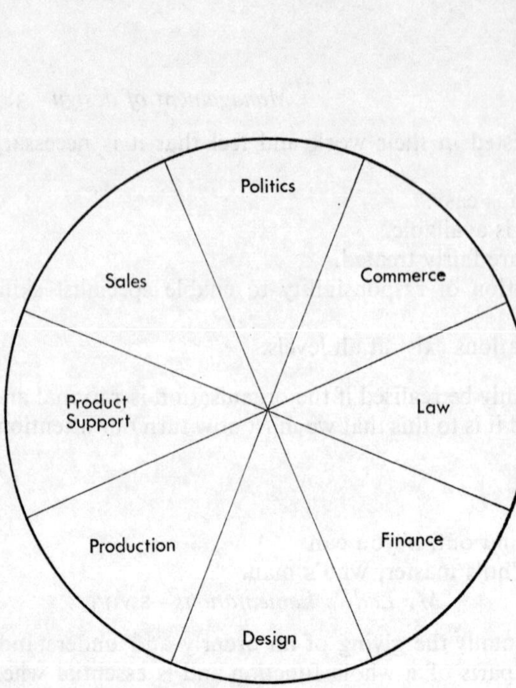

Figure 2 The original engineer

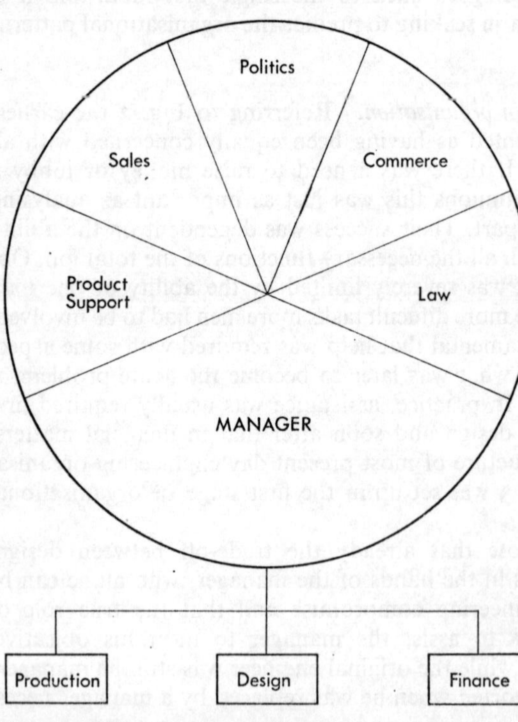

Figure 3 The first specialisation

sarily with the kind of abilities that did not exist in the specialist departments that had grown up. In this way, it was perhaps inevitable that the mantle of the original engineer should fall on the design department, no longer in a position, nor indeed with the competence, to replace him.

Following now the development of a typical design organisation as it grew in size under the Chief Designer, we find that organisational distinctions based on separate phases of the project, e.g. Project office and Design office, made their appearance. These were followed closely by the specialist disciplines of Stress and Aerodynamics.

This second stage of development led to the type of design organisation illustrated in Fig. 4 which was typical just before the Second World War. Points to note are that, as with any growing organisation, individuals in charge were in some cases totally unsuited to manage the larger departments as they grew. The introduction of organisational distinction based on phases of the work led to overlapping of skills and types of work between departments, for example, draughtsmen were involved in

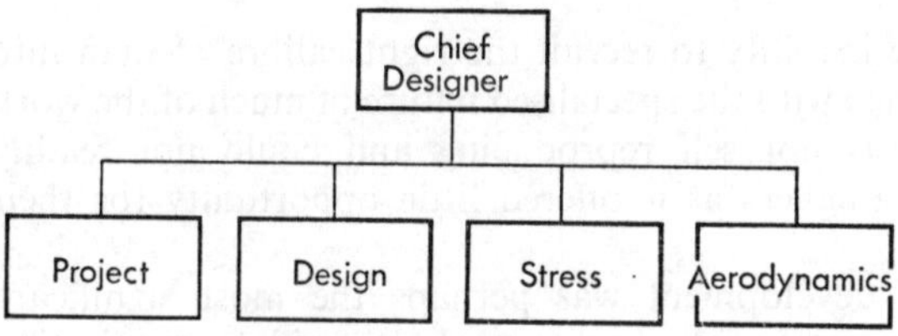

Figure 4 A typical early design organisation

project and design offices. In addition, the Chief Designer was beginning to have to cope with a number of specialist departments.

The third stage of development which took place during and immediately after the Second World War was characterised by a rapid proliferation of specialisations, as many functions grew to become departments in their own right. Aeroelastic, Avionics, Systems of all kinds, Materials, Flight Test and Instrumentation are just a few examples. During this period, partly as a result of the educational system, but also because of the rapid expansion, most of the formally trained and qualified engineers were recruited into specialist departments. This tendency undoubtedly had an unwholesome effect on the design teams. Designers became more and more dependent on specialist advice which they became less and less able to judge. The qualifications of, and demand for, the specialist tended to introduce a pay differential in his favour and the designer in the drawing office was made to feel an inferior animal. At one point it became fashionable to refer disparagingly to drawing office work as the 'ironmongery'. This situation unfortunately was a vicious circle and grew worse as fewer able recruits were prepared to take jobs in the design office. Surely it was predictable that the results of this period would be impressive technical achievements combined with poor detail design, leading in turn to unreliability, difficult maintenance, high costs and poor service to the customer.

In this position, illustrated in Fig. 5, the Chief Designer was the only man with an overall view and he was stretched to the limit in an attempt to co-ordinate all the specialist advice into a wholesome design. Also his grip on what needed to be done in every specialist area was lost and he became reliant on the good sense of his specialists for the direction and depth of the work.

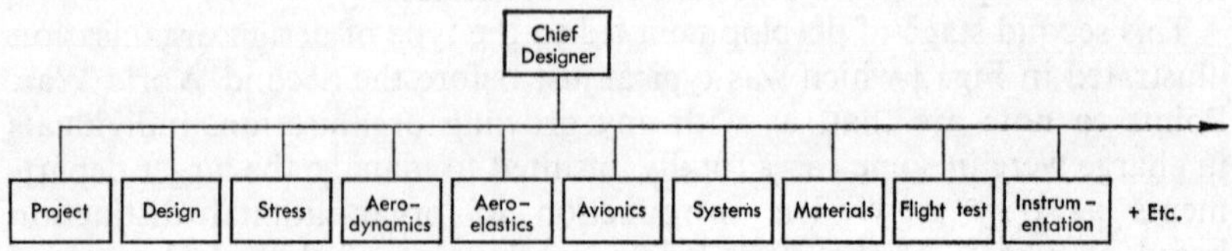

Figure 5 The growth of specialisation

One side effect of the inability to recruit the right calibre of man into the design office, combined with the specialised nature of much of the work was that the system was not self reproducing and could not readily produce future Chief Designers as it offered little opportunity for their training.

The fourth stage of development was perhaps the most significant because it was the point at which the classical pyramidal organisation proved entirely inadequate and new organisational thinking was essential to continued growth. This crucial development was the advent of the multi-project situation in which one team was attempting to handle a number of different complex projects.

Referring to Fig. 6 it is clear that, although individual departments could usually cope with the situation, the Chief Designer had to provide co-ordination of all the specialist departments for all the projects, which was an impossible task. Obviously the Chief Designer had to delegate part of his co-ordinating responsibility, but as already mentioned the system

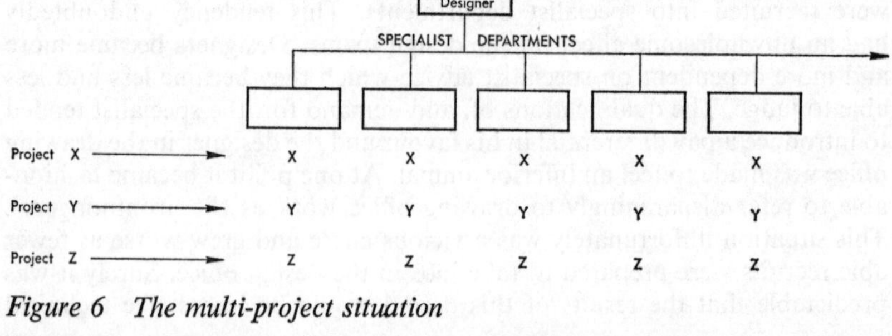

Figure 6 The multi-project situation

was not one which readily allowed individuals to acquire a broad experience, and for this reason it was extremely difficult. In practice some measure of responsibility was often delegated to individuals in the design office. In this way, to some extent, the mantle of the original engineer finally descended to the design office, where not only had the ability to balance market cost, production and other considerations been lost in the earliest stage of development, but also the grasp of a wide range of engineering specialisations was hopelessly inadequate to judge effectively the correct engineering compromise. One might almost think that Milton had a design office in mind when he wrote in *Paradise Lost*:

The will
And high permission of all ruling Heaven
Left him at large to his own dark designs
That with reiterated crimes he might
Heap on himself damnation.

This situation was, I believe, the nadir of design organisation and I think it was predictable that it would lead to a wide range of activities aimed at improving an untenable state of affairs. The Feilden Report recommended improving the status of the designer to attract better men into design. Educationalists began to talk about teaching design. Customers complained about the maintainability, reliability and cost of the product, advocated the use of maintainability check lists in design and sought reliability and cost guarantees. Production engineers were moved into design offices and value engineers made their appearance loudly proclaiming that this 'new?' technique had to be a company-wide activity. Of course it should be, but even today it is not difficult to find people who believe this should be a design department function.

The matrix organisation. It was inevitable that out of this situation would grow recognition that the Chief Designer would have to delegate his authority to individuals capable of taking a broad view of the whole design of a project (Project Designers) and that company-wide the managing director was steadily being forced in the same direction. This in turn led to the concept of a two-dimensional or matrix organisation which is illustrated in Fig. 7.

This type of organisation recognises to some extent the multi-dimensional nature of the problem and faces it using a two-dimensional solution. Perhaps the most important feature of this kind of organisation is that it allows the project designer to run a team on his project (within the dotted line of Fig. 7) in much the same way as the Chief Designer ran his team on a single project. That is, section leaders within design and specialist departments have much the same relationship with the project designer as department heads had with the Chief Designer. The essential difference is that, if important questions involve serious disagreement at project level, then the process of taking the matter up at higher level, both by the project designer and the specialist, naturally involves the most senior people. In this way, a number of projects can be run simultaneously with the most senior people involved in the most important problems.

Unfortunately, this approach entails a new kind of thinking and a new understanding and for this reason is not easy to establish. The one dimensional or pyramidal organisation has roots going back thousands of years and is based on ancient precepts, such as 'No man can serve two

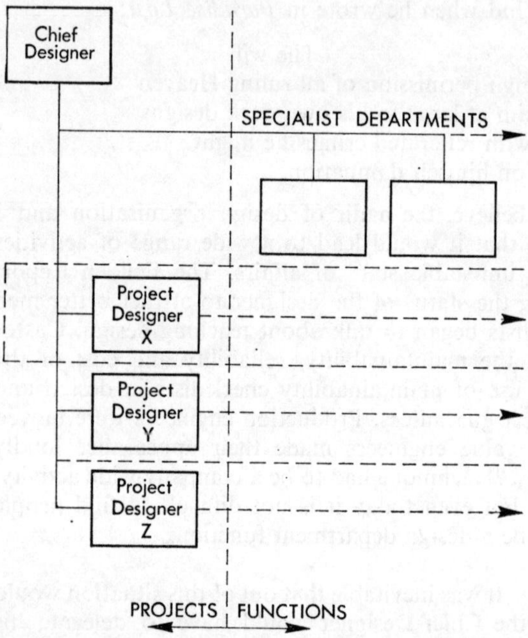

Figure 7 The matrix organisation

masters'. Well rooted ideas of this kind cannot be changed quickly and in my experience to convince anyone that he must now serve at least two masters and to develop an understanding of the relationship requires a great deal of time and patience. The logical positivistic approach of 'Is there a real situation in which you would have any difficulty in knowing to whom you should go for advice or to whom you should report?' is, I think, the most effective, otherwise it is easy to be drawn into hours of sterile debate about principles.

The other major problem involved in setting up an organisation of this kind is the difficulty of finding men of the right calibre and experience for the project designer jobs. I have mentioned previously that in the past many of the most able men became specialists and, therefore, often did not have the breadth of experience necessary. For this reason, at the present time, it is sometimes the case that project designers are little more than programme controllers and lack the essential grip on the design compromise. Unfortunately, this enables the concept of project management to be discredited particularly by senior specialists who may resent people whom they regard as less able having a wider influence on the design as a whole.

This situation, however, is transitional because the basic structure of the matrix organisation allows for the first time since the beginning of specialisation the natural and progressive training of broadly based designers. In the course of time, therefore, young men with an aptitude for overall design, after training in a number of specialist departments, will be able to assist on the project management side before being given a small job of their own to control. Progression through jobs of increasing size and responsibility will result in project designers who carry the respect of the specialist functions, and project management will be able to function in the right way.

A major virtue of this kind of organisation, I believe, is that it places specialisation clearly in correct perspective with the design as a whole and unequivocally places the leadership with the project designer. It may seem absurd that so obvious a fact should need emphasis. That it does need emphasis is a direct result of historical development that I have described. Leadership comes from those who can provide it and in the past many of those who had the ability to provide it were specialists.

This, then, is the present position, the concept of matrix organisation is in use and project management has started to grow. For historical reasons and because it involves understanding new kinds of relationship it is not being readily accepted and it is rarely to be found functioning in a proper fashion.

Therefore, the immediate future development that I foresee is a growing understanding and acceptance of this kind of organisation, the effectiveness of which will steadily improve as project designers of appropriate ability and experience develop within it. The balance of specialisation with respect to the whole design will be restored and it will be possible for one team organised in this way to handle a number of projects in a straightforward and businesslike way.

The matrix organisation concept, however, did not emerge as a result of some special peculiarity of the design problem but as an answer to a complex management situation. In the future, therefore, I believe this approach will be widely adopted across companies as a whole, for precisely analogous reasons to those which have forced its development in design.

Figure 8 is a diagrammatic representation of the principle applied to the whole company. Within the dotted envelope shown the project manager will be able to operate with the delegated authority of the managing director as though he were operating his own company, except that the administrative pattern will be set for him and any important disagreements will automatically be settled at a higher level. In this way, just as for Chief Designers, a progressive training of potential managing directors is made possible, thus solving a problem of the past which has resulted in managing directors being recruited from people having wide experience of administration and leadership, but often with very little knowledge either of the company or its products.

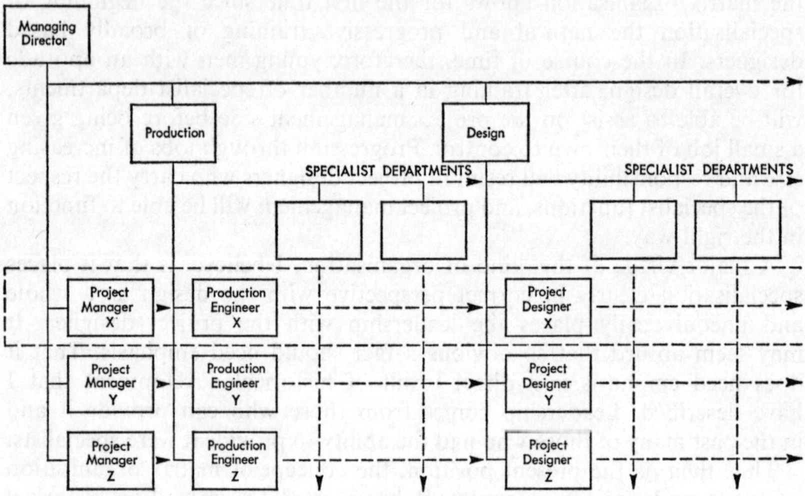

Figure 8 Company-wide matrix organisation

When this point is reached the evolution of organisation will in an overall sense have gone through a complete cycle. The original complete engineer was limited by his personal ability and efforts. This forced him to seek assistance in order to take on larger tasks and inevitably the process of specialisation was begun. He was, in course of time, often naturally replaced by an individual with abilities which were not contained within specialised areas. Some element of the total objective of the original engineer fell to design departments who were by now no longer in a position or competent to carry this responsibility. The system created failed to be naturally self reproductive in terms of generating overall leadership and this in turn led to increasing distortion of the *modus operandi* and company rationale. This in turn gave rise to inefficiency and a poor product. The system finally failed under the weight of the multi-project situation which generated the matrix organisation concept. The prospect that I have outlined will restore the balance of the whole operation and will also provide a framework for the natural development of new leadership at all levels. In this way the future managing director will replace the complete engineer of the past in terms of overall objective and leadership, but he will be equipped with an organisational framework which will enable him to deploy satisfactorily large teams on a number of different projects. Lest I should be misunderstood, I must emphasise that I am not implying that the future managing director must be a designer or engineer. In fact, on the contrary, I am insisting that he must possess a wide range of abilities and whether he starts life as a lawyer, accountant, engineer or salesman is immaterial, provided that, as managing director, his grasp of a wide range of subjects reflects in large measure their importance to his company.

Organisation as a multi-dimensional problem. Within the overall organisation pattern that I have discussed, further differentiation of function is necessary in the management of complex tasks. This is essential in order to provide an understandable framework within which clear communications can enable different parts of the total job to be carried out by different people with a minimum of confusion and duplication.

In the past organisational distinctions have naturally reflected ways of looking at the work to be done and these depend on deeply ingrained patterns of thought. The fundamental concept of time gave rise to distinctions based on phases of the work such as research, project, design, flight test, and development. The concept of space led naturally to distinctions based on the separate parts—wing, fuselage, etc—of a project and on work carried out at different geographical sites. Other distinctions owe their existence to the concept of quality and we have departments dealing with strength, weight, cost, etc.

For this reason we should consider whether or not, in the course of the next hundred years, there will be any change in these fundamental concepts. For the first time in man's history he is about to break out of his natural environment into space and figuratively to look back at himself. I believe that this is likely to have a profound effect on man's thinking on a cosmological scale but that it is most unlikely to change any concepts that are relevant to the organisation problem. Meanwhile 'Man is what he thinks' and organisation will continue to reflect the fundamental ways in which he thinks.

It is important, then, to recognise that organisation is a truly multi-dimensional problem. Organisational distinctions are not mutually exclusive and suffer from the imprecision of the words used to describe them and of the concepts underlying them. It follows that they cannot be added in any arithmetic sense to arrive at the whole and must be regarded simply as ways of looking at the work. Because there are an infinite number of ways of doing this it is important to determine which are the most significant and useful in setting up an organisation.

In the future I think organisation will be treated as a multi-dimensional problem, to which the two-dimensional matrix organisation is a better approximation than the single-dimensional pyramidal type. Individual functions will be seen as falling into different logical categories and will be grouped and treated accordingly. Communications will be recognised as a major factor in determining the choice of functional interfaces and will become the subject of measurement and analysis to enable organisations to be created in which both lack and ease of communication become measures of the quality of the organisational structure.

I started by defining organisation as giving an orderly and understandable structure to related parts of a whole function. If the structure is well understood then in a sense there is no need for formal organisation and many examples spring to mind of highly efficient small teams which are not formally well organised. Most large teams need to be formally organised in order to make the structure understandable to people working within it. In the future, with the development of more rational and, therefore, more easily understandable structures and with the growing realisation that responsibilities cannot be unidirectional, the need for formal organisation structures may well decline. Individuals and groups will see themselves more as functions with a general responsibility within the organisation for the efficient execution of their functions, rather than as experts in particular fields responsible to one manager for whatever he chooses to define as their task.

Analysis of organisational structures. In the past I believe that too little attention has been given to the analysis of organisational structures. The obvious fact that organisations should be arranged to make the best use of individual abilities has, I think, often been allowed to obscure the fact that such a course can lead to the creation of awkward functional interfaces and distortion of the whole *modus operandi* of an organisation. In the future I believe that emphasis will be placed on developing managers to carry out functions rather than tailoring organisations to suit individuals.

In the general context of analysis of organisational structures I suggest that much could be done to throw light on their properties by mathematical modelling. As an example, in very simple terms, if we consider a simple pyramidal organisation, the total staff in the organisation can be expressed as:

$$\text{Total staff} = \frac{N^{L+1} - 1}{N - 1}$$

where L Number of management levels
N Number of staff reporting to individuals at the next higher level.

The total staff given by this expression is shown for a range of values of N and L in Fig. 9 which illustrates that for practical purposes, in this kind of organisation, the number of management levels is relatively insensitive to the size of the design team. However, for a given size of team, say 1500 people, three management levels imply approximately eleven people reporting to each manager and five management levels would reduce the number of people reporting to four. Bearing in mind that the difficulty of communication is some function of both the number of people controlled by one man and also of the number of levels through which to communicate it is hardly surprising that a team of this size is typically run by four levels of management, Chief Designer, Department

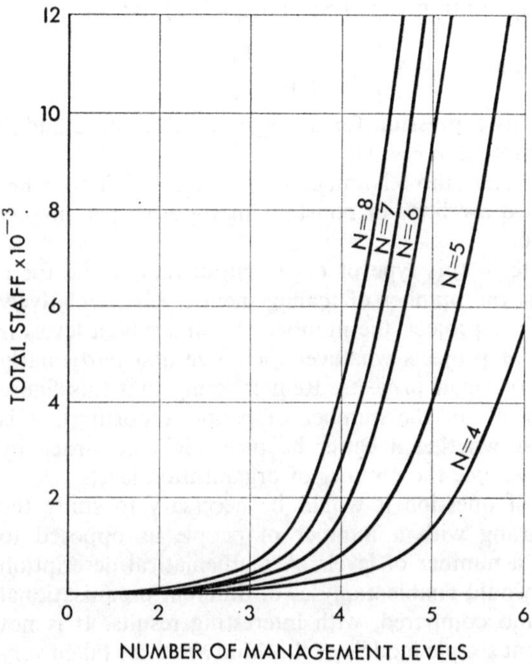

Figure 9 Total staff in a pyramidal organisation

Head, Group Leader, and Section Leader, with approximately six people controlled by any one manager. One may speculate as to the relevance of this analysis to the classic management precept that a manager should not have more than six people reporting to him, but if it throws any light at all on the problem then I believe the approach is worth pursuing.

The next obvious step is to apply this approach to a matrix organisation in which at every level a manager runs his department on a project management basis. In this case the manager has reporting to him not only specialist functions but also project engineers concerned with all the department's specialist functions on his project. In this way a model (represented diagrammatically in Fig. 10) of an organisation consisting of continued layers of project management can be built.

If as before we assume that:

	L	Number of management levels
	N	Number of staff reporting to individuals at the next higher level,
then if	P	Number of projects
it follows that	$(N-P)$	Number of specialisations reporting to any one manager.

The total staff in the organisation can then be expressed as:

$$\text{Total staff} = 1 + N\left[\frac{(N-P)^L - 1}{(N-P) - 1}\right]$$

the total staff given by this expression for a range of values of L and P and for $N=6$ are shown plotted in Fig. 11.

In this case because Total staff=Number of projects × Number per project, it is possible to draw lines of constant man-power per project shown dotted in the figure.

It is interesting to note in this type of organisation that, while for a given number of projects the number of management levels is relatively insensitive to the size of the project, the number of management levels is sensitive to the number of projects whatever their size and particularly so with a large number of small projects. Remembering that this figure is drawn for a value of $N=6$ (the number of people reporting), it is perhaps worth wondering whether a better balance might be struck by increasing N in order to reduce the number of organisation levels.

To resolve this kind of question it would be necessary to study the efficiency of communicating with a number of people as opposed to communicating through a number of levels. A mathematical description of factors such as these would enable analyses of different organisational structures to be made and compared, with interesting results. It is not my purpose to pretend that simple models of this kind are to be taken very seriously. It is to point a direction and to suggest that mathematical techniques already developed in connection with topology, random theory, information and communication theory, etc, could be harnessed with advantage to the study of properties of organisational structures.

Control

Turning now to the subject of control, we must first refer back to the earlier discussion on the nature of the design process. In this discussion it was concluded that the phases of the design process are essentially of two different kinds. One is creative and involves speculation, ideas, judgment and decision; the other is deterministic and is concerned with the execution of ideas using known engineering practice. These two

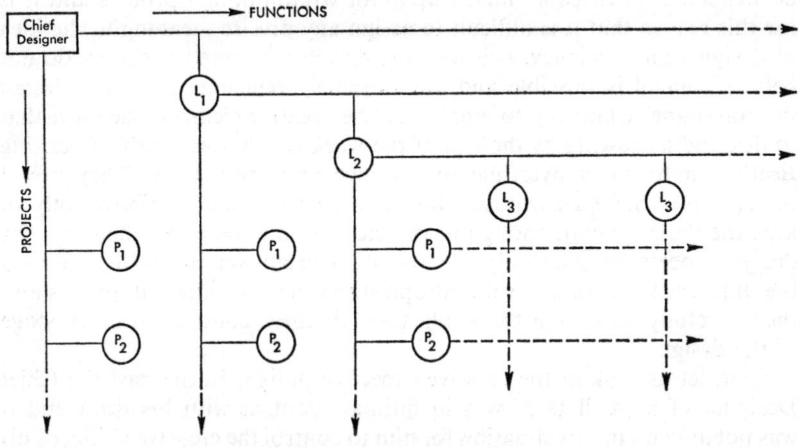

Figure 10 Model of continued project management

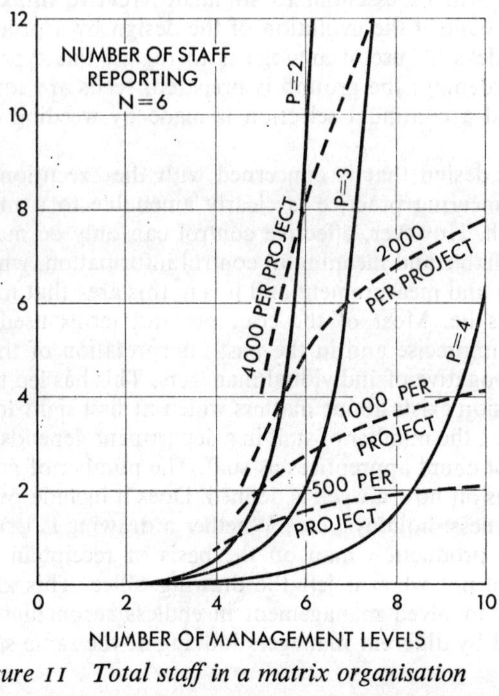

Figure 11 Total staff in a matrix organisation

elements are inextricably mixed up in the overall design process and it is for this reason that it is difficult to assign any precise meaning to the idea of design team efficiency. It is also the reason why some designers do not believe control is possible and are extremely reluctant to accept design programming. Choosing to emphasise the creative element, they feel that to demand a drawing by the end of the week is as unreasonable as asking Beethoven to work overtime to produce another sonata. They are, I believe, as mistaken as those who take the view that you have only to kick the designer hard enough to get the job on time. How, then, can the design process be effectively controlled? The answer lies in determining the different kinds of control appropriate to the two different aspects and then carefully balancing the application of these controls at every stage of the design.

First, let us look at the creative aspect of design. In the past the Chief Designer of a small team was in intimate contact with his team and it was not difficult in this situation for him to control the creative ability of his individual designers by direct discussion with them. In the larger teams of today there exists the danger that in exercising control the Chief Designer places undue emphasis on his own ideas, while remaining insulated from much of the thinking of his team. This kind of repressive control fails to realise the full potential of the design team and invariably leads to low morale. In the future it will be essential to stimulate creative thinking at all levels and then to control the evolution of the design by a process of careful selection of ideas. A useful analogy is to regard this type of control as similar to gardening; the ground is prepared, seeds are sown, growth is stimulated and a continual selection is made by weeding out unwanted plants.

The second aspect of design that is concerned with the execution of ideas using routine engineering practice is clearly amenable to a more straightforward approach. However, effective control can only be maintained on the basis of reliable and meaningful control information, which in turn implies definition and measurement and it is in this area that most of our present problems lie. Most of the concepts and terms used in management are highly imprecise and in the past interpretation of these has usually been the prerogative of individual managers. This has led to a situation in which confusion exists about matters which at first sight look quite simple. For example, the number of staff in a department depends on whether you do or do not count apprentices as staff. The number of man-weeks a job takes depends on how a week is defined. Does it include overtime, allowances for sickness, holidays, etc? Whether a drawing is issued or not is judged by the production man on the basis of receipt in his planning department and not when it left the drawing office. This kind of thing has in the past involved management in endless reconciliation of information generated by different managers looking at the same situation in different ways.

A further factor which has been a fruitful source of confusion has been the traditional concept of overheads which involves the belief that high overheads mean bad management. As traditional skills have given way to new, essential functions involved in new methods have often been classed as overheads. In this way a high overhead has come to be an indicator in some cases of efficient modern methods and in some cases of bad management, depending on the circumstances. The result has been the overhead neurosis which is common today and a great deal of ignorance concerning the true state of affairs.

More recently the advent of high-speed electronic computers has begun to force a discipline on managers which is long overdue. At first, ill-considered input data resulted in meaningless outputs and as a result many were quick to condemn the computer as useless for management applications. Today it is recognised that to make use of these machines it is necessary to analyse management information and to define precisely the meaning of data to be processed and the resulting output.

In the future the increasing complexity of engineering projects will require more accurate prediction and better control of costs and time scales. This can only be done by definition and measurement in order to build up a meaningful statistical background for prediction and to provide the necessary error signals for control. Current systems analysis is forcing this development and integrated systems have led to the concept of a data base from which all management information will be generated to suit individual managers' requirements.

In parallel with this development all the essential functions involved in the execution of jobs will be identified and control will be extended over them; governed by the simple criterion of minimum cost to carry out the work in the time allowed and to a given standard. In this way the manager in future will not have to spend time on reconciliation of information and the confusion currently surrounding overheads will disappear.

Managers will then have the system, the time and the information available to allow them to make a much needed scientific study of the nature and mechanisms of management control. Interest in cybernetic principles applied to management will grow rapidly and repressive types of control will give way steadily to feedback control, as the fundamental value of this type of control in probabilistic situations becomes widely recognised by managers.

Many interesting and useful analogies have already been drawn between management and the control of systems of all kinds. An impressive and rapidly expanding background of theory and experience of control already exists and I believe that the computer systems that are currently being developed for management purposes will provide the unambiguous quantitative data that will allow in the coming century the development of scientific management control.

CONCLUSIONS

It is perhaps not surprising that, in considering the development of design management over the next century, one is led to the conclusion that it will be dominated by the development of the computer.

In terms of the overall industrial environment the advent of the computer is seen as a parallel to the introduction of machine tools in the past and leading to what can only be described as a second industrial revolution. The potential productivity of computer-based methods will lead to an intensification of the growth of powerful industrial units, capable of sustaining the high cost of facilities and of selling in international markets.

Turning to design, the computer will have far reaching effects on design activities as they are currently understood. Approaches to the changing pattern of markets will be revolutionised by the ability of computers to predict the logical consequences of taking into account factors which today would be regarded as opinions. The distillation, codification and validation of existing knowledge that has been stimulated by the computer, will grow and, together with developments in rapid access, will have a profound effect on our ability to bring relevant knowledge and experience to bear on design problems.

The design team of the future will consist of highly trained and experienced engineers engaged in a continuous and intimate dialogue with extremely powerful computers. Software covering a wide range of design and calculation routines will be available to them and the design definition, continuously updated as the design proceeds, will be held in the computer store. Control instructions to the computer will be given by spoken word and via visual displays. Outputs from the computer will include numerical data, graphs and automatically generated drawings and will provide the designer with the necessary material to scheme or perform calculations when these operations are outside the capability of the system or not worth the effort of programming. In addition, to enable designers to visualise the design as it proceeds, the computer will generate three-dimensional visual displays and machine tool control tapes for model manufacture.

Throughout the design the team will be working on a data base provided by the computer store and consistency and compatibility will be ensured by continuous automatic re-running of routines affected by data changes. When the design is complete the computer will provide the data necessary for manufacture and operation, including control instructions for automatic machine tools. Hard copy of selected visual displays will also provide outlines that can be worked up for technical illustrations.

One of the most significant effects of these future developments will be on the overall time involved in working out the consequences of design changes and decisions. By today's standards this will become vanishingly small and as a result the future design team will develop a much more intimate and creative feel for the design. Indeed, one may imagine the future Chief Designer, in almost continuous session with his senior colleagues, developing a design with the same creative immediacy that an artist feels for his work.

The effect of these developments on the structure of the design team will obviously be far reaching. The highest levels of creative engineering ability will be required to develop and control the design. In addition specialist engineers capable of working at the frontiers of knowledge will be needed to advance design methods and to break new ground in engineering techniques. No longer, however, will large numbers of people be employed carrying out routine calculations, making detail drawings or keeping records. Instead there will be a need for more electronic engineers and technicians, specialists in software development, programmers, etc.

An inevitable result of these changes will be a drastic re-appraisal in the future of our approach to the education and training of engineers. Not only will the balance of numbers of people with training in particular disciplines change, but also, I believe, the present emphasis on analysis will give way to a wider approach in which analytical thinking, creative or synthetic thinking and the teaching of experience are regarded as equally important.

In the field of management the impact of the computer will be equally far reaching. In the past management has been largely instinctive and often bad; managers looking at the same situation in different ways have themselves generated a good deal of the confusion that they have then sought to control. At the present time the computer is beginning to force a logical approach and a large effort is being deployed on the analysis of management systems. In the future the computer will provide the unambiguous quantitative data upon which, for the first time, a serious scientific study of the nature and problems of management control can be based.

The computer will also significantly affect organisational thinking. The design of company-wide systems to provide management control information and the need to build up statistical bases for prediction, will steadily strengthen the functional concept of organisation. This in turn will emphasise the need to train managers for functions rather than to tailor organisations to suit people.

As far as design management is concerned one would conclude that, for tasks of a given size and complexity, the problem would become more simple in the future. However, I believe that the complexity of tasks that we now undertake are rarely limited by scientific and engineering knowledge but by human effort and man's ability to comprehend, organise and control situations. In the future I see no reason to suppose that design tasks will not continue to be determined by the same criteria and if so design management will be as difficult for future managers as it is today.

Finally, I must emphasise that the future I have predicted for management of design involves enormous advances in computer technology and in our ability to make practical use of them. The way ahead is clear, but the rate of advance will be determined, not only by available technology and the cost and effectiveness of new methods, but also by the rate at which an industrial environment can evolve capable of sustaining design teams of the high cost and productivity envisaged. It is for this reason that, in my judgment, I have looked forward far enough and indeed to the many people struggling with the practical problems of computers today the future that I have outlined may seem like a bad dream. Nevertheless, we face the prospect of continued rapid development and looking back in one hundred years' time, I believe today's difficulties will have been forgotten in the revolution that has taken place.

ACKNOWLEDGMENT

The author acknowledges the permission of Hawker Siddeley Aviation Limited to publish this chapter and points out that the views and opinions expressed are personal and not necessarily those of the Company.

18 Law for aerospace activities 1966–2066

H. Caplan

PERSONAL PROLOGUE

The purpose of this prologue is to outline how I have approached the arrogant and impossible task of surveying an unborn century of law. I may also be able to illustrate that the nature and quality of the task is completely different from that attempted in the preceding chapters.

In this chapter I have done little more than infer repeatedly, in different ways

(*a*) that the shape of the future so far as law is concerned will be determined by the methods of communication adopted between sectors of the aerospace community and between the aerospace community and society at large, and

(*b*) that the search for effective methods of communication is urgent.

I take it as axiomatic that within a century some changes in the law will take place. Just as in other chapters—the question of 'choice' has been emphasised—so also, in the field of law, the question of 'choice' is paramount. If future systems of law are to be any improvement on the present, then this can only be done by the conscious efforts of many thousands striving towards similar ends.

The task which appeals to me is to improve the chances that whatever changes occur will be both rational and beneficial to the community at large and the aerospace community in particular. Of course, this is not the only possible strategy—but in the field of law it is impossible to be neutral about the future and I make no pretence of objectivity.

My colleagues who have contributed far-seeing chapters on scientific and technological subjects know that even within a century they have limits set by the known laws of nature—even though there are many laws and phenomena to be discovered in that time. The realm of possible technology and engineering is circumscribed completely by the state of scientific knowledge at any given time.

Although man-made laws for men are not subject to these limitations,

Originally published in the *Aeronautical Journal of the Royal Aeronautical Society*, April 1969.

the two major characteristics of the forces involved in legal change should be noted:

(*i*) the superficial simplicity of the mechanism which approves or enacts the change—the legislature.

(*ii*) the inherent and necessary resistance to change which exists in every stable society.

Traditionally lawyers would say that the second characteristic tends to be dominant and this is certainly the popular image of how difficult it seems to be to change the law. Thus, until 1966 I would have thought that any work attempting to forecast trends in law for a century ahead would have to be the result of prodigious research and study into the various mechanisms at work for legal reform in the leading Air Transport nations. Thereafter, an impressive study in comparative law, social trends and forecasts of technological change would enable a working draft to be circulated for comment to leading experts and in a few years perhaps one would have an intelligent prognostication for at least half a century if not a full century ahead. Indeed, this type of analysis—followed by a synthesis using the latest techniques of technological and strategic forecasting, might be necessary and productive.

On the scale I have indicated above, a 100-year forward survey is obviously beyond the resources of one man and may even be beyond the resources of a single private research institute. But one event in 1966 persuaded me that, however thorough such a study might be, there was still no way of calculating, in advance, the possibilities of such a revolutionary, cataclysmic and widespread change as that which occurred when just over 30 of the world's airlines agreed that for passengers travelling to, from, or by way of, the USA, they would accept liability for death or injury *without any fault on the part of the airline* up to a limit of $75,000 (inclusive of lawyer's fees) or $58,000 plus fees.

A stranger to the airline world might not think this curious even if he admitted that international air passengers enjoyed a legal liability system far in advance of that available to surface travellers. For the moment it may suffice to say that the concept of liability without fault is foreign to several hundred years of development in the common law, and the limit of liability previously in existence for such travellers was $8290. In a period of less than six months the shape and size of the airline liability world was changed. True, indeed, that 'absolute liability' as it is sometimes called, was not unknown to the common law systems—but it was linked to such special circumstances or to such minute sums of money as to be a negligible source of inspiration for widespread legal development. True, also, that for some years it had been apparent that the old limit of $8290 was so low as to be a threat to the continued existence of international air law. But no-one had the means to foresee even the approximate shape of the future—even one year before the event.

In the following year, the basic international treaty governing peaceful uses of outer space was launched, and this embodies rational and helpful principles which were scarcely even in the dreams of lawyers ten years earlier when Sputnik I achieved orbit.

In these respects 'Law for aerospace activities' may appear to be dragged screaming into the twentieth century—a long way behind the scientific or economic facts of life, but usually some way *ahead* of other laws.

Law is not an abstract topic of interest only to lawyers. In general terms the law is those rules of conduct by which the community, by its chosen legislators, consents to live, work and play. Today there are laws governing nearly every aspect of human behaviour.

In 1969, therefore, one should no longer be surprised that bursts of unpredictable human activity render the tasks of prediction hazardous—but I have already confessed that it was not until 1966 that I thought that any area of the youthful Aviation Law was capable of such radical or rapid change. This forces me to adopt an entirely speculative rather than analytical approach to the Society's Second Century.

Furthermore, I have had the advantage of reading a majority of the chapters in this book before finalising my own. I have read each one in apprehension—wondering whether the possible scientific or technological developments will give rise to unheard of categories of law. So far I have failed to find new territories—instead I see many of the present-day problems writ larger and in more terrifying detail—particularly in the field of liability for defective products. Thus, I am encouraged to think that the field of my legal speculation is comparatively untrammelled by the advances elsewhere.

The law for *this* century at least, is likely to remain law for *human* activities and it will not be necessary to consider new departures unless, for example, we encounter non-human intelligences in the course of exploring space or extra-sensory perception, or if self-reproducing machines capable of adaptive learning are created. . . .

Hence it may be enough if the law can provide the machinery to keep pace with development and not outstrip it. I do not dissent from the general conservatism of the legal profession: I applaud it provided that there is an adequate Research and Development effort. In this country, a considerable structure for R and D is beginning to emerge—based upon the work of the Law Commissioners, and the widespread activity which it stimulates in all sectors of the legal community.

Changes in the law can be effected if sufficient people at the right level can be persuaded that change is necessary. Thus, merely to state a desirable or possible future goal before the right audience may be enough to start the long processes of reform necessary to reach that goal. This is a process which has its counterpart in the scientific and engineering aspects of Aerospace—but with one important difference—the cost of reaching a given goal in engineering hardware, may be a significant fraction of the whole nation's purse. *The cost of effecting a change in the law, by comparison, is minute—but the benefits can be enormous* ranging from improved compensations for the widows of air accidents to increased business opportunities for aircraft manufacturers.

It may be going too far to say that *any* prediction, however far-fetched, must influence the future pattern of events either positively or negatively. But that is the challenge and those are the stakes for any who wish to alter the future 'Laws for aerospace activities' in a more rational or profitable direction. It is not a task that any sector of the aerospace community can embark upon alone: lawyers, scientists, engineers, civil servants, politicians, aircraft operators, industrialists and educationists all have a part to play, sometimes separately, sometimes collectively. The Royal Aeronautical Society embraces all categories and is uniquely qualified to play an influential role in developing future 'Law for aerospace activities'. It is my hope that this book will prove to be part of an active process which will survive into the Third Century.

The Society in its earlier history was fully alive to its responsibilities: 'By explaining what is being done, by discussing the conditions and possibilities of aeronautics, by securing co-operation between civil scientists and the Services, and by reminding Governments of their responsibilities, I believe this Society is doing very valuable and patriotic work'.[1]

A distinguished lawyer, Mr. Roger Wallace, KC, speaking to the Royal Aeronautical Society on 19th November 1913, described the task which is peculiar to the law:

'. . . we have to consider many theories governing a different kind of law and suggest means by which they can be reduced to practical regulations. We have not the same opportunities as the physicists, no official laboratories or opportunities for making experimental tests, and have to leave this to the cumbersome machinery of the administrative authorities of the governing political powers'[2].

There is still no means of providing the experimental facilities, and the 'cumbersome machinery' still exists. The opening words of a paper which marked the Society's 95th birthday (and included the above two quotations from the 1914 *Journal**) ended by recommending that the Society should promote and establish a 'Law Reform Committee or a Law Group with the general duty of applying Aeronautical and other Sciences to the evolution of law'. The Council has in fact established an Air Law Group which has from the beginning attracted distinguished English and foreign lawyers interested in aviation. Under the leadership of Sir Richard—now Lord Wilberforce (who remains as Honorary Chairman)—the Society has become the principal national focus for public discussion of air law topics, and the *Journal* has published more articles on the subject than any other periodical in the UK.

**The Aeronautical Journal of the Royal Aeronautical Society.*

But the Air Law Group has still to find effective ways of 'applying Aeronautical and other Sciences to the evolution of law'. Its prospects were never brighter than under the current working Chairmanship of Sir William Hildred, a notably un-retired Civil Servant and Director General Emeritus of IATA. He is *not* a lawyer but has a remarkable grasp of what lawyers should be doing and he played an active role in launching the final stages of the 1966 cataclysm. In its Second Century, therefore, the Society is well-poised to play an active role in influencing, if not creating, a great deal of the future 'Law for aerospace activities' —thereby continuing the tradition referred to by Lord Sydenham[1] and started by Roger Wallace[2].

1 INTRODUCTION

This chapter attempts to outline with broad brush, possible courses of development of the law as it will affect the principal sectors of aerospace activity in the years 1966–2066. No attempt has been made to illustrate a full spectrum of possibilities—the choice is dictated by subjective feelings of what is probable and in some cases it may be obvious that personal feelings of what is desirable have been allowed to conceal practical difficulties. The only universally applied test is 'Is it foreseeable?'

The following general definitions are adopted:

'The Law' may be defined as those rules of human conduct which are ultimately enforceable in the Courts.

'Aerospace Activities' will be used to denote the present and future scope of activities with which the Royal Aeronautical Society is concerned.

The scope of the chapter is international and is not meant to be a description of possibilities within any one jurisdiction—but, as my experience is rooted in the common law—my predictions will be on shaky ground elsewhere.

There are two basic messages I offer for all of my colleagues who are not lawyers:

1 Law is not a subject which concerns lawyers alone. It is part of the formal structure of civilisation and can only be improved or altered to suit the circumstances of aerospace development if sufficient people involved in that development are prepared to take a hand in modelling the appropriate laws.

2 Although law for aerospace activities can (and has already) made outstanding advances on its own, the general law applicable to other activities in each jurisdiction will place limits on unrestricted development solely for aerospace purposes. So, for example, Air Transport interests have succeeded to a large degree in simplifying the documentation and procedural requirements for Customs and Immigration for air passengers and (to a somewhat lesser extent) for air cargo in a large part of the world —but the ultimate limitation on further improvements is the Customs

and Immigration laws in each country—which are derived from legitimate national and international social and economic protectionism.

In part this chapter must be based on my understanding of developments in aeronautical law preceding 1966, but space does not permit an adequate synopsis of this era. The well-equipped reader is referred to that excellent brief book *An Historical Survey of the Law of Flight*[3] to which I am greatly indebted. Specialist readers will turn to the early chapters of that magnificent work of reference—*Shawcross and Beaumont on Air Law*[4]. Non-specialist readers may wish to make do with section 2 of 'The Law versus Science in Aeronautics'[5] which contains a somewhat uneven sketch of historical development in air law.

2 THE GENERAL DEVELOPMENT OF LAW

2.1 *General*

While major wars accelerate scientific or technological progress in specific directions, their main effect on the law seems to be to suspend or slow down progress. Science fiction writers and political leaders seem to be agreed that a Third World War would be the nearest approach to Armageddon imaginable and I find it impossible to speculate what form of society or law would emerge thereafter. Therefore, my speculations have to exclude that interesting possibility, and assume that in the next century we will continue to see a steady succession of wars which do not develop into global nuclear spasm. Non-global wars will have little effect on the development of general law.

No-one need doubt that so long as nations retain the right of self-defence (as under Article 51 of the United Nations Charter) all methods of waging war and of countering threats will be explored by all nations with the necessary technology and resources. Rules for war will continue to be made and discussed, but real wars will be limited more by military and retaliatory considerations than by ethical principles. There can be little doubt that with or without treaties intended to limit the proliferation of nuclear weapons, the membership of the nuclear club will grow and means will be found for using so-called tactical weapons in localised conflicts.

The theory of current disarmament talks is that the major powers will submit to inspections to permit verification of their progress in implementing any agreed pattern of disarmament. Even if the technical problems are solved for short periods of time (no-one could believe that controls can be satisfactory for long periods)—this does not by itself remove the sources of tension between nations.

Within the first half of the century I would expect the major powers who possess a deterrent (nuclear, biological, chemical or physical) to agree to comprehensive mutual inspection programmes. Thus aircraft and missile manufacturers who are already accustomed to teams of their own Government or Service inspectors in residence at their factories and offices will have to make room for teams initially of *neutral* observers and subsequently teams formed by potential or actual *enemies.* Secrets are at present the food of espionage and diplomacy—but we now seem to be moving into an era where the most effective form of counter-espionage (allied to mutual inspection) will be to deluge potential enemies with accurate information on a comprehensive scale and to offer generals, admirals, air marshals and chief designers full facilities for mutual *or even unilateral* inspection at all times. It is difficult to imagine unilateral successful military action if, for example, a high proportion of key military, scientific and engineering personnel from two or more enemy nations are in residence at each other's centres of activity.

The future shape of law is also likely to be comparatively little changed by whether economic growth in the developed nations is dominated by the gargantuan conglomerates (foreseen by Prof. Galbraith and others) or by massive State-owned corporations developed from current British or Soviet models. The modern nation is already a financial conglomerate corporation and several of the biggest American corporations already spend more money than small nations. The large corporations and the nations have already learned to live with a variety of national laws: the forces tending towards a harmonisation of the laws of international trade, or company law, or the reduction of tariff barriers—come from international motives for economic union or co-operation, e.g. the Common Market grouping, the European Free Trade Area, proposals for a North Atlantic Free Trade Area. This suggests that economic theory is a poor guide to the development of law, but economic or commercial co-operation can be a strong stimulus.

A corporation or a nation can attempt to protect itself from external or internal domination by means of anti-monopoly laws, anti-cartel provisions, laws restricting foreign investment, tariff barriers and so on —but this is a short-term victory unless the protected corporation or nation has the resources to be scientifically and technologically creative without outside help. As new nations rise to economic or technological maturity there will inevitably be new forms of protectionism developed and this is why I would still be pessimistic about the prospects for *very large* Free Trade Areas in the coming century. The message therefore, seems to be to fight for adequate resources in science and technology rather than seek support or protection for specific sectors of industry. The former is the seed of the future, the latter can survive only a few harvests.

There are two features of the present whose long-term development will, I suggest, have a considerable influence on the future shape of law and society, and will determine a large part of the background for aerospace laws:

(*a*) Student revolt
(*b*) Computer facilities.

Student revolt and the Fulton Report on the Civil Service have at least one feature in common—they are both likely to lead to forms of government in which 'participation' of the governed in the process of government will become more clearly defined and essential. No doubt it has been true for centuries that rebellion is a characteristic of young men and young nations. On the national scale, memories are short and the revolutionary origins of Magna Carta, Parliament, the United States and the USSR are easily forgotten. So long as rebellion is in the past it is safe and respectable. But student unrest is today a nearly universal phenomenon and although the causes espoused are as manifold as life itself, I suggest that as a symptom of dissatisfaction with the *status quo* it cannot be ignored. Succeeding generations of young scholars are being given access to the thrill of research, the tools for intellectual adventure, and the keys to understanding existing institutions and the nature of man. Many may fail to understand what they have learnt, others will be drunk with discovery and a few will, in their turn, become leaders for the next assault on human or scientific understanding. While this may be a slightly romantic version of what can happen in higher education—I suggest it contains sufficient clues for the understanding of student disquiet. It is impossible to give the tools for critical inquiry to generations of younger people without expecting that they will turn their critical apparatus on the society which has given them birth. Nor is it surprising that they may find it wanting, because society and the democracy which we cherish survives as much on apathy as on democratic activity. The student revolt is a challenge to the stable, conservative forces.

I suggest that the challenge can only be met by purposeful and genuine 'participation' by the challengers in the system they challenge. This is something more than giving the student president a seat on the governing body, or the trades union activist a seat on the management committee, or the creation of advisory councils for the benefit of a Department of State. I am not clairvoyant so I make no pretence that I can foresee the forms this 'participation' will take, but I forecast that within the century under study, genuine working methods will be found to promote and encourage:

(*i*) Government by open discussion and peaceful persuasion.

(*ii*) Full publication (or ready access to) all information and reasons relevant to the creation and implementation of government policy.

(*iii*) Participation, by all who are competent, in the processes of discussion and decision.

Hardly a revolutionary list!

The 'government' involved may be running anything from a local drama group to a Federal Union of States such as the USA (in which a great deal of the apparatus of publication and discussion already exists).

In theory, the apparatus for (*i*), (*ii*) and (*iii*) exists in all but totalitarian states, but it is the gap between theory and practice which is close to the heart of student unrest. Participation is not merely a fashionable word, nor is it a theory which implies an abdication of leadership—instead it calls for stronger, more able and convincing leadership capable, for example, of speaking the language of science and technology when necessary. Participation could lead, for example, to an active dialogue between scientists generally, doctors, schoolteachers and taxpayers (as well as political leaders) in any debate which called for an allocation of further resources to nuclear physics: to be weighed in the balance with other equally important projects such as greater facilities in a national health service or the educational services. It may lead to specialised debates in which the traditionally elected representative will not be heard unless he has taken the trouble to study the language and content of the subject. It may generate a new breed of multidisciplinary politicians, or yet a new breed of indifference. But it is going to happen within the next fifty or sixty years: slowly, if at all, in the primitive nations: quickly in the USA; somewhere in between in England. The significance for those involved in aerospace is that if they wish to have some say in the allocation of national resources to aerospace—from primary schools to advanced research or multi-billion pound production lines—then they must seek to educate, encourage and promote those who can speak and act for them in the disturbing new era which will be characterised by increasing participation at all levels of government by all those who are competent. Participation implies communication of all kinds from television and inter-continental computer links, to rapid transit systems (including air transport). Television, for example, has only just begun to influence electoral and deliberative processes. Its future impact may be as profound as the invention of printing. But government, even when opened up to theoretically widespread and competent participation, will continue to be a minority interest and occupation.

The growth of computer facilities is generally seen in terms of increased problem-solving capability, process control, strategic prediction, aids to automation, etc—but the more straightforward use of large computers for information storage and retrieval will lead to an equally great revolution in the work of professional men and in the processes of participation already outlined.

Already microfilm and microcards can be used to store such a large amount of easily reproducible data that every participant in a debate, for example, concerning the growth of the national economy could be given a copy of all the relevant facts—no matter how voluminous. This will do no more than assure an equality of access to information—but this is a considerable advance and is preferable to publication by older methods.

By the beginning of the twenty-first century, if there is the will to do it, there is no reason why access to nationally-gathered (or internationally-gathered) data should not be given freely to all competent applicants—

just like a ticket to a reading room of the British Museum. This means that all potential participants in the government of any organisation—from universities to industries and Departments of State—will have an equal opportunity to master, analyse and manipulate the data necessary for policy decisions. Concurrently, the input to such data systems would need to be regularly reviewed and monitored subject to legal safeguards giving the right to request regular input and operational checks. The 'Right to Free Data' might very well become a slogan to rival the 'Right to Free Speech'.

Sections of the legal profession in the UK and the USA are already studying various aspects of computer technology such as admissibility in evidence of computer output, legal methods of safeguarding privacy and data, and the use of information storage as a tool of legal research and practice. Within the next twenty-five years it should be possible to place the whole of the current law of England into computer controlled memory stores and to give access to all prepared to pay for the service. Legally-trained skills will still be necessary to formulate the retrieval programs, to interpret the outputs and to apply the results to a particular case. A whole generation trained in such facilities will trust the system sufficiently to abandon manual research and will thereby take the first steps to liberate the legal profession from unnecessary drudgery.

The preparation of legal texts for computer storage and retrieval will also have a profound and beneficial effect in encouraging logical legislative thought.

The cost of supplying computer services on a national scale will undoubtedly encourage many countries to develop the concept of a national legal service.

If this involves any sacrifice of professional standards or professional independence then such moves will naturally be opposed, with good reason, by the legal profession. Only if the profession can be persuaded to take a lead in adopting the benefits of computer technology will it be possible to avoid solutions imposed by society as the price for supplying computer services.

A great deal needs to be done before one can claim 'equal justice for all before the law' and before one can claim that all activities of the modern state are subject to supervision by the Courts. In France and in the USA, systems of administrative law are a well-developed concept: in England, the subject is dealt with in an incomplete patchwork composed partly of the work of the Council on Tribunals and that of the Parliamentary Commissioner. The coming century must see a considerable change and it may well be that participative government of the kind referred to briefly above is more suited to the English temperament than creating a system of courts or tribunals to handle economic, business, industrial, sociological, planning and similar problems. In other words, if the participative system of government effectively contains its own system for remedying grievances, then the role of the law should be that of safeguarding the system rather than operating a dispute machinery.

2.2 *International*

In 1961 I wrote in the *Journal of the Royal Aeronautical Society* that '. . . individual national sovereignty . . . is an ancient concept which may not survive the twentieth century . . . The increasing economic interdependence of nations, the 'explosion' of world population and the abuse of sovereign power—may ultimately lead to a dismantling of the extreme apparatus of individual sovereign power'[6].

I think I was wrong. There are no signs of any disappearance of national sovereignty within the next century—but there is evidence that economic interdependence reduces sovereign independence to harmless proportions.

The United Nations Organisation, through its specialised Agencies, Commissions and Committees is one of the most powerful and beneficial influences at work in the field of international law and in the harmonisation of differing national laws. This contrasts with the comparative ineffectiveness of the UN as a method for resolving conflicts or keeping peace across the world.

The 1967 'Treaty on Principles governing the Activities of States in the Exploration and Use of Outer Space, including the Moon and Other Bodies' (Appendix A) emerged from the work of an *ad hoc* UN Committee on the Peaceful Uses of Outer Space. There is little doubt that this Treaty and the subsequent agreements which flow from it will set the pattern of peaceful law in outer space for at least the next half-century and possibly more.

The ICAO, founded as a result of the 1944 Chicago Convention, is a specialised Agency of United Nations and is responsible, among other things for developing Standards and Recommended Practices for adoption by contracting States in international air transport. The Legal Committee of ICAO has been responsible for the text of five international agreements on air law since 1946.

By the end of this century there may very well be about a dozen principal international agreements affecting international air transport and a similar number relating to space activities. All of these agreements may be expected to influence or, to the extent necessary, may become part of the national law of contracting States. United Nations' agreements will not become truly universal until, as a minimum, the People's Republic of China is admitted to membership: hopefully this will occur in the next decade.

The USSR has not, so far, become a party to the 1944 Chicago Convention and China is unable while she is excluded from United Nations. As the USSR expands its own international air services and permits those of other nations to transverse her territory, the need for the USSR to become a member of ICAO will gradually seem more attractive and the USSR has already indicated agreement with ICAO Standards and Recommended Practices so far as possible. Thus the general prospects for a harmonisation of air law between East and West are good at the moment. The People's Republic of China, the Byelorussia and Ukrainian Soviet Socialist Republics, the Mongolian People's Republic as well as the USSR are all parties to the Warsaw Convention of 1929 dealing with the legal liability of air carriers. In fact the USSR has long ago ratified the 1955 Hague Protocol which *inter alia* doubled the limit for passenger death, injury and delay. The USA remains unrepentantly at the old limit—having secured for a majority the benefit of the 1966 cataclysm.

By the mid-century point, the co-existence of between 20 and 30 major international air and space agreements may give rise to renewed efforts for a set of revised and co-ordinated agreements covering the whole spectrum of air and space activities. There are many idealists (including myself) who think that such a task should have been possible before the end of the twentieth century, but the die is now cast into two moulds: one set of agreements governing international air transport and the other governing spaceactivities and it seems unlikely that there will be any serious efforts (if at all) for a co-ordinated series within the Society's Second Century.

In more general fields of law, developments must be expected in connection with an extremely large range of important topics such as the establishment of basic human rights, multilateral agreements concerning relief from multiple taxation, the control of tax-free havens, the reciprocal recognition and enforcement of court judgments, uniform laws for international trade and companies, the protection of intellectual property such as designs and literature, and the extradition of accused persons over a wider range of crimes. Wherever possible the trend seems to be towards multilateral agreements to succeed, reinforce or replace bilateral arrangements.

The international law-making and law-stimulating activities of United Nations and its specialised agencies has a considerable cross-fertilising effect not only from one branch of public international law to another (the law dealing with relationships between nations) but also in the domestic law of individual states. I would expect these trends to continue throughout the whole of the century under review. It is becoming more likely, in effect, that international law is 'the common law of mankind in an early stage of its development' (C. W. Jenks).

But there are many topics which do not seem likely to be dealt with by, or as a result of, international agreement and will develop within the municipal or domestic law of each nation. Some of these are referred to later.

2.3 *Domestic*

Court procedures, the execution of judgments, family law and the organisation of the legal profession are obvious topics which, for the whole of the foreseeable future, are likely to be dealt with on a purely domestic basis rather than by international agreements.

One of the most interesting topics which many people feel should be the subject of international agreement is given the legal name of 'damages' or financial compensation for loss. Within the period under study I see no real chance of widespread international agreement on the subject because it is so intimately bound up with the social theories of each nation. However, regional groups which share a similar economic or social philosophy (such as the Scandinavian countries) may already have a great harmony of law on the subject of damages. Nevertheless wider groupings will emerge only slowly.

Thus, one of the greatest disparities between law in one nation and another is the rules relating to the assessment of damages for death or personal injury—not only are the legally admissible factors different, but also the absolute amounts available in similar economic circumstances are quite different. So far, no agency or committee of United Nations seems to be examining the problem: partly, I suspect, because comparatively few people are aware of it.

There are some areas such as the common law provinces of Canada, South Africa and New Zealand where the proceeds of certain insurance policies are wholly or partly taken into account so as to reduce the claimable damages. In the UK and the USA this practice has been abandoned by law as it is said to penalise those who have been provident. In Scandinavia where social insurance is highly developed, the assessment of damages is based more upon need than compensatory loss. In the UK one half of the social security over 5 years may be deducted from damages in personal injury cases, but nothing is deducted in death cases.

But what is the purpose of damage awards? Generally it seems to be a system whereby those who have suffered as the result of another's wrongdoing may be compensated for their loss so far as money can do it. Academic discussion favours the concept of each business enterprise accepting liability for accidents occurring regardless of fault—this is the so-called concept of absolute enterprise liability. If you are killed at work or on a bus or railway train, so the theory goes—your dependants should not have to prove negligence in order to recover damages. The business enterprise is in the best position to bear the loss, to spread the cost by means of insurance if the enterprise cannot afford the direct costs out of its own budget, and so the cost of enterprise accidents is reflected back in charges to the consumer. So far as the limited scale of compensation for accidents and disease at work is concerned, this is already a feature of law in many countries.

It may be significant that only in aviation has this theory been accepted in any large-scale sense, first, in relation to liability for damage caused on the surface by an aircraft, and secondly, in the 1966 air carrier agreement referred to in the Prologue. It is certainly scandalous that the dependants of those who are killed on the roads or railways have to prove negligence before they can recover damages—even though in many cases they can be helped by application of the legal rule of evidence *res ipsa loquitur* which effectively shifts the burden of proof to the carrier to demonstrate that there was no negligence. If that is scandalous, then it is a crime of unparalleled magnitude that shipowners should still attempt to exclude all liability whatsoever by means of microscopic print on their voluminous tickets. Compensation from a shipowner is made to appear an act of grace rather than a legal right.

If absolute liability gains wider currency in air transport then within one or two decades it is certain that a clamour will arise for similar treatment in other forms of transport. Until other forms of transport have caught up with air transport in matters of legal liability—further extensions of the doctrine of absolute liability in air transport may be held back—but this is the only inhibitory influence I can foresee.

But even if all forms of business enterprise (including transport systems) were to accept absolute liability in respect of visitors or passengers (and so far there has been no suggestion of applying the doctrine to cargo) would this be the end of the road? Surely not.

Further anomalies are already visible. Why should the bereaved family secure compensation for the loss of a breadwinner only if he is killed or injured by some enterprise which accepts absolute liability? What if he kills himself in a car accident or suffers heart failure at a picnic? Everyone knows the answer to that one—he should take out his own personal insurance for such contingencies. But if there is personal insurance in force for such contingencies—should it not also be in force for enterprise accidents? In practice the answer is that the basic insurance is in force at all times—but travellers often take out additional cover on a trip basis. The prudent man will always effect sufficient insurance to ensure that so far as money can do it, his death or injury will not deprive the family of their standard of living. But the rule in most jurisdictions is that insurances are not taken into account in any damage award—so the result can be that an amply-insured widow is properly advised to secure further compensation by means of a lawsuit. On an objective view this is wasteful if the widow is truly compensated by the personal insurance arrangements.

Does this not suggest that a legitimate aim of social research might very well be to evaluate the total cost of litigation to the community and to see how the same ends can be more effectively served at lower cost and greater expedition?

Surely the true social evil is the person who, in an affluent society, does not make adequate personal insurance arrangements? If this view is only approximately correct then within the first half of the century there is likely to be considerable research devoted to this end. Steps are more likely to be taken in Western Europe than in the USA, and more likely in New Zealand[7] than anywhere else.

By the end of the century—led by socialist countries—I would expect the following views to be gaining wide currency:

1 The State will lay down minimum standards of personal insurance to cover death, injury or incapacity from any cause whatsoever including accident, sickness, disease and old age.

2 In liberal socialist economies the individual will be allowed to purchase the minimum insurance from private sources. In strict socialist economies the State will be the monopoly provider of the minimum insurance.

3 The State will supervise the administration of funds in both the State and the private insurance systems to ensure that proper compensation is paid.

4 Litigation for the recovery of compensation from anyone who has caused an accident resulting in death or injury will be abolished (criminal penalties may remain as the only consequence of wrong-doing) and the insurance funds (State or private) will not be allowed to reimburse themselves from those who may have caused or contributed to such an accident. (This latter feature is already a widespread principle of law relating to life insurance.)

These are integrated steps and there would seem to be little purpose in steps 1, 2 and 3 unless accompanied by step 4 and vice versa.

Litigation for the recovery of car damage claims has to a large extent been abolished already by private agreement between British motor insurers. In Iceland, for example, insurers who have paid a damage claim of any kind are not allowed to litigate for the recovery of sums paid except in the event of criminal negligence. Each of the four steps already exists in one or more countries. All that would be new is the integration of the four steps and their widespread adoption as an integrated principle.

Thus it seems probable that the first half of the century will be characterised by the growth of so-called enterprise liability and the latter half may very well see the general spread of State-supervised comprehensive personal insurance accompanied by the abolition of all litigation for personal injuries and death.

Finally there is no reason why the abolition of litigation should stop at personal injuries and death where the emotions can be deeply engaged in any debate. There should be fewer emotions involved in loss or damage to property—which, if it is property insured for full value, need not be the subject of further litigation for the benefit of the 'All Risk' Insurers who presumably have received a proper premium in consideration of the known risks.

If the day dawns when all litigation is abolished for accidental personal injuries and death, and loss of, or damage to, property, then I am quite certain it will not be in my lifetime—but it is foreseeable as a trend in the latter half of the century—particularly if 'participative' forms of government have developed and decisions are based on better and more comprehensive data than is possible in this pre-computer age. It is not a cheerful thought to forecast the possible demise of part of the insurance industry in which I work!

3 COMMERCIAL AIRCRAFT OPERATORS

3.1 *Introduction*

This section sketches some possible developments in a limited number of topics relating to commercial aviation. The potential field is extremely wide—so the selection has been based on those topics which have wide commercial or financial significance.

The rapidly rising cost and size of new aircraft will probably be the most pervasive single influence on the developing law. Among the principal effects will be:

(*a*) international consortia for financing, insuring, operating, maintaining and leasing,
(*b*) multinational ownership and operating rights,
(*c*) new methods for improving annual utilisation,
(*d*) aggressive and novel marketing.

Each of these topics may be expected to modify the legal environment.

It is not surprising that all of the major international agreements relating to air law are of significance to aircraft operators. The keystone of the post-war world of civil aviation—the 1944 Chicago Convention—is unique because it contains comprehensive obligations for contracting States which ultimately affect all components of the civil aviation industry air traffic control authorities; airworthiness, licensing and registration authorities; designers and manufacturers; those engaged in aircraft operation, maintenance and repair; chart makers; telecommunications and airport operators.

So far, there are no conventions dealing specifically with the legal liability of air traffic controllers, aircraft manufacturers and designers, or governments exercising control over civil aviation.

3.2 *The right to fly*

By comparison with the freedom accorded to international shipping, the right to fly scarcely exists and yet air transport has long ago become the principal means of conveying passengers across the oceans. In the 1944 Chicago Convention, the right to operate scheduled services was left to individual bilateral negotiations with the result that the principal scheduled air routes are the subject of hundreds of bilateral agreements—most of which reserve rights for the carriers designated by the governments concerned.

Non-scheduled traffic was nominally given greater freedom in the Chicago Convention, but in practice the concessions have been strictly controlled for the purpose of protecting scheduled carriers: so much so, that in Europe, new multilateral agreements have emerged with the object of restoring more freedom to limited classes of non-scheduled traffic—but again the object has been, so far as possible, to protect the scheduled carriers.

The rules, fare-concessions and intensive operations by so-called non-scheduled operators have already blurred the once sharp distinction between the two categories. It is becoming increasingly clear that the distinctions cannot survive for much longer as the distinctive rules become more complex and difficult to enforce. At the same time, larger and more expensive aircraft are being introduced which require more intensive utilisation and more simplified procedures if they are even to pay off the financing or leasing burden within the useful life of the aircraft.

Therefore, I forecast that within two decades the distinctions between scheduled and non-scheduled services must disappear on the busiest air routes. When this happens, bilateral agreements which regulate scheduled capacity will be eligible for re-negotiations and at this point, the 'freedom of the air' may be re-opened for a fundamental re-examination.

The only surviving purpose of route and capacity agreements between governments will then be to give a fair and equal opportunity to their respective designated carriers. There is no fundamental reason why this cannot be left to commercial competition and if the result is that carrier *X* swamps the market with more capacity than country *A* is prepared to tolerate on behalf of its carriers, there is no reason why country *A* should not levy a surcharge in respect of each passenger carried by carrier *X* in excess of the capacity which country *A* is prepared to tolerate.

Poorer nations may even find that it pays them to live off the proceeds of carrier surcharges rather than provide air services of their own. If the net result is to improve trade and communications with the territory, there can be no real disadvantage to the economy, while the field would always be open to the national air transport entrepreneur to enter or re-enter the market at any time.

Today governments seem to distrust market forces or economic pressures to regulate the flow and development of air transport. This may have been a correct philosophy before profitable aircraft were available to the infant industry—but it can no longer be justified with the third, fourth and later generations of money-making subsonic jets. Thus I expect to see economic forces producing a gradual relaxation and even an abandonment of the restrictions on air services over the next fifty years.

Whatever the market is for supersonic aircraft, I venture to prophesy that for the next century at least, there will always be a market for large and small, short-range and long-range subsonic jets and that ultimately they will operate as freely in the regular and irregular air routes as ships once did over the high seas in the heyday of shipping.

By today's standards, such a vision may be regarded as Utopian—but very soon it will become a harsh economic necessity if the civil aviation world is to make full use of the aircraft production lines in the UK, USA and Europe. In no more than fifty years I would hope to see the present bilateral agreements being disregarded increasingly. In 100 years I hope that the right to fly will be studied only by historians.

3.3 *Legal liabilities*

There are two basic international agreements which at present deal with two separate aspects of the air carrier's legal liabilities:

The Convention on Damage caused by Foreign Aircraft to Third Parties on the Surface [1952 Rome]

The Convention for the Unification of Certain Rules relating to International Carriage by Air [1929 Warsaw].

The 1952 Rome Convention has been accepted by over twenty States (not including the UK, USA or USSR) and does not seem likely to develop much further. This Convention deals with the liability of the operator of an aircraft registered in one contracting State which causes damage to persons or property on the ground in another contracting State. Compensation is payable on proof that 'damage was caused by the aircraft in flight or by any person or thing falling therefrom'. The liability is often referred to as absolute because the operator is given very few circumstances in which he can escape liability. The operator's liability is limited in accordance with a graduated scale related to the certified gross weight for take-off. Contracting States are entitled to ask for evidence that the aircraft operator can pay compensation in accordance with the graduated scale of liability—this may be in the form of certificates of insurance, guarantees or cash deposits.

It is not easy to discover or to summarise all the reasons why the 1952 Rome Convention is unlikely to make any further progress without substantial amendment. Initially, for example, the USA was opposed to the principle of absolute liability—but that objection has been withdrawn following the 1966 carrier agreement on passenger liability. The limits of liability are quite small by modern insurance standards. Recently it has become more certain that the Convention is not really suited to deal with supersonic aircraft because it is by no means clear whether sonic boom damage would be covered by the Convention: there is no right to compensation 'if the damage results from the mere fact of passage of the aircraft through the airspace in conformity with existing air traffic regulations'. So far as insurance is concerned, the 1952 Rome Convention is an improvement on an earlier Rome Convention (1923) which made insurance (or equivalent guarantee or cash deposit) compulsory.

There seems no purpose in a modern convention unless it can deal clearly with all the damage which can be caused by foreseeable aircraft. This means that sonic boom damage must be dealt with clearly one way or the other and the limits must be greatly increased. But international requirements for liability insurance by themselves will not solve the problem of securing compensation for innocent victims because of the wide variety of methods and markets for providing such coverage in various parts of the world. For example, in many cases, a State certificate asserting that a local insurer is in a sound position may be irrelevant if the local insurer has substantially reinsured the risk with another market in another country.

Moreover there is no evidence that innocent persons on the ground have not received compensation because the aircraft operator was unable to pay or his insurers refused to pay. Therefore, it is doubtful whether there is any need to include this aspect in an international convention—but if there is, then it is more certainly accomplished by means of mutual guarantees between States, to be called upon *only* if the aircraft operator fails to pay.

Until such time as there is any radical change in the law relating to compensation I would expect a new Rome Convention to emerge within the next two decades containing the following features:

(*i*) Retention of the principle of absolute liability and its application to sonic boom damage.

(*ii*) Increased limits of liability—with provision for local or general increases at the request of individual States or groups of States so as to deal with local circumstances, nuclear-powered aircraft, etc.

(*iii*) The operator's ability to pay backed by a guarantee from the State of Registration (leaving that State free to make such rules as it pleases in relation to its own aircraft and the recovery of any guarantee payments).

(*iv*) An obligation on contracting States to make the same minimum rules for its own aircraft as well as foreign aircraft.

The above four features of law relating to civil aircraft operations are also relevant to the development of law for space activities—for which there is, as yet, no international agreement concerning liability although several drafts are in an advanced state of readiness. One of the basic questions in space law which at present divides East from West is whether it is appropriate for the State concerned to assume legal liabilities for the consequences of injury or damage done in space activities. Point (*i*) places primary liability on the operator and point (*iii*) makes the State a financial guarantor for the operator's ability to pay. Thus if work on an agreement for space activity liability can be delayed long enough, and work on a new Rome Convention for aircraft can be accelerated it is conceivable that the two sets of draftsmen would be persuaded to pool their ideas, and if this should happen then there is some faint hope that the two problems could be dealt with in a single international agreement.

The 1929 Warsaw Convention has been accepted by approximately 100 States including many representatives of common law and Civil Code countries as well, linking East and West. By an enormous margin it is the greatest international success in solving problems of 'lawyer's law' e.g. which system of law applies in an international accident where several choices could be made depending upon which courts consider which problem? The Warsaw Convention (like the Rome Convention) provides a uniform code of law which can be applied in each contracting State with a high degree of probability that similar results will be reached in the courts of each State when considering similar factual problems.

The Warsaw Convention places a presumption of liability on the air carrier for death, injury or delay to passengers and for loss of, damage to or delay of cargo and baggage. Thus the plaintiff has very little to do to establish his case and it is for the defendant carrier to use the defences provided by the Convention. The carrier's principal defence is if he can prove 'that he and his servants or agents have taken all necessary measures to avoid the damage or that it was impossible for him or them to take such measures'. In return for this simplification of the plaintiff's task, the carrier is given a limit of liability—worth approximately $8300 for death, injury or delay and $16·60 per kilogram for loss of, damage or delay to cargo and registered baggage. The carrier cannot rely upon the limits if he is guilty of 'wilful misconduct' or in the event of certain documentary defects.

By modern standards the death limit is very low and in 1955 several changes to the basic Convention were agreed at the Hague. One of the main alterations was a doubling of the death limit to $16,600—which was still less than the $25,000 which the USA sought in 1955. In the USA, most of the domestic law does not contain a limit of damages for wrongful death although there are still ten States which have limits ranging from $25,000 to $110,000. Thus every time the USA administration approached Congress for approval of the 1955 limit of $16,600, there was renewed discussion of the Convention and fresh opportunities for its opponents to urge US withdrawal from the Convention. These pressures reached a peak in 1965 when, in November of that year, the US gave the necessary six months' notice of withdrawal but said that the notice could be withdrawn if sufficient international carriers could agree, as an interim measure, to accept a liability limit of at least $75,000 for each passenger travelling to, from or by way of the USA. At the 11½th hour, in May 1966 following the initiative of Sir William Hildred (then Director General of IATA) the minimum necessary group of the world's airlines acceded to the US Government's demands which by that time had *increased* by the inclusion of a new condition that the airlines should, *in addition to the $75,000 limit*, agree to waive their principal defence under the Convention (quoted above). This is usually referred to as the Montreal Agreement because the detailed terms were settled in Montreal on 4th May 1966.

The result was that the USA withdrew its notice of withdrawal the day before it would have taken effect, and the world's airlines took the fateful step of accepting what is loosely called 'absolute liability' (although the carrier can still use the defence of the passenger's contributory negligence in appropriate cases) and a limit of $75,000 (inclusive of lawyer's fees) or $58,000 in those countries where lawyer's fees may be *added* to the Court award of damages (e.g. England). Today the Montreal Agreement has well over 100 signatories and there are many more carriers who have accepted the terms but have not yet signed the agreement (e.g. Aeroflot).

There is no doubt that the limits and the principle of absolute liability are widely acceptable. But there is by no means universal agreement on methods for translating the Montreal principles into an inter-Governmental agreement with a stature comparable to that of the original Warsaw Convention. There are many anomalies to be cured, e.g.:

(*a*) The Montreal Agreement only covers passengers travelling to, from or by way of the USA (BEA alone has accepted the limits for all its international passengers. The UK and New Zealand have extended the $58,000 limit *by law* to non-Convention passengers. The West German Government has requested a similar limit to be adopted *by contract* by carriers on the German domestic routes.)

(*b*) In the USA passengers can secure unlimited damages if they can prove that they received inadequate notice of the Warsaw limits.

(*c*) In the original Warsaw Convention the carrier is not protected by the limits if he fails to deliver a passenger ticket, and under the Hague amendments, the ticket must contain a notice about the possible application and effect of the Convention.

(*d*) The carriers have accepted liability without fault even though a third party may have caused or contributed to the damage, e.g. in an accident caused by errors of air traffic control or aircraft design, or if the aircraft is shot down by an unfriendly State or by a terrorist.

Furthermore, comparatively few States have adopted the 1961 Guadalajara Convention which supplemented the original and the amended Warsaw Convention principally by enabling the passenger to sue either or both the carrier with whom he concluded a contract and the carrier who actually performed the contract—but he could only recover one set of damages limited by the Conventions.

International airline passengers today may be governed by any one of five limits of carriers liability—$8300, $16,000, $58,000 or $75,000 and all could be found travelling on the same aircraft!

Under the auspices of ICAO, the principal Warsaw States are examining the question of a further Convention or an amendment to the existing ones which would have the effect of re-unifying the code of law in the post-Montreal era. The United States is uninterested in these processes until such time as there is a prospect of a liability limit exceeding $75,000. In September 1969, the US indicated willingness to accept a limit of $100,000 and, for the first time, was ready to abandon the requirements for giving warning to passengers about liability limits. A recent limited survey of passengers arriving from the USA (conducted on behalf of the Board of Trade) showed that despite larger notices to passengers in 10-point size print (given in accordance with the Montreal Agreement) only 19% had read the notice and only 4% had, as a result, purchased any additional insurance.

The result at the moment of writing in 1969 seems to be that even if other nations produce what they think is an adequate successor amendment to the Warsaw Conventions, there is little prospect that the USA will be a party except on terms which will be unacceptable to the others. The impasse is a hopeful sign because without it, there seems little likelihood of a fundamental re-examination of the issues involved. Unfortunately the mere fact of an impasse is, by itself, insufficient to generate the necessary re-examination, but if, by some miraculous series of negotiations, a further compromise is reached between the USA and the rest, then there will be little chance of a fundamental inquiry for at least another generation.

The most likely result in the next decade is the emergence of a successor (or a comprehensive amendment) to the existing Warsaw system with the object of removing as many anomalies as possible and raising the limit of liability for death to at least $75,000 and possibly $100,000 (with automatic periodic increases) on a basis of 'absolute liability'—modified to take account of some of the factors noted above. If the result involves an abandonment of the requirement to give notice to passengers, then the USA is unlikely to be a party even if the limit of liability is satisfactory.

In an era which opens with the requirement to handle three or four hundred passengers in a single aircraft (probably without tickets and using credit cards as the sole check-in document) it is clear that the carrier's liability limit should no longer depend on whether a ticket has been issued or whether a particular form of notice is given. It is also out of tune with the times that the carrier's liability for baggage should depend upon the actual weight of individual baggage: a single universal limit per passenger should suffice. At the same time, it is essential, with such large numbers of passengers that some form of liability limit for death and injury should exist. (There seems no need to increase the limit for delay.) When personal accident insurance is so cheaply available on a world-wide basis there is really no cause for compassion for wealthy widows who might be able to secure no more than say $250,000 from the airline on whose aircraft her husband was killed. There is a cause for compassion in the case of those passengers who survive serious accidents and are severely incapacitated for the rest of their lives, because adequate personal insurance is difficult to arrange at present, particularly for young people.

There is great interest, even in the USA, in a Convention which might have a reasonable single limit accompanied by 'absolute liability' but without any method of breaking the limit. At the moment, the USA government is unwilling to examine the consequences of its insistence that every passenger should be given notice of liability limits so that he can, if necessary, purchase additional personal insurance. If this is an honest and humanitarian motive, then the answer is surely a nation-wide publicity campaign aimed at

(*i*) educating the public to insure against every conceivable event (and not merely when they travel by air) and

(*ii*) encouraging a change in the law in each State and in the Federal law whereby it would be a criminal offence for each adult not to have minimum personal insurance in force.

The USA will undoubtedly be the last place in the world to abolish legal actions arising out of personal injury and death, but they may be among the leaders in national legislation towards universal insurance—despite the socialist implications. Life and medical insurance *per capita* in the USA is the highest anywhere in the world. US agencies as diverse as the Civil Aeronautics Board, the Military Airlift Command and the California Public Utilities Commission already call for the most comprehensive and expensive *liability* insurance to be effected by relevant air carriers. In not more than three decades from now, I prophesy that someone in the USA will inaugurate a nation-wide cost-benefit survey of the various life and medical insurance programmes, the costs of administering the legal system and the court awards and the policy of the regulatory agencies towards minimum insurance requirements—all in relation to the basic question 'What is the quickest, fairest and most economical way of dealing with the financial consequences of accidental death and injury in the population?' It would be very much to the credit of the American Bar Association, for example, if they were able to take the first steps in stimulating such a far-reaching and costly survey.

My own forecast of likely answers to these questions was outlined in section 2.3, and while I think the USA may be among the first to ask and answer many of the questions, they are likely to be the last to accept the consequences if their present and foreseeable attitude to the Warsaw Conventions provides any guidance.

Damage awards in the USA seem to be escalating faster than anywhere else and a court in Chicago still holds the world's record for a jury award in respect of the death of the head of a family—$2 million in 1965 (settled for less prior to appeals). There are cases known which could beat this record if they are not settled before trial.

Therefore, I think it likely that with or without the Warsaw Conventions (and any successor amendment) personal injury and death litigation in the USA is continuing to accelerate away from the rest of the world faster and faster, and must ultimately be reflected in a very serious community cost burden. It is possible that within 50 years the federal Government will be moved to intervene and the most obvious and economical first step would be a return to the position whereby the proceeds of life insurance policies would be applied to mitigate the damages recoverable from wrong-doers. This would not really be a fundamental step (although no doubt it will be denounced as retrograde) and it would not involve the sacrifice of too many sacred principles—but it could slow down the flood of damage settlements and awards for a decade ro more. In the course of the slow-down, more fundamental solutions could be analysed and discussed.

What is the future of cargo litigation? It would be easy to hope that this might be abolished within a few decades because nowadays litigation is solely for the benefit of cargo 'all risks' insurers and they would have no principles to lose if those they insured had no rights to further recovery. It is possible that the 'all risks' insurance rates would rise over a long period of time to compensate for the absence of any net recoveries.

The difficulty of such a facile piece of wishful thinking is that, at present, there are no visible pressures to produce any great change in the legal position of cargo shippers/consignees in relation to air carriers. In many parts of the world tariff limitations on the carrier's liability are accepted and enforced by the Courts. Again it is the Warsaw Conventions which provide a means for making claims in excess of the limits.

It is possible that the growth of containerised cargo or indivisible pallets will give rise to increased cargo litigation because of the possibility which exists for the individual cargo shipper to sue not only the person who makes up the container but also the carrier. When this tide of litigation is in full flood over the next two or three decades, then pressures may arise to seek an overall examination of the purposes of the litigation. If the only object of the litigation process is to transfer the burden of accidents and thefts from one sector of the cargo industry to another by very costly and inefficient means, then, at that stage I would be hopeful that rational means will arise for minimising or abandoning litigation.

The final question under the heading of 'Legal Liabilities' is 'Whatever legal liabilities are borne by the air carrier, what are the comparable liabilities of others who may be involved?' The position of designers and manufacturers is considered in the next section but the position of governments and air traffic control authorities is far from clear.

If the air carrier bears absolute liability by operation of law, then there is some hope that either expressly or by implication, the carrier will have rights of recovery against others who might truly be responsible. Provision for this is contained in the 1952 Rome Convention and in equivalent United Kingdom legislation. Thus if an aircraft crashes by reason of some air traffic control error—both the 1952 Rome Convention and the UK legislation would require the aircraft owner to pay compensation to those injured on the ground without prejudice to the aircraft owner's right to seek recovery from the air traffic control authority. Hence 'absolute liability' does not bring an end to litigation—it merely facilitates, as it is meant to, the original plaintiff's task.

Not all governments who supply air traffic control facilities can be sued in the Courts. But in the USA, for example, where the FAA has been successfully sued for air traffic control errors (and even for alleged errors in the airworthiness certification of aircraft) the Government is interested in some legislation to give them a limited form of liability. There is also international interest in a convention which would limit the liabilities of air traffic control authorities—but this topic will take a long time to settle—first of all because of the difficulties of setting suitable limits of liability and second, because in many countries, air traffic control is a military responsibility and the local law is not yet such that military agencies can be sued in the civil courts.

Many governments might still be in a position to request and enact legislation which would effectively give them immunity from litigation arising out of air traffic control activities, or alternatively give them strictly limited liability. But the general trend of legislation everywhere seems to be away from special categories of immunity. Thus if a government felt impelled to grant its air traffic control authorities an immunity, that government might find it necessary to do so only in the context of prohibiting *all* litigation arising out of an air accident. But that is unthinkable unless the general law is ready for such a change, and that would be predicated upon widespread requirements for personal insurance as outlined in section 2.3. Thus if sufficient governments are worried about their potential liabilities arising out of a mid-air collision between two jumbo jets fully laden with expensive Americans (the total bill could be around $150 million)—then this may give some impetus to fundamental inquiries into the purpose and cost of litigation as a means of dealing with the financial consequences of accidents.

Fortunately the number of accidents attributable to air traffic control errors is very small and hence the question of legal liabilities arising therefrom is scarcely the most important problem that governments will have to solve in the next century even though some of their lawyers may have sleepless nights imagining what *could* happen.

Until such time as there are any changes in the general law of a large number of air transport nations, I do not foresee any curtailment of the air carrier's rights of recovery from others who may be responsible for accidents such as air traffic control authorities. Therefore, there will be a gradual evolution towards some form of international agreement to limit the liabilities arising out of air traffic control. The history of the past twenty-five years illustrates that if the relevant limits of air carriers' liability (or any limits which may in the future apply to air traffic control) fall seriously out of step with damage awards then, litigants will not only seek to destroy the limits by every legal means, but will also seek supplementary targets such as aircraft manufacturers.

The problems to be faced by governments in liabilities arising out of air traffic control, or airworthiness control activities are similar in legal nature to those which they will have to assume in relation to space activities. Again there is a theoretical opportunity for a co-ordinated and comprehensive international agreement to cover air and space activities, but again it seems unlikely that the opportunity will be seized within the coming century.

3.4 *Miscellaneous*

The widespread use of computers for storage and retrieval of information should facilitate international and national record-keeping for all vital purposes: for example aircraft registration, mortgages, leases and all other financial information including the existence of insurance, cash deposits or guarantees if required. There need be very little gap between international agreement and implementation on these subjects provided that suitable assistance can be given to nations unable to afford their own national data banks. Again economic forces will shape the necessary agreements.

For example if the 1948 Geneva *Convention on the International Recognition of Rights in Aircraft* proves to be useful in the long term, comparatively simple data stores would be sufficient on an international basis to contain all the relevant information for all commercial aircraft everywhere. Agreement would be needed for the location of the central memory, and for the rules governing input and access to the data, together with a compendium of laws of all the contracting States on topics relevant to the Convention. The Convention already provides that 'Reasonable charges may be made for services performed by the authority maintaining the record'.

If ever interest rates come down nearer to the cost of insurance, the way should be open for commercial agreements which will link financing and insurance in a single package over the first working life of an aircraft. This would be the beginning of a considerable reduction in operating costs which can be facilitated and monitored by computer-assisted systems containing data on aircraft registration; financing; mortgages; leases; insurance; operating costs; revenue; payments to creditors.

If the technical problems of securing confidential input and access to the various categories can be solved, the remaining steps will be legal, i.e. developments in the law to protect the operation of a multi-category data store with security requirements which vary from one category of data to another. Less than a decade may be necessary to solve the technical problems, but the legal problems may take the remainder of the century.

Customs procedures may continue to be a source of irritation to air carriers and cargo shippers/consignees for the whole of the foreseeable future, and some form of Customs control is essential so long as each nation has specific tariff barriers or trading problems to solve. Many notable advances have been made in recent years (e.g. the SPEED and SPEAR systems at Heathrow) and computers are being introduced to deal with import documentation. But the fundamental system in all countries still seems to involve personal examination of relevant documents and cargo by Customs officials. With the predicted volume of air cargo growth (unaccompanied by a comparable growth in Customs staff or facilities) I suggest that this ancient system cannot continue throughout the coming century. In the United Kingdom and elsewhere, the Customs authorities have begun to trust shippers in various ways, e.g. by not requiring advance payments of Customs duty if there are suitable cash deposits. This trust must be expanded into a comprehensive licensing system whereby suitably qualified shippers/consignees become licensed agents of the Customs authorities—charged with the statutory duty of enforcing all applicable regulations, and keeping the necessary records for Customs verification at any time. The Customs authority would, therefore, become a supervisory body only and would collect a revenue from the licensing fees as well as the duties payable. Implementation is impossible without a period in which shippers/consignees receive education and are able to recruit suitable staff. In many areas we are halfway to such a system and one hopes it will not take more than a decade to plan a comprehensive licensing system so that in not more than three decades it is working well in Europe or the USA and may spread elsewhere by the second half of the century.

Immigration problems may only be of a temporary nature, but it is equally conceivable that they will assume a greater world-wide importance if the gap between under-developed areas and highly industrialised ones grows larger instead of smaller. This is another problem whose solutions will be more apparent when there is better communication and data transfer between consular officials in the country of origin and immigration authorities in the country of destination. At the moment, airlines in London, for example, are being asked to pay for the cost of detainees and repatriation.

In the late 1960s airlines should be able to rely on government departments to satisfy themselves before passengers are embarked at points of origin. It may not be long before airlines refuse to pick up certain categories of passenger unless they have written confirmation of acceptability direct from the country of destination. There are still governments who do not have adequate Telex facilities available to immigration and consular officials in important parts of the world. It may, therefore, take another fifty years before the fruits of existing communication technology come into general use and to speed up the government side of immigration regulation. The airlines can accelerate this process by formulating their own requirements for passenger embarkation.

Aeronautical regulations and operation rules need to evolve in such a manner that they do not inhibit technical or economic progress. In an earlier paper I argued the case for a considerable simplification of aeronautical regulations on safety grounds. I think the approach can be justified on the broader ground of avoiding technical detail in operating and airworthiness rules and concentrating instead on the essential philosophy of regulatory activities—leaving each aircraft design as a specifically 'acceptable means of compliance', and each aircraft's operator Manual for Flight or Ground Crew as a specifically acceptable method of carrying out the safety philosophy of the regulatory authorities.

Aircraft noise will continue to be a problem of balance between the industry and the resident population near airports and air routes for the whole of the foreseeable future. No-one has promised to eliminate noise completely—merely to reduce it to acceptable levels by means of aircraft or engine design or operational procedures. The 'acceptable level' will always be a matter of opinion and there is still no news of what will prove to be acceptable in relation to supersonic aircraft either in the airport vicinity or *en route*. I am confident that ways will be found to make supersonic aircraft acceptable to the resident population—but it is possible that the first generation of supersonic aircraft may be forced to accept economically unattractive routes, altitudes or speeds to make them acceptable to the relevant community. West Germany, for example, has announced a prohibition of supersonic flight over her territory. In an earlier paper[5] I concluded that 'rules of law can often act as a serious limiting operational parameter—particularly for supersonic aircraft'—this still seems to be a reasonable prediction of things to come. Aircraft and airport operators (and designers) will no doubt continue their efforts at noise reduction—but equally great efforts will be needed in each country for aeronautical enthusiasts to become familiar with the communities they serve and to understand the law-making processes. Failure to do so may mean that the air transport and aerospace manufacturing industries are being denied opportunities for expansion because they have not taken sufficient trouble to communicate with the society in which they live.

4 DESIGN AND MANUFACTURE

4.1 *Introduction*

This section is intended to cover some of the legal liabilities which may be expected to arise out of design and manufacture of *any* aerospace product—personal air-taxi, supersonic jets, or vehicles for planetary exploration.

Forward speculation is based on the primitive notion that what has happened in the USA will develop elsewhere, and that the manufacturer's position will get even worse in the USA.

4.2 *The tort of manufacture*

Several of the states forming the USA are now classified as 'strict liability' states from the point of view of a manufacturer's legal liability for defects in his product—often referred to as 'Products Liability'. That is to say in these states it is enough for an injured party to show that he was injured by some defect in the product for which the manufacturer was responsible, and that will be enough to require the manufacturer to pay compensatory damages. Shorn of technicalities this liability extends to users of the product such as airlines and their passengers. This means that if a product defect causes the total destruction of the aircraft and all on board, the manufacturer will be responsible for compensating everyone: the airline for loss of the aircraft, dependants of the crew and passengers, those with cargo interests and any third parties on the ground. If the airline is one to whom the manufacturer sold the aircraft, then it is possible for the manufacturer and airline to have agreed in the sale contract that no claims will be made for loss of the aircraft. Although theoretically the agreement could also deal with all the other claims, in practice it does not.

In most of the United States, contributory negligence is an absolute bar to recovery—but again this is only likely to affect the airline. So if the manufacturer can show, for example, that negligent handling of the aircraft was just as much a cause of the accident as any fault in the aircraft, this would defeat the airline's claim but it would leave unaffected all the other claims except those in respect of crew who were involved in the negligent handling. There are moves in the United States to introduce English concepts of contributory negligence—in which the Court assesses the degree of each party's negligence and instead of defeating the claim completely, damages are reduced in proportion to the degree of negligence involved. Thus it seems possible that 'comparative negligence', as it is sometimes called, will become a universal rule in the common law countries largely because it affords some degree of natural justice.

It is worth noting that in the strict liability States, the plaintiff does not have to prove negligence on the part of the manufacturer, and there is no real opportunity to refer to the state-of-the-art at the time of design, nor is compliance with airworthiness standards a good defence. Furthermore, the manufacturer can be made responsible for defects in parts which he neither made nor designed (e.g. an altimeter). Lawsuits may be successful ten and fifteen years after the design was completed.

Those who are engaged in codifying the American Law of Torts (civil wrongs giving rise to damages) have already proposed that strict liability should be the universal rule for products in the United States and although that view is strongly opposed, I have little doubt that strict liability will become the universal rule throughout the USA in not more than a decade. Moreover the rule will spread to other common law countries in not more than two or three decades because there is a dearth of legal precedent outside the USA, and other common law courts will turn naturally to the USA for guidance even though the present law elsewhere (e.g. in England) does not recognise strict liability in general for defective products. Parliament has now created an exception: employers are absolutely liable to their employees in respect of defects in any machinery or vehicles (including aircraft) supplied for use by the staff. There is no recorded case in the English courts dealing with products liability specifically for aircraft—but there have been many legal actions started because the legal principles are well enough understood, and all aircraft cases so far have been either abandoned or settled out of court.

The litigation which can arise out of a single aircraft accident in the USA is truly prodigious because claims may be brought in several of the constituent states and in the Federal districts concurrently. If there is any likelihood of a product defect being involved, it will be customary for plaintiffs to sue both the air carrier and the manufacturer for one of two main reasons:

(*i*) if there is a limit of liability for the benefit of the carrier (as in the Warsaw Convention, the Montreal Agreement or any of the new proposals for limited liability) — then the plaintiff may succeed in recovering only a part of his damages from the air carrier

(*ii*) if there is no limit of liability, but a risk that *neither* the air carrier *nor* the manufacturer might be 100% liable due to the operation of the rules about contributory negligence, the manufacturer will be joined as defendant.

If the manufacturer is liable by reason of a defect in some part of his product which was supplied by a sub-contractor, then the manufacturer may seek recovery against the sub-contractor. Once more the rules of contributory (or ultimately, comparative) negligence may be applied—subject to the terms of any contract between the parties which, in some cases will actively permit recovery action, in other cases all recovery may be precluded by the terms of the contract. It can be ten years or more before all litigation arising out of a particular accident is ended. Another feature of litigation in the USA is that it is not confined to civilian aircraft users and passengers. Dependants of Officers of the Defence Forces who are killed or injured as a result of manufacturer's defects do not hesitate to bring suit and I have little doubt that this will happen elsewhere in similar circumstances.

It is not easy to see what the long-term future of Products Liability will be. It must certainly become a great deal worse before there can be any real prospect of ameliorative measures. In the short term, the manufacturer can deal intelligently with those whom he has contractual relationships—the airlines, the government purchaser and the sub-contractors.

For example it would be consistent with 'fair play' if sub-contractors agreed contractually to indemnify the manufacturer for any consequences of a defect in the sub-contract product, or if manufacturers and their sub-contractors were brought within a common Products Liability insurance scheme to which all contributed rateably and by which all were indemnified.

In a sale contract, the manufacturer has a series of choices according to his negotiating strength: for example—offering to accept only a limited liability (as in the so-called warranty clauses) or accepting some greater liability based on negligence in workmanship, materials or design. It is even possible to imagine that the sale contract dealing with the situation which can arise where, following an accident, *both* are sued by the dependants of deceased passengers.

The modern manufacturer's responsibilities are not even confined to the design and production of aircraft: they also accept voluntary responsibility for initial crew training; instruction of maintenance and overhaul personnel; issue of maintenance and operating instructions; investigation, analysis and rectification of all reported defects occurring in service; general economic advisory service, etc. All of these activities are a source of potential legal liabilities if errors occur due to negligence.

In the long term it is difficult to believe that the law would be altered so as to favour manufacturers because most legislatures have become consumer-orientated. But there is a chance that legislatures will be disposed to differentiate between the innocent consumer who cannot protect himself (such as the user of food or drugs, or the airline passenger) and the commercial and knowledgeable customer such as an airline —whose technical staff may know nearly as much about the aircraft as the designers and manufacturers. This is one area where manufacturers might seek relief from the legislature. Another area is the sharing of responsibility with the government certification authority.

In the USA cases have already succeeded in the first round of demonstrating that the airworthiness authority is liable if it has wrongfully certificated an aircraft as safe which is subsequently discovered to have been unsafe. At the same time the manufacturer in the USA has so far been unable to escape liability by pleading that the aircraft meets the necessary airworthiness standards. A first step in injecting realism into the manufacturer's position would be to allow him the defence that his product met all the relevant airworthiness standards and the second step would be to make it clear that airworthiness authorities can and should be prepared to stand by the results of their own negligence just like any other enterprise. The law in England has yet to be tested on these topics.

The only long-term development which would begin to remove the necessity for these steps would be any moves towards the abolishing of litigation in general arising out of personal injury and death as outlined in section 2.3. So the general outlook for Products Liability in the next century is not bright, but there are several rational steps which could be taken following an initiative from manufacturers.

4.3 *Patents and design*

Prior publication can defeat a claim for patent protection: so also can lack of novelty. Both these topics are ripe for conversion to digital information stores equipped with programmes for pattern recognition and other features to analyse the content of a patent application.

Today the searches for prior publication and the examinations for novelty are carried out to a large extent manually, at great expense separately, in each country where patents are filed. There is no such thing as an international patent registration which is valid in all contracting States merely by being accepted in one State. The result is that comprehensive searches for possible patent infringement are impossible in any new aircraft which may contain thousands of patentable ideas, and may be exported to dozens of countries each with their own patent register.

I therefore look with confidence to computer developments adapted for patent searches which will yield rich dividends as a tool for engineering research as well as simplifying the processes of patent application (which can take several years in one country alone) and facilitating the truly international registration of patents.

It is likely to be at least fifty years before sufficient computing solutions are available to deal with the analysis of patents and the resultant searches. From that era onwards, the patent laws may develop into an item of positive assistance to practical designers. It can happen sooner if enough engineers and inventors are able to combine in seeking the necessary research facilities.

5 COMMERCIAL SPACE

5.1 *Introduction*

No-one knows at the moment whether manned exploration of the Moon or planets will yield rich dividends either in a commercial or a military sense. No-one doubts the dividends in prestige or scientific knowledge or the tremendous stimulus and spin-off in technology which is given by both the American and the Russian space efforts. Yet one does not have to travel further than the first space treaty (See Appendix A) to see something of what the nations fear in outer space:

Appropriation of the Moon or other celestial bodies.

Placing nuclear and other weapons in orbit or on celestial bodies.

Establishing military bases or testing weapons.

But at least there is an agreement, now being widely ratified, which came into operation from 10th October 1967.

Because I regard this as the most important landmark in space law and one which I hope will not be disturbed for at least a century (although it may be amended and supplemented) and because I hope that nations will use it, rather than analyse or destroy it, I propose in this concluding section to speculate as little as possible and quote the text of the Treaty (Appendix A).

5.2 *Commercial space*

It will be noted that in their wisdom, the draftsmen did not attempt to define outer space and indeed this task is still on the agenda of the UN Committee on the peaceful Uses of Outer Space. Neither the USA nor the USSR are pressing for a solution at this stage possibly because they recognise that any demarcation must be arbitrary and may endanger the present freedom which they both enjoy—to put up satellites without complaint from any nation.

Commercial activities such as the operation of a communication satellite system for private profit do not seem to be permitted by the UN Space Treaty unless it can be argued that satellite orbits are not in outer space as the term is used in the Treaty. Existing communication satellites appear to be regulated by agreement outside the Treaty. It seems likely, therefore, that the space available for strictly commercial exploitation may very well be limited already to orbiting satellites and there are no opportunities for the Space Travel Companies so beloved in science fiction.

The existing consent of all nations to orbiting satellites suggests that no nation regards its sovereignty as being disturbed. If that is correct, then the proponents of commercial satellites would be well advised not to seek any clarification on the subject of national sovereignty or where outer space begins.

There are many urgent and serious problems to be solved at the level of orbiting satellites—e.g. the allocation and control of broadcasting frequencies; the control of access to reflecting or transponding satellites; security control of coded communications, etc. The problems, however, seem to be primarily technical.

In the coming century I would expect that the franchises to operate satellites for commercial purposes will be guarded as closely as the privileges of Postal Administrations and perhaps subject to a similar type of international agreement.

5.3 *Non-commercial space*

I have not attempted any discussion of national sovereignty partly because I do not think it is a topic of practical importance and partly because I have found that to non-lawyers the subject is as wild and improbable as some of the more lurid types of early science fiction. All I can foresee in the next century is the possibility of international agreement to the effect that nations will not seek to exercise exclusive sovereignty in the airspace above a specified height, say fifty miles (eighty km)—which will be well below all practical satellite orbits. I cannot imagine that nations would sacrifice their right to exercise jurisdiction over astronauts or cosmonauts—because if they did so, they would lose their practical ability to adhere to the Space Treaty and any subsequent agreements. It has yet to be determined, for example, how a State will discipline an errant astronaut on the Moon: perhaps it will be by denying him access to communication or navigational facilities controlled from Earth.

The non-lawyer reader may also be surprised to learn that the word 'peaceful' as used in the Treaty has no generally agreed meaning. To the Eastern bloc 'peaceful' means 'non-military' whereas in the West we believe it means 'non-aggressive'. I do not foresee wars on the meaning of peaceful—but I do foresee serious and continuing disputes on the subject. To illustrate the problem I devised a light-hearted game some years ago. This is reproduced in Appendix B.

In the course of time the various articles of the Space Treaty will be amended, and the scope of international agreement will be extended by means of supplementary treaties such as the 'Agreement on the rescue of astronauts and the return of astronauts and the return of objects launched into outer space'—which is already in existence. But it is possible to continue for a very long time without any amendment or further Treaties—merely by applying 'good faith' to the intentions and operation of each Article in the basic Space Treaty. If good faith is insufficient, it is naïve to suppose that further definitions and agreements will remove the basic threats to world peace, e.g., those which the Treaty already seeks to control.

Therefore, I hope that members of the Royal Aeronautical Society of all grades will be able to play their part in using the Space Treaty as a passport to intelligent co-operation in space exploration, and I hope that lawyers in particular will refrain, in public, from too vigorous an analysis of the language of the Treaty. It has already been proposed that a United Nations Space Agency should be created to supervise and administer the various agreements relating to Space Law, and to act as the instrument to which nations might delegate their sovereign authority in space exploration. While I do not think that such an Agency is essential, it has the attractive feature of focusing national and international efforts on multinational lines and should, therefore, be encouraged.

Provided that United Nations on earth becomes truly universal and acquires some ability to control rather than debate events, one can, therefore, look forward to the creation of a United Nations Space Agency which would be authorised by the space powers to regulate and perhaps even to operate their space activities. This event is entirely contingent upon further progress on earth with particular reference to disarmament and it would be unwise to be too optimistic.

The important fact is that the Space Treaty with all its possible defects, contains virtually everything necessary for the peaceful control of space exploration for the greater part of the coming century. If all my other speculations fade into dust, this is the one prediction I would like to survive and, fortunately, I believe the Space Treaty has good long-term chances of survival.

References

1 LORD SYDENHAM. *Aeronautical Journal of the Royal Aeronautical Society.* Vol **XVIII**, No 72, October 1914.

2 WALLACE, R. The Right to Fly. *Aeronautical Journal of the Royal Aeronautical Society.* Vol **XVIII**, No 69, January 1914.

3 SAND, P., PRATT, G. and LYON, J. *An Historical Survey of the Law of Flight.* Institute of Air and Space Law. McGill University, Montreal. 1961.

4 KEENAN, P. B., LESTER, A., MARTIN, P, *Shawcross and Beaumont on Air Law.* Butterworths and Co. Ltd, London, 1966.

5 CAPLAN, H. The Law versus Science in Aeronautics. *Journal of the Royal Aeronautical Society,* Vol **65**, p 453, July 1961.

6 CAPLAN, H. Practical and Legal Problems of Disseminating Air Safety Information Part 2. Possible Solutions to some of the Problems. *Journal of the Royal Aeronautical Society,* Vol **71**, p 781, November 1967.

7 In New Zealand, a Royal Commission of Inquiry under the Honourable Mr. Justice Woodhouse DSc, was asked to investigate the law and practice relating to Workmen's Compensation claims. In December 1967, the Commission recommended that in place of the common law and current rights under the Social Security Act and The Workers' Compensation Act, there would be, in effect, nationally-financed '24-hour insurance for every member of the work force, and for the housewives who sustain them.' This report, with nearly 170 pages of well-reasoned argument and 12 Appendices, is a social and legal document of outstanding importance. The most recent similar enquiry in England [the Winn Report of July 1968 Cmnd. 3691] was given such limited terms of reference that it could not consider a comprehensive insurance scheme, but as this book goes to press there are reports of fresh efforts to set up a more comprehensive Royal Commission. [*The Times* 5th July 1969].

APPENDIX A

Annex to Treaty on Principles Governing the Activities of States in the Exploration and use of Outer Space, including the Moon and other Celestial Bodies

The states to this treaty

Inspired by the great prospects opening up before mankind as a result of man's entry into outer space,

Recognizing the common interest of all mankind in the progress of the exploration and use of outer space for peaceful purposes,

Believing that the exploration and use of outer space should be carried on for the benefit of all peoples irrespective of the degree of their economic or scientific development,

Desiring to contribute to broad international co-operation in the scientific as well as the legal aspects of the exploration and use of outer space for peaceful purposes,

Believing that such co-operation will contribute to the development of mutual understanding and to the strengthening of friendly relations between States and peoples,

Recalling resolution 1962 (XVIII), entitled 'Declaration of Legal Principles Governing the Activities of States in the Exploration and Use of Outer Space', which was adopted unanimously by the United Nations General Assembly on 13th December 1963,

Recalling resolution 1884 (XVIII), calling upon States to refrain from placing in orbit around the Earth any objects carrying nuclear weapons or any other kinds of weapons of mass destruction or from installing such weapons on celestial bodies which was adopted unanimously by the United Nations General Assembly on 17th October 1963,

Taking account of United Nations General Assembly resolution 110 (II) of 3rd November 1947, which condemned propaganda designed or likely to provoke or encourage any threat to the peace, breach of the peace or act of aggression, and considering that the aforementioned resolution is applicable to outer space,

Convinced that a Treaty on Principles Governing the Activities of States in the Exploration and Use of Outer Space including the Moon and Other Celestial Bodies, will further the purposes and principles of the Charter of the United Nations,

Have agreed on the following:

Article I

The exploration and use of outer space, including the Moon and other celestial bodies shall be carried out for the benefit and in the interests of all countries, irrespective of their degree of economic or scientific development, and shall be the province of all mankind.

Outer space, including the Moon and other celestial bodies, shall be free for exploration and use by all States without discrimination of any kind, on a basis of equality and in accordance with international law, and there shall be free access to all areas of celestial bodies.

There shall be freedom of scientific investigation in outer space, including the Moon and other celestial bodies, and States shall facilitate and encourage international co-operation in such investigation.

Article II

Outer space, including the Moon and other celestial bodies, is not subject to national appropriation by claim of sovereignty, by means of use or occupation, or by any other means.

Article III

States Parties to the Treaty shall carry on activities in the exploration and use of outer space, including the Moon and other celestial bodies, in accordance with international law, including the Charter of the United Nations, in the interest of maintaining international peace and security and promoting international co-operation and understanding.

Article IV

States Parties to the Treaty undertake not to place in orbit around the Earth any objects carrying nuclear weapons or any other kinds of weapons of mass destruction, install such weapons on celestial bodies, or station such weapons in outer space in any other manner.

The Moon and other celestial bodies shall be used by all States to the Treaty exclusively for peaceful purposes. The establishment of military bases, installations and fortifications, the testing of any type of weapons and the conduct of military manoeuvres on celestial bodies shall be forbidden. The use of military personnel for scientific research or for any other peaceful purpose shall not be prohibited. The use of any equipment or facility necessary for peaceful exploration of the Moon and other celestial bodies shall also not be prohibited.

Article V

States Parties to the Treaty shall regard astronauts as envoys of mankind in outer space and shall render to them all possible assistance in the event of accident, distress, or emergency landing on the territory of another State Party or on the high seas. When astronauts make such landing, they shall be safely and promptly returned to the State of registry of their space vehicle.

In carrying on activities in outer space and on celestial bodies, the astronauts of one State Party shall render all possible assistance to the astronauts of other States Parties.

States Parties to the Treaty shall immediately inform the other States Parties to the Treaty or the Secretary General of the United Nations of any phenomena they discover in outer space, including the Moon and other celestial bodies, which could constitute a danger to the life or health of astronauts.

Article VI

States Parties to the Treaty shall bear international responsibility for national activities in outer space, including the Moon and other celestial bodies, whether such activities are carried on by governmental agencies or by non-governmental entities, and for assuring that national activities are carried out in conformity with the provisions set forth in the present Treaty. The activities of non-governmental entities in outer space, including the Moon and other celestial bodies, shall require authorization and continuing supervision by the appropriate State Party to the Treaty. When activities are carried on in outer space, including the Moon and other celestial bodies, by an international organisation, responsibility for compliance with this Treaty shall be borne both by the international organisation and by the States Parties to the Treaty participating in such organization.

Article VII

Each State Party to the Treaty that launches or procures the launching of an object into outer space, including the Moon and other celestial bodies, and each State Party from whose territory of facility an object is launched, is internationally liable for damage to another State Party to the Treaty or to its natural or juridical persons by such object or its component parts on the Earth, in air space or in outer space, including the Moon and other celestial bodies.

Article VIII

A State Party to the Treaty on whose registry an object launched into outer space is carried shall retain jurisdiction and control over such object, and over any personnel thereof, while in outer space or on a celestial body. Ownership of objects launched into outer space, including objects landed or constructed on a celestial body, and of their component parts, is not affected by their presence in outer space or on a celestial body or by their return to the Earth. Such objects or component parts found beyond the limits of the State Party to the Treaty on whose registry they are carried shall be returned to that State Party, which shall, upon request, furnish identifying data prior to their return.

Article IX

In the exploration and use of outer space, including the Moon and other celestial bodies, States Parties to the Treaty shall be guided by the principal of co-operation and mutual assistance and shall conduct all their activities in outer space, including the Moon and other celestial bodies, with due regard to the corresponding interests of all other States Parties to the Treaty. States Parties to the Treaty shall pursue studies of outer space, including the Moon and other celestial bodies, and conduct exploration of them so as to avoid their harmful contamination and also adverse changes in the environment of the Earth resulting from the introduction of extraterrestrial matter and, where necessary, shall adopt appropriate measures for this purpose. If a State Party to the Treaty has reason to believe that an activity or experiment planned by it or its nationals in outer space, including the Moon and other celestial bodies, would cause potentially harmful interference with activities of other States Parties in the peaceful exploration and use of outer space, including the Moon and other celestial bodies, it shall undertake appropriate international consultations before proceeding with any such activity or experiment. A State Party to the Treaty which has reason to believe that an activity or experimental planned by another State Party in outer space, including the Moon and other celestial bodies, would cause potentially harmful interference with activities in the peaceful exploration and use of outer space, including the Moon and other celestial bodies, may request consultation concerning the activity or experiment.

Article X

In order to promote international co-operation in the exploration and use of outer space, including the Moon and other celestial bodies, in conformity with the purposes of this Treaty, the States Parties to the Treaty shall consider on a basis of equality any requests by other States Parties to the Treaty to be afforded an opportunity to observe the flight of space objects launched by those States.

The nature of such an opportunity for observation and the conditions under which it could be afforded shall be determined by agreement between the States concerned.

Article XI

In order to promote international co-operation in the peaceful exploration and use of outer space, States Parties to the Treaty conducting activities in outer space, including the Moon and other celestial bodies, agree to inform the Secretary-General of the United Nations as well as the public and the international scientific community, to the greatest extent feasible and practicable, of the nature, conduct, locations and results of such activities. On receiving the said information, the Secretary-General of the United Nations should be prepared to disseminate it immediately and effectively.

Article XII

All stations, installations, equipment and space vehicles on the Moon and other celestial bodies shall be open to representatives of other States Parties to the Treaty on a basis of reciprocity. Such representatives shall give reasonable advance notice of a projected visit, in order that appropriate consultations may be held and that maximum precautions may be taken to assure safety and to avoid interference with normal operations in the facility to be visited.

Article XIII

The provisions of this Treaty shall apply to the activities of States Parties to the Treaty in the exploration and use of outer space, including the Moon and other celestial bodies, whether such activities are carried on by a single State Party to the Treaty or jointly with other States, including cases where they are carried on within the framework of international intergovernmental organizations.

Any practical questions arising in connection with activities carried on by international intergovernmental organizations in the exploration and use of outer space, including the Moon and other celestial bodies, shall be resolved by the States Parties to the Treaty either with the appropriate international organization or with one or more States members of that international organization, which are Parties to this Treaty.

The remaining Articles, XIV to XVII relate to procedures in connection with the operation of the Treaty.

APPENDIX B

A lighthearted lunar duologue for spacemen

A game for any number of players

1 Take any simple conventional greeting (call this g_1 and its conventional response (gr_1).

2 Put the words into the mouths of two spacemen meeting for the first time at some remote space rendezvous, say on the Moon. So, for example, an American greets a Russian—denoted by g_1AR, and the Russian replies—gr_1RA.

3 Then affix the interpretation by specific nationalities which would be given to the exchange of greetings. e.g. $gr_1RA(A)$ denotes the American interpretation of the Russian response.

4 Continue in any number of combinations until translators and interpreters are exhausted.

Hypothetical example:

g_1 = 'Peace be to you'

gr_1 = 'And to you also, peace' (Ancient greeting)

g_1AR = 'Peace be to you' (says American astronaut to Russian cosmonaut).

$g_1AR(A)$ = American interpretation of g_1AR.
'I am stating my non-aggressive intentions positively whilst reserving the right to defend myself and notify United Nations of any exercise of the aforesaid right; I hope you are able to respond on my terms'.

gr_1RA = 'And to you also, peace' (replies the Russian cosmonaut).

gr_1RA(A) = American interpretation of gr_1RA.
'Your bourgeois capitalist approach is transparent: by way of counterproposal I assure you of my *renewed* quasi-non-military intention to continue seeking modes of co-existence—that is to say the ultimate world-wide acceptance of our classless socialist society without resort to overt force. Besides we might need each other when China acquires long-range nuclear capability'.

Contrast these with the corresponding hypothetical Russian interpretations:

g_1AR = 'Peace be to you'.

$g_1AR(R)$ = 'I wish to disarm you with my apparent frankness and therefore challenge you to meet me on equal terms: the visible portions of my hand of friendship are innocent of weapons and if you will allow me to inspect your secret installations I will let you see my well-publicised ones which you already know about from my propaganda'.

gr_1RA = 'To you also, peace'.

$gr_1RA(R)$ = 'In return I offer you the opportunity of non-violent coexistence until such times as our ideologies coincide'.

Because all the above examples have been invented by an Englishman, they should, strictly speaking, have been shown thus:

$g_1AR(A)\ (E), gr_1RA\ (R)\ (E)$, etc.

The combinations are endless—for example, the series g_1RA has not been explored, nor the series (*A*) (*R*) or (*R*) (*A*). This is best left to more competent scholars.

In contrast to these complex interpretations it may be worth noting the relative clarity and concision of British greetings—e.g.

g_2 = 'How do you do'
gr_2 = 'How do you do'
$g_2EE(E)$ = 'Hello'
$gr_2EE(E)$ = 'Hello'

[This game was inspired by *Wordmanship—Semantics as a Communist Weapon* —A study prepared for the Sub-Committee to investigate the Administration of the Internal Security Act and other Internal Security Laws of the Committee on the Judiciary, United States Senate. 87th Congress 2nd Session, Washington 1961.]

19 Airports and air transport; growth and transformation

Alan H. Stratford

I INTRODUCTION

At the first meeting of the Royal Aeronautical Society in January 1866, James Glaisher advised us to 'exclude any matter that may seem useless in forwarding the science, or that has the slightest chance of giving rise to ridicule ...' this warning was timely and relevant to this theme, for to consider the long-term future with as much rationality as is required for short-term forecasting and planning studies is a valuable exercise in restraint as well as in imagination. So many avenues of possible development are technologically interesting, but in economic terms are sterile and unrewarding. One emerges with a deep conviction that certain areas of research and development may be in a dominant position in their influence on future aviation growth. Two of these are noise control and the evolution of composite materials. Although fairly obvious it seems important that they and some others be given the financial support which they deserve.

Perhaps a little basic thought might have led to the recognition of the true limitations of lighter-than-air ships in the 1930s or could the picture still be changed with controlled thermo-nuclear power and with new composite materials? Safely controlled nuclear power associated with static lift could provide silent access from the air to the heart of our great cities. But this now seems unlikely because of the excessive volumetric demands of lift with displacement vehicles and the sheer vulnerability of large vessels in adverse ambient conditions. But imagination is after all the source of most human improvement, and the stimulation of the imagination was undoubtedly one objective of the sponsors of these chapters.

A deeper consideration of the long-term future has strengthened a conviction, which was previously held with some reservation, that much more serious attention should be paid to the many alternative possibilities for the last years of this century and beyond. More especially is this true

Originally published in the *Aeronautical Journal of the Royal Aeronautical Society*, May 1968.

in the fields of technological forecasting and facility planning. In this way we might more effectively prepare our institutions and our communities to accommodate the wide range of events which may be foreseen to offer the most probable options to succeeding generations of decision-takers.

The long-term planning of ground facilities for aviation, especially at the present time when extensive land use and transportation studies are likely to lead to critical decisions on urban systems, is essential, if sterilisation of available land is not to prove a constraint in due course to air system development. There is a remarkably persistent long-term influence exerted by all public planning decisions and the need to incorporate a high degree of flexibility into all land use plans becomes more important every day. A study of the longer term possibilities of air transport and the provision of facilities, makes clear the essential need for continuous institutional studies of the range of futures possible to us. Moreover, the desirability of taking due account of these in the sums of cost and benefit which today weight and favour so heavily the short-term interests of the community, must be fully considered.

The 1960s have been notable for a considerable extension of the field of air transport research especially in the so-called 'interface' areas where a number of disciplines converge. These have not been notable previously for their contribution to the development of aviation. Such areas are economics, geography, demography, environmental research (e.g. noise), international finance and human engineering. All the present indications are that the impact of these factors (new only in their more recent application to aviation) will become increasingly significant in the future. Indeed, as technology develops so as to master the economic requirements of man for movement in the air over the densely populated parts of the earth, we may well expect the deliberations of the Royal Aeronautical Society to be increasingly concerned with human inter-actions and the sociological impact from the extended use of near aerospace. Is there not already a notable trend in this direction, to be recognised in the increasing demand for university places in the various faculties of the Arts and Sociological Sciences, at the expense of the Physical and Chemical Sciences and Engineering? Education to meet the swift changes in technology, if only to allow people to change their career more than once in a lifetime, is already encouraging the flexible degree course based on a rigorous training in law or mathematics. Intuitive processes may become no less important in the coming decades. A study by Colin Leicester of Cambridge University published in 1968 indicated that between 1960 and the year 2000 the percentage of professionally qualified and technically trained occupations in Great Britain would increase from 8 to 20% of the total gainful occupations. In terms of the total number employed the figures suggest an increase of about three to one. When considered in conjunction with management appointments (which may change little as a percentage of total jobs), a major change in the pattern of work and employment and of the demand for transport services, is likely to emerge.

This situation should be seen in conjunction with the changing situation of industrial organisation which will determine the employment environment and prospects for advancement to the individual scientist and engineer in the decades ahead. A recent ICI study[1] has concluded that the economic advantages of large-scale production and marketing will inevitably cause industry in the long term to merge into progressively larger groups so that by the year 2000 world production may well be dominated by 100 or 200 companies. The implications of such a changing pattern of industry upon organised social life in all countries of the world will be enormous. And what will its influence be upon the transport system of the world and man's need of transport for business and pleasure and for the conveyance of goods? Professor Gabor[2] has gone so far as to suggest that in a future period a small minority of high IQ, probably less than the 8% of the population who today enjoy in the UK a university education, will be able to support the economic life of the whole community. Perhaps this is preferable to the graded activity and congenital breeds of *Brave New World*, but in these two approaches the scientist and the artist have both recognised the basic problem of the future which will in due course need to be faced by industrialist and government itself. The industrial society of the next century and its needs for transport of all kinds will be dependent upon the solution found to these problems. Deep unrest which might destroy society would result from government inaction, that easy way out which views a forthcoming election as more important than the long-term human and social need.

2 PRINCIPAL FACTORS IN GROWTH

One is tempted at this point to be over-confident, to extrapolate the favourable trends, and to seek a single and obvious solution which naively ignores the alternative futures which are open to man. However, some recent work, especially that of the Hudson Institute[3], has developed methods of approach which are together more modest and more rational. A preliminary consideration of these ideas may be of interest.

The Hudson method is in effect to establish a basic, 'long-term, multifold trend' in western society which can be recognised as the thread upon which can be strung the complex and interacting elements of our society since the early and middle ages. The principle trends which may now be recognised are:

1 Increasingly bureaucratic, secular, utilitarian and hedonistic societies.
2 Rapid accumulation of scientific and technical knowledge and the institutionisation of research.
3 The extending industrialisation, affluence and leisure of nearly all human societies.

4 Increasing literacy, and the gradual decay of secondary occupations.
5 Increasing population and urbanisation.
6 Increase in power available for production, transport and destruction.
7 Increase tempo of change since the eighteenth century.

Some liberties have been taken with the approach in the above summary but the trends can be seen to be dominant factors in any recent study or analysis of society, or of any aspect such as transport which is closely linked to advances in technology.

From the basic premise of the long-term trend a 'Surprise Free' projection and a standard world forecast can be made to meet the needs of a realistic study covering a range of futures. 'Canonical Variations' are proposed which are essentially arranged so that the pattern of change in comparison with the extrapolated standard world is provided as a basis for discussion.

Following a brief survey of the factors which seem to determine air transport and the need for facilities, we may then attempt to establish the most promising lines of technological development in the commercial use of aviation in the context of the probable changing pattern of world demand.

3 AIR TRAFFIC GROWTH

As A. C. Clarke has reminded us, any successful attempt to assess the future permits no failure of nerve or of the imagination. We are so close to the recent and current changes in aeronautical technology, and to a growth in the scale of activity in the air which has defied the most alert prophets in the post-war era, that it may be our nerve that proves wanting rather than our imagination. However, working firmly from the present situation and from the trends of development which may be recognised over recent years, we can, nevertheless, suggest in this chapter some likely patterns of growth and development. Over the next fifteen years, using the break point which has been in current practice in airport projects in the late 1960s, we could well claim that our picture is that of the probable situation, and over the subsequent eighty years who could claim that it is more than a possible pattern of alternative 'futures'.

It has been apparent in the past five years that the terminal and the contiguous air space is posing the most critical problems in air transport development and growth. This has arisen primarily in the great national airports such as New York, Chicago and London where passengers and aircraft movements, especially at peak periods, has raised congestion as a public issue and public enquiries have only served to fan the flames in the week-end press. No easy solutions at a reasonable cost have been found in the case of New York or London, although they could be found at considerable expense in environmental deterioration if access to the expanding national and international air services were a sole criterion.

It seems clear that a view on the long-term future of civil airports and ground facilities must depend as much on a sound assessment of the most successful lines of air vehicle development as upon our analysis of air transport growth *per se*. Clearly the two are inextricably connected, for recent history has shown that technological advance has allowed increased capability in performance and size to be matched by improved economy. The powerful contribution of these three critical parameters in vehicle development, supported by the advanced industries in six or seven countries in the world has provided the air transport capacity which built up air traffic to its present remarkable level and with a rate of growth which itself shows no sign of abating.

All those concerned with air transport have access to extensive statistics which need not be reprinted here. The debate continues as to whether the growth will persist or whether there are forces already at work which may neutralise the factors which have over more than twenty years been dominating the air systems of the world. Fig. 1 shows the total passenger and freight traffic of the ICAO States over the past ten years and the 'most probable' curves of growth up to 1980. A possible range of futures indicated by other forecasting bodies are shown on this figure.

We may view the immediate-, medium- and long-term time scales as shown below. Each period is in length approximately twice that of the preceding period:

1.	Immediate Future	1970–1985
1.1.	Medium-Term	1985–2015
1.1.1.	Long-Term	2015–2070

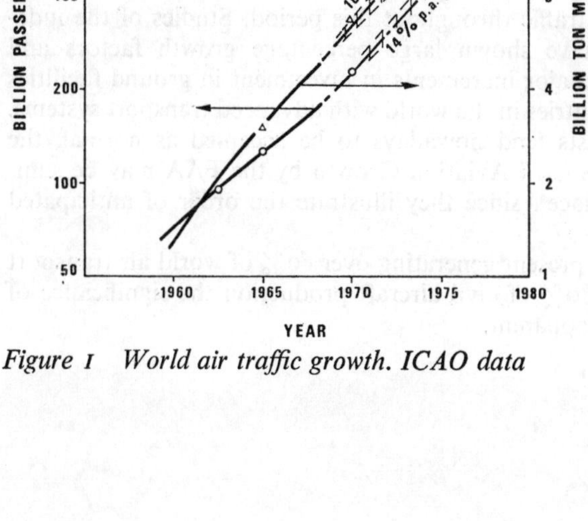

Figure 1 World air traffic growth. ICAO data

In this section we shall consider only the period up to 1985. Periods I.I. and I.I.I. are discussed in subsequent sections.

In the immediate future we shall be likely to experience a continuous development of air transport in the form and pattern which is to be recognised already today. Many studies have recently become available of the air traffic in the major world areas up to the end of the period 1985, which are almost certain to be within an acceptable order of accuracy and are indeed based upon the predominance of vehicles, systems and technology which will be under development within the next five to ten years. It is not difficult, therefore, to establish the airport pattern, the likely demand for facilities and the broad range of economic problems arising for operators, industrialists and airport terminal authorities through that period. Dominant factors in air transport development in the period will be:

An accelerating tempo of technological change.

A steady increase in the air traffic, passenger and cargo on the principal world routes.

A massive increase in investment in flight and ground equipment as well as in tourist and business amenities dependent upon air transport.

An intensive and probably accelerating growth in tourism (principally inclusive tours).

A competitive struggle with other transport modes on medium-haul stages except on over-water sectors.

Dramatic increases in the size and capability of the transport vehicle, with likely confirmation of the commercial feasibility of SST.

Increasing population, urbanisation, GNP* *per capita* and leisure.

Decreasing comparative unit cost of air transport.

The time-scale of the period, however, is too short for entirely new factors to be influential. Such events as the integration of the USSR or China into the world air system or the emergence of new traffic routes in Africa or Asia may need to await a later period. There is almost certain to be a steady increase in air traffic throughout this period. Studies of the independent authorities have shown large percentage growth factors and these are linked with major increments in investment in ground facilities especially in those countries in the world with advanced transport systems. Although such forecasts tend nowadays to be accepted as normal, the recent estimates of US Civil Aviation Growth by the FAA may be summarised in the first place[4], since they illustrate the order of anticipated growth.

Since the USA is at present generating over 60% of world air transport production and over 80% of civil aircraft production the significance of these figures is at once apparent.

*Gross National Product.

TABLE I *FAA aviation forecasts for the USA*

	1967	*1979*
Airline Passengers (millions)	126	444
Revenue Passenger Miles	86	343
Airline Fleet Aircraft (Jan)	2272	3860
General Aviation Fleet (Jan)	104,706	203,000
Jet Fuel Consumption (million gallon)	4697	16,890
Civil Aircraft Production (Aircraft Units)	15,171	32,480
Civil Transport Production (Aircraft Units)	372	280
General Aviation Production (Aircraft Units)	14,799	32,200
Airport Movements (FAA) millions (ATC)	48	167
IFR Flights (millions)	5·8	18·7

Within this period a remarkable growth in US airport movements, through airline activity with increasingly large transport aircraft, and through extensive general aviation movements can be recognised. The ambitious airport development programme supported by the FAA in the USA, therefore, will come as no surprise. Restrictions on airport operation at New York, Chicago and Washington, promulgated September 1968, and the urge to promote STOLports and STOL aircraft operations stem from the earlier recognition of these problems well before the critical weeks in July and August 1968.

Other comparable studies have been undertaken by the international organisations such as ICAO, IATA, ITA, and the EARB (Brussels). The forecasts of ICAO show that by 1980 the most probable total world air traffic volume (on scheduled airlines of the ICAO States) would reach 1500 billion passenger-kilometres compared with 275 billion in 1967 (930 and 170 passenger-miles). (Fig. 1). (1 billion$=10^9$)

The tonne-kilometres (tk) of freight between these dates is expected to grow from 6·7 to 48 billion. These estimates are based on the widespread introduction of high capacity jets and SSTs in the 1970s and the achievement of an overall reduction in operating costs from an average of 17·5 cents per tonne-kilometre (ctk) in the mid-1960s to 13·5 ctk in the mid-1970s due to the inherently improved operating economy of more advanced aircraft of large size. The last decade of experience in the introduction and operation of the first and second generation of jet transport has illustrated clearly the long-term economic advantage of the high productivity of fast and larger aircraft, for from the mid-fifties to the mid-sixties the world overall unit costs fell from 24 to 17·5 ctk. Such global traffic estimates, of course, incorporate the widely differing growth rates to be expected in the immediate future in various world areas as shown by the differing rates experienced in the past.

The factors determining growth have been recognised for at least ten years and the forecasts made by the airline industry in the USA and UK in the late 1950s assessed very accurately the passenger traffic on the trunk routes in the period up to 1970. Nor have the estimates of the aircraft manufacturing industry been greatly in error.

Methods and techniques in current use by the manufacturing and air operating industry were analysed in a recent paper prepared by Peter Smith for the Centre for Air Transport Development in Rome[5]. A wide range of methods of approach and techniques were identified.

These comments offer support to the assumption that we are at present in a position to estimate fairly precisely the scale and type of airline and airport operations in the 1970s and even up to 1980–5, as well as the demand for transport aircraft which this entails with the ground facilities which will be required. ICAO, and most other forecasting groups now at work, make certain basic assumptions clear in any forecasts of this kind. Typical of these would be:

(1) That no major world conflict arises in the next decade.
(2) That population and income per head will continue to grow at about 2·5% per annum.
(3) That inflation will continue at 2% per annum.
(4) That the capacity to be provided by world airlines will be such as to generate adequate services at load factors of 50–55% with lower values during the introductory phase with new large aircraft 1972–4.
(5) That high productivity and improved unit costs of operation (except for SST) will allow revenue rates for passenger and cargo to decline at 1·5–2·0% per annum.

In the relatively short period of growth up to 1985 it seems unlikely that great changes will occur in the form and origin of the demand for air services. The tourist element and air freight may well take an increasing share in the total traffic demand. Low rates in relation to surface transport will ensure that this occurs. It seems most improbable that the now under-developed countries will be able to do more than continue a steady increase in GNP and GNP *per capita*.

This would allow a regular increase in their demand for air transport, especially on international services to and from the principal industrial nations of the world. The GNP differentials between the rich and poorer countries will, in fact, continue to increase so that the drive towards tourism will be stimulated from both ends. In this period the growth of domestic traffic within the less well developed countries in North and West Africa, for example, and in South America, will be likely to continue but will not be dramatic and will rarely justify the acquisition of aircraft larger than the B707 or the DC8. Thus, the development of air terminals will follow along the now well established lines with few special requirements which have not already been encountered by the major airport authorities in their plans, well advanced in the late 1960s, for the advent of the giant jets (1971). What may well be called for is a greater dissemination of the 'know-how' and experience already gained by the airport operators, airlines and consultants in the problems of airport and air terminal design for the benefit of the small airport operators who may be in need of help.

The fundamental design requirements of the new capital city airports will be dictated by the giant jets. If the aircraft on order by the major carriers get no larger than the Boeing 747 and the Lockheed C5A, these aircraft will continue to serve as a criterion. In due course this type and size, however, must be superseded and beyond 1985 maximum gross weight must surely exceed one million pounds. Nevertheless, the more intensive operation of aircraft in the medium size brackets must be the mainstay of civil aviation.

Characteristic changes to be incorporated must be:

(1) The handling of large passenger flows, essentially sub-divided and grouped to allow physical control and personal comfort and convenience.
(2) The scientific management of baggage and freight with containers.
(3) The improved in-flight management of theatre size groups.
(4) The design of runways, taxiways and aprons to allow for the reduced manoeuvrability of giant aircraft.
(5) The first intensive operations of public transport links between urban centres and airport terminals.

Runway length and strength requirements will differ little from earlier long-range jet aircraft (e.g. DC8/55 and B707/320). A comparison of physical differences which may effect airport design are given in Table 2 for the DC8 and the DC10–Intercontinental.

Increases in the size and weight of transport aircraft have, over several decades, controlled the characteristics and specification of airports and

TABLE 2 *Physical differences between the DC8 and DC10-Intercontinental*

	DC8–55	*DC10 Intercontinental*
Gross Ramp wt (lb)	328,500	493,000
Maximum Seats (economy)	189	326
Number abreast	6	9
Maximum Cargo Load (cu ft)	1,390	3,040
Main Tyre Pressure (psi)	178	170
Max Landing Approach speed (kt)	143	139
Max Take-off Field Length (ft), (ISA+15 c)	10,400	10,100
Max Design PNdB, 4 nm from start of take-off roll	datum	-8 to -10

aerodromes. The limiting conditions have been first encountered at the capital city airports between which the new types have first been operated. The increasing demands for runway lengths, runway and taxiway strength and capacity of terminals and accessory facilities have incurred immense investments in capital works and services.

Certain problems will arise on airfields which may superficially be able to receive the largest of the new generation aircraft. For example, taxiways for the Boeing 747 (ramp weight 713,000 lb (324 Mg)) may require modification to provide adequate clearance between wheels and the pavement edge. New fillets may be required in the turning areas to maintain continuous movement. The 747 cannot make 180 degree turns on 150 ft (46 m) runways. Apron areas need to be carefully reconsidered.

In the immediate future period two major factors, Cargo and Charter Services will increase their influence substantially. Air transport growth can no longer be considered only as passenger traffic on scheduled services. Inclusive tour traffic at present of small significance outside Europe will become an increasingly important element on the North Atlantic and in all the dominant air transport regions and will increasingly influence the economy of some less well developed countries.

4 THE DEVELOPMENT OF INCLUSIVE TOUR AND CHARTER TRAFFIC, 1970–85

In the decade and a half up to 1985 we can confidently expect to see the inclusive tour market extend from the European-Mediterranean region and make powerful impact on world-wide routes.

Development on this market will be assisted by the following major influences:

(*a*) Increasing levels of prosperity, and of disposable income, in the more advanced countries.

(*b*) The desire of the less advanced countries to attract tourists to help their economy.

(*c*) The improved seat-mile costs of the large capacity jets will become a factor of even more significance to the whole-plane charter than it is to the scheduled operator. New generations jets will be introduced in considerable numbers on IT and charter work in the mid-1970s.

(*d*) The progressive lowering of inclusive tour charges on the longer-haul routes, where most savings can be made, will be aided by IATA's further recognition of IT as a solution to some of the problems occasioned by the introduction of the larger new generation jets.

Under these influences, IT and charter can be expected to develop along the following lines:

(*a*) A steadily increasing growth on longer stages.

(*b*) A concentration on the routes leading to the countries most attractive to the tourists, from consideration of cost, history, scenery and climate.

These efforts will lead, by the late 1980s, to inclusive tours penetrating into almost every part of the world's air transport system. With permitted numbers in inclusive tour groupings inevitably growing smaller rather than larger, and with the comprehensive schedule service system capable of carrying these groups to almost any destination in the world, the 'packaged tourist' will then have the world at his feet.

The dramatic increase in Western European tourism is well illustrated in the growth of foreign holiday traffic from the seven principal countries over the period 1962–65. Of the total population exceeding three hundred millions, over twenty five million took a foreign holiday in 1965. Of the five million British tourists, 508,000 took a holiday outside Europe. Table 3 indicates the recent trend of foreign holidays.

TABLE 3 *Recent trend of foreign tourists from Western Europe*

Country of Origin	*1962* (M)	*1965* (M)
West Germany	6·5	10·0
United Kingdom	3·5	5·1
France	2·0	3·0
Holland	2·0	3·0
Scandinavia	1·0	2·0
Switzerland	1·0	1·5
Italy	0·5	0·8
	16·5	25·4

Based on IUOTO *statistics.*

The projection estimated by IUOTO* of future holiday trips abroad from Britain for 1970 and 1980 are 6·6 and 9·4 millions respectively. US visitors overseas (excluding Canada and Mexico) are recorded as 1·77 million in 1962 and 2·30 millions in 1965. Forecasts of 2·73 and 3·83 millions have been made for 1970 and 1980.

Europe and the Mediterranean area still draws more than 50% of US overseas visitors. This proportion may decline in the last decades of the century as the tourist centres of the Middle East, Africa, Asia, and South America extend their facilities and the tour promoters, exploiting the economic wide-body jets, offer increasingly attractive holiday opportunities to the more affluent populations of North America and Europe.

5 AIR CARGO

The recent development of air cargo has been adequately recorded and ref. 6 gives one excellent account of the current status and the major problems in the short-term future as seen by the intra-European operator.

*International Union of Official Travel Organisations.

In this field, growth will depend on the capability of the airlines to continue to sell the rapidly increasing capacity provided by the larger passenger jets now coming into service on the principal passenger routes as well as the capacity of the all-cargo versions of the B707 and DC8, which are already providing a large proportion of the space on the key long-haul routes. On the North Atlantic in 1962 for the first time cargo loads on all-cargo aircraft flights exceeded that on passenger flights. Rates of growth are still high on the North Atlantic and in Europe. The potential for growth in the immense total freight market with the reduction in rates, which the new generation jets will soon allow, is the principal basis of hope for continuing air freight growth. Studies of growth have in the past been decried because they are generally over-optimistic. Methods more recently introduced which are based on the methods of total distribution costs are providing more consistent answers[7].

Air cargo growth, however, has proved to be even more variable than air passengers. In many world areas the scale of movement still does not justify the use of large enough aircraft to allow economic rates to be scheduled. Attempts to forecast air cargo movement even in the short term must take account of the wide variation in the need for, and the price that will be paid for, the higher speeds of air delivery. Air cargo growth is likely to be highly variable with immense growth in the domestic USA, on the key European routes and on the principal intercontinental sectors through the Middle East and the Pacific to Japan and Oceania, but moving up far more slowly in most other areas.

6 SHORT-TERM FORECASTS OF AIRPORT REQUIREMENTS

In the present period the airport situation is dominated by aircraft and passenger congestion at the world's major airports. Runway and terminal restrictions, however, will soon be seriously felt at the far larger number of airports where aircraft type development has overtaken the late development planning of many owners and operators.

In a number of countries some form of a national airport development programme has been inaugurated. This has the immense advantage of eliminating the less desirable elements of competition and redundancy in the overall airport situation but may arouse criticism in some quarters since local enterprise and knowledge of the special needs of a regional situation might be discouraged and lost. National Airport Planning has been accepted in the USA for many years, and has worked with such effect that it has been widely proposed in the United Kingdom[8].

The US Airport Plan programmed by the FAA is of particular interest in the context of this paper since it is likely to be a form of airport organisation which will be more widely adopted in the future in the case of nation states with extensive airport systems (which may have grown up from diverse origins) especially where a controlled and non-redundant

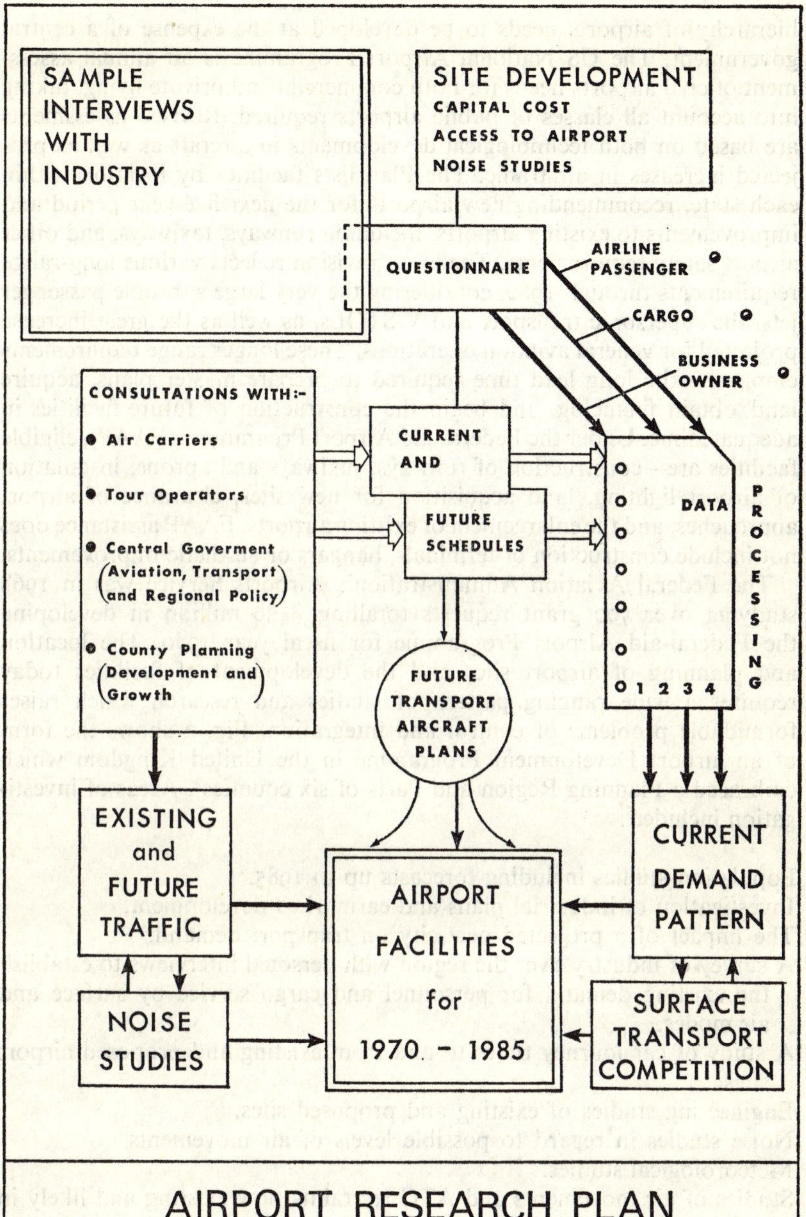
SAMPLE INTERVIEWS WITH INDUSTRY
SITE DEVELOPMENT
CAPITAL COST
ACCESS TO AIRPORT
NOISE STUDIES
QUESTIONNAIRE
AIRLINE PASSENGER
CARGO
BUSINESS OWNER
CONSULTATIONS WITH:-
Air Carriers
Tour Operators
Central Goverment (Aviation, Transport and Regional Policy)
County Planning (Development and Growth)
CURRENT AND
FUTURE SCHEDULES
DATA
PROCESSING
0 1 2 3 4
FUTURE TRANSPORT AIRCRAFT PLANS
EXISTING and FUTURE TRAFFIC
NOISE STUDIES
AIRPORT FACILITIES for 1970 - 1985
CURRENT DEMAND PATTERN
SURFACE TRANSPORT COMPETITION
AIRPORT RESEARCH PLAN

Figure 2

hierarchy of airports needs to be developed at the expense of a central government. The US National Airport Programme is an annual assessment of civil airports needs for both commercial and private flying, taking into account all classes of public airports required. Revised assessments are based on both technological developments in aircraft as well as projected increases in air traffic. The Plan lists facilities by location within each state, recommending new airports for the next five-year period and improvements to existing airports, including runways, taxiways, and other airport safety requirements. The latest revision reflects various long-range requirements through 1980, considering the very large subsonic passenger jets, the supersonic transport and V/STOLs, as well as the great increase projected for general aviation operations. These longer range requirements emphasise the long lead time required to prepare master plans, acquire land, obtain financing, and begin the construction of future facilities in adequate time. Under the Federal-aid Airport Programme (FAAP), eligible facilities are—construction of runways, taxiways and aprons, installation of airport lighting, land acquisition for new sites, clearance of airport approaches, and the enlargement of existing airports. FAAP assistance does not include construction of terminals, hangars or aesthetic improvements.

The Federal Aviation Administration's Airports Service was in 1968 studying over 700 grant requests totalling $450 million in developing the Federal-aid Airport Programme for fiscal year 1970. The location and planning of airport sites and the development of facilities today requires a wide ranging pattern of studies and research which raises formidable problems of control and integration. Fig. 2 shows the form of an airport Development Programme in the United Kingdom which embraced a Planning Region and Parts of six counties[9]. Areas of investigation included:

Population studies including forecasts up to 1985.
Investigation of industrial plans and earmarked development.
The impact of a projected new city on transport demand.
A survey of industry over the region with personal interviews to establish the existing demand for personnel and cargo service by surface and air modes.
A study of car journey times to and from existing and proposed airport sites.
Engineering studies of existing and proposed sites.
Noise studies in regard to possible levels of air movements.
Meteorological studies.
Studies of air movements and ATC restraints now existing and likely in the period of study.
Operating costs and revenue estimates for the alternative sites with due regard to all sources of airport site and adjacent property revenues.
Cost/benefit studies of the alternative proposals.

These areas of investigation are likely to arise in any extensively developed region in North-West Europe, North America, Japan or Oceania. Some simplification may be introduced in less industrialised zones. It is at last being realised that airport location and development is a critical and central factor in the planning of new regions, and if for no other reason than to minimise the nuisance to the community, the aviation facility must be correctly located at the start and the key functions correctly disposed in relation to them. It can now be said fairly that the principal planning authorities in Europe and the USA are well aware of this.

7 CHANGES IN AIRPORT REQUIREMENTS

Studies of airport development in the short-term future are naturally initiated by the airport owners or authorities with a keen sense of need for expanded facilities to meet the coming generation of transport aircraft. In the less developed countries the urge to attract air services, more especially to and from the USA and Europe has stimulated an airport building programme on a massive scale. Nearly every major city in the world is today planning a further extensive development of its international airport (first extended probably in 1960–65 to suit the early four-engined jets), or is seeking a new site to meet the expected traffic growth of the 1970s.

According to the 1968 edition of the US National Airport Plan (FAA November 1968) more than 800 new airports are to be built in the USA in the next five years to relieve present congestion and to accommodate the forecast growth. Covering fiscal years 1969 to 1973 inclusive, the plan recommends twenty-two airports accommodating airline aircraft, and seven hundred and forty-eight new airports for general aviation including twenty-five to be used as STOLports in the heavily congested New York, Washington corridor and on the Los Angeles–San Francisco axis.

Fig. 3 shows the increase in area provided by the major world airport

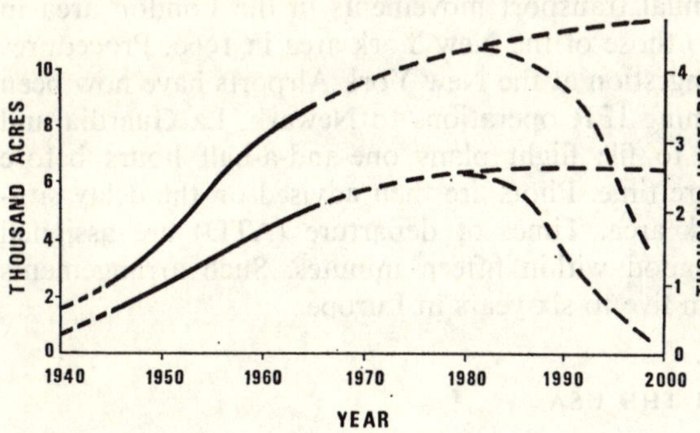

Figure 3 Growth of capital airport size. Upper curve: area controlled. Lower curve: aviation area

authorities over the past twenty years. The upper curve shows the total land area required for airport facilities including the adjacent land for aviation linked activity such as access roads, car parking and air-orientated industry. The lower curve indicates the actual land in use for airport services. The extension of these curves to the right is in question. Although runway length requirements have apparently reached a limit, parallel runway systems are now viewed with favour for new major projects, and the acreage still grows with 15,000–18,000 acres (60–70 km^2) for airports designed in the mid 1970s, seeming not unlikely. The British Airports Authority has proposed 5700 acres (23 km^2) for an idealised Mark 3 Third London Airport in 1976[10], but this may require major reconsideration in the light of an actual site, the approach road and rail systems and the regional situation.

The Palmdale site proposed for Los Angeles requires the acquisition of 17,000 acres (69 km^2) and the Dallas International Airport now under construction has involved planning control over 18,000 acres (70 km^2). The scarcity of land and high land values has stimulated interest in off-shore airport sites when a situation is available within an acceptable distance from the centres of industry and population. Siting of airports on off-shore locations is also under active consideration because of air traffic control conflict with existing major airports near the great cities and because of increasing public pressure to reduce noise from air traffic.

A study recently commissioned by the FAA will investigate new methods of construction for off-shore airports projects. Present interest centres on airports built on piles, those built within polder dams and protected by dykes, and floating airport projects. A Lake Michigan off-shore site is favoured by Chicago, as is the Foulness-Maplin Sands by two project promoters as a site for the Third London Airport. Plans for a Seadrome project 10 miles (16 km) off Los Angeles were published in late 1967.

The Airport Programme in the USA offers to us in Europe a foretaste of the demands which we will ourselves have to face in ten to fifteen years. Indeed the level of annual transport movements in the London area in 1967 were equivalent to those of the New York area in 1960. Procedures for the reduction in congestion at the New York Airports have now been called for. Pilots, planning IFR operations to Newark, La Guardia and Kennedy are required to file flight plans one-and-a-half hours before their proposed departure time. Pilots are then advised on the delay situation in the New York area. Times of departure (ATD) are assigned which must be made good within fifteen minutes. Such arrangements must be expected within five to six years in Europe.

8 STOL AIRPORTS IN THE USA

The recent stimulus to establish STOL services in the USA has arisen largely through the need to ease airport congestion at the principal air centres on the East and West Coast. No less than nine STOL airport sites are now planned in the New York area, including four in Manhattan. At present the city's only STOL facility is one runway at La Guardia Airport. FAA plans include $126 million expenditure in New York State. Washington National, Dulles International and Friendship International Airports, are already linked with regular STOL services offering sixty flights daily with Dornier Skyservant aircraft. At San Francisco approval has been sought for a 1000 ft × 50 ft (305 m × 15·2 m) site above an existing pier on the waterfront. The Los Angeles airport master-plan calls for VTOL/STOL metroports to relieve Los Angeles International and the proposed satellite airfields at Van Nuys, Ontario and Palmdale.

These plans are an important part of the pattern of airport development for the immediate future in the USA. Relief from congestion, noise and regional planning obstruction will be sought in the near-central or satellite airports to promote STOL services until such time as the next stage of a major international airport programme can be forced through or until VTOL aircraft on an airline scale can be introduced.

Guidance material already issued by the Aviation Administration has recommended runways 1500 ft (457 m) long and 100 ft (30·5 m) wide, 60 ft (18·3m) taxiways and pavement criteria for 150,000 lb (68 Mg) ramp weights. 150 ft (45·5 m) over-runs at each end are proposed. Recent trials with the 53,000 lb (24 Mg) McDonnell-Douglas 188 (Breguet 941) have been successful and the steep climb-out has been favourably noted in urban areas in comparison with conventional jet aircraft. The trial operations have been planned to give routings clear of airways and congested terminal areas, offering segregated and more direct tracks, and favourable flight profiles with minimum delay. The development of

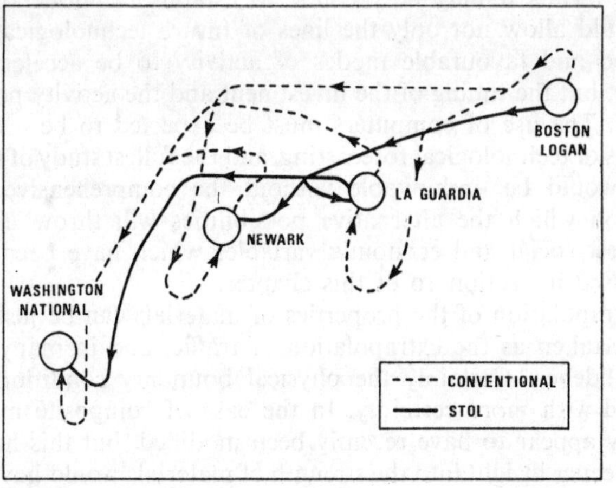

Figure 4 Illustration of STOL operation patterns in North East USA
(Shell Aviation News No 367)

navigational aids is of special significance and Decca Omnitrac pictorial display equipment with data-link has been in use with good effect. The routings planned on the Washington–New York–Boston legs of the Eastern Airways Inter-City STOL trials recently concluded are shown in Fig. 4. This illustrates the direct routings which have been shown to be possible in the Eastern Airways trials. This is likely to be a pattern for extensive operations of this type in the USA and in Europe in the 1970s.

9 TECHNOLOGICAL FORECASTING

From our previous consideration of the current trends of air transport it would not be possible to draw conclusions as to the further development of this industry without some indication of the likely changes in the key parameters of future world demand, illustrated perhaps as a series of the most probable situations with or without major unforeseen events. This would be equivalent to a 'surprise-free' future with alternative options to include dire events such as atomic war or a collision with Icarus on the one hand, and the economic evolution of an anti-gravity device or intellectual fertilisation from extra-terrestial visitors on the other.

Cost-benefit and cost-effectiveness studies may be inadequate to deal with the longer term future without a fairly complete picture or scenario of the environment within which the technology must be integrated. These studies of the basic trends in social activity, (population and urbanisation), and the first order derivatives (distribution of industry, concentration of investment, and economic growth), become essential before the key factors determining transport promotion and use (GNP *per capita*, leisure and disposable income) can be estimated.

By this means it may be possible to establish a mode of feed-back which would allow not only the lines of future technological change to be selected and favourable modes of activity to be accelerated in the time-scale, but the timing of the investment and the activity phasing to be optimised. The use of computers must be expected to be accelerated in these areas of technological forecasting, and the fullest study of the multiple 'futures' would be unthinkable without the comprehensive control of information which the alternative possibilities will throw up. Some of the simplest social and economic variables which have been mentioned are discussed in section 10 of this chapter.

The extrapolation of the properties of materials can be just as reasonably undertaken as the extrapolation of traffic, and in many cases with more confidence. Certainly the physical boundary conditions could be established with more certainty. In the case of composite materials old limits may appear to have recently been modified, but this has not been so, for a deeper insight into the strength of materials would have indicated, for example, the potential for multiple materials extending the envelope of specific tensile strength and specific modulus. The achievement of new design limits by means of computerised design methods should allow a still closer approach to the physical limits of materials.

Erich Jantsch[11] has stated that about six hundred American firms do their own technological forecasting on a regular basis. On it they spend about 1% of their total research and development budget. However, governments are now concerned with the long-term technological forecast in the civil field, and techniques originating in defence projects are being progressively developed to establish national goals, and the planning of programmes for the national research institutions. It is now becoming more clearly understood that inter-disciplinary teams are needed to pursue such research. The needs of the future must be met by education and training so as to fertilise joint activity in mixed teams, although it may take several decades for institutionalised education to reorganise itself. Even without success in the control of the gravity field, advances in the control of weight would have a very direct economic effect on the operation of transport aircraft, and thus hold out the prospects for a major comparative advantage in relation to surface transport modes. Herein lies the immense significance of the current work on new materials of all kinds.

A concentration of effort on noise reduction is already apparent. This has arisen through excessive noise from jet aircraft in the vicinity of major airports and from the intensive motor traffic in city areas. The effort must be expected to increase as the peripheral STOL sites become necessary in the vicinity of the great cities and as the VTOL sites, if not likely to be central, then very 'close in' to the major cities, are established in the last decades of the century. Physical limits to the reduction of noise may well be a determining factor, but the wide impact of noise from all sources upon the increasing population in urban areas makes it likely that a complete mastery of the objectionable elements of public noise will be achieved within fifteen–twenty years (short–medium term).

Air Traffic Control is going to be dealt with in the next chapter by Dr. A. Stratton. A fully automatic system of control of the movements of vehicles in the air space must be anticipated in the medium-term future.

Effective world-wide weather forecasting, and the regional control of weather may have an immense influence upon the future of air transport services. The former must have a significant bearing upon the safety of flight, and the latter an influence upon the distribution of tourist destinations in all parts of the world.

Other technical advances in the air vehicle and its power or thrust units in respect to power/weight and power/volume ratios will be likely to accrue as continuing economic progress is made in VTOL, STOL and conventional aircraft designs. Increase in size must be expected to continue in the interests of operating economy and air as well as ground space congestion, so that the ground handling of passengers, baggage and cargo will become of increasing significance.

Thus the investment in ground facilities will keep step with the investment in aircraft themselves, so that an increasing inertia in the heavily capitalised air transport industry will become more apparent. There is already an obvious danger that with the continuing heavy investment in airport facilities for the new generation of jet transport there will be a reluctance on the part of the major world air carriers to consider in their earliest stages the first VTOL transport projects. However, such is the energy and competitive spirit of the air transport industry that it would be most unlikely that such inertia could be sustained in the face of so significant a transport development. It is possible, however, that in the next century after two or three decades of VTOL experience with parallel massive developments in other new technologies not associated with aviation, the inertia may become more intense and a less alert airline attitude may inhibit the growth and the zest for improvement which is so apparent today.

General aviation

The development of VTOL techniques and the wider provision of ground facilities will be likely to lead to a considerable growth in the use of private and executive aircraft with VTOL capability. If control of the weather could be effectively achieved in the vicinity of the coasts and airfields a great extension in the use of general aviation would follow. At the present time the reduction in aircraft utility owing to uncertainty in future weather conditions deters many potential owners. This probably has a greater effect on business than on private aircraft use.

The pattern of ground facilities in the medium and long term must take full account of the needs of the private and executive user. Moreover the capability of the aircraft available for private use must be expected to be greatly augmented. Long-haul aircraft of four–eight seat capacity with transatlantic range may be in demand in the 1980s. Supersonic performance would surely be a requirement in a high-grade class with a limited market in the medium-term future period. VTOL capability may indeed be a requirement for all general aviation types by the turn of the century.

10 AIR TRAFFIC IN THE MEDIUM-TERM FUTURE 1985-2015

In the longer term we are unlikely as yet to be able to forecast with accuracy the detailed technology which will be available to us. This is patent to anyone who takes the most superficial glance at the very recent (1963–68) developments in fields such as VTOL, Air Cushion Craft, composite materials or supersonic aerodynamics, so that a forty–fifty year prognosis might at first sight seem pointless. Some guide lines for thought are, however, readily available to us. We are organised as human beings in social communities which may change to some considerable extent in five or ten decades but it is unlikely that we will be re-orientated emotionally or politically in any dramatic matter short of global disaster or thermonuclear war.

The pattern of world population change can be forecast with a fair degree of accuracy for half a century and published attempts at this are widely available. The continuing urbanisation of the human population (metropolisation) and the growth of world population within a century to well over 15,000 million are widely accepted by specialists in the field[12]. The re-distribution both of population and of the industrial centres of the world will inevitably cause changes in the pattern of demand for transport which will most probably become recognisable by the end of this century. By that time over 60% of the world's population are likely to be living in cities exceeding 100,000 and growing industrialisation of the countries now considered to be under-developed will inevitably increase the demand for air services.

In the development of air services most governments already spend a considerable amount of effort, and some of the least progressive in other ways incur costs which may be grossly disproportionate to the value of the services obtained. Nor should the new regions under development in the USSR (to the East) and in the less accessible parts of countries such as China, Brazil, or Australia be ignored.

The improved economics of the transport aircraft must be expected to contribute notably to the increase in air transport demand up to and beyond the end of the century. If a careful analysis is made of developments in aviation over the past forty years it seems clear that the principal advances have been achieved in the fields of increased size of vehicles, increased efficiency in the installation of power or thrust and the technological development through research into moving surfaces[7]. In the next forty–fifty years it seems likely that two factors will dominate the form of development. These are (*a*) the capability for vertical take-off and landing which may dramatically change the form of the vehicle and the utilisation of the air space, and (*b*) the development of materials for engines and air frames to allow very much more efficient payload/gross weight ratios. Thus, increased utility of the air vehicle, and at the same time improved economy of operation are likely to be available to a world of changing industrial and social form which will undoubtedly emerge in the medium-term future. It would be possible to consider these primary factors in far more detail than our long-term glance at the future would now justify. We shall return again to this future period in an attempt to assess the likely future demand for air transport services, airports and the terminal facilities.

Studies on the currently developing technologies have shown that a high proportion of these are highly relevant to the improved economy of air transport and the provision of facilities. In recent work by the Rand

Corporation on the likely time-scale for the advent of future technological achievements as measured by the consensus of expert opinion, eighteen of the twenty-one principal achievements are most likely to be complete by the end of 2015, that is, the medium-term future now under discussion. Of these eighteen, no less than seven would be of direct economic value to the air transport industry and three would be of important indirect value.

Of direct economic value to air transport

Increase by a factor of 10 in the capital investment in computers more especially in the now developing regions of the world.
Synthetic materials for ultra-light construction.
Regional weather control.
Automated motorways.
Automatic decision-making.
Central data storage of information.
Automatic language translation.

Of indirect value

—more especially in the now developing regions of the world.
Use of sophisticated teaching machines.
Economic mining of the ocean floor.
Economic desalination of sea water.
The control of gravity was judged to be outside this time period, and the control of noise was not apparently a subject of examination.

The redistribution of population is likely to be a key factor in the overall growth of air transport in the medium-term period (1985–2015). While population growth and increases in GNP and disposable incomes are dominant factors today and in the short-term future, the longer and medium-term periods must be strongly influenced by the wide ranging forces emerging from the immense changes in these basic parameters most of which we can only just recognise today. Some of the principal parameters may be briefly discussed.

The increasing population of the world

This is now believed to have increased at an accelerating tempo since the first widespread use of agriculture, and from the present level of 3500 million it is forecast to increase to about 6300 million in the year 2000. The more developed parts of the world today including Europe, North America, Oceania and Japan support a population of 1100 million or about one third of the total world population.

In the year 2000, the developed countries have been estimated to be likely to support a population of 1612 million or about one quarter of the world population at that time. The population must be expected to exceed 15,000 million within one hundred years. This must surely be the dominating factor in the demand for air transport in the new century.

The redistribution of population over the land and sea areas of the world

This redistribution is more striking and more significant in its impact on the prospects for world air transport when the remarkable growth of population in the, so far, less developed areas of Asia are considered. For the increase in Asia (less Japan) from 1965 to 2000 is likely to be no less than 1787 millions, almost precisely a doubling of the existing population in that area. The population of Africa and South America would more than double in this period and in relation to Europe and North America the increases would be such as to affect considerably the flow of trade and transport and the influence of nations in world affairs. However, population itself may be more a deterrent than a stimulus to the economic development of a state without large resources and sources of natural wealth, and the economic 'take-off' may require an immense investment in ideas and capital from friendly states before growth becomes significant. Recent history records the slow rate of progress in this field.

The increasing wealth of states is likely to be as potent a factor as population itself, and the GNP, or Gross National Product, is a significant parameter in the assessment of trade growth and of the demand for transport. While a five-fold increase in GNP is forecast for the less developed continents from 1965–2000, Europe, North America, Oceania and Japan are likely to experience a gain of similar order with an increase in the resultant differential GNP which will be dramatic in its influence. Thus by the year 2000 both North America and Europe may exceed by a factor of between 2 and 3 the combined GNP of all the less developed countries in the world.

The increasing contrast in the wealth of states

It is in the GNP *per capita* that the increasing contrast between States in the next fifty years can be most clearly identified. Fig. 5 shows selected estimates of GNP *per capita* based on the work of the Hudson Institute[3]. Medium rates of increase in GNP have been used, but these can be no more than very broad guesses at future development. It may be seen that the less developed States seem even within the time-scale of the chart, to have hardly begun their acceleration let alone achieved take-off. Serious political problems would seem to exist if this disparity in the individual wealth and in the wealth of States is allowed to persist or to increase.

The increase in leisure and disposable income

There seems little to indicate that the level of *per capita* wealth in the now undeveloped countries would allow extensive overseas travel by the end of this century. Using the medium or most probable estimates of the Hudson Institute and other sources, the *per capita* wealth in India in the year 2020 would be little more than one half of Italy in 1965. We would conclude that by the early years of the century a remarkable development

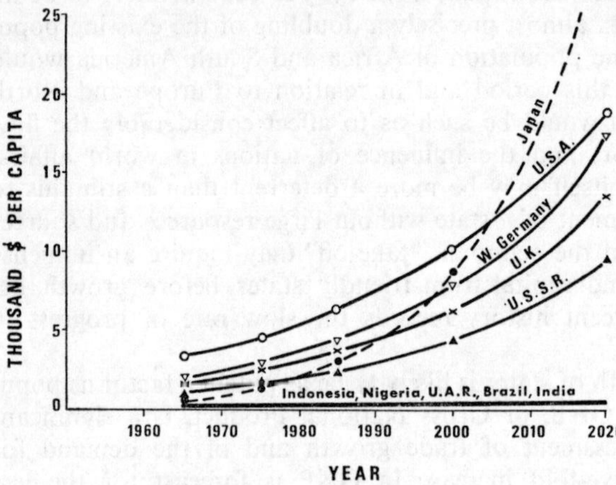

Figure 5 Possible growth of GNP 'per capita'—as indicated by studies of the Hudson Institute

in air transport is to be anticipated, but this will occur primarily in the already well-developed countries with a steady but slower growth (on the scale we have witnessed in recent years on European domestic services) in those other parts of the world such as South America, Africa, and Asia, (except Japan), which will be only gradually increasing their *per capita* wealth. This would in no way exclude the probability that the rapidly expanding activity of international business and the ramifications of the large international organisations, whose increasing influence we discussed earlier in this chapter, will generate an ever-growing traffic on the principal international air routes. Thus an increasing tempo in the development of the great international air services is foreseen, and a slow but steady increment in the domestic air systems of the world outside the most advanced industrial states in North America, Europe, Oceania and Japan.

The increasing urbanisation of population

These influences will surely be encouraged by the increasing urbanisation of the world's population. Figures estimated of the population living in urban areas differ somewhat because of the different definitions of the urban area. From study notes prepared for *Europe since 1945* by Professor A. G. Lehmann and Dr. R. Pryce (1968), it is recorded that urban populations have increased in the case of six European countries, the USA and the USSR to the following values:

Population living in cities exceeding 100,000

Belgium	1947	62·7%
Denmark	1955	69·0
France	1954	55·9
Germany	1961	76·8
Sweden	1962	51·9
UK	1961	80·0
USA	1960	69·9
USSR	1959	47·9

A remarkable increase in these ratios has been occurring over the past fifty years and Fig. 6, based on the figures of Professor Kingsley Davis[12], shows the likely trend into the early decades of the next century. A close

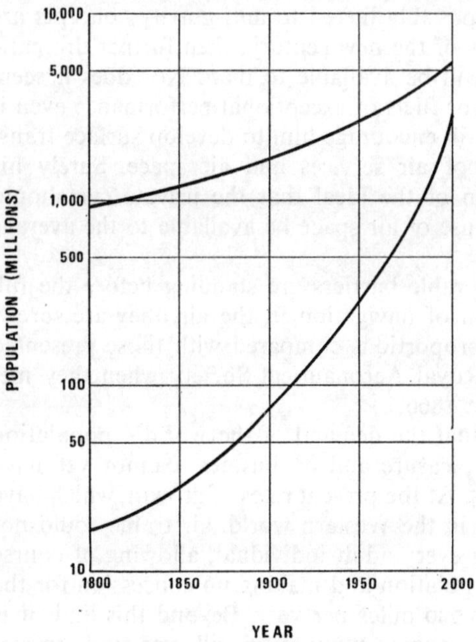

Figure 6 Total world and urban populations, estimated to Year 2000

correlation between GNP *per capita* and the percentage of urbanisation of a country has been established and the relation to air transport and the demand for air services seems self-evident. However, the probable contrast in the wealth of States which we suggested as a deep and dangerous issue in our medium-future period, may well offer extended opportunities for travel as GNP *per capita* slowly arises in the now undeveloped countries (in 2015 more fully developed countries), who will be able to offer their citizens the chance of wider transport experiences in the known world.

II LONG-TERM 2015–2070

It may be thought time wasting to push even further ahead than the second decade in the new century to consider the air transport system, its vehicles and terminals, but even in this period some basic concepts seem to be justified. The technological development of the vehicle will continue to determine the use of the air medium as emphatically and continually as in the medium-term period considered above. Automatic landing and take-off VTOL aircraft as well as automatic ATC will in the long term allow a very high intensity of air movement. If the high grade composites such as carbon or boron, reinforced aluminium or resins, are fully incorporated into engine and structural design by the 1990s and a new generation of new materials, possibly linked to anti-gravity concepts are introduced in the first decades of the new century, then further dramatic potential for the air vehicle will be available to man. Nor does it seem likely that the new materials (or fuels of exceptional performance even if available at reasonable cost) will encourage him to develop surface transport systems at the expense of air services and air space. Surely his period should see the fruition of the ideal that the private car should take to the air and the fuller use of air space be available to the average citizen.

At the present time considerable barriers are standing before the full development of such a system of navigation in the air, they are surely, however, of quite negligible proportions compared with those presented to the first members of the Royal Aeronautical Society when they met on that celebrated occasion in 1866.

It seems almost inevitable that the demand of the world's population for travel, both for personal pleasure and on business cannot yet have been satisfied by the year 2015. At the present rates of growth, which have not been difficult to maintain in the Western world, air traffic could not have reached a level at which every adult individual, allowing of course for the expected growth of population and making no concession for the 'pill', would have travelled 30,000 miles per year. Beyond this limit it is thought most unlikely that the human population will ever wish to go. Nor are we convinced that circumnavigation of the Moon would be a popular pastime, any more than visits to the Antarctic ice-cap have been in the 1960s. Journeys into far space may in that age more usefully and enjoyably be undertaken in the inner mind of man. Other bands on the spectrum of communication will be available to the Fellow in 2069 who may then abhor the congested but relatively quiet VTOL transport system which he helped to perfect in the early decades of the century.

12 AIRPORT DEVELOPMENT IN THE MEDIUM TERM (1985–2015) AND LONG TERM (2015–2070)

From the STOL airports discussed in section 3, it is a major step to the VTOL ports which are already in the planning stage for the 1980s and beyond. The essential requirements of the STOL site differ very little from the local airfields which have been provided in most countries since the First World War. The requirements of the VTOL site, however, are not yet in any sense rigidly defined. The VTOL vehicle itself may yet appear in many different forms for the technology is in no sense agreed and a number of lines of attack are being pursued in the USA and Europe at the present time. Initially the support of defence budgets both here and in America must be forthcoming if accelerated development is to be sustained. Although the UK has had a lead so far in direct lift vehicles, more recently based on high by-pass ratio 'advanced technology' turbofan engines, this lead may soon be lost if action is not taken to pursue the projected VTOL designs into hardware of commercial significance. The new technology offers improvements in specific weight, consumption and noise level which will undoubtedly be exploited in the transport of the 1980s.

The case for VTOL transport does not need to be made here. The advantage in journey times is most dramatically to be shown on short-haul direct city centre trips, and may become even more striking in future years as traffic congestion on the roads makes airport-town centre times unacceptable to the air traveller. If the value of the time saved by various sectors of the community is computed as an effective cost saving for the fastest form of transport, an economic case can be presented for VTOL in comparison with most forms of surface transport and with conventional air services operating between existing airports which are so often remote from the true beginning and end of an air journey.

The theoretical city centre VTOL site may not, however, prove to be a practical proposition. The recent history of discussion on the London Helicopter sites and current studies on sites for VTOL operations into London have shown that major obstacles to the true city centre site exist. Foremost of these are the problems of noise, conflicts of land use (residential, government or parliament), cultural centres, terrain and tall structures. The most favoured sites for 'near-central' London are now in the Surrey Commercial Docks and St. Catherine's Dock area. Earlier proposals for Nine Elms and over the railway termini such as King's Cross, have now been discarded for various reasons including noise, hazards of navigation and congestion in the ground approach areas. In Paris it seems hardly likely that true central sites will be acceptable and the same is probable in New York and most other major cities of the world. In Manchester it was found that four potential sites were possible to meet the scale of movements forecast for the 1980–90 decade but little disadvantage arose from peripheral sites and a favourable one was located with good access to the city, with a wide natural catchment area and with existing general aviation facilities. Interline connection with conventional aircraft services are also likely to be a major advantage of peripheral sites at least in the medium-term future period of airport development, and this trend is almost certain to be emphasised by the active development of STOL services to reduce congestion in the principal city airport zones in the next five years.

In any case decisions on VTOL city centre sites will need to be taken at a very early date (early 1970s) if they are to be preserved for the medium-term future period since far-reaching planning decisions are now in train in all the principal cities of the Western world. In the UK we may mention the Manchester rapid-transit system and, the London Airport (Heathrow)/Victoria rail link which will create nodal points in metropolitan transport well into the long-term future of these cities (45–100 years).

In the urban growth which we have discussed is to be found an increasing inflexibility in our social organisation. To meet this the fixed surface transport links and airport terminals surrounded by high value property, are unsuitable solutions. The true value of air transport should lie in a high degree of flexibility, and VTOL technology linked to automatic ATC surely offers this in the highest degree. It seems inevitable that in the long-term future sites for the VTOL ports will be distributed peripherally to the urban zones with adequate points of access to allow short road times to all air terminals from the key residential, business and industrial zones. The future of urban development in many regions lies in linear, cluster, or rand type growth rather than in uniform circular growth about a hub or nucleus; examples of this lie in the Netherlands (Randstad, Holland), West Germany (Ruhr Plan, SVR), and in the cities now under development in England (Central Lancashire and South Hampshire). VTOL ports peripheral to the linear conurbations seem the most likely long-term future development to meet the needs of this growth. Fig. 7 illustrates a possible development of the type described. As now, the smaller and self-contained communities will need close-in peripheral airfields to meet the combined needs of VTOL, STOL and conventional aircraft of all types. Nor must it be forgotten that the costs of operation of all airports will need to be met by all the available revenue sources taken together. Such costs are not greatly sensitive to land values. Not only are these lower in the peripheral area but these locations encourage many secondary revenue sources of great importance, such as general aviation, maintenance and overhaul facilities, training functions, air orientated manufacturing industry and cargo storage/container depots. Air cargo reception and handling would be extremely difficult in the central city areas. In the short, medium, and probably the long-term future only a small number of airports in countries such as Great Britain, Germany and France can expect to be able to ignore the secondary sources.

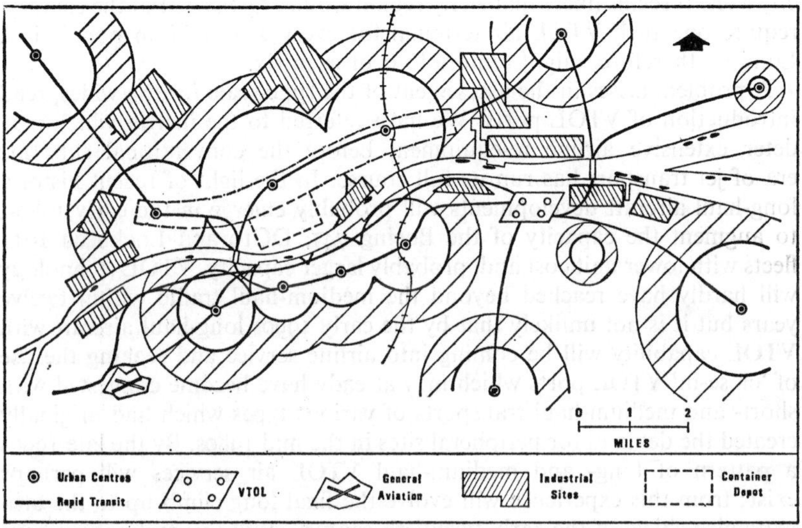

Figure 7 VTOL and general aviation sites. European City, Year 2000

The requirements for VTOL ports which may determine the main elements of choice in site planning in the medium-term future may be briefly stated as follows:

1 Landing and take-off pads which allow unobstructed approach and climb away paths. 15–25° approach angles. Noise sensitive areas to be retained outside a three-quarter mile (1·2 km) radius movement funnel.
2 90–95 PNdB noise level to be acceptable within 500 ft (152 m) of the vehicle approach pattern although initially 95–100 PNdB may be necessary.
3 Good access to and from surface transport in all significant directions.
4 Catchment within thirty minutes of half a million population to justify first serious consideration for airline passenger operation.
5 Scale and design of facility would be determined by multiple factors including scale of population, industry, other competitive transport, tourist characteristics, GNP and economic factors.
6 Adequate capacity of surface transport to meet peak air demands.

Figure 8 illustrates a possible solution to the VTOL city centre or inner urban site which might be suitable for passenger movement rates of 30 million per annum.

Careful consideration of all the factors which can be identified in the medium-term future period within which it seems most likely that the first VTOL ports will be inaugurated, suggests that they will be peripheral to the cities and conurbations and probably four to six miles (6–10 km) from the central city areas or a similar distance from the metropolitan city axes. It seems likely that a city region would, by the end of the century, require one such VTOL air terminal for every 2–2½ million population. London, therefore, might require four or even six.

A problem arises in the assessment of the time-scale for the widespread introduction of VTOL ports. We have referred to the inertia which may deter extensive airline re-equipment before the conventional subsonic era of jet transport has run its full course. In the light of recent history, long-haul aircraft developments will probably emerge in the early 1980s, to augment the capacity of the Boeing 747, DC10 and Lockheed 1011 fleets with lower unit cost and, probably larger capacity. VTOL technology will hardly have reached beyond the medium-haul traffic within twelve years but it is not unlikely that by the early 1990s long-haul aircraft with VTOL capability will be coming into airline service and seeking the use of 'close-in' VTOL ports which may already have become congested with short- and medium-haul transports of various types which had originally created the demand for peripheral sites in the mid 1980s. By the late 1990s a pattern of long- and medium-haul VTOL air services will perhaps exist; from this experience will evolve the final long climb up to the ultimate air vehicle of the early twentieth century. Will this technology have achieved the economic mastery of supersonic flight? It seems inevitable that it should be so, although the 750–1000 seaters of the last decade of this century may be subsonic in the interests of operating economy. However, it is unlikely that the 1000-seat VTOL–SST long-haul transport will need to wait until the Twenty-first Century, since no clearly defined constraint can at present be identified. Surely ground handling systems for passengers, baggage or cargo could be designed in the 1980s to meet the requirements of a vehicle which in this respect only differs in scale from the giant transports of 1970.

If the sonic boom should offer restrictions, then the North and South Atlantic and the Pacific routes are a rich field for activity by SST and hypersonic transport. If financial resources should be lacking, the constraint may lie far deeper perhaps even in a growing disenchantment in the need, at such a time, for a new generation of transport aircraft.

13 LESSONS FOR THE 1970S AND BEYOND

The more critically one analyses the growth factors in civil aviation the

Figure 8 VTOL site. An artist's view Year 2000
(Hawker Siddeley Aviation and Brian Colquhoun and Partners)

more convinced one becomes of the vast increases in air movement which will be a main characteristic of advanced societies in the early decades of the next century. The geography of the earth is such that the more constrained systems of surface transport will not meet the major part of world transport as it must develop at the turn of the century. Other forms of communication such as the Television Phone are likely to become competitive to air services for business, but tourism should be little affected by this. Tourism is more likely to be seriously eroded in due course by the familiarity of exotic scenery and resorts through colour television. The increase in population, the growing urbanisation of human societies in both the technologically advanced and in the less well developed states, the increase in gross national product *per capita* in all countries, and the extension of hours of leisure will surely all combine to stimulate the demand for travel, for pleasure and for business.

The improving technology of aircraft design, matched by the great opportunity which an increased scale of operations offers for the provision of air services at reduced unit cost, will combine with increased world purchasing power to promote a continuing massive expansion of commercial aviation. To define its direction would be most difficult beyond the end of the century, since the increase in individual purchasing power and discretionary income might stimulate the use of general aviation to the point when control of the weather and automatic air traffic control might encourage the virtual domination of the air space in the latitudes 30–60° N by privately owned and operated high performance VTOL vehicles. Widespread weather control and climate controlled resorts in northern latitudes may widen the geographical demand pattern for tourist and personal air travel in the early decades of the new century.

A warning note must be sounded with regard to the now undeveloped countries of the world which, while increasing their GNP, will increasingly lag behind the advanced countries (Fig. 5). Except where tourism or key mineral products may generate a special demand for business and cargo air services, the growth of communications and the provision of ground facilities cannot be expected to grow at the rate of the well developed States. The development of consortia for airline operation on the model of KLM, SAS, Swissair, or the more modest proposals of the Maghreb airline for the North African States, will be likely on an increasing scale, since the operation of the giant aircraft necessary for economically competitive international air services call for an investment which will often be beyond the resources of individual States. We may look to the growing demands for extending the political and economic communities in Europe and in Africa to be pressures wholly favourable to the sound commercial development of air transport.

It seems most likely that the use of STOL aircraft will increase very considerably within the immediate future (1970–85) if only to ease congestion at the great capital airports and to activate use of the now passive air space close to our major conurbations. Reduced noise levels must be anticipated before this stage of development can be intensively programmed with direct lift VTOL aircraft. Probably the VTOL era for short/medium-haul transport must await the last decade of the century for its full impact on our society. It seems inconceivable that the revolution in convenience and economy which VTOL will bring can be delayed beyond then. Intensive research into noise and materials might step up the rate of progress even faster than this, for the existing ATC problems and congestion increasing annually at the main terminals might create sudden political pressures to stimulate VTOL services and operations if they could be viably sustained.

The great era of long-haul air transport with giant subsonic and supersonic VTOL jets may not wait until the end of our medium-term future period, 2015. By the mid 1990s the VTOL technology will surely be proved, and the burden of conventional take-offs from our existing remote capital city sites may no longer be acceptable. This will be the era of the universal VTOL ports in or close to the urban centres of our great cities and adjacent to the axes of our extended metropolises. Access to the smaller city must become possible in that day for the short-haul and the long-haul air liner no less than to the small vehicles of the business man and private owner. That will be a great day for the civic planners and consultants who should be grateful to us if we can preserve for them the widest possible choice of site and opportunity. Present and future planning decisions must be taken with boldness and with an imaginative sense of the future.

It would be foolish to spend time on a pointless examination of the future which posterity can report on so much better than we can ourselves, however, we would be wise to encourage the art of technological forecasting for we should then be in a much better position to improve our own future and to invest more wisely in those areas of our activity which will be the most beneficial to our descendants. The need to build flexibility into our planning should be more fully appreciated. Such planning effort which may seem greater in the short term, may recognisably ease the constraints within which the future has to work.

What industry should be better able to plan the future than the aviation industry. Even if it is the unaltering human characteristics of man which may prove the most unmanageable factor in the future, it is our industry alone which might lift him, within his own life-time, to the stars!

References

1 LEVY, J. C. Engineering our Future. *Journal of the City University*. December 1968.

2 GABOR, D. *Inventing the Future*. Pelican, 1964.

3 KAHN and WIENER. *The Year 2000*. Hudson Institute, Macmillan, New York 1967.

4 FAA Aviation Forecasts, 1968–79. TAD-484-3 Washington, 1968.

5 SMITH, P. Methods of Forecasting in Air Transport. ASA 75. Alan Stratford & Associates report.

6 MILWARD, SIR ANTHONY. The Development and Future of Air Cargo. *The Aeronautical Journal of the Royal Aeronautical Society*, Vol **72**, No 695, November 1968.

7 STRATFORD, A. *Air Transport Economics in the Supersonic Era.* Macmillan, London, 1967.

8 STRATFORD, A. H. UK Airport Planning in the 1970s. Address to the Aerodrome Owners Association Conference 1966.

9 Airport and Air Service Development in Lancashire. Lancashire County Council. ASA 73.

10 MASEFIELD, P. G. The Modern Airport and its Future. *Journal of the Royal Society of Arts.* August 1968.

11 JANTSCH, E. Forecasting the Future. *Science Journal.* October 1967.

12 DAVIS, K. The Urbanisation of the Human Population. *Scientific American.* September 1965.

13 Transport in the Year 2000. Institution of Mechanical Engineers, 1965.

14 CLARKE, A. C. *Profiles of the Future.* Gollancz, 1962.

15 JOHNSON-MARSHALL, P. *Rebuilding Cities.* Edinburgh University Press 1965.

16 SEALY, K. R. Air Transport Facilities and Regional Planning. *The Aeronautical Journal of the Royal Aeronautical Society*, Vol **73**, No 703, July 1969.

17 BRIGGS, A. Technology and Economic Development. *Scientific American.* September 1963.

18 DAVIES, H. Some Thoughts about the Future of European Aeronautics. *The Aeronautical Journal of the Royal Aeronautical Society*, Vol **72**, No 689, May 1968.

19 HEITMEYER, R. Future Air Traffic and Implications for Airports. IATA. 17th Technical Conference October 1967.

20 Studies of the Site for the Third London Airport. ASA 49. July 1966.

21 Report into the Development of Stansted as the Third London Airport. HMSO. 1967.

22 Airport Requirements and Sites in the Metropolitan New York—New Jersey Region. Port of New York Authority, 1966.

23 Forecasting Future Markets for Air Transportation. Lockheed Aircraft Corporation. Report 7185.

24 GREENWOOD, GABRIELLI and CLARKSON. The Prospects for New Types of Aircraft in the Decade 1970/80. AICMA, Symposium, September 1967.

20 Air traffic and navigation in the Second Century

A. Stratton

INTRODUCTION

Navigation has been defined as 'the business of conducting a craft as it moves about its ways'[1]. This broad definition of navigation encompasses the whole process of handling, steering, control, guidance and operation of the craft in a safe and economic manner. Navigation has a much longer history than aeronautics. As man first began to explore the world and to develop trade, the need for safe and predictable operation of the craft was a spur to the application of the most advanced science and technology of the day.

The primary threat to the craft was natural hazards, which increased greatly on approach to land and much attention focused on the problem of making a safe landfall. For the open sea the problem of the navigator was the timely arrival of the craft at its destination.

As the number of craft increases a new problem arises, that of traffic. The dictionary definition of traffic is 'vehicles, pedestrians, etc, collectively using a thoroughfare'. The essence of traffic is thus the interaction of the craft with each other and the prime risk of collision that is generated by this interaction. The collective use by vehicles of the thoroughfare (the term thoroughfare will be taken in the broad sense to apply to the air) can only be tackled by establishing certain requirements for collective behaviour. These requirements extend both to the characteristics of the vehicle itself in terms of its operational performance and capability and rules of procedural behaviour for the operator of the vehicle. To be effective all vehicles and operators must conform to a certain specified minimum standard and a system of control has to be established to ensure that these standards are met.

For the early aviator, the problem of navigation was similar to that of the earlier mariner, namely securing his personal survival by avoidance of collision with natural obstacles and on over-sea flights securing a safe arrival at his destination within the fuel capacity of his aircraft. To this end, marine techniques were directly applicable, dead reckoning by

Originally published in the *Aeronautical Journal of the Royal Aeronautical Society*, June 1969.

compass and log (in the case of the aircraft operating on air pressure) and position fixing by astronomical observation using the sextant and chronometer. Francis Chichester's first East-West solo crossing of the Tasman Sea in 1931[2] using these simple aids to make an accurate landfall at the limit of endurance of his craft is a classic example of the navigational skill of the pioneering aviators.

Although aviation started with free movement of the craft in three dimensions, avoidance of other craft being by visual observation and a 'collision avoidance' manoeuvre, increasing speed, numbers of craft and the need for predictable operation under all weather conditions brought, as in other means of transport, a need for rules and procedures to ensure safe conduct of traffic.

Such rules and procedures take two forms:

(*a*) rules for manoeuvre to avoid collision when two craft are in close proximity and

(*b*) planned routing, or traffic lanes, to reduce the number of occasions in which craft are brought into conflict.

When the number of craft is small it may be argued that reliance can be placed solely on collision avoidance rules, thus ensuring the greatest freedom of movement. Even when only two craft are involved and their relative velocity is low (as at sea) the formulation of unambiguous rules is a matter of great difficulty[3, 4]. With the much more severe consequences of air collision, compared with sea, air traffic has thus long been forced to rely for safety primarily on planned routing in traffic lanes and other controlled aerospace, 'free flight' only being allowable outside these controlled regions. Use of air thoroughfares necessitates for safety the fitting of special equipment to agreed international standards and a regulated level of training and experience by the pilot.

In forecasting the future, it is tempting to assume that the need for such regulation should not be considered as one of the factors determining scientific and technical progress. Furthermore it is tempting to assume that economic and social considerations can also be set aside. At any one time, however, the method of operating air traffic—and indeed of any traffic—is a complex balance between technological, social, political, economic and regulatory factors and any examination of the technical future must take account of the influence that these factors will have on the achievement of technological ideals; for example, in considering the introduction of new technological developments allowance has to be made for evolutionary introduction with the new operating at the same time as the old. The long-term forecaster, however, enjoys a major advantage over the current designer of new systems in that the former can start from an ideal and consider how it might be implemented, whereas the latter is heavily constrained by the real and practical difficulties of the most immediate future. To be credible the forecaster nevertheless has to define the framework or model within which his forecast is made.

The requirements of such a model for aviation are formidable. The fundamental input is a demographic forecast of the future growth of world population; this has then to be applied to a sociological forecast of the way in which future society will organise its living and occupy the globe. The latter will be influenced by and determine the future means of transport and communication. For any state of social development, forecasts have then to be made of the modal split between air and other forms of transport which results from the interaction between technological, economic and social factors.

The above in combination will determine the nature of future air traffic and lead to an operational model against which the application of future science and technology to navigation and traffic control can be considered. For this chapter the demographic and social forecasts will be culled from other writers to produce a simplified transport model against which some postulates will be advanced on future transport needs. The fundamental factors determining the flow of traffic will then be discussed and the future role and special characteristics of air traffic examined.

This will provide an operational model defining objectives for future science and technology to fulfil. Navigation always has been in the forefront of exploiting new scientific discovery. The application may be limited by technological problems of implementation either in performance or in reliability or by some performance limitation that is inherent in the natural phenomena utilised. An example of the latter is the magnetic compass which has reached the limit of its performance due to fluctuations in the earth's gravitational field. This chapter will examine the possibilities for navigation in the future, unfettered by technological problems, in an attempt to define what should be possible against predicted operational requirements; it will be restricted nevertheless to existing scientific knowledge and fundamental limitations.

THE DEMOGRAPHIC AND SOCIAL MODEL

The growth and distribution of world population for AD 2000 as forecast by the United Nations is shown in Table 1. It provides a more than

TABLE 1 *Predicted distribution of world population*

	Europe	*USSR*	*North America*	*Latin America*	*Asia*	*Africa*	*Oceanic (Pacific)*	*Total*
	$\times 10^9$	$\times 10^9$	$\times 10^9$	$\times 10^9$	$\times 10^9$	$\times 10^9$	$\times 10^9$	$\times 10^9$
1963 (estimated—UN)	0·44	0·22	0·22	0·22	1·75	0·29	0·02	3·16
2000 (forecast—UN)	0·59	0·40	0·33	0·65	4·25	0·66	0·03	6·9
Growth Factor	34%	82%	50%	195%	140%	130%	50%	120%

two-fold increase to $6{\cdot}9\times10^9$; it is characterised by a much larger proportionate growth in Latin America, Asia and Africa than in Europe and North America and is dominated by growth in Asia.

Another author in this book[5] uses for his model a two-fold increase in the next century, and a recent author on future transport systems[7] uses a similar distribution to the UN forecast with an overall figure of $8{\cdot}7\times10^9$ for 2016. A recent longer-term forecast[6] suggests a minimum of 20×10^9, a medium of 35×10^9 and a maximum of 50×10^9 in 'one or two centuries from now.'

The important factor in a transport model, however, is the way in which the future population, be it 7×10^9 or 50×10^9, will be living and, in particular, the form of urban development. One such recent study[6]. will be taken as an example in which the forecast of a world population of at least 20×10^9 is linked to an urban population of at least 10×10^9, a growth of 15 times over the present. Such a large increase could only occur if the greater part of the urban population lived in interconnected cities forming large urban areas of many tens of millions of people living at a high density. This trend is already evident in many areas of the world, for example in the eastern seaboard of the USA from Boston to Washington DC. The shape and form of these complexes will depend to a major extent on existing urban areas, natural features and resources and transport and communication methods will play an increasing part in shaping urban development. It is not possible to forecast the number and absolute size of such areas; in many regions[6], however, the forecast shows a tendency to an urban complex of elongated form. Such large areas of course, would not be evenly populated and would have a fine structure of linked urban development. Population densities of up to 50–200 per hectare* are forecast (the density of the Greater London Council area averages 50 per hectare). For the want of a clearly defined terminology the title 'Urban' (capital U) will be used in the broadest sense to describe areas with populations calculated in tens of millions and 'urban' to describe the smaller units within the Urban area. The concentration on the Urban population must not allow, however, the needs of the many thousands of millions outside Urban areas to be neglected.

TRANSPORT MODEL

A formalised model of the Urban concept would look like Fig. 1 with variation in the number and size of the individual urban units. Some models of the urban units themselves are discussed by Legg[7], and they in turn will have their own fine structure. As Table 2 demonstrates, a population range of over 100:1 could be accommodated within a 5–15 km radius of urban units, leading to a separation between urban units of a few tens of kilometres.

*1 hectare $=10^4\text{m}^2=2.47$ acres.

TABLE 2 *Size and population of urban units*

Radius km	*Population Density* per hectare	*Population* × 10⁶
5	30–100	0·2– 0·7
10	50–200	1·5– 6·0
15	50–200	6·5–26·0

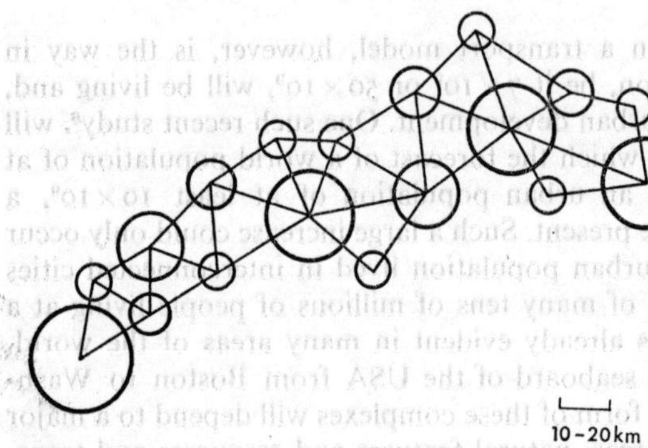

Figure 1 *Transport model of Urban development*

In present day development, the lower limit corresponds with the new city of Milton Keynes; at the upper end the GLC occupies an area of about 30 × 50 km with a population of 8 million.

The length of the 'chain' will be taken in the hundreds of kilometre band (Boston-Washington ~650 km; London-Manchester ~300 km; Tokyo-Osaka ~500 km).

Thus the following transport system requirements will be postulated.

(*i*) intra-urban; within the urban areas— ~10 km
(*ii*) inter-urban; linking urban areas— 10–50 km
(*iii*) intra-Urban; within Urban areas— 200–1000 km
(*iv*) inter-Urban; between Urban areas— say, 500 km upwards
(*v*) non-Urban; within outer areas— from a few kilometres up to many hundreds.

Such a model can be applied only in a gross way to indicate the nature of future transport requirements. The amount of travel will depend very much on the future form of society and, in particular, the extent to which new methods of audio and visual communication will avoid the need for travel. Within the total transport scene the modal split between the air and other components will be determined by a complex interaction between technological, economic and social factors. There are, however, certain basic properties of traffic flow which provide a unifying framework within which the properties of different modes of transport can at least be compared. With the evolutionary nature of transport, an examination of current trends and problems will also indicate those factors which will play the greatest part in shaping development over, say, the first half of the next century and these will now be considered.

CURRENT TRENDS

When considering the usage of airspace, it is customary to divide aviation into military, transport (scheduled and non-scheduled), training and general aviation. The special needs of military aviation are dictated by defence requirements and, in particular, rapid free movement is vital. In the past the higher performance characteristics of the military aircraft has provided considerable natural separation in height, but this no longer applies. We are already in the situation therefore where, for traffic purposes, military and transport aviation have to be considered as a whole. Furthermore, by virtue of its task, military aviation is equipped with the most advanced navigation technology which gives it the capability of highly accurate free movement both within and outside controlled airspace. General aviation may be divided into two categories—the company and other private aircraft which are fully equipped both in type of equipment and crew certification for operation in controlled airspace and the so-called 'private flyer' who is not equipped (the cost of equipment may be considerable) and not certified for operation in controlled airspace. Growth in free movement of general aviation that has little or no navigation capability creates an increasing hazard outside controlled airspace and if this lack of capability persists it can only lead to increasing restraint on free movement of this type of aviation.

In determining the extent of controlled airspace a major conflict exists between allowing for the legitimate requirements for free movement of military and general aviation and making the most economical and safe use of airspace by control. Furthermore, even within certified use of controlled airspace, conflict arises over economic use of airspace due to diversity of aircraft flight characteristics. The number of possible interactions between freely moving aircraft increases as the square of the number of craft and it is inconceivable that any technological advances could allow even today's density of traffic to be operated on a completely free movement basis. It will be taken, therefore, as axiomatic that any traffic system for the future must be based on controlled airspace and routing but the volume of controlled airspace must be minimised.

The greatest interaction between aircraft occurs at terminals; this used to mean a relatively small area around the airport itself. The increased traffic, the relatively inflexible nature of the approach path of present conventional aircraft, the need for all weather operation have all led to an increase in the size of the 'terminal area' of interaction. With traffic densities requiring many runways and several airports to serve major population centres, complex interactions occur between the individual airports serving one centre and the interaction rapidly extends to include other population centres, even though they may be separated by hundreds of kilometres. In addition, as the need for more civil airports grows there is increasing conflict between the terminal area requirements of military and civil traffic.

Considerable air traffic delays accumulate under peak operating conditions. There is in addition the major problem of matching air and surface traffic. The noise and other social penalties of airports are making it difficult to find acceptable sites near large centres of population with an adequate surface traffic distribution system. Surface movement and waiting time is taking an increasing part of the total journey time; for example, the block speed from the centre of London to the centre of Paris is only 30 per cent of the flight speed and the block speed from the centre of London to the centre of New York on introduction of the Concorde will be less than 50 per cent of the cruising speed.

Any consideration of a system for the future, therefore, must be based on the assumption that the 'terminal area' of today becomes a volume of airspace proximate to the Urban area of the future, within which there is intense interaction between air traffic. Furthermore, it will no longer be possible to concentrate solely on the airborne part of a journey, ignoring the linking surface movements and the interface problems. Transport systems will be judged on the overall cost-benefit of the complete journey, taking into account both direct costs and social cost-benefits including time, convenience and environmental factors such as noise and pollution, and social amenity.

In considering future growth, aircraft design and development and marketing focus on the growth in revenue miles. However, the total traffic flow through airports is determined by the number of aircraft movements (take-offs and landings) and the number of passengers carried. The rate of traffic flow will depend on monthly, daily and hourly variations in flow, the peak rate being reached for a few hours of a few months in the year. Interpreted in terms of current passenger flow at Heathrow[8], the peak hourly rate of passengers is about 3·5 times the average annual hourly rate and the peak hourly movement rate about 2·5 times the average annual hourly rate[9]; the greater increase in passenger movements corresponds to a higher load factor at peak.

The world air traffic situation is shown in Fig. 2 which gives the 1965 distribution of passengers and aircraft movements for the first 25 world airports and 40 other major world airports likely to handle high capacity jet aircraft and supersonic transport aircraft in the early 1970s (IATA). It clearly demonstrates the domination of European and North American traffic which, in proportion to the population of the areas, is 10–50 times greater than that for other regions of the world; 55 per cent of the passengers pass through 12 airports. Fig. 3 shows the (1967) total air traffic

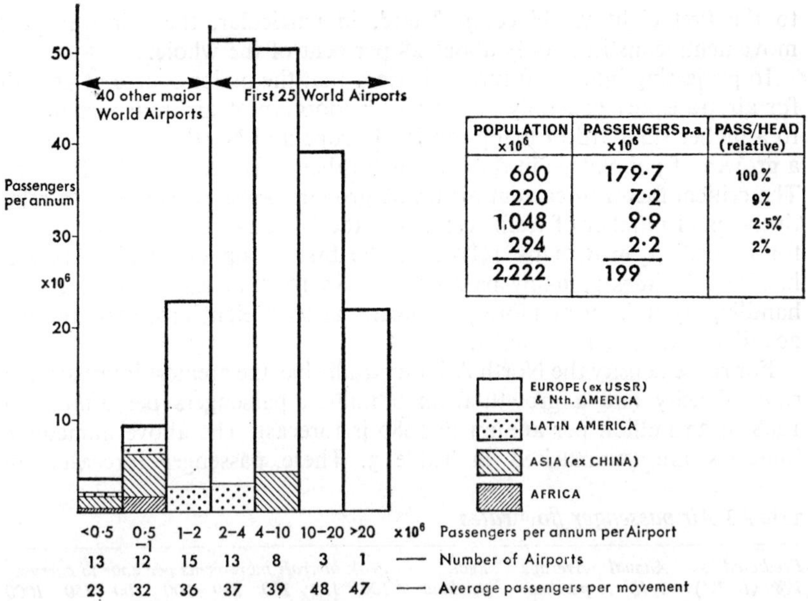

POPULATION x10⁶	PASSENGERS p.a. x10⁶	PASS/HEAD (relative)
660	179·7	100%
220	7·2	9%
1,048	9·9	2·5%
294	2·2	2%
2,222	199	

Figure 2 Distribution of air traffic at major world airports (1965)[15]

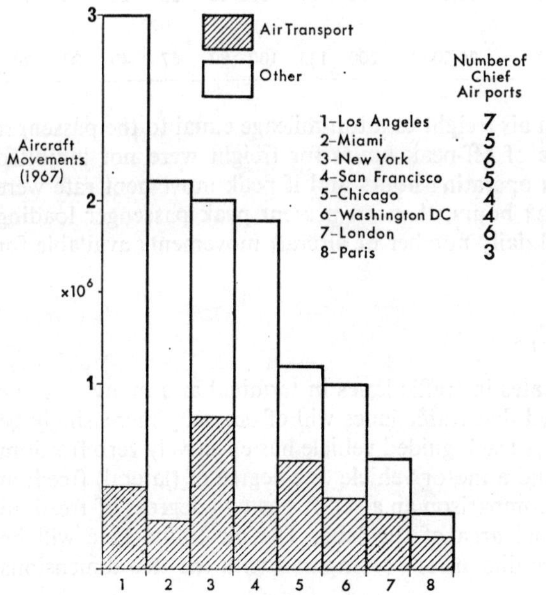

Figure 3 Chief centres of world air traffic (1967)[9]

to the first eight world centres[9] and, in particular, that air transport movements constitute only about 28 per cent of the whole.

In projecting into the future, it is clear that the major source of growth for air transport generally lies in wider adoption of air transport linked to a higher standard of living outside Europe and North America and to a growth of population in those areas (in themselves competitive factors). The critical factor in current air traffic growth, however, is that of a relatively small number of traffic centres in the USA and Europe. Forecasts[15] for immediate growth by 1980 are: the first group of 25 airports will handle 4–8 times as many passengers as in 1965, at least four* airports handling 45 million or more passengers, at least eight airports handling 20 million passengers annually.

For route density the North Atlantic traffic has the greatest international route density and a growth from 6 million passengers per annum in 1968 to 32 million per annum in 1980 is forecast. The above immediate forecasts are summarised in Table 3. These passenger forecasts are

TABLE 3 *Air passenger flow rates*

Predicted 1980 (IATA)	*Annual* $\times 10^6$	*Average per hour*	*Peak per hour*	*Peak aircraft movements per hour to carry* 100	150	200	250	300	500	750	1000 *passengers per movement*
Route (North Atlantic)	32	3600	10,000	100	67	50	40	33	20	13	10
Airport (Terminal area group)	45	5800	20,000	200	133	100	80	67	40	27	20

linked with a growth in air freight to a ton-mileage equal to the passenger ton-mileage. If the use of off-peak hours for freight were not inhibited by noise restrictions in operating hours and if peak movement rate were sustained throughout 24 hours, then on present peak passenger loading there is about an equal daily number of aircraft movements available for freight.

THE FLOW OF TRAFFIC

Air traffic already operates in traffic lanes in terminal and many *en route* areas and it is postulated that traffic lanes will of necessity increasingly be the mode of operation. A track guided vehicle has effectively zero freedom within the traffic lane and a motor vehicle one degree of (lateral) freedom within a road lane; by comparison an aircraft has two degrees of freedom within the cross-sectional area of the lane. The air traffic lane will be defined in route in three dimensions (compared with the two dimensions

*The official[10] 'most likely' forecast for the London area as a whole is 44 million passengers annually by 1980.

of surface vehicles). The problems of defining such a lane in space and of ensuring that the aircraft travels within the lane will be considered quite separately from the flow of traffic along the lane; the cross-sectional area of the air traffic lane has to be related to the certainty with which it can be forecast that the aircraft will not be outside the lane.

In a traffic system arrangements must be made for lanes to cross and for traffic to join and leave a lane. It is self-evident that for maximum capacity there should be separation in one of the two cross-sectional dimensions where lanes cross.

Having defined such air traffic lanes, an identity with surface transport is established in the study of traffic along a lane. Fig. 4 gives the space

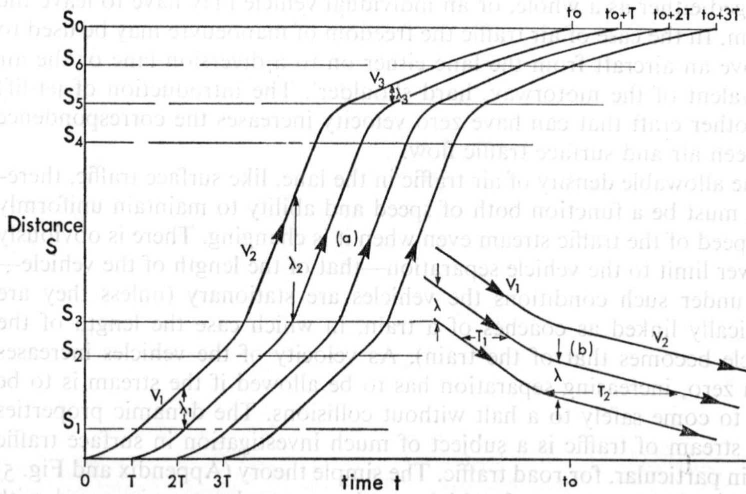

Figure 4 Flow along traffic lane

/time relation for traffic flow (the distance travelled S is measured along the lane which may be any curved path). Vehicles depart at regular intervals T. The first vehicles leaving at $t=0$ accelerates to V_1 covering a distance S_1, maintains uniform speed to S_2 where it accelerates to V_2 over S_2S_3 and so on through deceleration to V_3 and arrival at S_0 at time t_0. The second vehicle departs at time T and maintains exactly the same space/time profile, delayed by T and arrives at t_0+T. A regular flow of traffic at frequency $1/T$ is thus established, vehicles passing any point at intervals of T (which is called the headway). The separation λ between vehicles is related to the velocity V by $\lambda = VT$.

Alternatively by writing $q=1/T$, the rate of flow (in vehicles per hour) and $k=1/\lambda$ the concentration of vehicles (in vehicles per kilometre), $q=kV$. It is clear that for uniform traffic flow q must be constant at all points, otherwise there would be an accumulation of vehicles; k, therefore, must be a function of velocity. This is illustrated in (*b*) of Fig. 4 where vehicles decelerate from V_1 to V_2 maintaining separation; there is a cumulative delay of $T_2 - T_1 = (V_1 - V_2)/V_1V_2$ per vehicle. It is also clear that to achieve safe flow of traffic along the lane, account has to be taken of possible variations in vehicle timing from the ideal postulated in Fig. 4. Any sustained difference in velocity will give conflicts and it is axiomatic, therefore, that all vehicles travel at the same mean speed. Separation between vehicles must be sufficient to allow for fluctuations of individual vehicles about the mean when in steady flow conditions and for varying response to change in velocity; furthermore, it is necessary to consider emergency conditions when the flow of the traffic stream may have to be changed either as a whole, or an individual vehicle may have to leave the stream. In the case of air traffic the freedom of manoeuvre may be used to remove an aircraft from the lane either on to a diversion lane or the air equivalent of the motorway 'hard shoulder'. The introduction of jet-lift and other craft that can have zero velocity increases the correspondence between air and surface traffic flow.

The allowable density of air traffic in the lane, like surface traffic, therefore, must be a function both of speed and ability to maintain uniformly the speed of the traffic stream even when it is changing. There is obviously a lower limit to the vehicle separation—that of the length of the vehicle—and under such conditions the vehicles are stationary (unless they are physically linked as coaches of a train, in which case the length of the vehicle becomes that of the train). As velocity of the vehicles increases from zero, increasing separation has to be allowed if the stream is to be able to come safely to a halt without collisions. The dynamic properties of a stream of traffic is a subject of much investigation in surface traffic and in particular, for road traffic. The simple theory (Appendix and Fig. 5) assumes that a stream of vehicles undergoes a decrease in speed with variable deceleration between vehicles and a time lag in successive vehicles commencing deceleration. The vehicle with the lower deceleration 'catches up' with the vehicle in front and there is a decrease in separation distance. The analysis gives a functional relationship between vehicle flow rate and vehicle density which has a maximum, the lane capacity, at a vehicle density one half of the density at which the stream becomes stationary. At the point (A) (Fig. 6) the flow has a maximum of 60 vehicles per hour at a speed of 600 km/h. To achieve this at a deceleration of 0·2*g* requires the difference in deceleration between successive vehicles to be less than 42 per cent of 0·2*g*; the stream could then be brought to a halt at a density of 0·2 vehicles per km, i.e. a separation of 5 km. If the density of traffic is increased beyond the maximum value of 0·1 vehicles per km (free movement separation of 10 km) then the capacity of the channel is decreased if the same deceleration and separation standards are maintained. For example at *D* the flow has reduced to 30 vehicles per hour at a speed

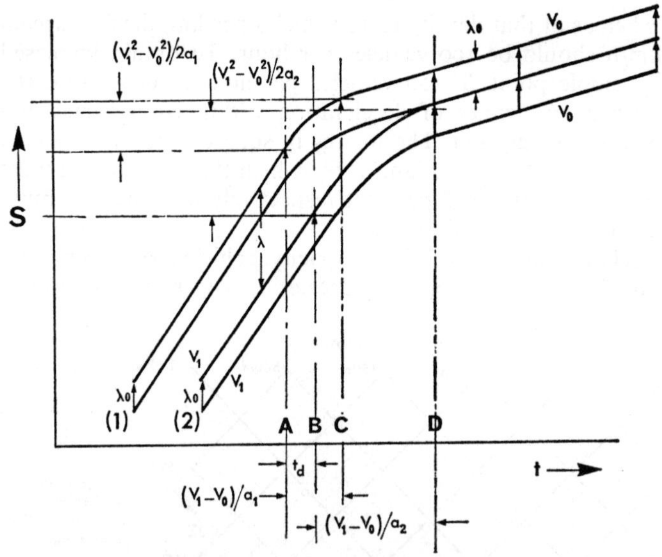

Figure 5 Relation between vehicle separation and speed

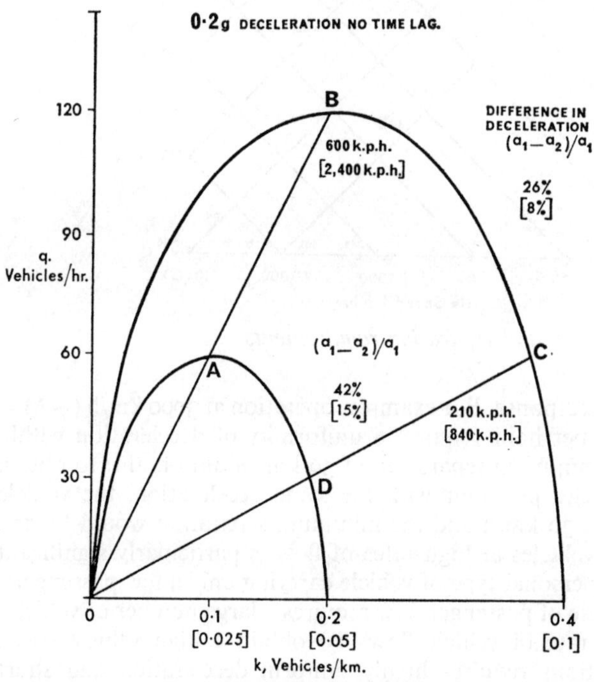

Figure 6 Variation in traffic lane capacity with vehicle density

of 210 km/h whereas at that density (0·18 vehicles per km) the free stream flow at 600 km/h should be 100 vehicles per hour. To obtain increased capacity, for example point *B*, 120 vehicles per hour at 600 km/h, the requirements are a uniformity of deceleration within 26 per cent and miminum separation of 2·5 km. The figures in square brackets relate to a speed of 2400 km/h (M=2) at *A* and *B* for which the deceleration must be uniform to 15 per cent and 8 per cent respectively and the minimum separation would be 20 km and 10 km.

The general relationship between lane capacity, vehicle speed, minimum separation and allowable difference in deceleration is shown in Fig. 7

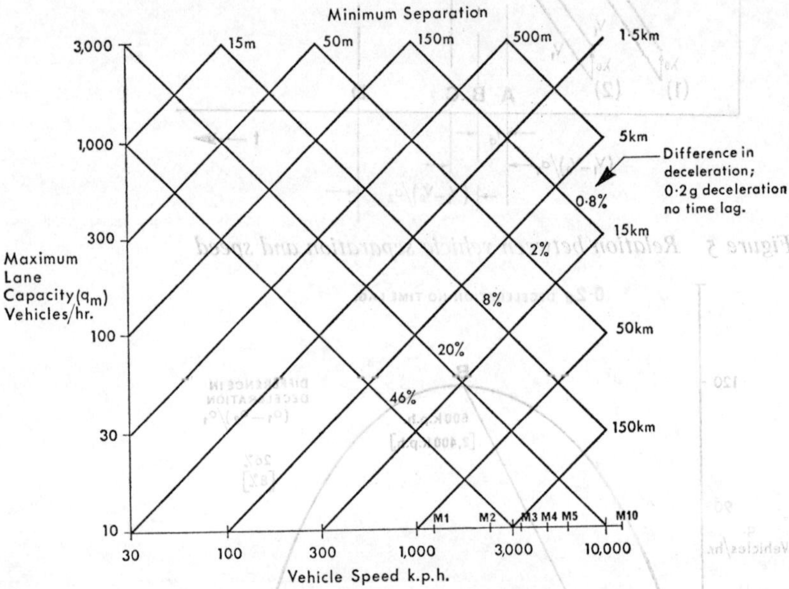

Figure 7 Deceleration and separation requirements

for zero time lag in response. For example, operation at 3000 km/h (~M=3) and 30 vehicles per hour requires a uniformity of deceleration within 20 per cent and a minimum separation of 50 km would obtain; to obtain a rate of 300 vehicles per hour with the same deceleration, the vehicle speed must drop to 30 km/h and the minimum separation would be 5m. The low speed of vehicles at high rates of flow is particularly significant in considering the personal type of vehicle carrying only a few passengers. To obtain a high rate of passenger flow requires a large number of vehicles and hence a high rate of vehicle flow. To obtain a high vehicle speed under these conditions requires highly uniform deceleration and short reaction times (a flow rate of 3000 vehicles per hour at 300 km/h requires deceleration uniform with 2 per cent at the 0·2g level; a time lag of just over a second would halve the flow rate).

The effect of a time lag in commencing deceleration is to reduce the lane capacity (Fig. 8), the allowable difference in deceleration and the minimum separation distance; for example, a time lag equal to the free flow headway ($1/q$) halves the flow rate.

Figure 8 demonstrates the increasing dominance of response time lag on flow at high densities and with a human operator controlling the vehicle

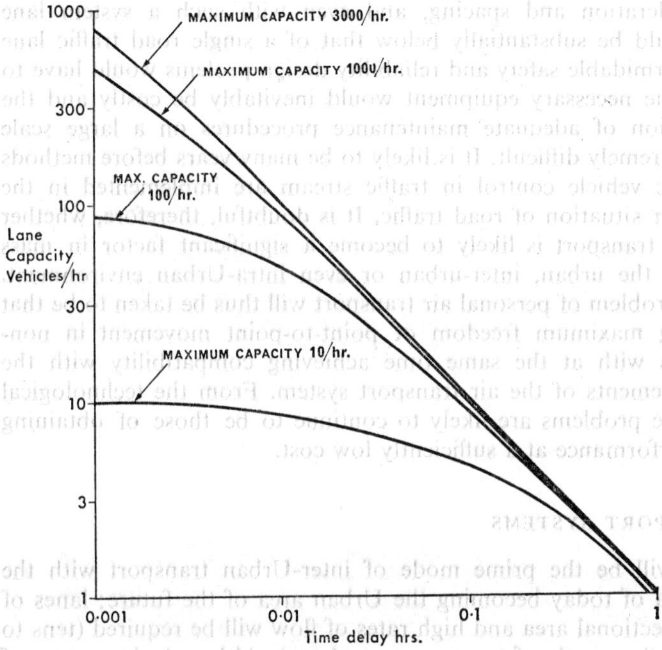

Figure 8 Effect of response delay on traffic lane capacity

reaction time will be a limiting factor. For example, for road traffic, even under free flow conditions, a maximum lane capacity of about 1500 vehicles per hour obtains at a speed of about 30 km/h.

High rate of passenger flow at high speed is thus best achieved by carrying a large number of passengers in a single vehicle (or directly coupled vehicles which behave as a single vehicle).

THE FUTURE ROLE OF AVIATION

Consideration of the more immediate forecast trends and the basic properties of traffic flow in relation to the long-term model leads to the following postulates.

PERSONAL AIR TRANSPORT

The freedom of individual point-to-point movement given by general aviation is analogous to the freedom that the automobile gives to surface transport and creates a similar, although at present less severe, conflict between private and public transport. Utilisation of the potential high speed of personal air transport under high capacity flow conditions requires a multiplicity of lanes combined with close automatic control of vehicle deceleration and spacing, and even with such a system lane capacity would be substantially below that of a single road traffic lane of today. Formidable safety and reliability design prolems would have to be solved, the necessary equipment would inevitably be costly and the implementation of adequate maintenance procedures on a large scale would be extremely difficult. It is likely to be many years before methods of automatic vehicle control in traffic stream are implemented in the much simpler situation of road traffic. It is doubtful, therefore, whether personal air transport is likely to become a significant factor in mass transport in the urban, inter-urban or even intra-Urban environment. The traffic problem of personal air transport will thus be taken to be that of obtaining maximum freedom of point-to-point movement in non-Urban areas with at the same time achieving compatibility with the traffic requirements of the air transport system. From the technological viewpoint the problems are likely to continue to be those of obtaining adequate performance at a sufficiently low cost.

AIR TRANSPORT SYSTEMS

Air traffic will be the prime mode of inter-Urban transport with the terminal area of today becoming the Urban area of the future; lanes of small cross-sectional area and high rates of flow will be required (tens to hundreds of thousands of passengers per hour). Although these rates of flow are high by current air transport standards they are not exceptional in terms of the fundamental capabilities of a traffic stream of high capacity vehicles (Table 4). Matching the air transport to the surface transport system and the location of the air terminals in relation to the Urban

TABLE 4 *Passenger capacity of traffic lane*

Lane Capacity	*Speed* km/hr *1,000*		*3,000*		*5,000*		*Passengers per vehicle* *300*	*500*	*700*	*1,000*
Vehicles/ hour	λ_0 (km)	$1-a_2/a_1$	λ_0 (km)	$1-a_2/a_1$	λ_0 (km)	$1-a_2/a_1$	*Passengers per hour*			
100	5	20%	15	8%	25	5%	30,000	50,000	75,000	100,000
50	10	34%	30	15%	50	10%	15,000	25,000	37,500	50,000
30	17	46%	50	22%	83	15%	9,000	15,000	22,500	30,000

structure and environment will be major problems requiring a much more integrated approach to transport and environmental planning than has so far obtained.

Intra-Urban transport will require high block speed (hundreds of kilometres per hour) over distances of hundreds of kilometres. This is the region in which high-speed surface transport is increasingly competing with air transport. The outcome will be determined by economic and social considerations with increasing advantage to surface transport as passenger numbers increase[7].

As a model against which to examine air traffic control and navigation, attention will thus be focused on the problem of high density flow, with particular reference to the terminal area.

FUTURE PROSPECTS IN NAVIGATION

The most significant scientific factors in navigation are gravitation and time[11].

The inertial properties of matter and knowledge of the gravitational field are the basis of inertial navigation. Knowledge of the vertical in a moving vehicle and time are essential to navigation by observation of the heavenly bodies. Measurement of the time of propagation of radio waves is fundamental to all radio navigation aids.

Inertial navigation has the unique property of being completely self-contained and unaffected by any external agency. It also provides a highly stable vertical reference for flight control that is unaffected by vehicle acceleration. Up to the present time, the attainable accuracy has been dominated by the technological limitations in sensor (primarily gyroscope) performance. By steady development and by the advent of new techniques such as the laser gyroscope[11] the instrument performance may improve by one or even two orders of magnitude over the present order of one to two kilometres per hour of flight. When combined with the reduction in flight times to 1–3 hours by the introduction of higher speeds, instrumentally an accuracy in plan position of 100 m or less over the whole flight may be achievable.

At this order of accuracy, however, the nature of the physical limitations in accuracy have to be examined; these are knowledge of the earth's gravitational field and the earth reference ellipsoid, the location of the earth's axis and fluctuations in the rotation of the earth and in movement of the earth in space. The most significant factors[12] are local variations in gravity, which in an extreme case could give a 1 km error in position relative to the standard reference ellipsoid, and limits in geodetic information on position. Great advances have taken place in our geodetic knowledge due to observations on artificial satellites[13] and knowledge of the basic factors which limit inertially derived position accuracy is likely to keep step with sensor development. For the navigation system it will thus become a matter of incorporating the appropriate corrections within the system computer.

From high-altitude aircraft stars are visible by day; thus there is the possibility of using stellar information to give a bound to errors in the inertial system. Automatic star tracking systems are expensive and taking optical observation from high-speed aircraft difficult. So far the advance in instrumental accuracy of inertial systems has kept ahead of the need for a stellar reference and this may well continue.

Inertial navigation, however, can only give position on a normal to the surface of a reference ellipsoid; height above that surface cannot be determined by inertial means. Altitude in aircraft navigation is normally determined barometrically with reference to a pressure level and traffic lane standards are established relative to a pressure level. Direct measurement of height above the surface of the earth can be made by a radio altimeter (which times the transmission of a radio wave). To convert this into information for general traffic purposes would require continuous correction for height of the surface with respect to a reference. In principle it would be possible to make this correction but a large computer capacity would be required (for the UK, storage of height at 10 m intervals for every 100 m square would require 4×10^9 bits) and this method is hardly likely to commend itself to the system designer except for restricted applications such as landing (where the radio altimeter is already in use as a prime aid) or possibly over sea where, at the most, a much simpler correction for tide obtains. Adequate height determination may be dependent, therefore, on other forms of radio navigation aid.

All forms of radio navigation aid basically use the predicted propagation path of the wave with a near constant velocity together with a time measurement[11]. The basic accuracy is limited by ability to predict variations in the propagation conditions and coverage is limited by the range over which predictable propagation can take place. The advent of artificial satellites has considerably extended the coverage area at the expense of uncertainties in knowledge of position of the stations. It is not the function of this chapter to engage in a detailed comparison between radio aids and the individual prospects that they offer for the future; it should be noted, however, that there is a considerable lag in general introduction of new technology to radio navigation aids for air transport due first, to the need to obtain international agreement to fit the aid and then to the consequent heavy investment that is made in the aids. For example, the present ICAO aid VOR (a directional beacon) was developed over 20 years ago and will be in service for at least another six years. However, one particular technique of considerable promise will be selected to illustrate future potential.

The simplest measurement of distance by radio propagation uses two clocks maintaining exactly synchronous time. If the time by one clock at point *A* is transmitted over a radio frequency carrier and received at *B*, the time delay in propagation and hence the distance *AB* can be determined by comparison of the time with the clock at *B*. Absolute stability of time is required, the error in distance measurement being 30 m for each tenth of a microsecond error in time.

If each aircraft carries a master clock keeping a standard atomic time to a high order of accuracy and broadcasts the time linked to an identity signal (the broadcasts would be in series, only one vehicle broadcasting at one time) then at a single receiver on the ground, equipped also with a standard clock, the range to each aircraft can be determined. The use of three spaced receivers (Fig. 9) enables the position of each aircraft in three dimensions to be determined by triangulation (in practice more than three would be used to resolve ambiguities).

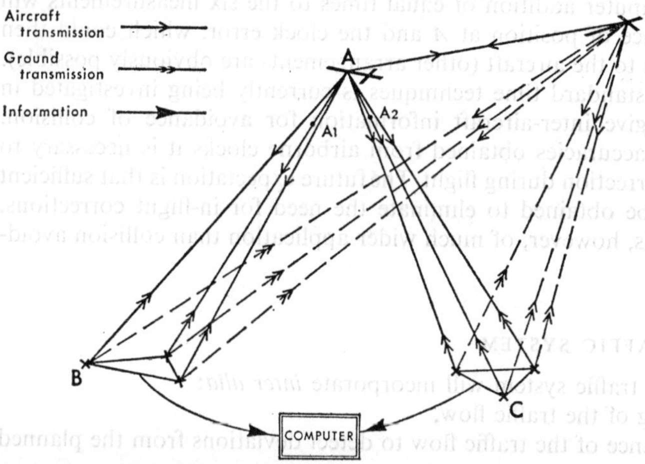

Figure 9 Navigation by time comparison

Multiple groups of receiving stations can be used to give area coverage and redundancy of information which is then correlated by computer. If the ground stations also broadcast time from the clocks then each aircraft can obtain position relative to the known position of the ground station (double arrows of Fig. 9). Finally each aircraft can determine the range to any other aircraft.

Multiple reflections from other aircraft would be resolved by identifying the minimum propagation distance.

Such a system has the following properties:

(*a*) three-dimensional information on aircraft position both in the aircraft and on the ground;

(*b*) given adequate redundancy in the airborne equipment it has a high degree of system redundancy;

(*c*) it gives relative separation of aircraft.

The basic accuracy will depend on radio propagation conditions and may be increased by having a shorter distance between groups of stations to reduce the propagation range to the nearest group. The system design can be adjusted to match the accuracy to the requirement for different phases of flight and accuracies from a fraction of a metre upwards can be expected for any phase of flight including landing. The system is equally applicable to satellite-based stations giving a wide area coverage at the expense of accuracy. The instrumental accuracy will depend on clock stability and the simplest system obtains when no correction to the aircraft clock is required in flight. The introduction of redundancy into the system does enable clock errors to be determined; for example, in Fig. 9, if the clock at A is fast, station B will determine the position as $A1$ and station C as $A2$. Computer addition of equal times to the six measurements will give coincidence of position at A and the clock error, which could then be transmitted to the aircraft (other arrangements are obviously possible).

The use of standard time techniques is currently being investigated in the USA to give inter-aircraft information for avoidance of collision. With present accuracies obtained from airborne clocks it is necessary to provide for correction during flight. The future expectation is that sufficient stability will be obtained to eliminate the need for in-flight corrections. The method, is, however, of much wider application than collision avoidance.

THE AIR TRAFFIC SYSTEM

The complete traffic system will incorporate *inter alia*:

(*i*) planning of the traffic flow,

(*ii*) surveillance of the traffic flow to detect deviations from the planned flow,

(*iii*) means for dealing with deviations from the planned flow both on a semi-routine and an emergency basis,

(*iv*) means for traffic to join and leave lanes, and

(*v*) means for maintaining stable flow along lanes.

Overall the system must be designed to handle the expected traffic at the required rates and at an acceptable level of safety. Absolute safety does not exist—only an acceptable accident rate. Although the engineer and system designer strive for as high a degree of safety as possible, the limit is set by a practical balance between the operation of a transport system and an accident rate that is not found unacceptable by society.

Aircraft in free movement are exposed continually to a collision risk which can be resolved only on an individual encounter basis. The planning of traffic flow in lanes establishes a controlled situation in which there is a considerably reduced risk of collision if movement is precisely determined and controlled.

In practice, allowance has to be made for deviations in traffic flow from the ideal and the volume of space required for the traffic flow is determined by the extent of these deviations. The traffic system has thus to effect a working compromise between the available volume of air space, density of traffic flow and the chance of an accident.

For air traffic moving in lanes, aircraft are exposed to risk of collision when they are separated by only one dimension, i.e. one of the transverse dimensions of the traffic lane or time along the lane. The operating separation standards in these three dimensions have to be set from an estimate of the statistical probability of the aircraft being outside that dimension of the lane either due to an error in position or time information or in controlling the aircraft to the desired position or time. At the low probability levels arising it is difficult to get information on the statistical distribution of the 'tails' of the errors—certainly it cannot be assumed that they are normally distributed[14].

Navigation accuracy (including errors in control) will determine the cross-sectional dimensions of the traffic lanes; the figures quoted as estimates of possible future navigational accuracy are expected values and the traffic lane dimensions will be greater by a factor depending on the estimate of error distribution. To obtain even an approximate estimate of separation standards, however, requires analysis of a specific system and is not appropriate to this paper.

It can be stated with certainty that a considerable reduction in the volume of individual traffic lanes from the present standards must occur if even the immediate growth of traffic is to be able to flow freely.

TRAFFIC CONTROL

Operation of a traffic system at or near capacity will require complete pre-planning of the traffic flow (by computer) and adherence to the plan. A standard time navigation system can give on the ground complete surveillance of the air traffic and automatic data links can provide exchange of flight data between air and ground. Provision of continuous self-control of the aircraft to the planned flight position within the traffic pattern and within a traffic lane is essential for the quick reaction required for high density flow and will enable ground surveillance to concentrate on the minority of deviations due to abnormal conditions and determine the necessary action. All normal functions will be performed under computer and automatic control, with human intervention being planning and decision at a high level. There will be no technological barrier to implementation of any system design in terms of computing and displays for the operator.

Operation within the traffic system will be limited to aircraft with the requisite performance characteristics and equipment.

The extent to which such a system could be applied to a special traffic system for personal aircraft would depend on cost. Transference of computing functions to the ground is possible even to the extent of determining the flight path errors in three linear dimensions and time and transmitting them to the aircraft. The response time of the system, however, would be considerably increased due to time sharing and lane capacities would drop.

TRAFFIC LANE GUIDANCE

The basic control loop within the area of the traffic lane is shown in Fig. 10. The centre co-ordinates of the traffic lane are defined in three dimensions as a function of distance travelled along the lane. This may be in the form of a programme stored in the aircraft or transmitted from

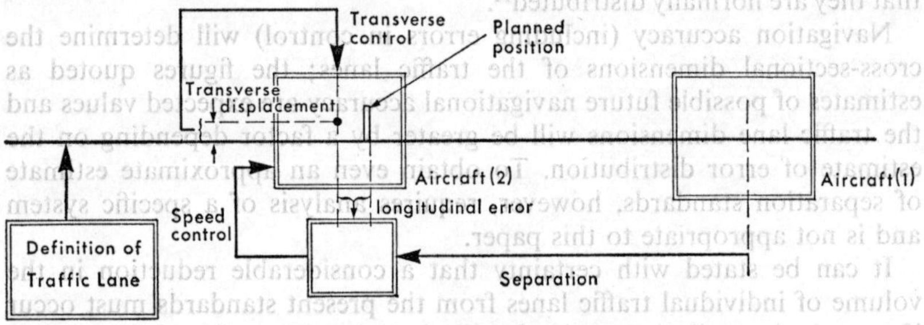

Figure 10 Guidance and control of aircraft in traffic lane

the ground. The navigation system determines aircraft position continuously as a function of distance travelled along the lane; the transverse displacement relative to lane centre can thus be determined as an input to control the aircraft to travel along lane centre.

The use of inertial navigation in the aircraft combined with a time-based radio aid giving, both in the air and on the ground, continuous aircraft position in three dimensions will give high accuracy and a high degree of redundancy together with continuous ground surveillance; this will allow a substantial decrease in lane dimensions compared with the relatively rudimentary system in use today.

Maintenance of longitudinal or time separation is, as the basic theory of traffic flow shows, a problem in dynamic flow. The basic system of longitudinal control (Fig. 10) is to a planned distance along the lane as a function of time. This planned position can be computed to allow for variations in meteorological conditions. The standard time radio aid gives a direct measure of vehicle separation (and also relative velocity by the Doppler effect), which is applied as a monitoring of separation. The overall system would have to be designed to ensure stable flow of the chain of coupled vehicles during acceleration and deceleration.

FREE MOVEMENT

The use of a standard clock gives inter-vehicle distances. While Doppler measurement derived from this gives relative velocity and, in combination with the range, the distance of nearest approach, the most that can be determined by a single receiving antenna on the aircraft is the distance of another craft and the angular direction relative to the axis of the receiving aircraft, i.e. the other aircraft lies somewhere on a cone. The use of three spaced antennae (wing tips and tail) will give a bearing and range, but four antennae are needed to determine if the other aircraft lies above or below the flight path. Although such a system will provide collision warning between two aircraft in the case of multiple threats it will be difficult, if not impossible, to resolve the collision conflicts without exchange of intent to manoeuvre. While some protection will be given by the fitting of such a system it cannot be regarded as a substitute for an air traffic system.

SUMMARY AND CONCLUSIONS

Air traffic is a product of the latter third of the First Century of Aviation; the rapid growth in traffic over the past few years and the predicted growth over the next few years are already posing most crucial problems for a few major air traffic centres in North America and Europe.

Penetration of mass air travel into other regions of the world and predicted growth of world population increasingly concentrated in large Urban areas would bring these traffic problems to many other traffic centres of the world.

Aviation can potentially offer the possibility of free point to point movement and this potential will increase with development of aircraft capable of short and vertical take-off and landing. A basic conflict exists, however, between such free movement and air traffic. Whatever science and technology may offer, the number of conflicts between large numbers of freely random moving craft would defy resolution and, except in low traffic density conditions, controlled movement of traffic becomes essential. The dynamic properties of a moving stream of traffic relates traffic flow, speed separation and uniformity of response to changes in speed. The flow of traffic is a maximum at the lane capacity which is simply related to speed and separation; for a given uniformity of response the speed of traffic flow decreases with increased lane capacity. Time delays in response become increasingly significant as the lane capacity increases and attempts to operate above the lane capacity must either result in an increased chance of collision or a decrease both in the traffic flow and speed, with accumulating delays.

The smaller the vehicle the greater the traffic flow to carry the same number of passengers and the more stringent the requirements in short time delay in response and of uniformity of response to changes of speed. As far as the personal air vehicle is concerned, whatever the vehicle developments of the future, like the motor car, its properties as traffic will remain in conflict with the basic requirements that obtain for high density flow.

Whatever new navigation and control technology may bring, it is difficult to visualise the double step of reproducing, say, the densities of road traffic at a speed of 5–10 times that of the present road maximum capacity of about 30 km/h and at the same time achieving the wider three-dimensional freedom of movement. Furthermore, although new systems and technology such as standard time radio aids will give some general alleviation of the collision risk under free movement conditions and provide universal navigation traffic surveillance, the problems of personal air transport are likely to remain those of ensuring compatibility with other air traffic at an acceptable cost and risk.

In comparison with future transport needs, air transport will remain unique in providing high-speed, long-haul movement between large Urban areas; the form of future Urban development, however, will increasingly favour high-speed, surface transport over short-haul routes that have no major geographic barriers.

For air traffic the critical factor is still seen as the terminal area of the future Urban complex. One of the major problems of today is the interaction between airports accentuated by the inflexible nature of the approach path of current aircraft. Thus, it is interesting to look at the possibility of a single air terminal handling a large volume of traffic.

Particularly with a possible future air vehicle that has complete flexibility in speed and flight path, the aim would be to accommodate a large number of high capacity terminal lanes in the traffic flow pattern (certainly much larger than the four runway major air terminal of the immediate future with its estimated peak operating capacity of 128 aircraft movements per hour[10]). Design of a traffic flow pattern would in many ways be simpler for a large single terminal.

Figure 11 shows a possible high capacity approach and landing system. The approach stream is 100 aircraft per hour at 200 km/h which enters a circular lane. Aircraft leave tangentially and land with a short or vertical landing on pads in sequence as numbered. When pad 20 is full the next aircraft goes around the circle landing on 1 which must have become vacant. Each pad can thus be occupied for 12 minutes, giving 200 movements per hour. On take-off a similar procedure is followed in to a circle outside the pads; this is separated in height when it crosses the approach stream. By interleaving, multiple approach paths can feed the landing circuit. With, say, five approach lanes and five times the number of landing pads, the traffic separation in the circuit is 0·4 km which from Fig. 7 requires a uniformity in deceleration of better than 20 per cent. The diameter is about 10 km (giving a radial acceleration of 0·1g) and with a movement rate of 1000/hr and say 250 passengers per movement, the traffic capacity on a continuous basis is 2000 million passengers a year or ten times the traffic of the 65 world airports of Fig 2. The pads would be integral with a high capacity air terminal unit. They would be linked with each other by an internal transport system and the whole complex would be linked to the Urban Area by a high capacity, high-speed surface transport system.

Whether or not there should be a requirement for such an airport, as an example it illustrates that high capacity depends on design for even traffic flow and close separation between vehicles, and that freedom to control speed and path in three dimensions will be the most valuable asset that future aircraft development can bring to air traffic.

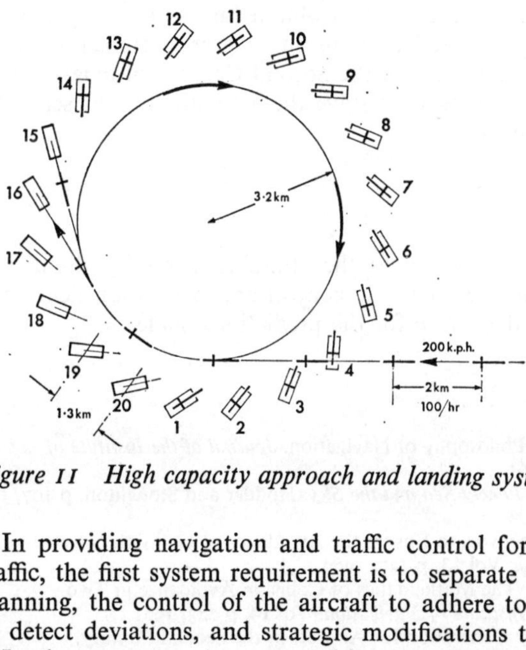

Figure 11 High capacity approach and landing system

In providing navigation and traffic control for such high density of traffic, the first system requirement is to separate the functions of traffic planning, the control of the aircraft to adhere to the plan, surveillance to detect deviations, and strategic modifications to the plan.

In the present system, ground-based equipment and controllers play a primary function in control of the aircraft and the relatively large response delays cause considerable reduction in traffic capacity. Given complete freedom of concept a system can be specified in which the aircraft has the capability of flying along a lane and maintaining separation from other aircraft in the lane without ground intervention. Technologically, advances in computing, control, inertial navigation and radio navigation (such as a standard time system) will enable such a system to be implemented at an accuracy that will give much higher rates of flow in a much smaller cross-sectional area of lane than obtains today. The ground system can be designed to perform the functions of traffic planning, surveillance and strategic modification to the plan. Technologically, advances in computing and display will allow implementation of the full aims of the system designer and human intervention would be limited to the higher level planning and decisions. The same standard time navigation system that is used for aircraft navigation can provide complete surveillance, and the same principle can be applied to all phases of flight at any point of the globe by a combination of ground-based and satellite aids.

There is nothing novel in this concept and although major advances in accuracy and technological refinements can be expected, it is basically a system which technologically could be implemented today with a dramatic increase in traffic capacity. Implementations of new concepts is constrained, however, by the need to obtain international agreement for use, by economic factors and by requiring a graded transition from the old to the new. The advances of the Second Century of aviation will be as much dependent on how we resolve these conflicts as on scientific and technological advances.

ACKNOWLEDGMENT

The author gratefully acknowledges the stimulus given by discussion with many colleagues in the air traffic control and navigation fields. He alone is responsible to the future for the predictions made.

References

1 ANDERSON, E. W. A Philosophy of Navigation. *Journal of the Institute of Navigation*, Vol **14**, p 1, 1961.

2 CHICHESTER, F. *The Lonely Sea and the Sky*. Hodder and Stoughton, p 107, 1964.

3 CALVERT, E. S. Manoeuvres to Ensure the Avoidance of Collision. *Journal of the Institute of Navigation*, Vol **13**, p 127, 1960.

4 HOLLINGDALE, S. H. The Mathematics of Collision Avoidance in Two Dimensions. *Journal of the Institute of Navigation,* Vol **14**, p 243, 1961.

5 THRING, M. W. Power Generation for Aircraft in the Second Century. Chapter 9.

6 DOXIADIS, C. A. *Ecumenopolis*: *Tomorrow's City*. Britannica Book of the Year, 1968.

7 LEGG, K. The Influence of Aerospace on Future Transport Systems. *The Aeronautical Journal of the Royal Aeronautical Society*, Vol **72**, No 693, p 771, September 1968.

8 British Airports Authority—Traffic Study Heathrow.

9 MASEFIELD, P. Airports ATC and Air Traffic. Seventh Convention of Guild of Air Traffic Control Officers.

10 The Third London Airport, Cmnd 3259, HMSO.

11 STRATTON, A. The Science and Technology of Navigation. *Journal of the Institution of Navigation*, Vol **22**, p 1, 1969.

12 KAYTON, M. Fundamental Limitations in Inertial Measurements. American Rocket Society Guidance Control and Navigation Conference, 1961.
13 KING-HELE, D. G. The Shape of the Earth. *Journal of the Institute of Navigation*, Vol **17**, p 1, 1964.
14 REICH, P. G. Analysis of Long-Range Air Traffic Systems—Separation Standards—I. *Journal of the Institute of Navigation*, Vol **19**, p 88, 1966; II p 1969, 1966.
15 IATA. Major Airport and Terminal Area Problems. 17th Technical Conference, October 1967.

APPENDIX

Equation of traffic flow

In Fig. 5 vehicles (1) and (2) are moving at a uniform speed V_1—the heads of the vehicles corresponding to the heads of the arrows, and separated by λ. At time A vehicle (1) decelerates at a_1 reaching a speed V_0 at C. Vehicle (2) begins deceleration at B after a time delay t_d and decelerates at a_2 reaching a speed V_0 at D; the separation is reduced to λ_0 (t_d+ve; $a_2 < a_1$). From the simple dynamics of the situation

$$\lambda + \frac{(V_1^2 - V_0^2)}{2a_1} + V_0\left(\frac{V_1 - V_0}{a_2} + t_d - \frac{V_1 - V_0}{a_1}\right) = \lambda_0 + \frac{(V_1^2 - V_0^2)}{2a_2} + V_1 t_d$$

i.e.
$$\lambda = \lambda_0 + (V_1 - V_0)t_d + \frac{(V_1 - V_0)^2}{2}\left(\frac{1}{a_2} - \frac{1}{a_1}\right) \quad . \qquad (1)$$

Substituting $1/K = \lambda$, $q/K = V_1$ eqn (1) may be written

$$\frac{q^2}{2}\left(\frac{1}{a_2} - \frac{1}{a_1}\right) + qKt_d' + K^2\lambda_0' - K = 0 \quad . \qquad (2)$$

where

$$\left.\begin{aligned} \lambda_0' &= \lambda_0 - V_0 t_d + \frac{V_0^2}{2}\left(\frac{1}{a_2} - \frac{1}{a_1}\right) \\ t_d' &= t_d - V_0\left(\frac{1}{a_2} - \frac{1}{a_1}\right) \end{aligned}\right\} \quad . \qquad (3)$$

Equation (2) establishes a functional relationship between q and K in terms of the parameters $\left(\frac{1}{a_2} - \frac{1}{a_1}\right)$, λ_0', t_d', which has a maximum $q = q_m$; $K = K_m$ given by

$$\left.\begin{aligned} \lambda_0' &= \frac{1}{2K_m}(1 - q_m t_d') \\ \frac{a_1 - a_2}{a_1 a_2} &= \frac{K_m}{q_m^2}(1 - q_m t_d') \\ q_m &= V_1 K_m \end{aligned}\right\} \quad . \quad . \quad . \quad . \qquad (4)$$

Equation (2) may then be expressed in simple parametric form

$$\frac{q^2}{q_m^2} + \frac{(K - K_m)^2}{K_m^2} = 1 \quad . \quad . \qquad (5)$$

which is an ellipse of semi-axes q_m, K_m and centre (0, K_m). By substituting from eqn. (4), eqn. (3) may be written

$$\left.\begin{aligned} \frac{a_1-a_2}{a_1a_2} &= \frac{(1-q_m t_d)}{q\,(V_1-V_0)} \\ \text{and} \qquad \lambda_0 &= \frac{1}{2K_m}\left(\frac{V_1+V_0}{V_1}\right)-(V_1-V_0)t_d/2 \\ &= \frac{1}{2K_m}(1-q_m t_d)+\frac{V_0}{2q_m}(1+q_m t_d) \end{aligned}\right\} \quad (6)$$

Equations (5) and (6) may be applied either to establish the necessary criteria in terms of uniformity of acceleration $(a_1-a_2)/a_1a_2$, allowable time delay t_d, and minimum separation λ_0 that must be met to achieve a given maximum flow rate q_m at a velocity $V_1=q_m/K_m$ or to determine the maximum flow rate and traffic velocity or density that will result from given time delays, deceleration and minimum separation.

Although this theory is usually applied to bringing the vehicle to rest it will be seen that the functional relationship between q and K is universal and applies to any change in velocity. For simplicity, however, all examples in this paper are for $V_0=0$, i.e. the stream is brought to a halt.

21 Looking ahead in aeronautics and astronautics—A US view

Ronald Smelt

I INTRODUCTION—SOME COMMENTS ON TECHNOLOGICAL FORECASTING

I would first like to pay tribute to the inspiration which suggested this series of chapters on The Next Hundred Years, the Second Century of the Royal Aeronautical Society. John E. Allen's introductory chapter set the stage for a remarkable series of individual chapters in which technology, philosophy and sociology have combined to give a vision of a challenging future, often exciting but occasionally sobering. Coming at the conclusion of this series, I have interpreted my task to be that of looking at this future through the eyes of a technical man in the aerospace industry of the United States. Before settling down to this task, it seems appropriate to make a few remarks on the general nature of technological prophesies; in doing so, I am following a pattern which has been set by a number of my predecessors in this book. Occasionally these general remarks have taken the form of a reminder of the uncertainty of the process of technical forecasting. Küchemann, in his very thoughtful chapter on the evolution of aerodynamics, for example, invokes the privilege of 'zero order' accuracy in his predictions. I must, of course, request the same privilege; and I would also like, in these opening paragraphs, to discuss the fact that I am using two quite different types of forecast in this attempt to assess the patterns of the future.

Technological forecasting is currently quite a respectable art, in both the United States and in Europe, dignified by a number of recent books[1–3] and even boasting its own journals[4, 5]. It has been stimulated by the realisation that some measure of planning of the longer-range affairs of government and industry, in fact, of most human activities, is possible and worthwhile. Most companies in the aerospace industry now engage in long-range planning to some extent, and usually have small corporate staffs of technical men, economists and political scientists reporting to

Originally published in the *Aeronautical Journal of the Royal Aeronautical Society*, August 1969.

the company president. These staffs are more or less continuously engaged in updating some aspect of the company's business projections; its future technical capabilities and needs; its markets; the economic or political constraints within which it will be expected to operate; etc. A few years ago, the extrapolations of these long-range planning groups extended about five years into the future; but recently there has been a general recognition that five years is not an adequate time for the maturing of major projects, and plans generally now extend ten to fifteen years into the future. The projection of technological change and growth looms large in this planning, possibly because (surprisingly!) it generally appears more definitive than the economic, political and other human factors which also have a part in defining the future scenario.

The word 'projection' is very descriptive of the process of extrapolation generally used in corporate long-range planning. We observe trends in a number of primary variables for the past few years, we follow the same continuous curve throughout the period of the long-range plan; and we then deduce the probable course of the items of most interest to us. As an obvious example, we project the current growth of free world passenger traffic or of airborne cargo (as was done by Allen in the first chapter) and we then predict on the basis of the probable demand in 1980 or 1990, the probable number and best size of commercial transport aircraft required. As a more sophisticated example, we look at developments in structural aircraft materials for the past few years, and link it with our growing ability to analyse the final detail of aircraft stresses, and we deduce that within the next ten to twenty years we should be able to build aeroplanes of over two million pounds gross weight before reaching the limit of current technology. Some of these projections can become quite detailed, and occasionally demand complex computer programs; but they are all based upon the same idea of a smooth continuous trend from the present.

In the simplest of these analyses, with very short-term objectives, the smooth trend line is generally exponential in character. For longer-term projections, measured in tens of years, a number of factors appear which can completely alter the nature of the extrapolation. The most readily recognised of these is the saturation effect, when continued growth along the exponential curve would ultimately exceed some obvious physical limit; the exponential is then modified to the well-known '*S*' curve*. A simple example of this can be seen in the projection of free world passenger traffic mentioned earlier. Since the current rate of growth of air passenger traffic greatly exceeds the growth rate of the population, we can show that well before the end of the next century if this growth rate is continued, every man, woman and child will be flying by air for a large fraction of his day. This obvious saturation condition is likely to be preceded by

*The author first encountered this term many years ago in marketing jargon in the electronics industry, descriptive of the saturation of the sales curve for domestic radio receivers after every interested household has purchased one.

several other saturation factors—the physical limitation to passenger access to airports, the limited number of aircraft which can be controlled with safety in the air space above major airports, etc. Some of these limits are already beginning to appear, and are forcing a demand for a discontinuous change in our techniques and procedures for traffic handling at major airports.

We see then, that this 'saturation' approach, applied across our aerospace world, can highlight those areas where our procedures or our technology must undergo significant—usually discontinuous—change in the future. Sometimes the basic nature of this change is clear; already in the United States, for example, planners are envisaging novel relations between the airport and the central city which might circumvent the problem we have begun to recognise. Regrettably, however, this approach often signals the need for a discontinuous change, but the mechanisms which may bring this about are not currently apparent. Probably the most disconcerting example of this occurs when we ask a question which is very germane to the future of the aerospace industry—how many engineers and scientists will there be in the next century to sustain the pace of technological change? Lord Bowden tried to answer this by extrapolation when he was Minister for Education and Science, and I cannot do better than quote his comments on the growth in numbers of scientists, since they apply equally well to engineers:

> 'If the rate of progress which has been maintained ever since the time of Sir Isaac Newton continues for another 200 years every man, woman and child on earth would be a scientist, and so would every horse, cow, dog and mule as well. It can't go on like this much longer. What will limit the growth of science? Always assuming that scientists are not buried under their own papers.'

Here we find ourselves in the position of prophets convinced of the inevitability of a major change in the foreseeable future, but without any idea of the forces which will bring it about.

The two different types of forecasts to which I referred in my opening paragraph will now be clear. The first or, 'evolutionary', forecast is the logical extrapolation which is the basis of most aerospace planning activities, encompassing about ten years. The second, much more difficult, 'revolutionary' forecast accepts the possibility, or even inevitability, of discontinuities in our technology having a major controlling influence over longer time periods. The examples above show how some of these discontinuities can be identified from extrapolation, by a *reductio ad absurdum* approach; this reveals the necessity for the revolution, but not the technological change which will satisfy it. There is an inverse group of discontinuities in the future, characterised by technological changes which we can foresee from current scientific advances, but whose effect upon the role of aerospace in society cannot currently be estimated with any certainty. As an obvious example of this category, the growth of communication technology is rapid and the development of low-cost, high-capability integrated circuits can conceivably provide every person with the means of immediate communication, by voice and picture, with any other person of his choosing. The saturation of available channels can be forestalled by the tremendous bandwidths offered by laser technology, and the switching problems are within the capability of the next generation of computers. Leaders in the science of communications, such as Dr. John Pierce of Bell Telephone Laboratories, already indicate this as a significant possibility, and it could clearly result in a very great reduction in the amount of travel, at least for business purposes, which could sharply reduce the projected airline passenger growth.

I have used the names 'evolutionary' and 'revolutionary' to describe the two types of forecast now identified. A. V. Cleaver's treatise in this book on the future of astronautics (chapter 5) makes a similar distinction between his projections for the first twenty years (1966–1986) and the following eighty years. Furthermore, he introduces two categories of change—'what could be done technically' versus 'what may actually be done'—in a manner which the reader will readily identify with the two indicators of revolutionary change discussed earlier; those foreshadowed by scientific advances versus those predicted from the growing needs of society. A. V. Cleaver's discussion of the changing relation between these two categories, and its dependence upon the size and cost of the disruption, is so complete a discussion of the essential points that further elaboration is unnecessary here. Instead, the reader is referred to section 3 of Cleaver's chapter. Since he introduced a time divider at twenty years, some discussion of the time scale of future events is appropriate. As mentioned earlier, technological forecasting by extrapolation is generally applied over five to fifteen years in the United States. The identification of potential revolutionary changes from current scientific advances raises the obvious question of the length of time typically required for the process which develops a scientific advance into a technically feasible engineering concept, and then into a generally accepted device, process or business. A very recent study by the US National Science Foundation[6] shows, as might be expected, a wide scatter in these time intervals, but in general the time periods required for the first phase (laboratory to engineering feasibility) cluster around twenty to thirty years, as shown in Fig. 1. The further period for complete realisation in practice takes another five to ten years. (A noteworthy exception is the transistor, where the practical value was immediately recognised and several laboratories within industry were in a position to accomplish immediate translation to the production line). New concepts in the aerospace industry have usually taken rather longer than these average figures for complete utilisation by society, although there has been some short-circuiting of the full development time for military applications. This is demonstrated, for example, by the early history of the aircraft itself, or by the period between first mention of the gas turbine and the first commercial jet aircraft.

On the basis of past history, then, we might expect on the average, a lag of thirty to forty years between a first 'laboratory' investigation of an aerospace advance and the commercial exploitation of its promise. We can also deduce that if we attempt to go further than about forty years into the future, we will be 'flying blind' in the sense that it is improbable that coming revolutionary events will have cast their technological shadows before them into today's laboratories. Perhaps the other approach, that of identifying a future saturation condition as a cause of revolutionary change, may enable us to look further into the future; but hopefully the gap in time between the first forecasting of an approaching saturation condition and the final satisfaction of the need is not quite so long.

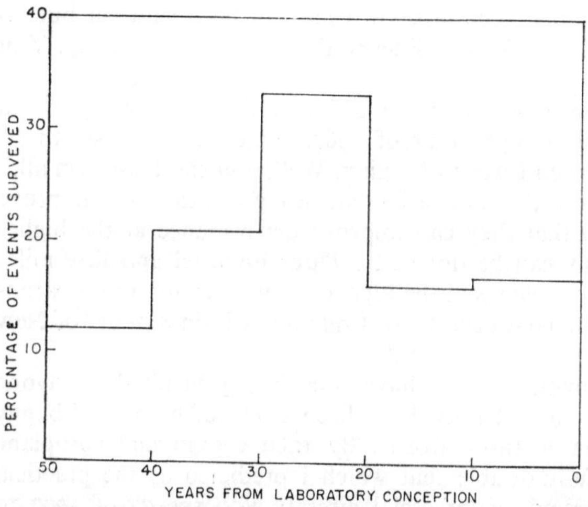

Figure 1 Time period required, laboratory to engineering feasibility

Of course there will be a number of revolutionary changes in the next forty years. The aerospace community at any time has a large number of concepts moving through this prolonged germination period. Some are barely starting as small programmes in the research laboratory; the forecaster must have both awareness and imagination to recognise them and to envisage their impact on society in the far future. Others, started many years ago, have almost completed the process. Again, taking an example from the aeroplane field, the supersonic transport is now coming down the home stretch about thirty-five years after the first signalling of the possibility, in the work of Taylor, Maccoll, Busemann and others in the early 1930s. Again we note how military exploitation of the same concept was achieved in a much shorter time period.

It has always seemed to me that this time scale is dismally long. I was therefore quite surprised to find that forecasters who have tried to go beyond simple extrapolation into the 'revolutionary' future have usually seriously overestimated the time requirement. The reader can verify this by reference to two books[7,8], both with the same title: *The Next Hundred Years,* and both with the distinction that their authors have had the courage to review them in print many years later, and check their prognostications against the facts[9,10]. There quotations from Dr. C. C. Furnas's 1967 review of his 1935 forecast will illustrate the point:

1935: 'A number of years ago the English physicist Soddy made the remark that inside the atom there is on almost limitless reservoir of energy. If we would only get at it the whole complexion of the world would be changed . . . Can the atomic physicist show us how to tap that reservoir profitably? The last word has not been said yet but do not buy any stock in an Atomic Energy Development Company. You will certainly lose.'

1967 Review: 'Looking at the signal calling for the 30 plus-year-old statement from the vantage point of 1967, someone obviously threw a touchdown pass when I wasn't looking. Well, you can't win 'em all.'

1935: 'It seems obvious that one task confronting aviation is to improve design of planes so that they can improve performance at the higher altitudes. It certainly can be done . . . Three hundred and fifty miles per hour for transport planes at the higher altitudes is not an unreasonable figure to expect. That brings San Francisco within 8 hours of New York and London within 10 hours.'

1967 Review: 'Although I would have been highly insulted if anyone had said so at that time, I must have been guilty of horse-and-buggy thinking when I wrote those words. By 1960 commercial aeroplane performance was fully double that which I predicted as the practical ultimate, and supersonic commercial aircraft with speeds of 1500 to 1800 miles per hour will certainly be with us before long. My 1935 crystal ball was obviously operating under a cloud.'

1935: 'I do not doubt that some day a great crowd will gather at some aviation field to watch a man be locked into a peculiar bullet-like machine on the first bonafide trip to the moon. I hardly expect it to be in my day . . .'

1967 Review—'How overly conservative could a young man be in the 1930s? Unless someone pulls the actuarial tables out from under me, I now *do* expect to be around when that man (probably an American) first lands on the moon. Further, I hope to have the privilege of flicking my fingers into the first sample of moon dust which he will bring back.'

Harrison Brown comments on the same conservatism in his review after ten years[10]:

'In 1954 the United Nations estimated that by 1980 the population of the world would be between 3·3 and 4·0 billion persons, with 3·6 billion given as the most likely figure. By 1965 the population had already reached 3·3 billion, and the United Nations demographers were estimating that the 1980 world population would lie between 4·1 and 4·5 billion, with a median of 4·3 billion. Thus, in the span of but a decade the estimate for the most likely population in 1980 has moved upward by about 20 per cent'.

With the weight of this evidence, I do not doubt that some of the group contributing to this book will read it again at the turn of the century and ask how we could possibly have been so shortsighted. In spite of this, I propose in the following paragraphs to look at a few of the major segments of aerospace activity, and to examine those evolutionary or revolutionary changes which can now be perceived, perhaps dimly, from the vantage point of US industry.

2 SPACE FLIGHT

It may appear to be a reversal of the logical order to discuss space flight first, and to leave air transportation to the next section. The developments in space transportation in the next one or two decades, however, may have a fundamental influence on air transport of the future. Furthermore, the unfolding of the space programme in the USA is an excellent example of the concepts of evolutionary and revolutionary change as indicators of the future, as discussed in the introduction.

The evolution of the rocket motor, from the early experiments of Goddard and others, to the 7½ million pounds of thrust provided by Saturn V for the Apollo lunar programme has progressed directly down the path pre-directed by pioneers such as Oberth in the early 1930s. It presents a classical example of an aerospace evolutionary pattern; after the first experiments, the recognition (in the late 1930s and 1940s) of a potential military application; next, the perfection of the device using military resources to step up its size and its reliability; and finally its conversion to non-military use, with continuing further increases in size and impulse. At various stages during this growth there has arisen a number of special applications which we can confidently expect to continue their evolutionary pattern for many years in the future:

(1) *The ballistic missile* itself, which will continue to evolve with payloads of increasing capability and complexity until one is used in earnest, or more hopefully, until they are outlawed by the world.
(2) *The communication satellite*, which is now firmly established as a vital segment of intercontinental communication, and will therefore evolve in consonance with other developments in communications and electronics. In particular, we can expect intercontinental picture-phone relay from home to home to be as common as the current telephone within one or two decades, and of course universal television relay via

satellite. The educational and political applications of these satellites can also be predicted.

(3) *The meteorological* and *navigational satellites* have also demonstrated useful application, and will grow in capability. Both are of interest to primarily aeronautical readers since they can be expected to contribute in the future to the solution of some of the problems of air transport. In particular, the application of navigational satellites has so far been confined primarily to ship position-finding; within the next decade it will be extended to air traffic control on a large scale, both as position-finding equipment and also as a communication medium between aircraft pilot and controller.

(4) *The earth resources satellites* are in a relatively early stage of development, where the potential of different forms of satellite sensors for monitoring of our natural resources is being explored. Workers in this area look forward to major contributions of such satellites in the fields of oceanography, geology, and agronomy—particularly in the detection of plant growth and disease on a large scale.

Following the philosophy of our introduction, it is logical to ask what factors will arise in the future to limit the evolution of these unmanned satellite applications. The most obvious is the possible saturation of available communication bandwidth; but there are already developments which will move this limit many years into the future. Some satellites already incorporate data compression devices, permitting all but the most relevant information to be filtered out at the satellite before being communicated to earth. The adoption of optical laser communication is an even more promising development, offering far greater bandwidth, provided that the satellite can be stabilised well enough to give the necessary pointing accuracy.

Absurd though it appears, a second limit to the evolution of these satellites is that we may run short of space! For the majority of communication and surveillance applications, there is a preferred position for the satellite—in 'stationary' orbit, 20,000 miles (32,000 km) high above the earth's equator. A further constraint on the position on this circle is the need for the satellite to see its specific points of interest on the earth—a transatlantic communication satellite, for example, is restricted to a small arc of the circle from which New York and London are visible. Since each such satellite currently wanders for a hundred miles or more around its stabilised position, a few tens of satellites in the most favoured position may represent uncomfortable crowding. The stationary orbit may become, like the Suez Canal, a strategic communication point and a focus for national rivalries in the next generation.

This, of course, is only an example of the critical part to be played by international co-operation in the future of satellite applications. Ultimately the pace of evolution of many of the applications which can now be foreseen will depend completely upon the forces of international good will versus national suspicion. For this reason, the timetable of unmanned satellite developments is extremely difficult to estimate.

The future of manned space flight presents a complete contrast, for it is already clear that it must undergo a revolutionary change in the very near future. Only a few years ago it was assumed that the sequence of rocket vehicles of steadily increasing size and power for manned space flight (Atlas, Saturn I, Titan III, Saturn V, with their space vehicles Mercury, Gemini, Apollo) would continue to evolve to still larger vehicles; indeed there were a number of studies of the next upward step, under the code name of Nova. It now appears that these studies neglected one of the most basic saturation factors, that of cost; it does not appear likely that the USA, or probably any other country, will commit to an investment significantly greater than that of the Apollo-Saturn lunar programme in the near future, without some potential economic return. We have, in fact, reached a saturation point at which further progress requires a revolutionary change—a change which will reduce the cost of space operations by one of two orders of magnitude.

This challenge of major cost reduction is currently engaging the attention of many aerospace engineers who regard it as the key to a successful future space programme. Two alternative directions of development are now apparent. One possibility is indicated by the observation that solid-fuel rockets are significantly less costly, per launch, than the liquid-fuelled vehicles; and further, that the technology of solid rockets has now advanced so far that it has displaced the liquid rocket for the vast majority of missile applications. This does not necessarily imply a conversion to solid fuel: the ingredients of the solid rockets' economy can be recognised (cheap case materials, pressure-fed fuel, etc.) and many of these can be incorporated into the liquid-fuelled vehicle. These thoughts have given rise to the concept of a 'big dumb liquid' launch vehicle.

The second possibility concentrates upon the tremendous waste which is involved in constructing and checking out a complete vehicle, with all its equipment, then discarding it after a single launching into space. Early in the manned space programme there was a detailed examination of the possibility of recovering the components of the launch vehicle—the booster and the upper stage—using deceleration devices such as balloons, or parachutes of various forms. It was found impossible to save any costs by such techniques; the cost of the recovery operation itself, and the detailed refurbishment of many of the expensive components, more than outweighed the potential saving. It was realised then that the approach to a recoverable vehicle must involve the design of a completely new vehicle, together with a fundamental change in the concept of launch which would bring the operation closer to that of a conventional commercial airline. The build-up in space technology outside the Apollo programme has now reached the point that such a novel manned space operation appears to be a real possibility. There are in fact members of the aerospace community who believe that we can now design a vehicle which will go directly into orbit without discarding any of its launching stages[11]. This vehicle can then return, land with sufficiently low accelerations that its equipment is not significantly deteriorated, and it can then be refuelled and re-used.

Other engineers, slightly more conservative, have evolved designs in which all of the valuable components of the entire vehicle (the engines, guidance system, communications, and, of course, the life support equipment for the passengers and crew) are located in a sub-vehicle which can be flown back to earth and landed on a conventional airport, so that only the cheap, empty fuel tanks are discarded. Aeronautical engineers will recognise close similarity between this concept and the familiar drop tanks of the conventional aeroplane. Either of these approaches appears to offer a reduction of one or two orders of magnitude in the cost required to put a man into orbit. More specifically, the current generation of vehicles can put a payload into orbit at a cost between $1000 and $10,000 per pound, the exact value depending upon factors such as the size of the vehicle. The recoverable vehicles are variously estimated to cost between $20 and $100 a pound, the variation here depending largely upon the number of space operations to be conducted per year.

Most space engineers now believe that the next decade will see the development of this new class of recoverable vehicles. With it, and with the reduced costs that should be possible after the initial development expense is invested, it should be possible to provide economical manned transport to an earth-orbiting station such as could be placed in position by the present generation of space boosters. The road is then clear to demonstrate man's capability to live and work for prolonged periods in space, and the literature is already full of suggestions for useful applications. The interested reader may refer to a conference on the subject in St. Louis several years ago[12], a recent Symposium at Huntsville on Zero Gravity Applications[13], and a still more recent discussion of the opportunities for manufacturing in space[14].

We can combine these speculations (and other predictions) into a rough time-table for future manned space operations. It appears that by the end of the 1970s, the revolutionary change from expendable manned space vehicles to re-usable vehicles will be apparent. There will be a few manned space stations in earth orbit, placed there by the present generation of boosters—Titan and Saturn. Their re-supply and crew change, however, will constitute the first task for the new generation of recoverable vehicles, launched vertically from existing spaceports but landing horizontally on conventional runways. In the decade of the 1980s these logistic operations will expand, and take on some of the aspects of airline operations, as the first specialised manufacturing activities begin in space. These are likely to be:

Crystal growth and refinement in zero-*g*.*

Manufacturing of ideal shapes of deformable bodies, such as spheres and other minimum-energy configurations.

Ultra-purification and refinement, particularly of refractory metals, in a vacuum.

As these activities grow, flight into space will become more commonplace. Additionally, the stresses imposed on the space traveller on ascent and re-entry will become less severe as the development of hypersonic lifting vehicles and configurations with changing geometry is stimulated by the requirements of the new generation of vehicles. Accordingly, before the end of the century, it appears likely that space excursions for thrill or pleasure will begin. At this point space flight will begin to converge with long-distance point-to-point earth transport by air, a subject to be discussed in the next section.

It is challenging to speculate when the first hotel will be constructed in earth orbit, or the first specialised hospital planned for beneficial use of the zero-*g* environment. Both of these developments are presaged in the literature, and become quite practicable if the developments outlined above reduce the cost to less than $20 per pound in earth orbit.

These revolutionary changes can be expected to have an impact upon the exploratory aspects of space flight, interrupting the sequence of exploration of earth's upper atmosphere, exosphere, moon, near planets, far planets. The unmanned lunar and planetary probes and orbiters already launched by the USSR and USA will undoubtedly be followed by similar unmanned explorations with more advanced equipment, and to planets other than Mars and Venus. An exciting and unique opportunity to reach four of the outer planets (Jupiter, Saturn, Uranus, Neptune) occurs in the 1976–1980 time period due to the unique alignment of these planets at that time. This 'Grand Tour' presents an opportunity, which will not recur for another 171 years, to minimise the launch energy requirements by swinging around each planet in turn. The mission has been investigated in detail by the staff of the Jet Propulsion Laboratory, and is described in a recent paper[16]. Although funding of this programme is not yet authorised, it has extremely high priority in future unmanned mission planning; it can be done with the present generation of launch vehicles.

In contrast, manned missions to the planets will probably be delayed during the entire period of the revolutionary change in manned orbital operations. This is evident from the fact that it is necessary to determine how the human being reacts to long periods in the environment of space, before he sets out on a two-year journey to cover the nearest planets. An established earth space station with its logistic support system is

*Aerodynamicists will be interested to note that this application relates back to a long-standing problem in aerodynamics, that of suppression of turbulence in the liquid in which the crystals are grown. The interested reader may obtain further details in ref. 15.

therefore a desirable precursor to interplanetary flight. It is further evident that such a station, if it can be operated economically as asserted above, could be the first staging point on an interplanetary journey. This leads to the consideration of an entirely new type of space vehicle which operates in space only, without any provisions for atmospheric flight or re-entry; its duty will be to shuttle between space stations in earth, lunar, or planetary orbit. Such a vehicle can be expected around the end of the century, if the lunar surface proves to have any economic value justifying frequent journeys. Its major new features will be its communication system and its propulsion (for acceleration and orientation), both optimised for extra-atmospheric application. In particular the propulsion system will not require high thrust, but must attain the highest possible specific impulse. Since it never returns to a planet, this is apparently an ideal application for nuclear propulsion. The solid-core reactor, now under development in the Nerva programme, does not provide sufficient improvement in specific impulse (compared with chemical rockets) for this purpose, but two other developments, currently in the early laboratory phase, promise a specific impulse of 2000 seconds or more. These are the liquid-core, or gas-core reactors, and the nuclear-electric generator applied to an ion or photon rocket.

Projection of man's space activities into the more distant future is straightforward but unnecessary, since the science-fiction writers have already found it a fertile area. The most significant point underlying the present discussion is that we now have the technology to take man to the planets, and we can see no basic technical obstacle to space travel. Whether and when man will go are questions which depend on non-technical matters—first political and national, then later primarily economic. We are entering an era which will stress the economic rewards for the exploration and exploitation of space, and will therefore put a premium on more efficient, lower cost space operations.

3 AIR TRANSPORT

3.1 *Airline growth and aircraft size*

The economic forces which may soon begin to shape the space programme have of course been acting upon the commercial aircraft for several decades. The evolutionary trends in air transport can therefore be projected by examining the nature of these forces, as the aeroplane responds to the rapid growth in the market. For the next few years the dominating factor in this growth will be the large rate of increase in commercial airline passenger traffic. This rate has been projected, with varying degrees of conservatism, in a number of recent papers, including John E. Allen's introductory chapter in this book; most writers suggest a rate of passenger-mile increase of about 6 per cent per year, about three times the growth rate of world population. Projection of the air cargo market is even more spectacular, most estimates indicating an increase of about 18 per cent a year in ton-miles for the free world's air cargo. It should be noted, however, that only a few per cent of the total inter-city cargo traffic goes by air in the USA, whereas the passenger aeroplane now claims 75 per cent of the total dollars for domestic inter-city travel[17], with the combined share for bus, rail, and other commercial carriers accounting for the remaining 25 per cent. At the present time, therefore, passenger travel is the major determinant of airline economics, but air cargo operations are growing in relative significance.

The influence of this growth on the size of transport aircraft is readily apparent. Although it could be accommodated by adding more aircraft of the current size, the economics of the situation forces the development of larger aircraft. In common with most forms of transport, the direct operating cost of an aircraft has two components—one roughly proportional to payload (or number of passengers) and the other substantially independent of payload, including such items as crew salary, maintenance costs for navigational equipment. The direct operating cost per unit of payload, therefore, necessarily decreases with increasing size. Results of a specific study for cargo aircraft are shown in Fig. 2. This implies that the larger aircraft will be more profitable, so long as there is enough payload to be carried; and we can therefore expect the size of transport aircraft to increase steadily in the future as passenger and cargo operations grow. The general case for the larger vehicle is reinforced in the transport aircraft by its improved aerodynamic and propulsion efficiency with increasing size and Reynolds number; and by the alleviation in air traffic

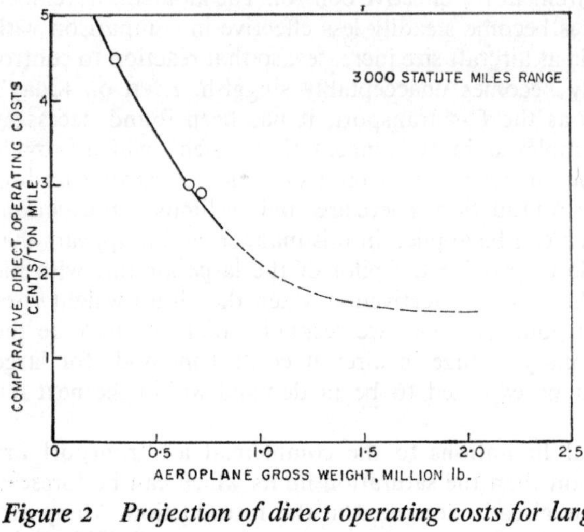

Figure 2 Projection of direct operating costs for larger subsonic cargo transports

control problems over major airports which is possible by the smaller number of larger aircraft.

As pointed out in our introductory remarks, however, this steady growth of transport aircraft size must ultimately reach a technological limit, and it is of considerable practical importance to know the probable magnitude of this limit. The classical 'square-cube' law has been quite misleading in any quantitative determination of such a limit in the past; the steady evolution in aircraft materials and in methods of detailed structural analysis has continually pushed up the maximum size. If we assume a steady continuation of the rate of past progress in these technological areas, and evaluate a future family of larger subsonic jet cargo transports, we obtain the rough curve shown in Fig. 2. This shows that the steady improvement in operating cost which we expect with increasing size will finally be offset by increased structural weight around an all-up weight of about 2 million pounds. This limit is unquestionably conservative; if we assume that a revolutionary change in structural materials will occur, reducing the ratio of structure weight to total weight from its present value of 30 per cent down to about 20 per cent, we find that the all-up weight of the aircraft can now increase to over 5 million pounds before we reach a turning point in the direct operating cost. We can invert this discussion, and assert that the approach to a limit size in transport aircraft, in the face of continuing growth in air travel, will strengthen greatly the impetus for use of revolutionary materials—the composites and the other possibilities described in some detail by A. J. Kennedy in chapter 14.

There is, however, another limitation in aircraft size, which can be seen to follow from the same dimensional arguments as the square-cube law, arising from the requirement for effective control. The moments developed by the control surfaces become steadily less effective in comparison with the moments of inertia as aircraft size increases, so that reaction to control movements ultimately becomes unacceptably sluggish. Even on today's largest aircraft, such as the C-5 transport, it has been found necessary to introduce fairly complex stability augmentation systems which provide anticipatory deflections in response to pilot demand, i.e. strong leading terms in the servo-control function. There are obvious limits to the amount of augmentation which can be applied in this manner, and it appears that it will not be possible to provide the pilot of the large aircraft with his customary 'feel' of the control effectiveness when the all-up weight rises above 1 or 2 million pounds. Again we recognise that this may be the signal for a revolutionary change in aircraft control methods for large aircraft, which might be expected to be in demand within the next ten to twenty years.

These technological limitations to the commercial air transport are probably less significant than the saturation limits which can be foreseen in ground facilities, as already described in the Introduction. Many airports are already experiencing peak-hour saturation, terminal facilities are strained, and airport access to the community is becoming increasingly difficult. It seems clear that the next decade will bring a new approach to the entire problem of airport design in relation to the cities, and a number of aeronautical engineering groups within the USA are already concentrating their efforts on this problem, as the most critical limitation which can currently be seen in the future development of air transport.

The problem of air traffic control may in fact turn out to give a more fundamental saturation limit. FAA forecasts, even though they envisage declining use by the military, predict sharply rising demands for air space by air carriers and general aviation. The total number of aircraft operations at FAA controlled airports is estimated to rise from about 45 million today to 140 million in 1977. Instrument operations at airports are predicted to rise as shown in Fig. 3[17]. When it is realised that this tremendous increase

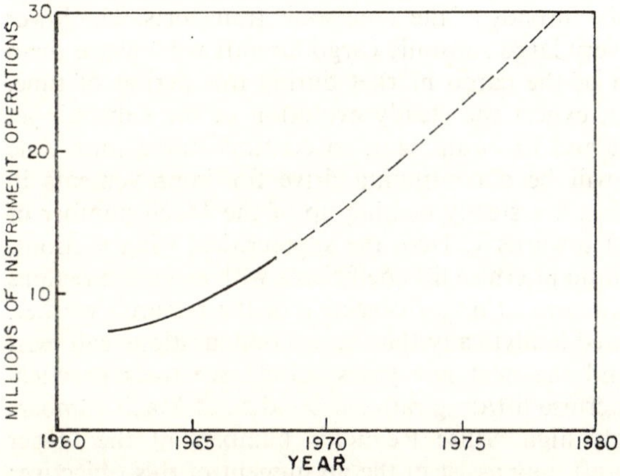

Figure 3 Rise of instrument operations handled by FAA

is superimposed on a system already showing some signs of saturation, and that it will be accompanied by a growth in the proportion of high performance aircraft such as the supersonic transports, the need for a major technical step forward is evident. This in fact is a good example of the type of saturation limit mentioned in the Introduction, which forecasts the inevitability of change but gives no clear indication of the new technology which will satisfy it. We shall discuss the air traffic control problem further, in relation to short-haul and personal aircraft.

3.2 *Transport aircraft speed*

The discussions of the preceding section make no mention of aircraft speed; our forecast of the steady growth in commercial aircraft weight up to several million pounds, under the economic pressures accompanying airline growth, appears to apply equally to subsonic and supersonic transports. However, speed has always been a critical factor in the economics of airline operation, and we can expect the urge for higher speed to continue into the indefinite future. This is not only because of the fact that velocity V enters directly into the aircraft's efficiency, as indicated, for example, by the Breguet equation*; additionally, the number of trips per day is clearly greater for the higher speed aircraft. The airlines recognise a further significant advantage in the preference of the travelling public for shorter travel time, although this factor may diminish in importance since we are now approaching the point where ground transport to and from the airport controls the total travel time.

Predictions made by Lockheed three or four years ago, and reproduced by John E. Allen in Chapter 1, would imply that the subsonic jet aircraft will carry a major fraction of the total revenue passenger miles at least until 1990, in spite of the inroads of the supersonic transports. The lower operating costs of the very large subsonic cargo aircraft will ensure them an even larger fraction of the cargo market during this period of time. In consequence, we can expect the steady evolution of the subsonic jet aircraft, its powerplant and its equipment, to continue for a long time in the future. There will be a continuing drive for improvements in transonic lift/drag ratio and a steady inching up of the Mach number of the subsonic jet aircraft towards 1. Here the supercritical wing sections, offering good lift/drag ratio at cruise lift coefficients with extensive regions of supersonic flow, are a current major objective of the research worker. It is already demonstrated analytically that these configurations can exist in frictionless flows, and the next few years should see their practical realisation, resulting in cruise lift/drag ratios around 20 at Mach numbers over 0·9. The extremely high cruise Reynolds numbers of the bigger subsonic transport aircraft may assist in the attainment of this objective; unfortunately the aeronautical world does not at present have experimental facilities capable of laboratory investigation in the Reynolds number range around 100 millions, so that the full development of supercritical wing sections may be delayed,

The evolution of the fanjet propulsion unit for the subsonic transport aircraft is by no means complete, and we can look forward to steady increases in performance for the next one or two decades. Future improvements will follow the trends of higher cycle pressure ratios (up to 40:1) and higher bypass ratios (perhaps up to 15:1), to lower the specific fuel consumption. The history of the gas turbine shows that new materials

*The Breguet cruise range equation in its simplest form may be written:

$$\ln\left(\frac{W_1}{W_2}\right)\Big/\text{range} = \frac{\text{sfc}}{V(L/D)}.$$

Where W_1 and W_3 are the initial and final cruise weights, sfc is the specific fuel consumption and L/D is the lift/drag ratio.

have been adopted by its developers somewhat in advance of the use of the same materials in aircraft; this pattern is still continuing in the use of composites, and we can expect the evolution of the gas turbine to emphasise further improvements in materials. As a result of this trend, together with improved cooling techniques, turbine inlet temperatures will steadily increase to keep pace with the higher pressure ratio.

Perhaps more exciting than these evolutionary changes, there is the possibility of a revolutionary change to nuclear propulsion in the largest subsonic transport aircraft, starting within the next decade. The earlier attempt by the US Air Force to develop such an aircraft was a failure, primarily because the aircraft considered at that period, about 400,000 lb (181 Mg) in all-up weight, was not large enough to permit complete shielding of the reactor to industrial standards. Indeed, current studies show that even with the improvements in shield weight which have taken place subsequent to the earlier programme, an aircraft of 400,000 lb (181 Mg) weight would be barely able to carry a fully shielded reactor without any payload. Fortunately, however, the biggest transport aircraft of today, such as the Lockheed C-5, are approaching 1 million lb (454 Mg) in gross weight and therefore permit full shielding to commercial industrial standards with a useful payload. Further, the requirement for a reactor shield of approximately constant thickness implies that the shield weight goes up quite slowly as the aircraft size and power increase, so that sketch designs indicate a payload of almost 50 per cent of the total weight when the aircraft reaches 2 million lb (908 Mg).

The biggest problem of the nuclear aircraft, and the one which is currently receiving most attention in development studies, is the danger of radioactivity in the event of a crash. A good description of the problem and the possible directions of solution has been given recently by workers in the NASA Lewis Research Center[18]. Following the traditional pattern for revolutionary aeronautical developments, it is expected that the nuclear transport aircraft will appear first in a military version, where the requirements for extreme range and for indefinite endurance are clear cut. It is of interest to note that the first studies of nuclear aircraft propulsion took place both in the USA and at the RAE in England around 1946; the comment in the Introduction regarding a thirty-year gestation period may apply to this development.

Although it appears that subsonic aircraft will continue to be the major form of commercial transport for the next twenty years, ultimate conversion to the supersonic and hypersonic speeds seems inevitable. A very general economic argument can be made for this assertion, based upon expected performance of aircraft and power unit in the higher speed regimes. As an example, the projection of cruise lift/drag ratio through the supersonic speed range, now and in the next fifteen years, is shown in Fig. 4. It will be noted that the product $V(L/D)$ actually increases somewhat with increasing Mach number. The efficiency of air-breathing engines is expected to increase steadily with speed, when the inlet compression is superimposed upon the pressure ratio of the engine mechanism itself, provided that an optimum design is selected. As is well-known, this optimisation involves steady decrease in the turbine compression ratio from the rotating machinery as the speed increases, until at hypersonic speeds the ramjet carries most of the task propulsion, with a change from subsonic to supersonic combustion as the speed increases still further.

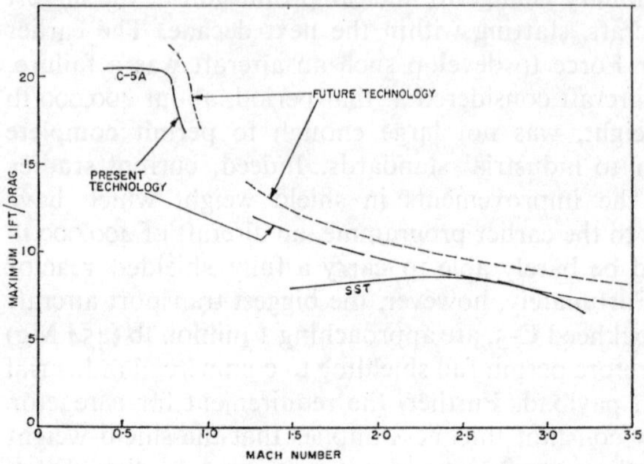

Figure 4 Cruise lift/drag ratio through the supersonic speed range

Direct insertion of these parameters into the simple Breguet relation gives results which are spectacular but misleading at the higher speeds of flight. Two additional factors appear which considerable modify the over-all efficiency of the hypersonic aircraft as a transport vehicle. First, it is apparent that the hypersonic vehicle must fly at extremely high altitudes by current standards; there is in fact a well-known corridor of flight between the low limit of excessive heating and the upper limit of inadequate density, as shown in Fig. 5. In consequence, the higher speed

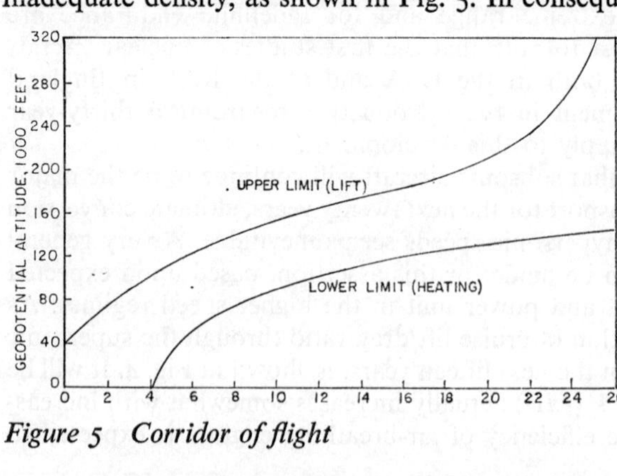

Figure 5 Corridor of flight

vehicle must expend a much larger proportion of its fuel in climb before it achieves cruise condition. The obvious limit of this is represented by the satellite at a speed of 26,000 feet per second (8 km/s), which requires no cruise fuel but requires its entire fuel supply and propulsion system to be designed for climb alone. This extended climb requirement must of course be taken into account in all evaluations of the operating effectiveness of hypersonic aircraft.

A second factor is the appearance of centrifugal 'lift' as the aircraft moves into the hypersonic speed regime. Again the satellite is the ultimate example of this, entirely sustained by centrifugal force. The influence of this component at lower speeds is shown in Fig. 6, in which the cruise range of a family of hypersonic vehicles obtained directly from the Breguet formula is compared with the actual range around the earth's surface.

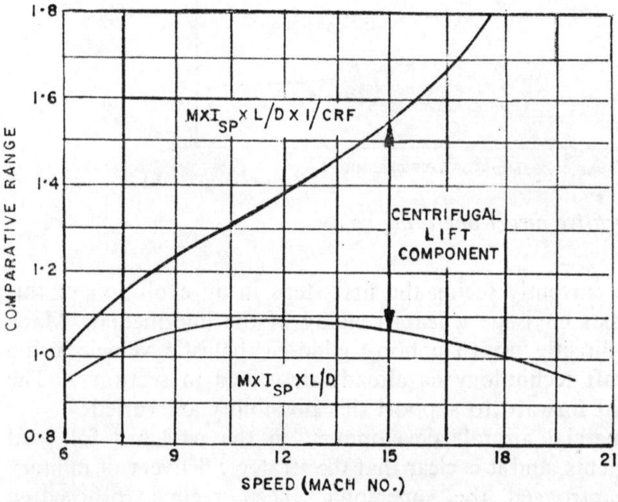

Figure 6 Favourable effect of centrifugal lift

If all of these factors are utilised in a few spot designs of supersonic and hypersonic aircraft, we obtain the rough comparative direct operating costs shown in Fig. 7. It is evident that there is no startling increase in operating cost to be expected in moving up the Mach number range towards hypersonic speeds; on the contrary, the increases in operating cost appear to be well within the acceptable regime for fast passenger traffic. Needless to say, this curve implies a considerable development programme for such higher speed vehicles, and our experience with the large subsonic and supersonic transport aircraft indicates that such developments represent major investments on the part of airlines, aircraft manufacturing companies, and even governments. In consequence, the exact direction which will be followed by the trend to higher speeds is difficult to predict. There are some projections which, recognising the need for higher temperature materials as speed progresses, make the assumption that the evolution of the higher speed aircraft will proceed in steps:

The aluminum aircraft to Mach number 2·3;

The titanium aircraft for a Mach number of 3 to 4;

The use of super-alloys up to a Mach number of 6 or 7.

Radiative cooling of columbium or tungsten surfaces for further Mach number increases.

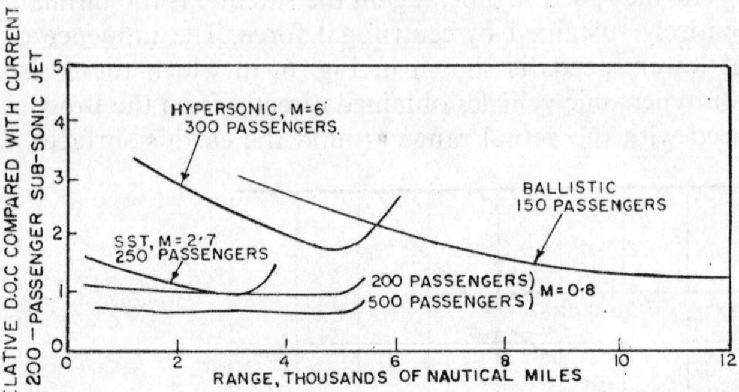

Figure 7 Comparative direct operating costs

Although we are currently seeing the first steps in an evolution of this sort, other forecasts envisage a leap-frogging of the intermediate Mach number ranges, directly up to a boost-glide or ballistic vehicle using re-usable spacecraft technology as already described in section 2. The arguments brought forward to support this possibility are varied:

(*a*) Much commercial aircraft development in the past has followed military developments, and it is clear that the strategic delivery of military payloads has leapfrogged the supersonic speed regime, proceeding directly to the ballistic missile at a Mach number of more than 20.

(*b*) The sonic boom problem of supersonic aircraft may not be readily soluble, and the pressures of the community may therefore restrict supersonic transports to overwater operation. If this is so, then the ballistic transport which has no boom, except in the immediate vicinity of launch, will clearly be more acceptable to the community.

(*c*) If we accept that the entire regime of flight speeds from zero to escape is now available to man for commercial use, then it would be logical to proceed as soon as possible to the speed which offers the highest operating efficiency. Studies on the lines indicated above imply that for moderate (domestic) ranges of flight, no strong case can be made for very high speed. On the other hand, for intercontinental ranges it appears that the most economical vehicle will use the highest possible speed, i.e., the speed for ballistic flight.

(*d*) The intermediate hypersonic speed requires considerable development of detailed technology, whereas the technology for boost-ballistic vehicles is already being developed, first in the ballistic missile and later in the space programmes. It might, therefore, transpire that the lowest development cost, as well as the lowest operating cost, would be associated with the highest speed vehicle.

Evidently the resolution of these questions is not possible at this time, although in another five to ten years some of the answers will be forthcoming. We shall know then, for example, whether the sonic boom is acceptable to the community, and whether it can be ameliorated sufficiently to permit universal use of SSTs. In addition, the next decade should see the start of recoverable space vehicle developments, which converge upon the commercial transports several decades later.

3.3 *Adverse aspects of air transport*

As the significance of air transport to the community grows, we can expect increased attention to some of its adverse aspects. Indeed, problems of engine noise near airports, smoke pollution, and the sonic boom are already the subject of major research and development programmes in connection with most current transport aircraft development, and we can expect even greater attention to these problems in the larger aircraft of the future. Projections from our current capabilities would suggest that it will be possible to hold down the airport noise and the local smoke pollution to acceptable levels; we cannot make a similar prediction for the sonic boom, in spite of the very large concentration of research on this subject.

The question of passenger aircraft safety in the future will in fact prove to be much more important to the aircraft designer than the alleviation of noise and smoke, and may present him with a larger problem. The subject has been discussed recently by the National Academy of Engineering's Aeronautics and Space Engineering Board[17] and a quotation from their report is appropriate:

'A high level of passenger safety must be maintained if civil air transportation is to continue growing. Simply holding today's safety levels may not be acceptable as larger aircraft are introduced. Although the safety record of scheduled air transportation is impressive and continues to improve when measured in fatalities per passenger miles flown, the accident rate has remained nearly constant for almost 20 years when gauged by accidents per 100,000 hours flown. This rate has varied very little despite significant increases in airplane size and passenger load. Tables 1 and 2 illustrate the record for the past seven years in terms of fatal accidents and passenger fatalities for all the member countries of the International Civil Aviation Organisation.

'Larger aircraft with a constant accident rate expose more passengers

TABLE 1 *Fatal passenger accidents*

Aircraft Type	*1960*	*1961*	*1962*	*1963*	*1964*	*1965*	*1966*	*7-year Total*
Turbojet	3	6	7	5	3	5	6	35
Turbopropeller	7[1]	6	7	5	6	2	5[2]	38
Piston-engine	23[1]	13	14	20[3]	15	17	14[4,5]	116
Total	33	25	28	30	24	24	25	189

TABLE 2 *Passengers killed*

Aircraft Type	*1960*	*1961*	*1962*	*1963*	*1964*	*1965*	*1966*	*7-year Total*
Turbojet	113	257	424	347	136	250	451	1978
Turbopropeller	264	192	100	47	252	35	143[2]	1033
Piston-engine	470	356	241	323	270	399	314[5]	2373
Total	847	805	765	717	658	684	908	5384

[1] Includes one mid-air collision between a turbojet and a piston-engine aircraft (counted as two accidents).
[2] Includes one helicopter with 20 passenger fatalities
[3] Includes one helicopter with three passenger fatalities.
[4] Includes a mid-air collision between two piston-engine aircraft (counted as two accidents).
[5] Includes one helicopter with two passenger fatalities.

to each occurrence. It follows that with the introduction of very large aircraft, the accident rate of the scheduled airlines must be reduced below the present average to keep the number of fatalities the same'.

We can evidently project that the larger aircraft of the future will need greatly enhanced safety features built into the design. The trend towards entirely fail-safe structures, for example, will continue. The problems of aircraft life-time and fatigue of structural materials will receive increased attention, and we are moving towards an era when the materials of aircraft construction will no longer be selected on the basis of ultimate load-carrying capacity per unit weight, but rather upon their endurance limit or fatigue properties after 60,000 to 100,000 hours of typical airline life.

The greatest steps in the direction of increased safety, however, can be expected to show up in the instrumentation, controls and general avionics of the aircraft. The larger size clearly permits the aircraft to carry more comprehensive equipment, so that multiple redundancy in critical areas can be expected. This trend will be assisted greatly by the reduced size and increased reliability of solid state components, and the developments in high reliability over long periods resulting from the space programme should have a significant beneficial effect. This entire subject has been discussed in considerable detail by P. A. Hearne in his discussion of future avionics in Chaper 13, and it is therefore unnecessary to go into further detail here. It should be emphasised, however, that the subject of safety extends beyond the avionic equipment of the aircraft, and imposes needs on the airframe structure, the propulsion units, and all other components which will force a continuing development over the next few decades.

4 V/STOL AIRCRAFT

Although I have followed the current custom in grouping V/STOL aircraft into a single chapter, it is important for purposes of projection into the future that we recognise the wide spectrum of requirements and classes of vehicle grouped under this heading. R. Hafner, in his excellent chapter 14 has summarised the many types of VTOL currently under development; he has followed the usual practice of grouping them according to 'disc loading' (weight divided by area of static lifting jet) in Fig. 6 of that chapter. In my Fig. 8 I have followed the same grouping as seen from this side of the Atlantic, and have included a rough curve showing the amount of hover time available if a fixed percentage of the all-up weight, actually 10 per cent, is expended as fuel for hover. It is of course

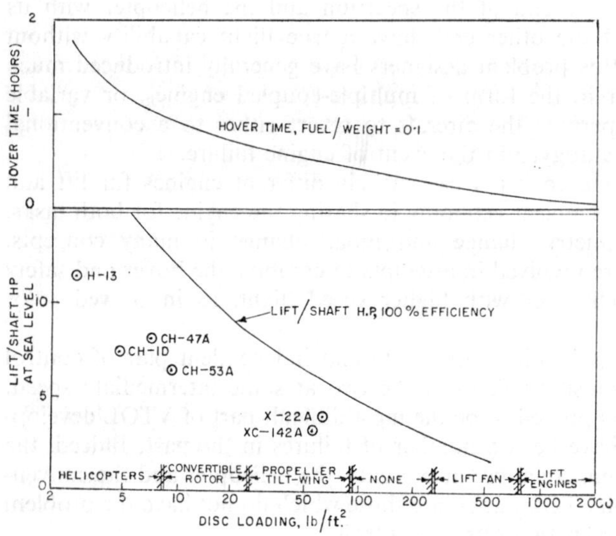

Figure 8 Classes of VTOL aircraft

necessary to supplement the vehicles represented in this chart by the STOL classes of aircraft developed from conventional aircraft by higher-lift flaps, blown flaps and the like. Evidently the progress from left to right on Fig. 8, with increasing disc loading, gives a steady decrease in hover time from over two hours in the case of the helicopter down to only five minutes or so for direct jet lift. In place of hover time, the higher disc loadings offer longer range and higher speed. The conventional helicopter is restricted to a maximum speed of about 300 knots (150 m/s) by the approach to unity Mach number on the advancing blade, and by the increasing vibration accompanying the reciprocating loads on each blade. The various types of propeller-VTOL aircraft can go higher in speed, up to 400–450 knots (200–230 m/s), before the propeller efficiency begins to fall off; perhaps in the future the successful development of supersonic propellers will extend this limit. The only forms of VTOL capable of supersonic cruising flight are currently those with disc loadings above about 300 lb/sq ft (15 kN/m^2), so that the combination of vertical and supersonic flight is only possible for those aircraft, both fan lift or direct jet lift, in which the hover time can only be a few minutes. It is, in fact, recognised that we must look to the helicopter for surveillance missions requiring extended hover time, and the configurations on the right of Fig. 8 are primarily for point-to-point transport in circumstances when the take-off or landing sites preclude the use of conventional aircraft.

Certain other basic characteristics are important for our future projection:

(*a*) Safety in the event of failure of an engine is critical. Only the conventional aircraft at one end of the spectrum and the helicopter with its autogyro mode at the other end, have a true flight capability without power. To meet this problem designers have generally introduced much complexity, either in the form of multiple-coupled engines, or variable geometry which permits the aircraft to revert either to a conventional aircraft or to an autogyro in the event of engine failure.

(*b*) Although some concepts use entirely different engines for lift and propulsion, there is a clear economy in sharing the engine for both tasks, which forces geometry change and mode change in many concepts. Similar changes are involved in attempts to combine the hover and safety aspects of the helicopter with higher speed flight, as in stowed rotor designs.

Thus many designs require a duplicate and independent pair of control systems, and a transition between the two at some intermediate speed. This transition has proved to be the most difficult part of VTOL development, and there have been a number of failures in the past. Indeed, the only classes in commercial use at present—the helicopter and the conventional aircraft with STOL flaps—are those which do not have the problem of transition between two sets of controls.

In all these classes of V/STOL aircraft we can of course envisage evolutionary improvements in response to specific technological advances discussed elsewhere in this series. In particular, we single out:

The steady reduction in powerplant weight as a consequence of improved design methods and materials;

Improvements in fatigue properties of materials, which have particular application to the oscillatory stresses of rotorcraft;

The application of supercritical transonic wing sections to helicopter and propeller blade tips;

The continuing improvements in the lifetime of engines, avionics and control equipment.

We do not get any clear indication of future V/STOL trends from this evolution in technology, since no revolutionary change is readily apparent. Perhaps a clearer picture of the future is obtained by examining the need for V/STOL aircraft. Here again the spectrum is very wide, but can be condensed into three categories: military, commercial and personal.

(*a*) *Military.* Military application to tactical warfare has been the chief stimulus for the development of VTOL aircraft during the past few years. The special requirements for surveillance of local ground action, and later of participation in such actions, generated a need for extensive hover time which only the helicopter could satisfy. In consequence, the helicopter has undergone rapid development, and has assumed a primary role in conflicts such as South east Asia. It has also performed supplementary tasks in point-to-point transport, but its restricted range and payload put it in secondary position relative to conventional transport aircraft. In other words, we can expect to see in the future a development of two different classes of tactical aircraft, one fulfilling the helicopter role of surveillance, and the other optimised for point-to-point transport to restricted airfields, the latter probably an STOL aircraft. The continuing recurrence of localised conflict will probably result in rapid development of these types of military vehicle within the next decade or two.

(*b*) *Commercial.* Chapter 14 by R. Hafner contains a very clear analysis of the developments in short-haul traffic which are forcing the introduction of V/STOL aircraft into airline service. It is only necessary, in this chapter to add emphasis to the growing need for air service to solve the problem of travel from the airport to the ultimate destination of the passenger. In fact, for many short journeys (and the majority of future travellers will still be expected to require journeys of less than 500 miles (805 km)) we must expect in the future that the passenger will by-pass the entirely conventional airport, using instead a network of city-centre or community-centre V/STOL ports. Table 3, compiled in 1966, compares the costs and times for a few major short-haul journeys in the USA, taking into account the entire time required to get from city centre to city centre, and including such items as limousine cost. If a similar analysis were to be made at the end of the next decade, assuming no drastic changes to alleviate the growing congestion in the air above the airports and on the roads leading to the airports, the time would be spectacularly greater for the conventional air shuttle.

TABLE 3

Routes	*Carrier*	*Time*	*Cost* $
Philadelphia—New York	Air Shuttle	2 hr, 25 min	15·05
	Rail (First Class)	1 hr, 20 min	9·35
	Rail (coach)	1 hr, 20 min	5·55
	Automobile	2 hr	
	V/STOL Carrier	35 min	10·50
Chicago—Detroit	Air Shuttle	3 hr, 45 min	22·75
	Automobile	5 hr	
	V/STOL Carrier	55 min	19·80
New York—Washington	Air Shuttle	2 hr, 35 min	20·95
	V/STOL Carrier	52 min	18·00

In all cases above, limousine time and costs at start and end of trip were included as far as appropriate.

The cost and convenience arguments in favour of short-haul V/STOL imply that a very significant commercial short-haul airline system will develop in the future. It is not clear at the present time whether VTOL or STOL types of vehicle will be used in this application. Economy and reliability considerations would emphasise STOL, at least for the forthcoming one or two decades, provided that the necessary runway length at the centre of population can be made available; for this reason, there is considerable interest in the possible development of STOL landing strips along city waterfronts—the majority of big cities in America are in fact located in such a way that the business centre is in close proximity to a waterfront. R. Hafner inclines towards the VTOL aircraft for city-centre operation and, indeed this is the preferred ultimate solution to the problem; the rapid current growth of STOL taxi service at major US airports suggests that the STOL aircraft may prove to be the preferred interim solution.

In either case, the size of such aircraft will grow with time in response to passenger demand, just as we have forecast for the longer-range conventional commercial transport. Probably the initial short-haul aircraft will be slightly larger in passenger capacity than the airport bus, with a full loading of 50–100 passengers. The size will steadily increase from this starting point, until, like the conventional aircraft, it reaches a physical limit. If the requirement develops as a VTOL vehicle because of urban landing restrictions, then the growth is severely restricted by the design requirements of large rotors. The STOL vehicle, on the other hand, can grow to a much larger size, the limit matching the corresponding limit for conventional aircraft.

(*c*) *Personal Aircraft.* Hafner does not subscribe to the idea of everyman's aircraft, and indeed any person contemplating this prospect must be deterred by the idea of tens of thousands of personal commuter aircraft converging upon a city from the suburbs at 8.00 am every morning. Yet there is a strong incentive for personal aircraft in the steady growth of road congestion, and all are agreed that the small VTOL aircraft is not much more complicated than the modern automobile and can be mass-produced at comparable prices, given sufficient demand. On second thoughts, it has seemed to me that our negative reaction to mass personal vehicles has some of the bias with which the early intrepid be-goggled motorist, peering through the dust of 1910 roads, must have contemplated the fearsome possibility of the little old lady driving to church on Sunday morning among many thousands of similarly qualified automobile drivers. It was then incredible but it is now commonplace, that people of all types and ages travel at speeds of 65 miles per hour (29 m/s) (relative speeds of 130 miles per hour (58 m/s)) within a few feet of each other, with decision time measured in fractions of a second. Regrettably, they do not all get where they are going, but the risk is accepted because of the benefit of rapid transport. The fact is that the ordinary human being is endowed with an amazingly complex and effective servo system, which is readily adaptable to situations of this sort.

Hafner's commentary on this, in juxtaposition with his delightful pictures of bird flight, lead me to comment that in the rise of a flock of thousands of birds following a gunshot, with only inches between wing tips, I have never seen a mid-air collision ensue. We could expect a human being to perform even better than a bird in flight*, provided that:

1 He performs such operations several times a day;

2 His input information is adequate, and is presented to him in a form which his control system can immediately assimilate and convert to actions. This comment applies equally to input information from a flight controller (the traffic policeman of the future), and to his sensor information regarding neighbouring traffic to permit collision avoidance;

3 A standardised procedure is established in three dimensions, equivalent

*Although the human being has a much more effective servo system than a bird, he has one obvious failing. Authorities have pointed out that animals fall into two categories, the predators, whose eyes look forward, and the hunted, with peripheral vision. The human being, of course, is in the predator category. As far as I have been able to ascertain, the birds which flock are all endowed with peripheral vision; the predator birds, such as the eagle and the owl, with forward vision, do not move in flocks. This implies that we must supplement the human being's vision with peripheral sensors as a collision avoidance measure. Some work on this problem for conventional light aeroplanes is already underway.

to the current traffic lanes on the ground, with traffic lights, on- and off-ramps, stop signs and all of the tools already found necessary for automobile traffic control;

4 His vehicle is able to execute all of the changes in direction and speed required by the above conditions No 2 and No 3,

The last requirement is, of course, imposed upon the vehicle designer, and there does not appear to be any insuperable obstacle to building vehicles which will satisfy it in the future. The real difficulty is contained in requirements No 2 and No 3, which require a complete revolution in the currently established methods of air traffic control. As we have already seen in section 3, this requirement is emerging from the growing operations of commercial aircraft. Comparisons with ground travel road traffic gives some indication of the nature of the change required. Evidently we cannot accept much longer the personal control of individual aircraft which characterises current traffic control; it must convert in the future to a fully automatic system, a three-dimensional equivalent of the traffic lane and stoplight, with clearly marked traffic lanes and prohibition of other regions, at least in the vicinity of the major cities. The great deterrent to this approach in the past has been the expense and uncertain reliability of electronic equipment. This situation is rapidly changing, and computer capability is growing to the point where the magnitude of the task of tracking and recording thousands of vehicles should not dismay us. I therefore conclude that a very large growth in personal aircraft will be a major feature of the more distant future.

References

1 JANTSCH, E. Technological Forecasting in Perspective, Organisation for Economic Co-operation and Development, Paris, France, 1967.

2 LENZ, R. C., JR. Technological Forecasting, 2nd edition, Aeronautical Systems Division, AFSC, ASD-TDR-62-414, Wright-Patterson AFB, Ohio, 1962.

3 BRIGHT, J. R. *Technological Forecasting for Industry and Government*, Prentice-Hall, Inc., Englewood Cliffs, New Jersey, 1968.

4 *Futures*. Quarterly beginning 1968, Iliffe Science and Technology Publications, Inc., Guildford, Surrey, England.

5 *Technological Forecasting*. Quarterly beginning 1969, American Elsevier Publishing Company, New York.

6 *Technology in Retrospect and Critical Events in Science*, prepared for the National Science Foundation by Illinois Institute of Technology Research Institute, Vol I, December 15, 1968.

7 FURNAS, C. C. *The Next Hundred Years; the Unfinished Business of Science*, Reynal Publishing Company, 1936.

8 BONNER, J., BROWN, H. and WEIR, J. *The Next Hundred Years*, CIT Press.

9 FURNAS, C. C. The Next Hundred Years—Three Decades Later, paper presented to The Thursday Club, 13th April 1967.

10 BONNER, J., BROWN, H. and WEIR, J. *The Next Ninety Years*. Proceedings of a conference held at the California Institute of Technology, March 1967, CIT Press.

11 CLAUSER, F. H. No Law Says Space Must be Expensive, *Astronautics and Aeronautics*, May 1969, Vol 7, No 5.
12 St. Louis Bicentennial Space Symposium, Fifth National Conference on Peaceful Uses of Space, May 1965.
13 *Manufacturing Techniques Unique to Zero Gravity Environment*. Proceedings of Conference at MSFC, Huntsville, Alabama, November 1968.
14 WAHL, B. W. *Analysis of Selected Opportunities for Manufacturing in Space*, Sixth Space Congress, Cocoa Beach, Florida, 1969.
15 UTECH, HARVEY P. and FLEMINGS, MERTON C. Elimination of Solute Banding in Indium Antimonide Crystals by Growth in a Magnetic Field. *Journal of Applied Physics*, Vol 37, No 5, April 1966.
16 LONG, J. E. Multiple-Planet Missions for Effective Exploration of the Outer Planets, *Astronautics and Aeronautics*, Vol 7, No 6, June 1969.
17 STEVER, H. GUYFORD How Should Civil Aviation Develop to Serve our Society Best? *Astronautics and Aeronautics*, February 1969.
18 ROM, F. E. and FINNEGAN, P. M. Will the Nuclear-Powered Aircraft be Safe? *Astronautics and Aeronautics*, March 1968.

22 Looking forward again—notes on the needs, probabilities and possibilities in aeronautics

Harold Roxbee Cox, Lord Kings Norton

INTRODUCTORY

In 1940 I had the honour of giving the 28th Wilbur Wright Memorial Lecture[1]. It was called 'Looking Forward', and it was about the future of British civil aviation. In his introductory chapter, Mr. Allen[3] was kind enough to include it among the prophetic writings he examined.

'Looking Forward' was produced under difficult conditions. Had I not been afflicted in the spring of 1940 with German measles, I do not think it could possibly have been written. I could not, moreover, consider in my predictions the influence of the gas turbine, with the development of which I was already closely associated, because the work of Whittle and his friends was necessarily shrouded in secrecy.

I do not mention these facts to excuse the paper, but to explain first, how time could be found to create such a mass of verbiage in the midst of war, and second, to explain what might at first seem a curious omission.

I delivered the paper almost exactly 29 years ago, but the occasion is clear in my mind. The Society's President, Mr. Griffith Brewer, was in the United States with his old friend Orville Wright, and in the chair was that wonderful character, Brab, at that time still Lt. Col. J. T. C. Moore-Brabazon. London was under attack, but the lecture was held in the great hall of the Institution of Electrical Engineers. The possibility of a less attractive venue was, however, in the Chairman's mind, for he announced the determination of the Society to continue the Wilbur Wright lectures in the War 'even if it became necessary to hold them, like the early Christians, underground in catacombs'.[2]

But it is not only the abnormal conditions surrounding the composition and delivery of that paper which impel me to refer to it. In addition, it is remarkable for its implied underestimate of subsequent progress and for its wholly misplaced dependence on logical argument.

I was, of course, too cunning to attach a time-scale to my predictions

Originally published in the *Aeronautical Journal of the Royal Aeronautical Society*, July 1969.

of developments to come. But the projections of aircraft speed and size and the ideas of frequency of transatlantic flight now look extraordinarily jejune. Development has taken place far more rapidly than I, in that paper, appeared to think possible, and I cannot recall that any of the friends who read the paper in draft or those who spoke about it immediately after it was published had any criticism to offer on my ideas for the future.

Indeed, I think one would have had to be clairvoyant to have foreseen the rate at which not only aeronautics, but space travel, nuclear energy and electronics have been developed since the war. And one matter I want to examine in this paper is the feasibility of the rate of progress we have witnessed continuing.

I have already indicated that my old paper depended too much on logic. The prophet who deduces the future by logical extrapolation from the past and his knowledge of the present is doomed to failure. It is difficult to suggest what other process except clairvoyance is available to him, but he would do better to consult the entrails of the appropriate fowl. I, quite logically I think, saw a future for the flying boat. But the rapid development of the large military aircraft, essentially a land-based vehicle, effectively stifled the development of the civil flyingboat; civil air transport was destined to be run by landplanes.

Aeronautical progress since the end of the war has certainly been extraordinarily rapid and we are apt to assume that this extraordinary rate of progress will continue. But we are already conscious of a change in the character of the progress and this change, in fact, may change the rate. If progress is thought to consist of steps upward, then the steps are getting higher and the pause before each is longer. Putting it differently, the quanta are becoming bigger and fewer and the gaps between them greater.

Size is today best measured in prototype capital cost. Quite apart from bigger jobs taking longer, decisions to commission them are taking longer to reach. This is not unreasonable. When a budget can be spread over a variety of projects, successes may cancel out the failures. But when a great proportion of the whole budget is concentrated in one project, it is reasonable to expect a long period of enquiry, planning, design of alternatives, costing and the multitudinous other considerations on which success depends, before deciding to proceed. Twenty years ago it would have seemed incredible that the USA should consider their supersonic transport project for six years and still not reach a decision, but this indeed has happened. And when these great and lonely projects are organised internationally, when even only two nations collaborate in one of them, delays are naturally longer still.

There is, too, another reason why decision is difficult and often delayed. There was a time when the future could be considered in compartments. There was indeed in the 1920s an admirable series of works among which I recall the titles of two—*Daedalus*, or *the Future of Flying*, and *Lars*

Porsena, or *the Future of Swearing*. Each area of activity—not, of course, only these two—could be considered in reasonable isolation. But now aeronautics is so intimate a part of the pattern of life, changes in it affect the pattern, and changes elsewhere in the pattern, affect *it*. The decisive improvements in UK railway communications following the electrification developments have reacted on our internal air services. Developments in, and foreseen in, overhead railway systems will influence airport siting[21]. We must indeed think not of air transport but of the whole field of communications, and our aim, sometimes conscious but generally subconscious, must be to make the overall system the best for the majority of people, insofar as politics, international jealousies, wars and rumours of wars will allow.

It is shocking to think how great is the influence on civil aeronautical progress, and indeed all civil progress, of what is euphemistically called 'defence'. I have already given an example of this influence from the last great war, and it would be childish to believe such influence was at an end. Indeed governmental, financial support today for technological progress in the name of defence is greater than governmental financial support for civil technological development.

Paradoxically, war, despite the appalling waste it means, is the great technological accelerator. Without the great wars of this century, our technological condition would be far less advanced. If we succeed in preventing another great war we shall have succeeded in hindering the rate of technological progress. It may be objected that the races for the Moon and the planets, with all the enormous volume of technological advances associated with them, are phenomena of peace. But they are really part of the aftermath of war and continuing enmity.

In this general consideration of the anatomy of progress, we have to remember that lines of development, like seams of coal, can be exhausted. The development of the railway steam engine is a case in point. It reached a stage when not only was technical innovation less and less frequent in its history, but it became an anachronism. There is perhaps development still to come in railway electrification, but even there one is conscious of approaching a horizontal asymptote, an asymptote which the conventional ocean-going ship has evidently almost reached. In the manufacture of ships there is probably a great deal to be done, but we cannot look for much improvement in the performance of conventional vessels, despite the influences of new propulsive systems based on the gas turbine and nuclear energy. Much of their work has been taken by aircraft, and the future may see some of the rest taken by hovercraft, hydrofoils and submarines[14]. Even the rotorcraft may not, on the time-scale of aeronautics' second century, have much further to go, perhaps not much further than half-way. It could develop as an admirable communting system: it has great possibilities of improvement in its control and operation: it might grow to accommodate 200 people: it could be—it is—a great weight lifter, a specially mobile crane. But in most visions of the more distant future, it has a relatively humble place[16].

I have already suggested that logical progress in civil transport is unlikely and have mentioned as disturbing influences on it war and preparation for war. If, as we all hope and pray, an era of peace descends upon the world in the Royal Aeronautical Society's Second Century, however, the main influences on technological progress will be deliberate planning, scientific discovery and economics. Scientific discovery is partly chance and partly planning, and I feel that if its influence is wrong in kind or in tempo the fault will be ours, the scientists' and engineers'. But to be sure that economics is operating fairly in its influence on progress is much more difficult. The costing and hence the comparison of the variety of solutions to communications problems is going to be intricate and I believe it is not too soon to begin the deep study of these matters. I will indicate the character of my doubts about our present economic assessments by two examples.

It is sometimes necessary to compare the costs of aeroplane and helicopter operations. In the former, the cost of providing the runways essential to operation is indeed generally partially reflected in the cost evaluation, but in evaluating the helicopter operation the helicopter is never given credit for the fact that the runway is, so to speak, built into the machine. With the advent of the civil VTO aircraft this matter becomes one of urgency.

My other example relates to the use of foreign equipment in government contracts. Insofar as decisions are made on cost comparison, then, except in times of full employment, when capacity thrown up in one activity theoretically can be absorbed by another, the estimated cost of home produced things should be reduced in the overall cost evaluation by the tax to be paid to the Exchequer by the staff concerned with their manufacture. This has never been done, but clearly it should be.

In both these examples, the error arises because of our habit of compartmenting. Costs in one government department apparently cannot be offset by receipts in another. This is surely absurd. In the future it is essential that our cost evaluation, inside and outside government, becomes more refined, and if we are to evaluate not only aircraft but whole communications systems accurately, there are complex problems for solution by our accountants and fresh ideas to be developed by our economists, and unless they and we collaborate, serious errors will be made.

There is another aspect of the economics of aircraft production which is much more the concern of the engineers and scientists than the accountants. I refer to the economic use of material. I have long been shocked by the production of high performance structural parts by machining away most of the material supplied. If the material is one with a high scrap value perhaps the situation is not as bad as it looks at first, but the correct evaluation of alternative processes, and the development of potentially economic forming processes, is obviously a matter of great importance for the future[9,17].

We are in a curious position at present in which, if medical science is left out of account, we could do without the discovery of new facts of science for quite a long time while we were applying to technological advance what we already know and while we are getting our economics straight. Flat statements of this kind always on examination need qualification and, in this case, the filling up of gaps in knowledge as an essential part of the pattern of development, of course, would continue to be necessary. But the assertion is broadly true, and while any attempt to convert it into a policy would be quite disastrous, it does at least suggest that a better distribution of effort might result from more deliberate planning than we now achieve.

Basic scientific discovery cannot be planned. Great and influential discoveries like that of electro-magnetic induction in the last century and the character of the atomic nucleus in this are always possible and when they occur they make nonsense of extrapolation; but they do not remove the need for planning. If we are to have any control of the technological future we must plan much more effectively than we have in the past.

When I say 'we', I mean, for two reasons, we here in the UK. I am not a xenophobe—far from it. But the world is not ready for global planning and even if Europe were ready for European planning, which manifestly it is not, we should still, in my view, need to have a national plan. It can be alleged that we have always planned, and the allegation needs examining. First, there is not much doubt that under the threat of war and in war, we did get down to the job of planning, and, having made plans, executed them with reasonable despatch. Our planning in peace, however, has been lamentably defective and depressingly so in aeronautics. The urgency of war unifies action. Party politics temporarily disappear. And it is politics that is the ruin of plans in peace.

In these remarks I am concerned with the future, but when one considers the string of bad decisions of the past in aeronautics, some of them stemming from the party in power changing, some from sheer lack of understanding by politicians of the issues at stake, we realise the need for a system which, in what should be the uncontroversial field of technological progress, would provide a coherent plan to which we could adhere.

Inevitably the national economy will affect plans, but a non-political continuing authority would be more likely to slow up or accelerate a programme than to initiate and cancel in the fantastically uneconomic way we have so often seen in the past, insofar as we have been able to see through our tears.

The planning of progress in communications or, if that seems to be too big a step all at once, of our progress in aeronautics, could in my view be the job of an authority independent of party, which did not *advise* Ministers, but was executive in its development of prototype systems and controlled only by its budget allocation. Because communication, and aeronautics in particular, are inevitably the concern of defence, the defence ministries would need to be associated with this essentially civil authority, if only to ensure no unplanned duplication of research and development effort and no unnecessary duplication of facilities. But the authority would itself be the chosen instrument for prototype communications development.

I put forward this idea of a new authority at a time when the proposal of the Edwards Committee for a Civil Aviation Authority concerned with safety and licencing is still new[23]. I make no comment in this chapter on an interesting proposal which at the time of writing is still under examination. And I do not suggest that the new prototype communications planning authority I have adumbrated should be added to it or put alongside it. But I should like the same consideration to be given to its merits and demerits as the Government, the Edwards Committee and the interested parties gave and are giving to the pros and cons of a central safety authority and the administrative alternatives to it.

One can scarcely make the point too often that the development of aeronautics in the next century depends on the development of the overall pattern of communications, on the development of other technologies and skills, as do these also depend on the development in aeronautics. This interdependence is basic and emphasises on the one hand the desirability of a single continuing non-political communications authority and, on the other, the overwhelming difficulty of predicting what is going to happen. The development of the video telephone, the possibility of conference between individuals in offices scattered over the world connected by such a device, must clearly make the need to travel less. One can indeed visualise, as E. M. Forster did in a story many years ago[24]. a world in which individuals are normally static in cells connected only by super communications, individuals to whom physical travel has once again become an adventure. But while all this makes the unwisdom of prediction manifest, it does not make the discussion of possibilities and probabilities uninteresting or indeed valueless. Indeed, did we think so, the whole series of chapters of which this is the last were misconceived.

The possibilities are indeed thrilling. It is difficult to see the limit of what could be done. One chain of possibilities flows from consideration of developments in structures. There is indeed a limit to what can be achieved, defined by our old friend the square-cube law[1, 9] and the use of the stoutest materials with the most economical set of physical properties developing their theoretical strengths. Had we stuck to the configuration of the subsonic transport, then with current materials we could confidently predict the economic aeroplane of maximum size. The square-cube law is a law of nature and, as the poet[25] said:

However great your strength or stature,
You can't repeal the laws of nature.

But, just as in the case of the law of income tax, there are ways round. One way, permitted—in fact, dictated—by the aerodynamicists is to change the shape[4]. Aspect ratio does not loom so large as it used to. The other, of course, is to develop materials with greater strength and stiffness and with the optimum combination, from the structural engineers' point of view, of these and other properties. Metals, as at present produced, are so far from their theoretically possible strengths, that alloys with twice the strengths now achieved are possible. Indeed, in the metallic field the alloys of steel and aluminium which have so long been the basis of aircraft construction have still a long course to run and the titanium alloys a still longer one[17].

But just as in the development of the gas turbine whole new fields of performance were opened up by the development of the Nimonic alloys, so do we begin, with the discovery of the carbon filament, to perceive the structural potentials of non-metallic materials. At present carbon is restricted because, to make it usable in anything but tension, it needs to be embedded in a matrix, and its own splendid characteristic of increasing in strength with increasing temperature is nullified by the more normal behaviour of the matrix resin. These difficulties will undoubtedly be overcome, perhaps by the creation of gigantic diamonds, but probably less dramatically.

In fact, it is more likely that ways will be found to exploit the extraordinary properties of the finedrawn wire, the filament and, most refined of these forms, the whisker. These have strengths and elastic moduli significantly greater than the materials they are made of have in their more conventional forms. It is remarkable that some fine whiskers have already been shown to have strengths near to the theoretical strength of the inter atomic bonds, so we are approaching the ultimate possibility in these cases. The potential that such achievements indicate is indeed prodigious[17].

But less dramatic properties than strength are of the greatest importance and we must hope for progress in two directions in particular. One is the improvement of fatigue properties of materials, the other is improvement in forming properties in such ways that the other desirable characteristics are not impaired.

In the case of fatigue failure prevention, the most obvious course is to decrease the ratio of working stress to ultimate stress in a vital area by increasing the latter, and the continual achievement of higher and higher strength materials makes this a hopeful route. Nevertheless, one still hopes for a more fundamental solution of the fatigue problem.

In the case of improved forming properties and hence the avoidance of uneconomic machinery and scrap creation, there is hope in the metallic field in the phenomenon of super plasticity and, in the non-metallic field, in mouldable materials such as glass. Mouldability could be influential in the elimination of joints and the avoidance of stress concentration, all of which is of increasing importance as desirable materials become more brittle[9,17].

Altogether one feels that the aircraft designer has ahead of him an immense choice of exploitable advances in material properties, properties which will enable him to reduce structure weight. How quickly these properties will be available to him, however, is quite another matter, and dependent upon the way government—or an independent authority—allocates its research and development finances and resources.

One way in which a considerable saving in structure weight could be made immediately, is by the elimination of windows, and it can be argued that, as in many large aeroplanes, many of the passengers cannot see out of them, and, further, that those who can are offered a depressing monotony of scene, they would not be missed. Whether the claustrophobes would be consoled by strategically disposed television screens showing what is going on outside is a matter which should be investigated. The other possibility is to develop moulded glass as the structural material for fuselages.

The development of materials and its effect on lightweight structural design gives greater flexibility to the designer in deciding upon aerodynamic configurations. He has, we are told, a whole spectrum of major types of aircraft within which to work[4]. It stretches from the infra-red of the classical aircraft to which we have grown accustomed, through the swept-wing aircraft to the slender-wing machine typified by the Concorde, to the hypersonic aircraft in the ultra-violet, the machine which rides on the shock wave it creates. With the development of the VTO slender wing aircraft and the possible advent of the wave-rider, new design problems loom large. One in particular is the choice of fuel for propulsion[11]. Today we use the fuel which gives the greatest calorific value per litre, kerosene. If we feel we need the greatest calorific value per kilogramme, however, we turn to hydrogen. We are then confounded by the fact that its calorific value per litre is only one quarter of that of kerosene. Aeroplanes as we know them are embarrassed by bulky fuels. However, the wave-rider may not be. It is a little early to predict, but if it is not, then it may be able to exploit the immense calorific value per kilogramme of hydrogen, nearly three times that of kerosene. If this can be done in connection with some form of boundary-layer propulsion, we shall have a vehicle, the shape of which I cannot at present visualise, capable of reaching the antipodes in a couple of hours.

But the most fantastic of our visualisations of the possible future in air travel pale into the commonplace beside the possibilities in space travel[7]. The materials and some of the fuels which I have just mentioned in their mundane context have their parts to play in the spatial context. But much more dramatic in their gigantic propulsive powers are nuclear devices depending for energy production preferably on nuclear fusion to nuclear fission. The development of earth orbiting satellites as military weapons is a regrettable present and future condition; their advantages for observational purposes, even if not used for offensive ones, is difficult for mankind in the present state of its great knowledge and great unwisdom to resist. But one can in the future see communications satellites developing from vehicles conveying programmes between nations to travelling broadcasting stations with programmes of their own, a possibility for propaganda alarming in its potential. This is only one of the manifold possible developments ahead but it alone is a compelling reason for this country to play a major part in space developments if we are not to be swamped by alien ideologies.

We can readily visualise the communications satellite of today developing in such ways that the need for business travel is reduced, the weather satellite developing from a data gatherer to a vehicle for prediction. Certainly satellites will increasingly be used as aids for navigation and traffic control for ships and aircraft[7].

Whether we need visualise the space vehicle as a means of travel, except for purposes of exploration, in the next 100 years, is doubtful. If, as Mr. Cleaver and Mr. Thwaites think, in AD 2066 there will be people living on the Moon, on Mars, in orbiting space stations and even on Jupiter and Saturn, then a certain amount of physical interplanetary transport will be required[6,7]. Even then, however, I feel it will be regarded as something of an adventure. It will presumably have meant, however, the achievement of the non-expendable vehicle capable of landing and taking off from both the earth and the other planet, and the extrapolation into space of terrestrial law[20].

It is in this latest and most fascinating and amazing field of human effort that prediction, prophecy, speculation—call it what you will—is most tempting. It is in this field too that the influence of some fundamental scientific discovery still utterly hidden from us, and leading perhaps to a completely unthought of source of energy, is likeliest to be. Conceivably, some understanding of the nature of gravity may develop, possibly from discovery in a scientist's laboratory, possibly from a discovery in the less conventional atmosphere of séance.

Possibilities for the next 100 years we can discuss, probabilities not. Probabilities for the next 20 years are hard enough to assess, but this year's moon and Venus achievements suggests that we should be cautious indeed as prophets if we did not see a landing on Mars in the 1980s.

However far our thought can reach, however bizarre our conceptions of supersonic aircraft, of wave-riders, of space vehicles, however refined the computers and their programs, we visualise man in control. Man may delegate more and more to machines he has designed as the vehicles of delegation, but unless he can delegate foresight, prediction, and the handling of the unpredictable, there appears to be a limit to the degree of his delegation. Moreover, the complexity of the machines over which he must still exercise his dominion will increase, and he has the problem of directing them through the functioning of his own physical and neural composition. We usually think of his having available for this control his senses of seeing and hearing, as the direct means of receiving essential information. No-one has so far suggested his senses of taste and smell for this purpose. But there are possibilities in his tactile sense and in the use of his sense of acquiring information about movement and attitude from the balancing faculty and his head[13]. It appears that man is capable of using these senses in more and more refined ways: that he may, for example, learn to use his parafoveal vision, given the right kind of signal for the purpose, in addition to his narrow direct vision, which requires him to scan. He may be assisted by three-dimensional displays projected in the line of his narrow forward vision. There are, too, the possibility of more precise use of hearing, through technological advance in the two-way reconstruction of original speech, the possibility of other languages, like the language of music, and which convey accurate information. There is the possibility of training and aiding the ears to advance beyond their already extraordinary direction finding abilities[13]. And if the extended abilities of vision and hearing are not enough there is the possibility of developing for control the tactile sense, which the blind reader of Braille has demonstrated is capable of great refinement in perception.

For the exercise of control as distinct from the receiving of data, man may possibly develop far beyond the wheel turning, button pressing phase he is now in. He is capable, as the players of musical instruments show, of translating a specialised language by great manual and oral dexterities, and faculties of this kind might be used for control. Man may, at the centre of an elaborate control system, influence action by small movements of his head relative to his body. In fact, for the refined control of elaborate apparatus, there may be almost unimaginable possibilities yet to be discovered in the elaborate neural and muscular complex which each of us is.

We have to remember, however, that this complex mechanism has so far operated in a 1*g* environment. If we are in the next decade to adventure into extra-terrestrial regions, we shall have to adapt our physiology to different gravitational environments, and the man-machine partnership will be varied accordingly[13].

The fantastic speculations possible about man's control of his machines and man's control of man relate, of course, to the whole social scene, of which aeronautics is but a small part. With its immense influence in physical communications, however, it is an important part, and we in aeronautics must realise that we are in the beginnings of a new phase in evolution, a phase in which man himself is beginning to influence significantly his own development[13]. Natural selection is no longer in control. We have a responsibility for the future which is terrifying in its magnitude, and we must pray for a wisdom commensurate with our knowledge and its incredible proliferation.

It is hidden from us still whether in the century ahead we shall still see man sitting in the midst of the vehicle he controls, controlling it by refined apparatus in intimate connection with his own anatomy, or whether we shall see the control exercised from afar by a computer, in communication with the vehicle, controlled by a program man has schemed. The rate at which the computer will develop in its manipulation of our affairs is determinable only by a computer operating to a program of incredible refinement and comprehensibility. But the increasing influence of the computer not only in aeronautics but in our lives is ineluctable. It is probably the most pervasive influence to which man has ever been subjected: it has indeed a potential universality beyond that of fire, of the wheel, of printing[6]. The possibilities indeed are so outrageous that it is a comfort to restrict speculation to aeronautics and think merely of computers controlling aeronautical transport, even though we extend its scope as far as the Moon.

The computer will be capable of almost any process which possesses a structure describable by the human intellect[6], and consequently we shall see it, in the coming years, more and more the instrument of aircraft design, more and more the instrument not only of calculation and record but of discovery by the logical integration of its store of knowledge. Design by drawing on paper will be with us for quite a while, I expect, but lines on paper as a process for storing knowledge is already *vieux jeu*. Journalists who for so long have loved to talk of blueprints for the future, unconscious that the blueprint went out of existence before the war, will at last have to talk of tapes for the future. But with the development of advanced systems of visual display, of instructions to the computer by the spoken as well as the typed word, we can visualise the design team of the future as highly trained engineers in intimate dialogue with powerful computers[19], a vision a little disconcerting to those who believe that the great designer is a supreme artist. Will the great designer of the future still be an artist? Will there still be room for gracefulness and imagination to manifest themselves through this incredible machine? We must hope so. But while in the course of considering my material for this chapter, I have imagined many things I have never imagined before, I have not succeeded in imagining the characteristics of the Sydney Camm of 2066.

He was, in every sense of the word, an individual, individual in the way he absorbed knowledge, individual in the way he used it. Will education, with the increasing use of televised and other visual programmes[18], computer controlled courses, and computer stores in place of human memory, impose such a pattern as to cause to proliferate the standard man? It is an awful thought. If we are all to be alike, I hope we shall all be tolerant, polite, kind and well spoken. Present trends suggest that this is unlikely and mankind needs to make a conscious effort to arrest his *descensus averno*. This is not the place to discuss social behaviour except in one aspect—language. If we are, as I have said, orally to instruct computers, if we are to have dialogues with computers, it is advisable that both ourselves and the computers use an accurate and compact language. Perhaps it should be Hebrew, which is such a language, but if it is to be English then for goodness sake let it be terse. Let us rid ourselves of the cumbrous transatlantic neologisms which are invading our language. Drop 'assignment', reinstate 'job'. Cut out 'elevator' and 'apartment' and restore 'lift' and 'flat'. Let us 'start' and not 'commence'. Let us cross out 'currently' wherever we find it—the sense will be unchanged. Believe it or not, I heard an American voice in a space programme say, 'Are you in visual contact with the earth?' He meant, presumably, 'Can you see the earth'. And in an estimable British technical journal I read the other day 'A VDR offers aircraft designers the capability of varying (such and such)'. Surely 'A VDR allows designers to vary (such and such)'. I believe we must arrest this tendency to sloppy verbosity and that early in the second aeronautical century we must reinstate English in educational curricula or we are going to have some damned longwinded computers talking to us later on.

I have made the point already that we shall be wise to think of the development of communications rather than of aeronautics by itself, and one of the most remarkable influences in recent years on the communications pattern is the air cushion vehicle, recently by law called the hovercraft[14]. The genus to which such craft belong is the surface effect vehicle—machines which are sustained in travelling across a surface because of their proximity to it. This sustentation could be achieved by means of an air cushion, positive or negative in pressure, through electromagnetic repulsion or just conceivably by other means designed to defeat gravity. So far the only practical vehicles have been of the air cushion type and, the first great inventive step having been taken (and as recently as a decade ago), there is little difficulty for the most conventional imagination visualising tremendous possibilities, particularly in the fields of tracked and marine vehicles.

The extrapolation of current technology already makes possible marine craft of 750 to 1000 tons with top speeds of, say, 75 kt, conveniently amphibious for docking purposes. Such craft would be propelled by airscrews like the 170 tons SRN4. Beyond 1000 tons, however, this kind of propulsion becomes difficult and the large hovercraft of the future, 10,000 tons in weight, 500 ft long and 150 ft wide will probably be driven by water jets. These, and the hydrofoil stabilisers which will probably be needed, demand underwater appendages which, from the viewpoint of the purist, seems a pity. Such a step does, however, effect an interesting hybrid craft, bringing together two competing technologies, the air cushion vehicle technology and the hydrofoil ship technology. If this is to happen, the synergy will need to be effected with great skill and recollections of the *Great Eastern* with its screws, its paddles and its sails. But perhaps it is too soon to assume that the marine hovercraft will not develop systems of stabilisation and propulsion of its own which do not require the help of the turbulent fluid below it.

The hovercraft has been visualised too as a vehicle which might help in the agricultural field, for example, for sowing paddy fields[5]. It is not likely, however, to oust the aeroplane from the position it has so rapidly developed in recent years in less specialised sowing, in spreading fertilisers, in pest control. In the world of agriculture there is not much doubt the aeroplane has an important part to play, particularly in developing farming of the great lands of the agriculturally less developed countries. But exactly what the aeroplane or helicopter or hovercraft does in future depends not only on its own abilities but on the progress in the rich field of agricultural chemicals and the combined operation could well develop wonderfully to the benefit of mankind, bringing prosperity and plenty where there is now poverty and misery[5].

And with the fantastic developments we see as possible do we see ourselves travelling as safely as today? I think we do. Take-off and landing are the most critical phases of air travel, and to achieve them vertically is potentially safer than charging along and approaching miles of concrete runway however refined our control becomes[22]. Furthermore, the development of quality and reliability techniques and the control of quality and reliability exhibited by the achievements of the Americans and Russians in space give considerable confidence in the safety of the high performance travel of the second century.

But as we peer into the mists of the future, first in this direction, then in that, we who are of a certain age see little to console us. The world we have, with its million faults, seems not so bad after all. Should we not be wiser to slow things up a bit? If so, who is to be bold enough to take the first step back. It is as big a responsibility as creating the atomic bomb. But if we never discover another fact, the vast inertia of advancing technology will take us a long way down the roads we have been contemplating to a greater uniformity. Probably, the only thing around us that we shall not succeed in changing is the weather, though we may get better at using it to our advantage and pleasure more efficiently[10], and better at predicting it, avoiding it and taking off, flying and landing through it[8]. Some of us may come to regard it with affection as the only thing we cannot control.

Envoi

Joseph Conrad and Ford Madox Hueffer wrote a book called *Romance*[26]. Somewhere near the beginning occur the words:

> '. . . romance. It lies a little distance before us and a little distance behind . . . One discovers that one has passed through it just as one passed what is today our horizon. One looks back and says "Why, there it is". One looks forward and says the same. It lies either in the old days when we used to, or in the new days when we shall . . .'

When I look back on the great things I was privileged to take part in—the building of R101, the creation of the jet engine—the scenes are illuminated with an heroic glow. I have always felt the romance of achievement and I think my contemporaries in aeronautics feel the same. But when I look forward, does the horizon glow with the same glory? I am not sure; the old eyes grow dim. But if it does not, however affluent we become, the future is going to be pretty grim.

References

1 ROXBEE COX, H. (later Lord Kings Norton). Looking Forward—Prolegomena for a detailed Study of the Future of British Civil Aviation. 28th Wilbur Wright Memorial Lecture. *Journal of the Royal Aeronautical Society*, Vol **44**, No 357, September 1940.

2 MOORE-BRABAZON, J. T. C. (later Lord Brabazon). Introduction to the 28th Wilbur Wright Memorial Lecture. *Journal of the Royal Aeronautical Society*, Vol **44**, No 357, pp 681–2, September 1940.

3 ALLEN, J. E. Looking Ahead in Aeronautics, Chapter 1.

4 KÜCHEMANN, D. An Aerodynamicist's Prospect of the Second Century, Chapter 2.

5 WHEELER, A. H. The Future for Agricultural Aviation, Chapter 3.

6 THWAITES, B. The Future Role of Computers with Special Reference to Aerodynamic Design. Chapter 4.

7 CLEAVER, A. V. Astronautics—Its Development During the Second Century of the Royal Aeronautical Society (1966-2066). Chapter 5.

8 O'HARA, F. and BURNHAM, J. The Atmospheric Environment and Aircraft—Now and the Future. Chapter 6.

9 MOLYNEUX, W. G. The World of the Structural Engineer. Chapter 7.

10 SHENSTONE, B. S. Unconventional Flight. Chapter 8

11 THRING, M. W. Power Generation for Aircraft in the Second Century. Chapter 9.

12 DAWSON, L. G. Propulsion. Chapter 10.

13 MELVILL JONES, G. Man-Machine Integration: A Long Term Look. Chapter 11.

14 STANTON-JONES, R. Hovercraft. Chapter 12.

15 HEARNE, P. A. Avionics—Towards the Twenty-First Century. Chapter 13.

16 HAFNER, R. The Future of Rotorcraft. Chapter 14.

17 KENNEDY, A. J. The Prospects for Materials. Chapter 15

18 BLACK, J. Education and Training in the Second Century. Chapter 16.

19 STAMPER, J. T. Management of Design. Chapter 17.

20 CAPLAN, H. Law for Aerospace Activities (1966–2066). Chapter 18.

21 STRATFORD, A. H. Airports and Air Transport: Growth and Transformation. Chapter 19.

22 STRATTON, A. Air Traffic and Navigation in the Second Century. Chapter 20.

23 British Air Transport in the Seventies. Report of the Committee of Inquiry into Civil Air Transport. HMSO Cmnd 4018, May 1969.

24 FORSTER, E. M. *Collected Short Stories*. Sidgwick and Jackson, 1965.

25 ORTON, K. N. *Little Songs for Tiny Scientists* (*unpublished*).

26 CONRAD, J. and HUEFFER, F. M. *Romance*. Dent, 1949.

Index